Handbook of Molecular Biology

Handbook of Molecular Biology

Editor: Erik Pierre

STATES
ACADEMIC PRESS
www.statesacademicpress.com

States Academic Press,
109 South 5th Street,
Brooklyn, NY 11249, USA

Visit us on the World Wide Web at:
www.statesacademicpress.com

ISBN: 978-1-63989-262-4 (Hardback)

Cataloging-in-Publication Data

Handbook of molecular biology / edited by Erik Pierre.
p. cm.
Includes bibliographical references and index.
ISBN 978-1-63989-262-4
1. Molecular biology. 2. Biomolecules. 3. Biology. I. Pierre, Erik.
QH506 .H36 2022
572.8--dc23

Table of Contents

Preface

This book aims to highlight the current researches and provides a platform to further the scope of innovations in this area. This book is a product of the combined efforts of many researchers and scientists, after going through thorough studies and analysis from different parts of the world. The objective of this book is to provide the readers with the latest information of the field.

The branch of biology that is concerned with the study of the molecular basis of biological activities within and between the cells is known as molecular biology. It encompasses molecular synthesis, mechanisms, modifications and interactions. It is closely related to the fields of biochemistry, genetics, computational biology and bioinformatics. Molecular genetics is one of the most prominent sub-fields of this area. Molecular cloning, gel electrophoresis, polymerase chain reaction, macromolecule blotting and probing, allele-specific oligonucleotide, microarrays, etc. are some of the techniques used within this discipline. This book brings forth some of the most innovative concepts and elucidates the unexplored aspects of molecular biology. The various studies that are constantly contributing towards advancing technologies and evolution of this field are examined in detail. Those in search of information to further their knowledge will be greatly assisted by this book.

I would like to express my sincere thanks to the authors for their dedicated efforts in the completion of this book. I acknowledge the efforts of the publisher for providing constant support. Lastly, I would like to thank my family for their support in all academic endeavors.

Editor

1

Crystal structure of *E. coli* arginyl-tRNA synthetase and ligand binding studies revealed key residues in arginine recognition

Kelei Bi, Yueting Zheng, Feng G"ao, Jianshu Dong, Jiangyun Wang, Yi Wang✉, Weimin Gong✉

Laboratory of Non-coding RNA, Institute of Biophysics, Chinese Academy of Sciences, Beijing 100101, China
✉ Correspondence: ywang@moon.ibp.ac.cn (Y. Wang), wgong@ibp.ac.cn (W. Gong)

ABSTRACT

The arginyl-tRNA synthetase (ArgRS) catalyzes the esterification reaction between L-arginine and its cognate $tRNA^{Arg}$. Previously reported structures of ArgRS shed considerable light on the tRNA recognition mechanism, while the aspect of amino acid binding in ArgRS remains largely unexplored. Here we report the first crystal structure of *E. coli* ArgRS (eArgRS) complexed with L-arginine, and a series of mutational studies using isothermal titration calorimetry (ITC). Combined with previously reported work on ArgRS, our results elucidated the structural and functional roles of a series of important residues in the active site, which furthered our understanding of this unique enzyme.

KEYWORDS arginyl-tRNA synthetase, amino acyl-tRNA synthetase, isothermal titration calorimetry, site-directed mutation, X-ray crystal structure, *E. coli*

INTRODUCTION

Aminoacyl-tRNA synthetase (aaRS) serves to generate the raw materials for protein synthesis—aminoacyl-tRNA in organisms of all complexities. aaRS catalyzes the esterification reaction between the amino acid carboxyl group and the 2′ or 3′ end hydroxyl group on the corresponding tRNA. The highly specific recognition of amino acid and its cognate tRNA is a critical aspect of the enzyme, the malfunction of which may severely impact cellular survival. Understanding the mechanism of amino acid binding is also beneficial to the ongoing efforts of protein expression with non-natural amino acids (Hendrickson et al., 2004; Wang and Schultz, 2005; Sinkeldam et al., 2010; Wang et al., 2012). Namely, natural aaRSs are ideal templates for creating new aaRSs that recognize side chain moieties beyond the pool of 20 natural amino acids.

Being such an essential enzyme for medicine and protein engineering, aaRSs received extensive research in the past two decades (Martinis et al., 1999; Woese et al., 2000). The binding and catalytic mechanisms of most aaRSs are now well understood, based on which the 20 aaRSs are categorized into two classes. As of biochemical functions, the 20 aaRSs differ only in the recognition of cognate amino acid side chain and tRNA anti-codon, whereas the primary and tertiary structures of aaRSs are highly diverse. Among aaRSs from different species, very low sequence similarity exists between organisms even closely related evolutionarily, indicative of convergent evolution.

A member of the class I aaRS, the arginyl-tRNA synthetase (ArgRS) possesses several distinct characteristics. Prior to the formation of the arginyl-tRNA bond, the activation of arginine by ATP requires the binding of tRNA to the enzyme, which is only observed in GlnRS and GluRS (Mehler and Mitra, 1967; Mitra and Smith, 1969; Kern and Lapointe, 1980). ArgRSs of many organisms lack the canonical KMSK motif which constitutes the active site of all other class I aaRSs (Zhou et al., 1997). There are six sets of codons encoding arginine, which poses considerable challenge to ArgRS insofar as identifying cognate tRNA via matching the anti-codons. Consequently, ArgRS utilizes several highly conserved basis on the D-loop region of tRNA as additional identity elements. To better recognize these identity elements, ArgRS possesses an N-terminal domain

Kelei Bi and Yueting Zheng have contributed equally.

of roughly 100 amino acids—the Add1 domain, which is unique among aaRSs.

Hitherto, several research groups have reported the crystal structures of ArgRS from three species. The 3D structures of yeast ArgRS-L-arginine complex and the ternary complex of ArgRS-L-arginine-tRNA^{Arg} shed considerable light on the mechanism of tRNA recognition in ArgRS (Cavarelli et al., 1998; Delagoutte et al., 2000). The recognition of anti-codons by the enzyme opens the active site of ArgRS, which in part rationalized the dependence on tRNA binding for arginine activation. The crystal structure of the *Thermus thermophilus* ArgRS (ttArgRS) apoenzyme, reported by Shimada et al., further accounted for the critical role of the identity element A20 on tRNA recognition (Shimada et al., 2001). Konno and co-workers determined the structure of ArgRS-tRNA^{Arg} complex from *Pyrococcus horikoshii* (pyroArgRS). The atomistic model of the ATP-bound complex, constructed based on the pyroArgRS structure, depicted the detailed mechanism of arginine activation in the presence of ATP (Konno et al., 2009).

While most research efforts focused on tRNA recognition and ATP-PPi exchange, the details of arginine binding are largely unexplored. Although Cavarelli et al. identified three highly conserved residues in the binding pocket of yArgRS, the thermodynamic contributions of these and other surrounding residues to the overall binding affinity remain elusive. In addition, ArgRS from *E. coli* (eArgRS), which shares only 29% sequence similarity with yArgRS, has no reported crystal structure. Herein, we report a 2.9 Å structure of the eArgRS-L-arginine complex (PDB ID: 4OBY) and a series of mutational studies of the enzyme active site. Our results showed that not all the conserved residues are critical for substrate binding. The thermodynamic details of the ligand binding reaction provided a mechanistic view of how ArgRS achieved high affinity and specificity towards L-arginine. As an enzyme from the model organism *E. coli*, an atomistic model of eArgRS would help further understand the structure-function relationship of ArgRS and provide structural guidance for future modifications of ArgRS in protein engineering.

RESULTS

The overall structure of *E. coli* ArgRS

The refined structure contains the whole sequence except residues Q178–A187, D203–E204 and the C-terminal M577 (Fig. 1A). It can be divided into five domains. The crystallographically invisible residues (178–187 and 203–204) are part of the insertion domain 1 (Ins1) as defined by Cavarelli and co-workers (Cavarelli et al., 1998). *E. coli* ArgRS contains the N-terminal additional domain (Add1, residues 1–112), the catalytic domain (113–382) including two insertion domains (Ins1, residues 164–221, and Ins2, residues 258–311), and the C-terminal additional domain (Add2, 383–577). A well-ordered L-arginine binds in the active site of the catalytic domain. Although 2.5 mmol/L ATP was always present in the crystal growth solution, ATP molecule is not visible in the electronic density map. The difficulty in obtaining a complex structure of ArgRS with ATP stably bound has been noted by several research groups and was attributed to the following two reasons: 1) the presence of both tRNA^{Arg} and L-arginine are prerequisite for ATP binding (Rath et al., 1998); 2) a lack of sufficient hydrophobicity and certain critical residues at the ATP binding site may have prevented the stable binding of ATP and ATP-analog (Konno et al., 2009).

L-arginine binding and recognition

In the refined structure, a well-ordered arginine molecule is present in the active site (Fig. 1A). The active site of eArgRS consists of a Rossman fold (α11, α12) and one strand (β5, Supplemental Figure). L-arginine binds tightly to the enzyme via a network of hydrogen bonds and salt bridge interactions. The main chain atoms of arginine are recognized by a series of residues strictly conserved in ArgRS, e.g. N123 and Q341. The β-bulge between the important strand β5 and helix α5 encompasses the main chain atoms of L-arginine. Three residues in the region (A121, N123 and H132) formed hydrogen bonds to the main chain atoms of L-arginine and adopted similar orientations as observed in yeast structures (Asn153 and Gln375 of yArgRS), independent of tRNA binding (Cavarelli et al., 1998; Delagoutte et al., 2000). Our structure therefore confirms the discovery from yArgRS structures that tRNA is not required for L-arginine binding (Fig. 1). The side chain recognition motif, which dictates the specificity of the enzyme, contains three conserved residues (D118, Y313, D317) making direct H-bonds (or salt bridges) to the guanidinium moiety of arginine. Another residue R324 formed two hydrogen bonds with D118, stabilizing the orientation of the latter.

We mutated the four aforementioned residues to alanine and generated four stable mutant ArgRSs (D118A, Y313A, D317A and R324A). The binding thermodynamics between ArgRS and L-arginine, determined via ITC, confirmed that these four residues were important for the recognition of arginine (Fig. 1B and 1C). Consistent with our structural data, the binding isotherms fit reasonably well to the typical one-to-one binding model (Fig. 2). Removing either of the two D317 and D118 residues completely abolished the binding reaction, while the mutation of Y313 and R324 introduced much less serious reduction in binding affinity (Table 1). In particular, the Tyr-to-Ala mutation at position 313 resulted in a merely two-fold reduction in binding affinity. On the other hand, both the Y313A and the R324A mutation greatly reduced the favorable enthalpy of binding, which were largely compensated for by the favorable change in binding entropy.

The A20 recognition pocket and the Add1 domain

The N-terminal domain (NTD, Add1), a unique structural feature of ArgRS, plays a critical role of recognizing the

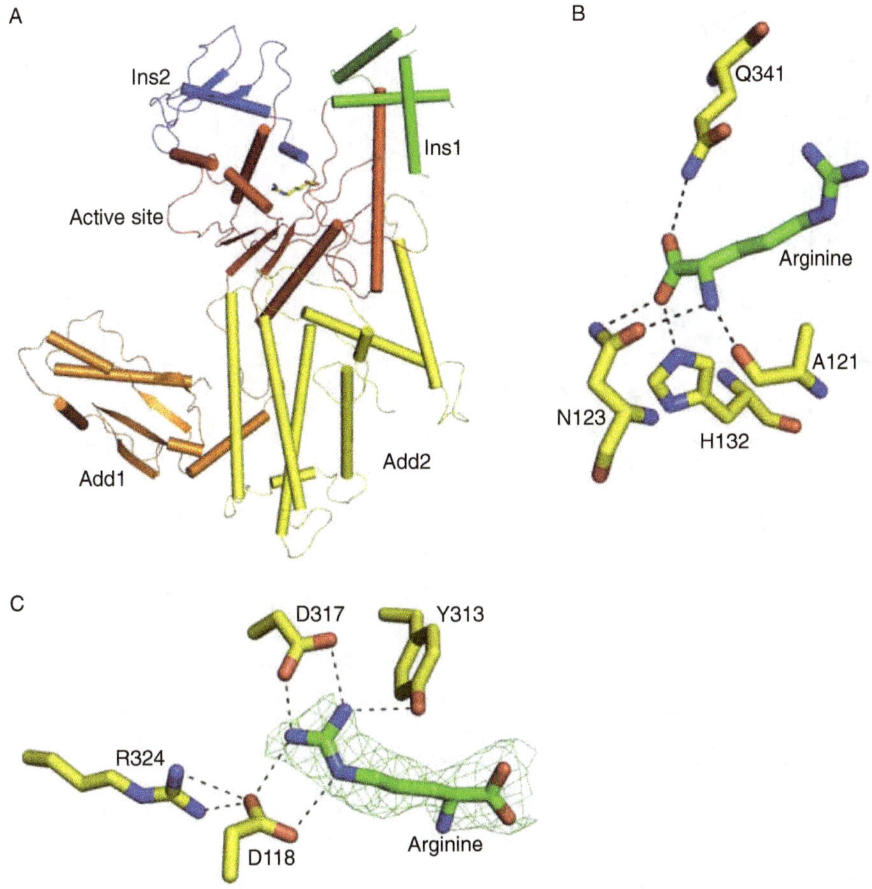

Figure 1. Overall structure of the eArgRS-L-arginine complex and the detailed structure of the binding site. (A) Overall structure of the eArgRS-L-arginine complex. The five domains previously identified by Cavarelli et al. are highlighted in different colors. (B) Four residues form hydrogen bonds with the main chain atoms of arginine. (C) Three conserved residues help recognize and hold the side chain guanidinium moiety. The omit map of L-arginine was generated at σ = 2.5.

identity elements of cognate tRNA. Phylogenetic analysis identified the nucleotide A20 as the major identity element (Liu et al., 1999), which is highly conserved in $tRNA^{Arg}$ of most organisms with only a few exceptions including yeast. The N-terminal Add1 domain accommodates the A20 recognition pocket as found in the structures of ttArgRS and pyroArgRS (Shimada et al., 2001; Konno et al., 2009). The bottom of the pocket is formed by the antiparallel strands β3 and β4. Cavarelli et al. (1998) and Shimada et al. (2001) proposed a Tyr/Phe-Asn/Gln motif in the Add1 domain for A20 recognition. Both the primary structure and tertiary structure analyses confirmed that in eArgRS this motif was present as Y84–N82 (Fig. 3), indicating that A20 recognition is highly conserved in ArgRS.

Although the NTD did not undergo significant movement (relative to the rest of the enzyme) upon tRNA binding (Delagoutte et al., 2000), Shimada et al. noted that the overall orientation of NTD with respect to the rest of the enzyme in ttArgRS was markedly different from that of yArgRS (Shimada et al., 2001). The NTD of eArgRS adopted a different orientation from that of yArgRS as well. When superimposed based on the active site Cα atoms of ArgRS, the L-arginine ligands in yArgRS and in eArgRS aligned nearly perfectly (Fig. 4A). However, when the NTD Cα atoms served as the superposition standard, we observed large deviation between the active sites of both orthologues (Fig. 4B), which is independent of tRNA and L-arginine binding (Cavarelli et al., 1998; Delagoutte et al., 2000).

Difference in the NTD orientations of various ArgRSs most likely reflected the difference between the relative orientations of the tRNA identity element region and those of the acceptor stem region. We picked helix α1 and the strand β5 as representatives of the NTD and the active site respectively, and calculated the angle between these two motifs (ζ). The three reported yArgRS structures had an average ζ of 125.2° ± 0.8°, while for eArgRS, ttArgRS and

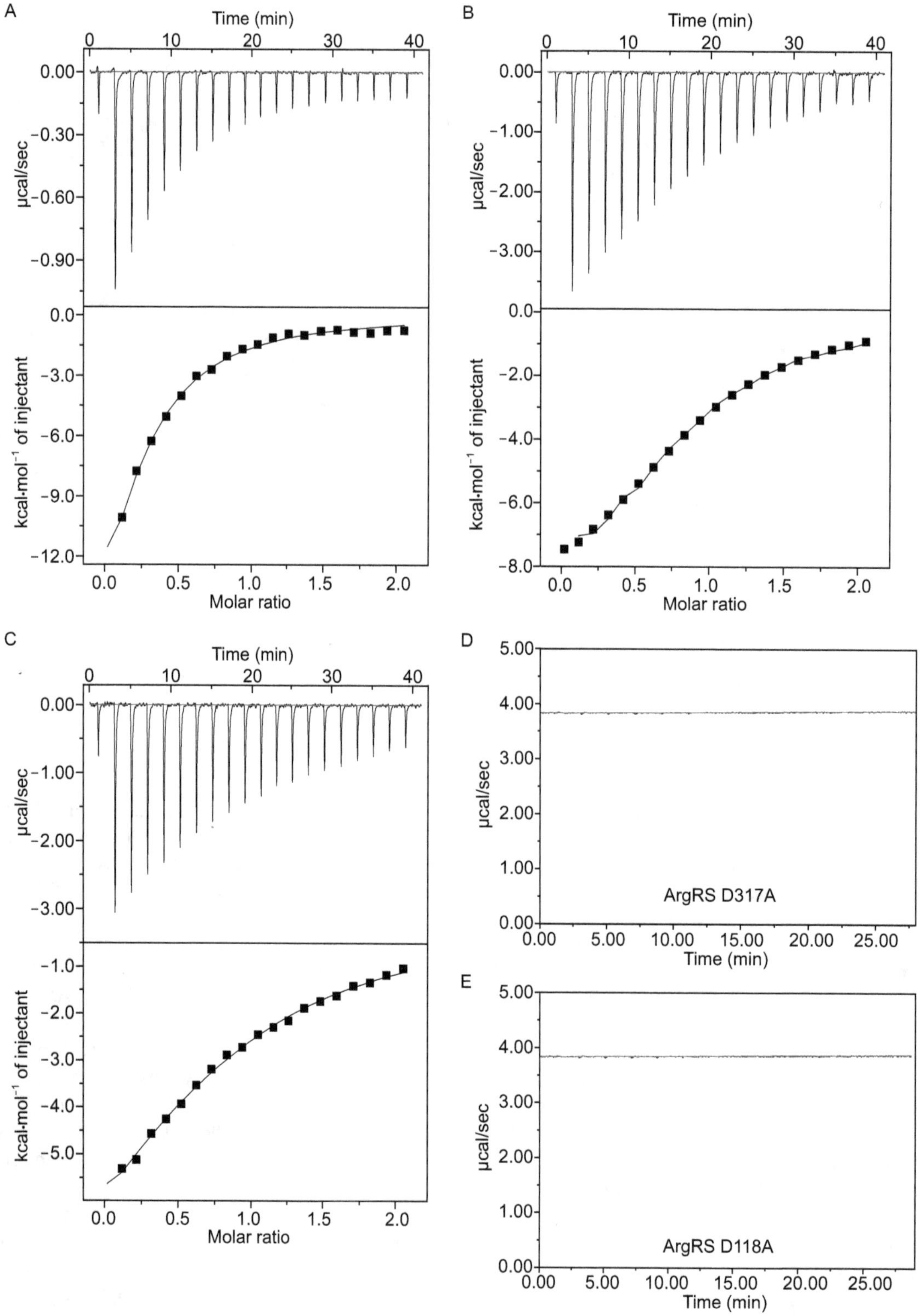

Figure 2. Binding isotherms of WT and mutant ArgRS with L-arginine. (A) Wild type; (B) Y313A; (C) R324A; (D) D317A; (E) D118A.

pyroArgRS, ζ were 120.0°, 122.1° and 113.9°, respectively. If we regard the archaic pyroArgRS as the most primitive enzyme in regard to evolution, it appears that ArgRSs of higher species tend to have larger ζ than those of lower ones in general. Nonetheless, the exact angle is highly species-dependent.

Table 1. Binding thermodynamics of WT and mutant eArgRS with L-arginine

ArgRS	Stoichiometry (N)	K_a (M^{-1})	ΔG (kcal/mol)	ΔH (kcal/mol)	-T•ΔS (kcal/mol)
WT	0.69	2.15E4	−5.9	−16.0	10.1
Y313A	0.86	1.24E4	−5.4	−10.5	5.1
R324A	0.82	5.23E3	−4.8	−11.3	6.4
D317A	N/A				
D118A	N/A				

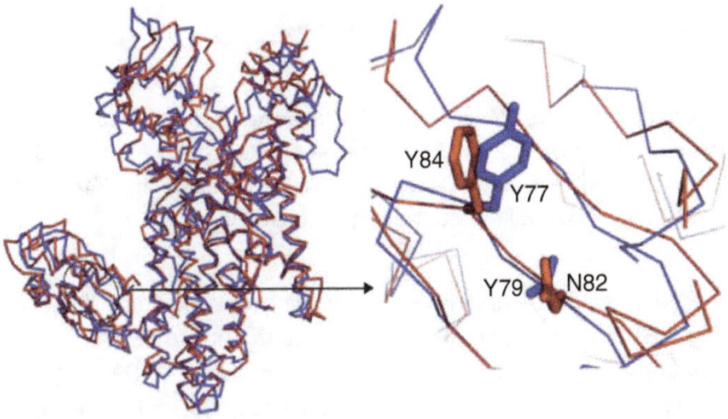

Figure 3. Residues critical for recognizing the identity element A20: yeast (blue, PDB ID: 1BS2), eArgRS (red).

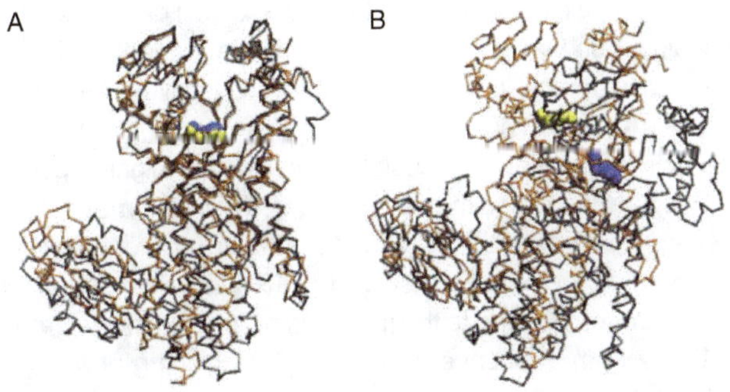

Figure 4. Different NTD orientations with respect to the catalytic domain in eArgRS and yArgRS. (A) yArgRS (gray, PDB ID: 1BS2) is superimposed to eArgRS (orange) according to the active site domain. The bound L-arginine molecules are depicted as blue (yArgRS) and yellow (eArgRS) vdW spheres. (B) yArgRS is superimposed to eArgRS according to the NTD.

The Ω-loop

The Ω-loop (A451–A457), connecting helices α15 and α16, acts as a molecular switch upon tRNA binding (Delagoutte et al., 2000). Using the residues spanning Ω-loop, α15 and α16 as superposition standard, the eArgRS Ω-loop structure aligned perfectly with the yArgRS-Arg complex (Table 2). The conformational change in this region upon tRNA binding manifests mainly as larger deviation of Cα atoms from eArgRS in α15 (Table 2).

Table 2. Cα RMSD between eArgRS-arginine structure and yeast structure

Secondary structure	yArgRS-tRNA-arginine	yArgRS-tRNA	yArgRS-arginine
H17(yeast)-α15	1.2 Å	1.3 Å	0.5 Å
H18(yeast)-α16	2.4 Å	2.4 Å	0.3 Å
Ω-loop	1.2 Å	1.2 Å	0.4 Å

Yao and co-workers measured the lifetimes of the tRNA-bound and free conformations of this α15-Ω-α16 region in eArgRS using ^{19}F NMR spectroscopy (Yao et al., 2003, 2004). The interconversion between the bound and the free conformations proceeds in the intermediate to slow exchange regime of ^{19}F NMR time scale. In the eArgRS structure, we observed a special structural feature that is absent in the yeast orthologue. R400 is in close contacts with both W446 and E396 (Fig. 5). While being stabilized by the salt bridge from E396, the guanidinium moiety of R400 adopts a conformation parallel to the indole ring of W446, where a cation-π interaction is likely in play. Adding to more constraints of W446, it is also surrounded by a series of non-polar side chains, which resembles the yArgRS structures. Sequence alignment shows that yArgRS also possesses these three residues on helices H15 and H17 (helices α13 and α15 in eArgRS): E424, K428 and W475 (Fig. 5). However, W475 of yArgRS locates too far from K428 to allow any

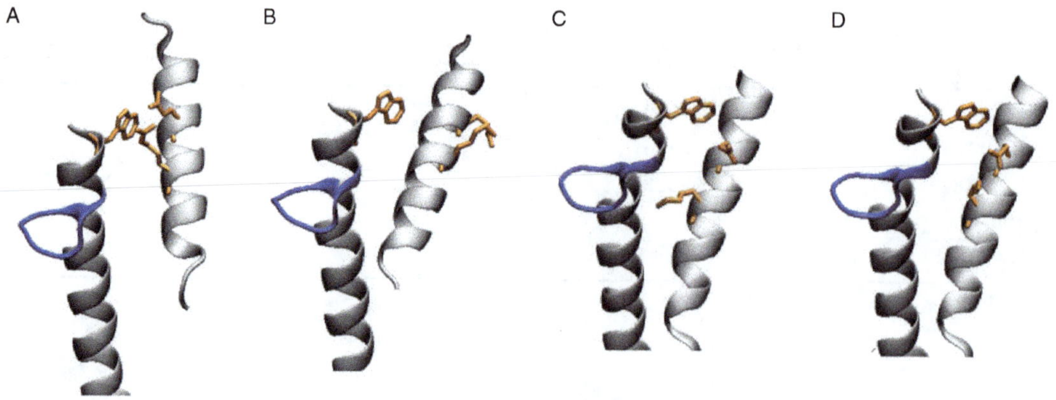

Figure 5. Structural comparison of the triad W446-R400-E396 in eArgRS and W475-K428-E424 in yArgRS. Comparison between the eArgRS-L-arginine structure (A) and the yArgRS-L-arginine structure (B–D, PDB ID: 1BS2, 1F7U, 1F7V) at the region of α15-Ω-α16 and α13 (H17-Ω-H18 and H15 in yeast). The Ω-loop, is highlighted in blue. The W446-R400-E396 triad of *E. coli* and the corresponding residues of yArgRS (W475-K428-E424) are highlighted in orange.

direct contact. yArgRS without tRNA has E424 and K428 pointing away from the Ω-loop (Fig. 5B). The binding of tRNA to yArgRS results in a ~90° rotation of helix H15 and a downward shift by about one helix turn (Fig. 5C and 5D) (Delagoutte et al., 2000). If such conformational change is maintained in eArgRS upon tRNA binding, the W446-R400 cation-π interaction would be absent in the tRNA-bound form, which might in turn rationalize the slow exchange kinetic feature measured in ^{19}F NMR experiment (Yao et al., 2004). That is, the W446-R400 interaction serves to lengthen the lifetime of the tRNA-free conformation. Enzymatic assays and preferably ^{19}F-NMR binding kinetics study of yArgRS, and the eArgRS-tRNA complex structure are required to further confirm the role of this W446-R400-E396 triad in controlling the conformation of Ω-loop and the binding kinetics of cognate tRNA.

DISCUSSION

In this study, we solved the crystal structure of *E. coli* arginyl-tRNA synthetase in complex with its substrate L-arginine, and carried out ITC studies of a series of mutant enzymes. In addition to providing more information regarding the mechanism of tRNA binding by ArgRS, our study shed considerable light on the less understood arginine recognition.

Although several residues in the arginine binding pocket are highly conserved among ArgRS, or even among many aaRS's, our mutational study showed that some residues may in fact be non-essential for achieving high binding affinity. The elimination of electrostatic interactions by mutating either D317 or D118 abolished the binding of L-arginine, while the enzyme without Y313 or R324 retained some of the wild type's ligand binding ability. In contrast to the small changes in binding free energy, a closer look at the thermodynamic data of the Y313A and R324A mutants revealed much greater changes in binding enthalpy and binding entropy from those of the wild type. The well-documented enthalpy-entropy compensation manifested in both mutants (Dunitz, 1995; Lemieux, 1996; Perozzo et al., 2004). R324 does not make direct contact with the bound arginine, but serves to lock the side chain of D118 at a favorable rotameric conformation (Fig. 1C). Without the salt bridge from R324, D118 was allowed to sample a much larger configurational space and might be adopting a less favorable conformation upon the binding of arginine, giving rise to a 4.7 kcal/mol drop in binding enthalpy. Such large decrease of favorable (negative) binding enthalpy was well compensated by a 3.7 kcal/mol favorable (positive) change in TΔS. That is, L-arginine is so tightly bound in the active site that any destabilization of the highly specific interactions increases the overall degrees of freedom.

A comparison with the yArgRS structure without arginine showed that the highly conserved Y313 was solvent-exposed and hydrogen bonded to the main chain carbonyl oxygen atom of W162 in the absence of arginine. The binding of arginine induced the rotation of the Cα-Cβ bond in Y313 and the Y313-W162 H-bond switched to the Y313-L-arginine H-bond. Mutation of W162 to alanine diminished the catalytic activity of eArgRS (Zhang et al., 1998). Using ^{19}F NMR spectroscopy, Yao et al. showed that W162 was involved in arginine binding (Yao et al., 2003). Since only the main chain atoms participate in the Y313-W162 hydrogen bond, the side chain of W162 may be involved in maintaining a favorable orientation of the backbone carbonyl group towards the Y313 side chain hydroxyl group. Presumably, the Y313A mutation should introduce negligible change in binding enthalpy, given that Y313 is always involved in hydrogen bonding prior and after the binding of L-arginine. However, the drop of ΔH in the Y313A mutant from that of the wild type was as large as 5.5 kcal/mol. A closer look at the active site of the eArgRS-L-arginine complex revealed a closed loop of hydrogen bonding moieties, consisting of L-arginine, Q341 and Y313 (Fig. 6A). Y313 accepts a hydrogen bond from the

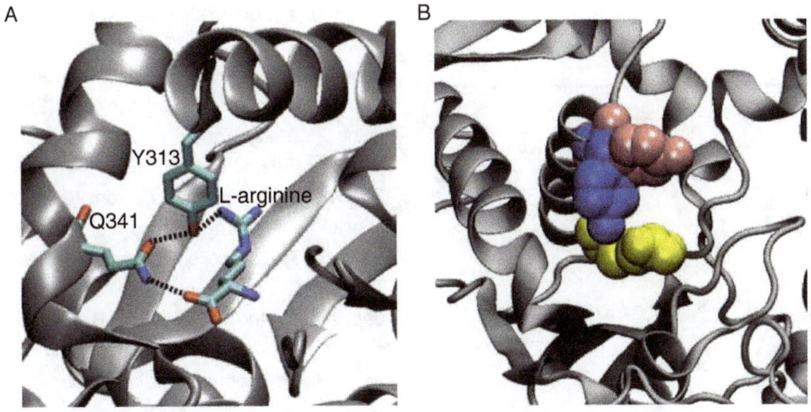

Figure 6. Thermodynamic and kinetic roles of residues in the L-arginine binding site. (A) Y313, Q341 and L-arginine comprise a cooperative ring of hydrogen bonds (black dashed line). (B) Y313 locks the bound arginine (yellow) in the active site. Upon the binding of arginine, the phenol ring of Y313 swings from a horizontal orientation (pink) to a vertical downward orientation (blue), which shuts the active site. The coordinates of the phenol ring in the absence of arginine (pink) were taken from the corresponding residue of Y347 in yeast structure (PDB ID: 1F7V).

guanidinium group of L-arginine and donates a hydrogen bond to the amide carbonyl group of Q341. Q341 in turn donates a hydrogen bond to the carbonyl group of L-arginine. The cooperative effect renders this hydrogen bond network highly favorable energetically. In the Y313A mutant however, Q341 would participate in only one hydrogen bond with the carbonyl group of L-arginine. It is not uncommon to have an enthalpy difference as large as 5 kcal/mol between the presence and absence of cooperative hydrogen bonding (Elrod and Saykally, 1994). On the other hand, the hydrogen bond loop of L-arginine-Q341-Y313 imparted restriction to the mobility of L-arginine. The removal of Y313 from the loop significantly increased the degrees of freedom for L-arginine. As a result, the favorable change in $T\Delta S$ nearly canceled the unfavorable change in ΔH, leaving the difference in ΔG at only 0.5 kcal/mol.

Although Y313 is strictly conserved among GlnRS, GluRS, TyrRS, TrpRS and ArgRS, this tyrosine residue assumes different roles in different aaRS's. In particular, Y313 of eArgRS acts as a gate-keeper: keeping the binding pocket open in the absence of arginine via the Y313-W162 H-bond, while locking the bound arginine in position after rotameric rotation. Y313 contributes more in the kinetics regime of arginine binding rather than in the regime of binding thermodynamics (affinity)—keeping the substrate in the active site sufficiently long for catalysis to take place (Fig. 6B). Further NMR binding kinetics study and enzymatic assays of the mutants are required to confirm our conclusion based on crystal structures.

MATERIALS AND METHODS

Gene expression and protein purification

We used the plasmid containing ArgRS gene derived from *E. coli* as a template to amplify the ArgRS gene, using the Primer5 software (http://www.uea.ac.uk/~e130/Primer5.htm) to design the up primer 5′-CGGGATCCATGAATATTCAGGCTCTTCTCTC-3′ and the down primer 5′-CCCAAGCTTTTACATACGCTCTACTGTCTC-3′. The gene was inserted into another plasmid vector pET-28a-SUMO. We transformed the pET28a-SUMO/ArgRS plasmid into the expression bacterial *E. coli* BL21 (DE3) (Novagen, Madison, WI). Protein expression was induced with 0.2 mmol/L isopropyl β-D-1-thiogalactopyranoside (IPTG) at an optical cell density (OD_{600}) of ~0.8, followed by incubation at 16°C for 20 h. Cell pellet was sonicated with lysis buffer (20 mmol/L Tris-HCl pH 7.5, 300 mmol/L NaCl, 10 mmol/L β-mercaptoethanol, 1 mmol/L PMSF). The lysate was centrifuged at 18,000 rpm for 30 min. The supernatant was incubated with Ni^{2+}-NTA affinity resin (Qiagen) for 3 h at 4°C, then the resin was eluted with an imidazole gradient up to 200 mmol/L imidazole and the target protein was eluted with 100 mmol/L imidazole. The protein was dialyzed overnight against a reservoir solution of 20 mmol/L Tris-HCl (pH 7.5) and 300 mmol/L NaCl in the presence of Ulp1 protease (made in house) which cleaved the N-His-SUMO tag from the target protein (enzyme:protein = 1:30 *w/w*). The released His-SUMO tag was removed by a second round of Ni^{2+}-NTA chromatography. The protein was further purified by size exclusion chromatography using HiLoad 16/60 Superdex 200 column (GE Healthcare) in gel-filtration buffer (20 mmol/L Tris-HCl pH 7.5, 300 mmol/L NaCl). The volume of the collected protein sample was reduced to reach a final concentration of 10 mg/mL.

Crystallization and data collection

Initial crystallization was screened by several commercial kit, such as crystal I/II and Index (Hampton Research) and crystals were obtained in the reservoir containing 100 mmol/L Tris-HCl pH 8.5, 200 mmol/L ammonium acetate, 25% PEG-3350. We mixed the protein sample with its substrate including 2 mmol/L L-arginine, 5 mmol/L ATP. At last, the best crystal was grown at 16°C by sitting drop diffusion method against a reservoir solution of 50 mmol/L HEPES (pH 7.2), 100 mmol/L sodium acetate, 22% PEG-3350. We mixed the 1 μL reservoir with 1 μL protein solution which contained 10 mg/mL ArgRS, 2 mmol/L L-arginine, 5 mmol/L ATP, 10 mmol/L $MgCl_2$.

Table 3. Statistics of data collection and model refinement

Data collection	
Space group	*C2*
Unit cell parameters (Å)	a = 118.60, b = 94.14, c = 62.72; α = γ = 90°, β = 110.5°
Resolution range (Å)	50–2.60 (2.69–2.60)
No. of total reflections	133676 (12309)
No. of unique reflections	20334 (2018)
I/σ	13.6 (4.8)
Completeness (%)	99.5 (99.4)
R_{merge} (%)[a]	12.0 (65.1)
Structure refinement	
Resolution (Å)	30–2.6
R_{cryst}/R_{free} (%)[b]	23.9/26.5
r.m.s. deviations	
Bond bonds (Å)	0.003
Bond angles (°)	0.750
Ramachandran plot	
Favored region (%)	93.5
Allowed region (%)	4.5
Outlier region (%)	2.0
No. of atoms	
Protein	4248
Water	30
Ligand	12

[a] $R_{merge} = \Sigma|I_i - I_m|/\Sigma I_i$, where I_i is the intensity of the measured reflection and I_m is the mean intensity of all symmetry related reflections.

[b] $R_{cryst} = \Sigma||F_{obs}| - |F_{calc}||/\Sigma|F_{obs}|$, where F_{obs} and F_{calc} are observed and calculated structure factors.

$R_{free} = \Sigma_T||F_{obs}| - |F_{calc}||/\Sigma_T|F_{obs}|$, where T is a test data set of about 5% of the total reflections randomly chosen and set aside prior to refinement.

Numbers in parentheses represent the value for the highest resolution shell.

Crystal grew in two days and its shape was rectangle with sharp edge. Ethylene glycol was used as cryoprotectant. We first soaked the crystal in the 10% ethylene glycol and then transferred the crystal into a higher ethylene glycol concentration up to 20%. Finally, the crystal was preserved in the liquid nitrogen.

For data collection, Diffraction data was collected at the beamline BL17U of Shanghai Synchrotron Radiation Facility (SSRF) using an ADSQ315 CCD detector. Data were processed with HKL2000 (Otwinowski and Minor, 1997). The crystal structure of ArgRS belonged to the space group C_2. The partial coordinates of yeast ArgRS (PDB 1BS2) (Cavarelli et al., 1998) as ensemble 1 and the structure 3GDZ as ensemble 2 were used as search models for ArgRS using the program Phaser from CCP4 package (Read, 2001). The initial model was built using the rigid body refinement with Phenix.refine in the Phenix program (Adams et al., 2010). Subsequent refinement was carried out with alternating cycles of manual refitting and building under 2Fo-Fc and 1Fo-Fc electron density map in Coot (Emsley and Cowtan, 2004) and using XYZ coordinates and individual B factors in the Phenix.refine program (Adams et al., 2010). The *R*-working and *R*-free dropped to 23.9% and 26.5% for all data from 30 Å to 2.6 Å. The final structure was checked for geometrical correctness with PROCHECK (Laskowski et al., 1993). Data collection and refinement statistics are summarized in Table 3. The atomic coordinates and the structure factor have been deposited in the Protein Data Bank (PDB code: in submission). All structural figures were generated using PyMOL (http://www.pymol.org) and VMD (Humphrey et al., 1996).

Isothermal titration calorimetry study of the wild type and mutant ArgRS

We designed four ArgRS mutants including D118A, D317A, Y313A, R324A based on ArgRS structure. The mutation was introduced into pET22b/ArgRS using the site-directed mutagenesis and purified these proteins with the same method as described previously. The formation constant and thermodynamic parameters for the inclusion of Arginine in ArgRS were measured by the titration calorimetry method by using an ITC MicroCal 200 (GE Life Sciences). All solutions were prepared in a 200 mmol/L Tris-HCl pH 7.5, 300 mmol/L sodium chloride. A solution (0.25 mmol/L) of ArgRS was placed in the sample cell, and a 2.5 mmol/L solution of Arginine was added in a series of 20 injections, the heat evolved was recorded at 25°C. The heat of injecting the Arginine into a neat buffer solution is nearly zero. The data were analyzed and the binding isotherm was fitted to a single-site model in the ORIGIN 7.0 software (GE Life Sciences).

ACKNOWLEDGEMENTS

This project was funded by the National Basic Research Program (973 Program) (No. 2010CB912301), the Chinese Academy of Sciences (KSCX2-EW-G-7-2) and the National Natural Science Foundation of China (Grant No. 91219202).

ABBREVIATIONS

AMP-CPP, adenosine 5′-α,β-methylene triphosphate; ArgRS, arginyl-tRNA synthetase; ATP, adenosine triphosphate; HEPES, 4-(2-hydroxyethyl)-1-piperazine ethanesulfonic acid; ITC, isothermal titration calorimetry; Ni^{2+}-NTA, Ni^{2+} nitrilotriacetic acid; NTD, N-terminal domain; PMSF, phenylmethanesulfonyl fluoride.

COMPLIANCE WITH ETHICS GUIDELINES

Kelei Bi, Yueting Zheng, Feng Gao, Jianshu Dong, Jiangyun Wang, Yi Wang and Weimin Gong declare that they have no conflict of interest.

This article does not contain any studies with human or animal subjects performed by the any of the authors.

REFERENCES

Adams PD, Afonine PV, Bunkoczi G, Chen VB, Davis IW, Echols N, Headd JJ, Hung LW, Kapral GJ, Grosse-Kunstleve RW et al (2010) PHENIX: a comprehensive Python-based system for macromolecular structure solution. Acta Crystallogr D 66:213–221

Cavarelli J, Delagoutte B, Eriani G, Gangloff J, Moras D (1998) L-Arginine recognition by yeast arginyl-tRNA synthetase. Embo J 17:5438–5448

Delagoutte B, Moras D, Cavarelli J (2000) tRNA aminoacylation by arginyl-tRNA synthetase: induced conformations during substrates binding. Embo J 19:5599–5610

Dunitz JD (1995) Win some, lose some: enthalpy–entropy compensation in weak intermolecular interactions. Chem Biol 2:709–712

Elrod MJ, Saykally RJ (1994) Many-body effects in intermolecular forces. Chem Rev 94:1975–1997

Emsley P, Cowtan K (2004) Coot: model-building tools for molecular graphics. Acta Crystallogr D 60:2126–2132

Hendrickson TL, de Crecy-Lagard V, Schimmel P (2004) Incorporation of nonnatural amino acids into proteins. Annu Rev Biochem 73:147–176

Humphrey W, Dalke A, Schulten K (1996) VMD: visual molecular dynamics. J Mol Graph Model 14:33–38

Kern D, Lapointe J (1980) The catalytic mechanism of glutamyl-transfer rna-SYNTHETASE OF *Escherichia coli*. Evidence for a 2-step aminoacylation pathway, and study of the reactivity of the intermediate complex. Eur J Biochem 106:137–150

Konno M, Sumida T, Uchikawa E, Mori Y, Yanagisawa T, Sekine S, Yokoyama S (2009) Modeling of tRNA-assisted mechanism of Arg activation based on a structure of Arg-tRNA synthetase, tRNA, and an ATP analog (ANP). Febs J 276:4763–4779

Laskowski RA, Macarthur MW, Moss DS, Thornton JM (1993) Procheck: a program to check the stereochemical quality of protein structures. J Appl Crystallogr 26:283–291

Lemieux RU (1996) How water provides the impetus for molecular recognition in aqueous solution. Acc Chem Res 29:373–380

Liu W, Huang YW, Eriani G, Gangloff J, Wang ED, Wang YL (1999) A single base substitution in the variable pocket of yeast tRNA (Arg) eliminates species-specific aminoacylation. Biochim Biophys Acta 1473:356–362

Martinis SA, Plateau P, Cavarelli J, Florentz C (1999) Aminoacyl-tRNA synthetases: a new image for a classical family. Biochimie 81:683–700

Mehler AH, Mitra SK (1967) Activation of arginyl transfer ribonucleic acid synthetase by transfer ribonucleic acid. J Biol Chem 242:5495

Mitra SK, Smith CJ (1969) Absolute requirement for transfer RNA in activation of arginine by arginyl transfer RNA synthetase of yeast. Biochim Biophys Acta 190:222

Otwinowski Z, Minor W (1997) Processing of X-ray diffraction data collected in oscillation mode. Method Enzymol 276:307–326

Perozzo R, Folkers G, Scapozza L (2004) Thermodynamics of protein-ligand interactions: history, presence, and future aspects. J Recept Signal Transduct Res 24:1–52

Rath VL, Silvian LF, Beijer B, Sproat BS, Steitz TA (1998) How glutaminyl-tRNA synthetase selects glutamine. Structure 6:439–449

Read RJ (2001) Pushing the boundaries of molecular replacement with maximum likelihood. Acta Crystallogr D 57:1373–1382

Shimada A, Nureki O, Goto M, Takahashi S, Yokoyama S (2001) Structural and mutational studies of the recognition of the arginine tRNA-specific major identity element, A20, by arginyl-tRNA synthetase. Proc Natl Acad Sci USA 98:13537–13542

Sinkeldam RW, Greco NJ, Tor Y (2010) Fluorescent analogs of biomolecular building blocks: design, properties, and applications. Chem Rev 110:2579–2619

Wang L, Schultz PG (2005) Expanding the genetic code. Angew Chem Int Ed 44:34–66

Wang KH, Schmied WH, Chin JW (2012) Reprogramming the genetic code: from triplet to quadruplet codes. Angew Chem Int Ed 51:2288–2297

Woese CR, Olsen GJ, Ibba M, Soll D (2000) Aminoacyl-tRNA synthetases, the genetic code, and the evolutionary process. Microbiol Mol Biol R 64(1):202–236

Yao YN, Zhang QS, Yan XZ, Zhu G, Wang ED (2003) Substrate-induced conformational changes in *Escherichia coli* arginyl-tRNA synthetase observed by F-19 NMR spectroscopy. Febs Lett 547:197–200

Yao YN, Zhang QS, Yan XZ, Zhu G, Wang ED (2004) *Escherichia coli* tRNA(4)(Arg)(UCU) induces a constrained conformation of the crucial Omega-loop of arginyl-tRNA synthetase. Biochem Biophys Res Commun 313:129–134

Zhang QS, Wang ED, Wang YL (1998) The role of tryptophan residues in *Escherichia coli* arginyl-tRNA synthetase. Biochim Biophys Acta 1387:136–142

Zhou M, Wang ED, Campbell RL, Wang YL, Lin SX (1997) Crystallization and preliminary X-ray diffraction analysis of arginyl-tRNA synthetase from *Escherichia coli*. Protein Sci 6:2636–2638

2

DNA replication components as regulators of epigenetic inheritance—lesson from fission yeast centromere

Haijin He[1], Marlyn Gonzalez[1], Fan Zhang[2], Fei Li[1✉]

[1] Department of Biology, New York University, New York, NY 10003, USA
[2] Clinical Translational Research Center, Shanghai Pulmonary Hospital, Tongji University School of Medicine, Shanghai 200433, China
✉ Correspondence: fl43@nyu.edu (F. Li)

ABSTRACT

Genetic information stored in DNA is accurately copied and transferred to subsequent generations through DNA replication. This process is accomplished through the concerted actions of highly conserved DNA replication components. Epigenetic information stored in the form of histone modifications and DNA methylation, constitutes a second layer of regulatory information important for many cellular processes, such as gene expression regulation, chromatin organization, and genome stability. During DNA replication, epigenetic information must also be faithfully transmitted to subsequent generations. How this monumental task is achieved remains poorly understood. In this review, we will discuss recent advances on the role of DNA replication components in the inheritance of epigenetic marks, with a particular focus on epigenetic regulation in fission yeast. Based on these findings, we propose that specific DNA replication components function as key regulators in the replication of epigenetic information across the genome.

KEYWORDS epigenetic inheritance, DNA replication, centromere, histone modification, heterochromatin, euchromatin

INTRODUCTION

In addition to the genetic code, which is defined by DNA sequence, cells contain another layer of regulatory information, referred to as the "epigenetic code". The epigenetic code consists mainly of chemical modifications of DNA and histone proteins, and plays an important role in gene expression regulation and genome stability through altering chromatin states (Strahl and Allis, 2000; Couture and Trievel, 2006). Like the genetic code, the epigenetic code can be faithfully transmitted for many generations. This phenomenon is referred to as "epigenetic inheritance" (Zhu and Reinberg, 2011). How genetic information is duplicated and transmitted to subsequent generations is well understood; it is achieved through a series of coordinated interactions between conserved DNA replication components during the S phase of the cell cycle. The process is initiated by the binding of the origin recognition complex (ORC) to replication origins, followed by assembly of a pre-replication complex (preRC), which includes Cdc6, Cdt1, and the MCM complex (Dutta and Bell, 1997; Li and Stillman, 2012). MCM helicase activity is required for unwinding the DNA duplex into single stranded DNA, which leads to the recruitment of proliferating cell nuclear antigen (PCNA) to the replication fork. PCNA functions as a platform for recruitment of a variety of replication factors, including DNA polymerase δ (Pol δ), and Pol ε. Pol ε is important for subsequent leading strand synthesis while Pol δ is required for lagging strand synthesis (Waga and Stillman, 1998; Burgers, 2009). During DNA replication, epigenetic information must also be transmitted to subsequent generations. How epigenetic inheritance is achieved remains poorly understood. In this review, we will discuss recent progress toward our understanding of the role of DNA replication machinery in epigenetic inheritance, particularly on the context of chromatin inheritance at centromeres in the fission yeast *Schyzosaccharomyces pombe*.

Chromatin consists of repeating particles of ~100 Å in size, termed nucleosomes. Nucleosomes are composed of

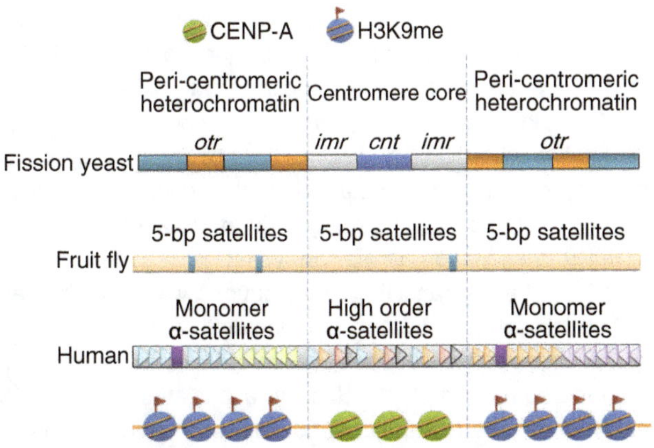

Figure 1. Schematic of centromere structure in fission yeast, *D. melanogaster*, and humans. Fission yeast centromere contains a central core region (*cnt*), flanked by *imr* and *otr* repeat regions. Each of the *otr* repeats can be divided into 2 smaller repeat units: *dg* (blue color) and *dh* (orange color). *imr* and *otr* repeat regions are heterochromatic. The sequence and size of DNA repeats in *D. melanogaster* and human centromeres are different from those in fission yeast (Schueler et al., 2001; Sun et al., 2003), but the general chromatin structure and epigenetic profile are conserved among these organisms. At core centromeres, the histone H3 variant, CENP-A, replaces histone H3 in nucleosomes, while histone H3K9 methylation is enriched at the peri-centromeric heterochromatin.

two copies of each of four canonical histone proteins H2A, H2B, H3, and H4. These barrel-like structures wrap around 146 base pairs of DNA, and are linked by histone H1. Strings of nucleosomes fold into higher order chromatin structures that are orderly packaged within the nucleus (Luger, 2003). The centromere is a specific chromatin structure responsible for correct segregation of chromosomes during mitosis and meiosis (Pluta et al., 1995; Amor et al., 2004). In most eukaryotes, centromeres are organized into two distinct domains: a centromeric core, and peri-centromeric heterochromatin that flanks the core (Fig. 1). The centromeric core is where the kinetochore assembles. The kinetochore is a multi-protein complex that mediates the attachment of spindle microtubules to centromeres during chromosomes segregation (McIntosh et al., 2002; Morris and Moazed, 2007; Bloom and Yeh, 2010). Defects in centromere regulation are catastrophic to cells and result in chromosome mis-segregation and an abnormal number of chromosomes, a phenomenon known as aneuploidy (Morris and Moazed, 2007; Weaver and Cleveland, 2007).

Centromeric core chromatin is defined by CENP-A, a centromere-specific histone 3 (H3) variant. CENP-A partially replaces canonical H3 histone at centromeres and is responsible for nucleating the formation of the kinetochore (Palmer et al., 1991; Black and Cleveland, 2011). In most eukaryotic centromeres, the centromeric core contains more than one CENP-A nucleosome, and is therefore bound by multiple spindle microtubules. Such centromeres are referred to as "regional centromeres" (Pluta et al., 1995). In contrast, budding yeast centromeres which consists of a single CENP-A nucleosome bound by a single spindle microtubule, are referred to as "point centromeres". Point centromeres are defined by the underlying DNA sequence, which consists of 125 bp of DNA wrapped around the sole CENP-A nucleosome (Cottarel et al., 1989). In contrast, in regional centromeres, although their architecture and function are conserved across eukaryotes, their underlying DNA varies significantly in size and sequence across species, suggesting that epigenetic mechanisms are important determinants in their regulation. CENP-A is considered the most likely candidate epigenetic mark for centromere identity and has been the subject of intense study (Henikoff and Furuyama, 2010).

Immediately flanking CENP-A-bound centromeric core is the peri-centromeric heterochromatin. This condensed and transcriptionally silent region is essential for gene regulation, genome stability and chromosome segregation. Peri-centromeric heterochromatin displays a protein binding profile and epigenetic environment that is clearly different from that of centromeric cores. The heterochromatic region is enriched in histone 3 lysine 9 (H3K9) methylation and devoid of H3K4 methylation. In fact, H3K9 methylation is considered an epigenetic hallmark of heterochromatin, conserved from fission yeast to humans (Fig. 1) (Rice and Allis, 2001; Carroll and Straight, 2006). This modification serves as the binding site for the highly conserved chromodomain protein, heterochromatin protein 1 (HP1). HP1 is a classic epigenetic "reader" protein, which can recognize specific epigenetic marks (Rice and Allis, 2001; Taverna et al., 2007). Histones in heterochromatin are also hypo-acetylated. This contrasts with active chromatin (euchromatin) in which H3K4 methylation and hyper-acetylation are enriched (Rice and Allis, 2001).

CENTROMERES IN FISSION YEAST

Fission yeast has emerged as an excellent model system for cell cycle and chromatin studies in the last several decades. Because of its similarity to mammalian cells at the most fundamental levels, it has gained the nickname of "micro-mammal" (Forsburg and Rhind, 2006). Fission yeast has a relatively small genome of 13.8 Mb harboring ~4800 genes, many of which exist as single copy. Similar to budding yeast, fission yeast is amenable to genetic and biochemical manipulations. Importantly, in contrast to the point centromeres of budding yeast, fission yeast contains "regional centromeres", which resemble those of higher eukaryotes (Carroll and Straight, 2006). These characteristics make fission yeast especially useful for dissecting the mechanisms governing heterochromatin and centromere structure and function.

Fission yeast contains three chromosomes that range in size from 40 kb to 110 kb. Like in multicellular organisms, centromeres in *S. pombe* consist of a core region (*cnt*, centromere core domain) and peri-centromeric heterochromatin. Centromeric cores span about 4–7 kb and share only limited homology. Immediately flanking the core region are the imperfect repeat regions referred to as innermost repeats (*imr*). *S. pombe*'s CENP-A homolog, Cnp1, is enriched at centromeric cores and *imr* regions. Flanking *imr* regions are the outermost repeat regions (*otr*) also composed of tandem DNA repeats (Chikashige et al., 1989; Allshire and Karpen, 2008). Both *imr* and *otr* regions are heterochromatic in nature. Exogenous marker genes inserted within these regions are silenced, although *otr* regions appear more strongly silenced than *imr* regions (Allshire et al., 1995). Each repeat unit within an *otr* region is composed of the two smaller repeat units, *dg* and *dh* (Wood et al., 2002). Distinct heterochromatin regions are also present in telomeric and mating type loci, which are all rich in tandem DNA repeats.

REGULATION OF PERI-CENTROMERIC HETEROCHROMATIN

Role of H3K9 methylation in heterochromatin formation

In fission yeast, like in higher eukaryotes, peri-centromeric heterochromatin is enriched with H3K9 methylation (Fig. 1). This modification is catalyzed by Clr4, a homolog of the mammalian histone methyltransferase, SUV39H1 (Nakayama et al., 2001b). The C-terminal end of Clr4 harbors a SET domain responsible for the enzyme's catalytic activity, while the N-terminal domain contains a chromodomain region that mediates binding of the enzyme to H3K9 methylation groups (Min et al., 2002; Zhang et al., 2008). Methylating H3K9 is pivotal for heterochromatin formation. Deletion of Clr4 results in loss of H3K9 methylation and heterochromatin silencing. In addition, artificially tethering Clr4 to euchromatic regions is sufficient to induce heterochromatin formation *de novo* (Kagansky et al., 2009).

Previous studies have identified Clr4 as a component of the multi-subunit Clr4 methyltransferase complex (ClrC). In addition to Clr4, the complex also contains Cul4, Rik1, Dos1 (also known Raf1/Cmc1/Clr8), Dos2 (Raf2/Cmc2/Clr7), and Lid2. Disruption of any of these components leads to severe reduction in H3K9me and silencing at heterochromatic regions (Allshire et al., 1995; Hong et al., 2005; Horn et al., 2005; Jia et al., 2005; Li et al., 2005; Thon et al., 2005). Cul4 is a core subunit of Cullin4-based ubiquitin E3 ligase (Jia et al., 2005). Rik1 contains a WD40 domain and is a functional homologue of human DDB1 (UV-damaged DNA binding protein 1) (Horn et al., 2005). Dos1 also contains a WD40 domain. Recent studies showed that Dos1 is a homolog of human DDB1-Cul4 Associated Factor (DCAF) (Buscaino et al., 2012; Kuscu et al., 2014). In humans, DCAF proteins interact directly with DDB1 and function as substrate-recognition receptors for the E3 ligases (Lee and Zhou, 2007). However, the substrate targeted by DDB1-Dos1-Cul4 in fission yeast is still unknown. Lid2 is a conserved H3K4 demethylase important for maintaining H3K4 hypomethylation in heterochromatin. It has been proposed that the ClrC complex may function in coordinating H3K9 methylation and H3K4 demethylation activity in heterochromatin to ensure proper establishment and maintenance of epigenetic states in these regions (Li et al., 2008).

H3K9 methylation in fission yeast is recognized and bound by the human HP1 homologue, Swi6. Like human HP1, Swi6 contains two distinct domains, a chromodomain and a chromoshadow domain. The chromodomain is the "reader" domain that specifically associates with H3K9 methylation, whereas the chromoshadow domain is involved in the dimerization of Swi6 (Cowieson et al., 2000; Bannister et al., 2001). It has been postulated that Swi6 may use a step-wised oligomerization process to generate a high-order multimer that in turn serves as a platform to recruit diverse regulators required for the assembly of chromatin into a less-accessible structure (Fischer et al., 2009). Swi6 also has the ability to recruit cohesin, a protein critical for sister chromatid cohesion (Bernard et al., 2001).

Histone deacetylation and heterochromatin assembly

Histone deacetylation is another highly conserved feature of heterochromatin in eukaryotes. Fission yeast contains all three subtypes of histone deacetylases (HDACs): Class I (Clr6), Class II (Clr3), and Class III (Sir2), all of which have been shown to be required for heterochromatin silencing. Clr6, a homolog of mammalian HDAC1 and HDAC2 and of Rpd3 in *S. cerevisiae*, deacetylate several lysine residues on histone H3 and H4. Clr6 is essential for viability and regulates heterochromatin silencing and gene activity at euchromatin (Bjerling et al., 2002; Nicolas et al., 2007). Clr3 deacetylates histone H3 on lysine 14, and is a component of the transcriptional silencing complex, SHREC. The SHREC complex physically associates with Swi6, and is important for heterochromatin spreading (Bjerling et al., 2002; Sugiyama et al., 2007). Sir2 is a conserved member of the Sirtuin family of HDACs that uses NAD^+ as a cofactor. Sir2 in fission yeast can deacetylate multiple lysines in histone H3 and H4, including H3K4, H3K9, H3K14, and H4K16. Recent studies have shown that Sir2 is critical for Clr4-mediated heterochromatin nucleation step and subsequent heterochromatin spreading (Shankaranarayana et al., 2003; Buscaino et al., 2013).

RNAi and heterochromatin silencing

RNA interference (RNAi), a RNA-induced gene-silencing phenomenon, was originally found to occur at the post-transcriptional level. Core components of the RNAi machinery include Argonaute, Dicer, and a RNA-dependent polymerase. An unexpected yet significant breakthrough was the discovery of a link between RNAi and transcriptional silencing in fission yeast (Volpe et al., 2002). A similar phenomenon was also later discovered in other organisms, including flies (Pal-Bhadra et al., 2004), *Arabidopsis thaliana* (Chan et al., 2004), and *Caenorhabditis elegans* (Guerin et al., 2013). Fission yeast contains a single copy of each RNAi component, namely, Ago1 (Argonaute), Dcr1 (Dicer), and Rdp1 (RNA-dependent polymerase). Remarkably, none of these proteins are essential for viability in fission yeast. However, deletion of any one of these components resulted in loss of H3K9 methylation and silencing at peri-centromeric heterochromatin, indicating the important role of RNAi in heterochromatin silencing (Volpe et al., 2002).

In fission yeast, Ago1, the chromodomain protein, Chp1, and Tas3, which contains a glycine-tryptophan (GW) domain, form the RNA-induced transcriptional silencing (RITS) complex. Like Swi6, Chp1 specifically binds to H3K9 methylation through its chromodomain, thereby anchoring the RITS complex to heterochromatin (Motamedi et al., 2004). Consistent with this, the level of siRNAs is reduced in mutants of the ClrC complex, presumably due to dissociation of the RITS from H3K9me-deficient heterochromatin in these mutants (Motamedi et al., 2004; Hong et al., 2005; Li et al., 2005). Tas3 serves as a backbone for the RITS complex, linking Chp1 and Ago1 (Schalch et al., 2011). The C-terminal end of Tas3 contains an α-helical motif which can assemble into a helical polymer that is essential for the *cis* spreading of the RITS complex at peri-centromeres (Li et al., 2009). Rdp1 associates with Cid12, a putative polyA polymerase, and the Hrr1 helicase, to form the RNA-directed RNA polymerase (RDRC) complex (Motamedi et al., 2004). Heterochromatin transcripts are processed into small single stranded RNA fragments through the slicer activity of Ago1 in the RITS complex (Irvine et al., 2006). These fragments are then used by RDRC to prime the synthesis of double stranded long non-coding RNAs, which are subsequently processed into siRNAs by Dicer (Motamedi et al., 2004). SiRNAs have been proposed to act as key nucleation factors for heterochromatin assembly by recruiting the ClrC complex to heterochromatin. Stc1, a LIM domain protein, interacts with both the ClrC complex and RNAi components to couple RNAi to H3K9 methylation at peri-centromeres (Bayne et al., 2010).

Cell cycle and RNAi

Previous studies have shown that phosphorylation of histone H3 at serine 10 (H3S10P) antagonizes the binding of HP1 to H3K9me, and that the binary phospho-methyl switch between the two modifications is crucial for heterochromatin assembly in humans (Fischle et al., 2003). In fission yeast, H3S10 phosphorylation increases during mitosis and results in dissociation of Swi6 from H3K9me sites during early S phase. At the same time, H3K9me is maximally reduced and H3K4me, the epigenetic mark for active transcription, increases at this stage. Correspondingly, heterochromatin transcription becomes active during early S phase. This coincides with the timing of replication of centromeres, which in *S. pombe* differs from the timing of replication of the other major constitutive heterochromatin domains, namely telomeres and mating type loci (Chen et al., 2008; Kloc et al., 2008). Heterochromatin in fission yeast is transcribed by RNA Pol II. Consistent with this, Pol II was found to be recruited to centromeric repeats during early S phase. Transcripts from both strands of *dg* and *dh* regions are immediately processed into small RNAs by RNAi machinery. Following the increase of siRNAs during early S phase, H3K9me begins to increase at late S phase, and reaches the highest level at G_2 phase (Chen et al., 2008; Kloc et al., 2008). During S phase, key ClrC components, including Rik1 and Dos2 are also recruited to peri-centromeric heterochromatin (Chen et al., 2008). These findings revealed that heterochromatin assembly is dynamic throughout the cell cycle, and pointed to the S phase of the cell cycle as a crucial stage for establishment and propagation of heterochromatic epigenetic marks.

DNA REPLICATION AND INHERITANCE OF HETEROCHROMATIN SILENCING

A role for the DNA Pol ε subunit, Cdc20, in inheritance of H3K9 methylation

Co-regulation of RNAi and the S phase of the cell cycle is intriguing. Recent studies provide clues into how RNAi and heterochromatin formation are coupled with DNA replication, thus providing mechanistic insight into how H3K9 methylation is inherited. Using mass spectrometry analysis, Li et al., demonstrated that the Dos2 and Rik1 components of the ClrC complex associate with the DNA replication factor, Cdc20, and the HEAT domain containing protein, Mms19 (Li et al., 2011a). Cdc20 is the catalytic subunit of DNA Pol ε (Durso and Nurse, 1997), and Mms19 is involved in diverse functions, including DNA repair, transcription, chromosome segregation, and the Fe-S pathway (Wu et al., 2001; Hatfield et al., 2006; Ito et al., 2010; Papatriantafyllou, 2012). Li et al. further revealed that the complex is crucial for assembly of heterochromatin during S phase. Disruption of Cdc20 results in severe loss of H3K9me and heterochromatin silencing. Specific association of Dos2 and Rik1 with peri-centromeric heterochromatin is also lost in a *cdc20* mutant, suggesting that Cdc20 may play an essential role in recruiting the ClrC complex to heterochromatin. Furthermore, the Dos2-Rik1-Cdc20 complex is important for the regulation of transcription of heterochromatic siRNAs during S phase (Li et al., 2011a). These findings demonstrated a key role for Cdc20 in RNAi-dependent heterochromatin assembly during S phase, and shed light onto how DNA replication, RNAi and the ClrC complex may be coordinated to process H3K9me and heterochromatin inheritance during DNA replication. Another study shows that during S phase of the cell cycle, RNAi-dependent processing of nascent heterochromatin transcripts promotes degradation of these transcripts and releases RNAPII from heterochromatin (Zaratiegui et al., 2011). This is important because it prevents potential collisions between transcription and DNA replication machinery, and allows the replication fork to continue advancing concomitantly with the recruitment of the ClrC complex by the DNA polymerase subunit, Cdc20.

Initiation of heterochromatin formation: RNAi or Cdc20-directed?

Small RNAs generated via the RNAi pathway have been proposed to serve as nucleating factors for heterochromatin assembly and to recruit the H3K9-methylating complex, ClrC (Verdel et al., 2004). However, H3K9 methylation is not completely abrogated in RNAi mutants and recruitment of the H3K9me-binding protein, Swi6, to heterochromatin is only weakly affected (Hall et al., 2003; Li et al., 2005). In contrast, H3K9me is completely lost, and Swi6 dissociates from heterochromatin in cells lacking components of the ClrC complex. In addition, in cells in which both Clr4 and the RNAi pathway are disrupted, overexpression of Clr4 bypasses the RNAi defect and results in the deposition of H3K9me at peri-centromeric heterochromatin (Shanker et al., 2010). In support of this, artificial tethering of Clr4 to a euchromatic locus was found to be enough to induce heterochromatin assembly (Kagansky et al., 2009), further confirming that H3K9me can occur independently of RNAi. These findings suggest the existence of a pathway for heterochromatin inheritance that acts independently of RNAi. In this regard, initiation of heterochromatin assembly by the ClrC complex has been proposed as an alternative model. In support of this view, components of the ClrC complex, such as Rik1, have been shown to be absolutely essential for H3K9 methylation and able to bind chromatin independently of RNAi to nucleate heterochromatin assembly (Reyes-Turcu et al., 2011).

However, how the ClrC complex is recruited to peri-centromeric heterochromatin in the first place remains unresolved. Here we propose that Cdc20, the DNA Pol ε subunit, as a central player driving the initiation of heterochromatin assembly. During DNA replication, Cdc20 recruits the ClrC complex to nascent heterochromatin resulting in the deposition of H3K9me at nucleation sites. Small RNAs generated from peri-centromeric transcripts may further facilitate and reinforce Cdc20-directed targeting of the ClrC complex to nascent heterochromatin during S phase by base pairing with heterochromatic DNA.

Other replication components implicated in heterochromatin silencing

In addition to Cdc20, other DNA replication factors in fission yeast implicated in heterochromatin silencing include, Orc and DNA Pol α subunits, Swi7, and Mcl1. Perturbing these factors results in loss of silencing at peri-centromeric heterochromatin. All of these factors were also found to interact with the homolog of human HP1 in fission yeast, Swi6 (Nakayama et al., 2001a; Natsume et al., 2008; Li et al., 2011b). Cdc18, the homolog of human Cdc6, also interacts with Swi6. Cdc18 is required for the association of Swi6 with pericentromeric heterochromatin but appears to be dispensable for heterochromatin silencing (Li et al., 2011b). The exact role of these replication factors in heterochromatin formation and inheritance of H3K9 methylation remains unclear.

DNA REPLICATION AND INHERITANCE OF CENP-A

Numerous studies have shown that CENP-A is essential for centromere structure and function, and acts as the epigenetic mark for centromeres. Unlike canonical histones, which are deposited into chromatin during S phase of the cell cycle, parental and newly synthesized CENP-As are assembled into core centromeres at different stages of the cell cycle: parental CENP-A is partitioned equally and incorporates into daughter centromeres following DNA replication during S phase, whereas newly synthesized CENP-A is loaded at centromeres only during later stages of the cell cycle (Jansen et al., 2007; Schuh et al., 2007; Black and Cleveland, 2011; Bui et al., 2012; Shivaraju et al., 2012). Assembly of newly synthesized CENP-A chromatin requires the histone chaperone, HJURP/Scm3 (Camahort et al., 2007; Foltz et al., 2009). Little is known about how the parental CENP-A is faithfully recruited to centromeric nucleosomes during DNA replication. In fission yeast, CENP-A loading during S phase requires Ams2, a cell-cycle-regulated GATA-type factor (Chen et al., 2003). A recent study demonstrated that Cdc20 interacts with Ams2 through Dos1/2 to mediate CENP-A incorporation during S phase. Disruption of the Cdc20-Dos1/2 complex leads to dissociation of CENP-A from centromeres and severe chromosome segregation defects (Gonzalez et al., 2013). This study suggests that epigenetic inheritance of core centromeres also depends on DNA replication components and provides mechanistic insight into how the parental CENP-A is inherited during S phase.

DNA REPLICATION FACTORS AS PART OF EPIGENETIC REPLICATION MACHINERY

Studies from various organisms implicate DNA replication factors in the inheritance of heterochromatin silencing. For instance, in budding yeast, mutations in components of the ORC complex affect transcriptional silencing (Foss et al., 1993; Bell et al., 1995; Fox et al., 1995). In *Drosophila* and humans, ORC2 interacts with HP1, and depletion of ORC2 in humans results in dissociation of HP1 from heterochromatin (Pak et al., 1997; Shareef et al., 2001; Prasanth et al., 2004; Prasanth et al., 2010). In budding yeast, MCM10 was found to interact with Sir2 and mutations in MCM proteins result in a reduction in heterochromatin silencing (Dziak et al., 2003; Liachko and Tye, 2005). In *Drosophila*, Mcm10 is known to bind HP1 (Christensen and Tye, 2003). In addition, disruption of PCNA in both budding yeast and *Drosophila* causes heterochromatin defects (Henderson et al., 1994; Zhang et al., 2000). In mammals, PCNA interacts with chromatin modification enzymes, including DNMT1, a DNA methyltransferase, and SET8/Pr-Set7, a histone methyltransferase for H4K20me, both of which are important for heterochromatin assembly (Chuang et al., 1997; Huen et al., 2008). Notably, DNA polymerases, including Pol α and Pol ε, have also been demonstrated to be involved in heterochromatin silencing in budding yeast and *A. thaliana* (Ehrenhofer-Murray et al., 1999; Yin et al., 2009; Liu et al., 2010; Hyun et al., 2013). In addition to the effect on heterochromatin and centromeres, DNA replication machinery has also been implicated in the epigenetic regulation of open chromatin: in human cells, PCNA was found to target ISWI, a chromatin-remodeling complex, to open chromatin during replication to mediate maintenance of active chromatin states (Poot et al., 2004).

Based on these observations, we propose that specific DNA replication components act as central regulators in the inheritance of epigenetic information throughout the genome. We envision such components acting in a coordinated fashion during S phase to recruit the appropriate histone modification or DNA methylation enzymes, required for the faithful inheritance of epigenetic marks. The advantage of sharing the same machinery for the replication of both genetic and epigenetic information is obvious since it allows effective coupling of DNA replication and epigenetic inheritance. However, given that epigenetic patterns vary throughout different regions of the genome, how can general DNA replication factors command the inheritance of diverse epigenetic states? One possibility is that DNA replication components are able to "sense" parental epigenetic states by interacting with specific epigenetic "reader" proteins. This, in turn, leads to recruitment of proper "writer" proteins that modify daughter DNA or histones (Fig. 2). Consistent with this idea, multiple replication components have been shown to be associated with the classic reader protein, Swi6/HP1 (Nakayama et al., 2001a; Christensen and Tye, 2003; Prasanth et al., 2010).

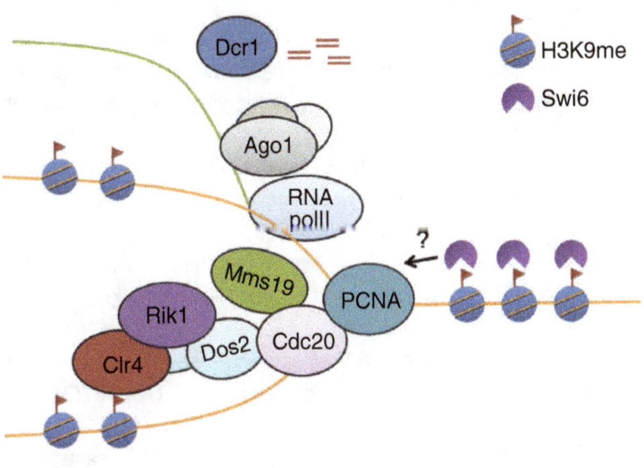

Figure 2. Model of inheritance of H3K9 methylation during DNA replication. During S phase of the cell cycle, Cdc20, the catalytic subunit of Pol ε, travels with the replication fork to synthesize the leading strand DNA. When it reaches heterochromatin, Cdc20 may sense the epigenetic state of parental nucleosomes by interacting, directly or indirectly, with the epigenetic reader protein, Swi6/HP1, associated with parental histones. This leads to recruitment of the ClrC complex to replication forks. Histone modification enzymes in the ClrC complex, including Clr4 and Lid2, modify the histones on daughter DNA to ensure hypermethylation of H3K9. Meanwhile, Cdc20 recruits the transcription regulator, Mms19, to heterochromatin, which in turn promotes heterochromatin transcription by RNA Pol II. These transcripts are processed by RNAi machinery into siRNAs, which facilitate the formation of heterochromatin after replication.

CONCLUDING REMARKS

One fundamental question in the field of epigenetics is how epigenetic states are faithfully transmitted through generations. Recent studies in fission yeast and other organisms provide significant insight into the role of DNA replication machinery in epigenetic inheritance. In the future, it is important to elucidate the mechanism for how different DNA replication components coordinate to mediate chromatin inheritance. Another important question is how DNA replication machinery guides the inheritance of various chromatin states. Furthermore, DNA replication is also known to provide a critical window of opportunity for changing epigenetic states, which can drive cellular differentiation during development (Probst et al., 2009). The molecular mechanism by which this epigenetic reprogramming takes place remains obscure. Future experiments are needed to further address these important questions.

ACKNOWLEDGMENTS

The work is support by the National Science Foundation (USA) (Grant No. MCB-1330557). We apologize to the scientists who made contributions to the field, but have not been cited due to space limitations.

ABBREVIATIONS

ClrC, Clr4 methyltransferase complex; HDACs, histone deacetylases; HP1, heterochromatin protein 1; H3K9, histone 3 lysine 9; ORC, origin recognition complex; PCNA, proliferating cell nuclear antigen; Pol δ, polymerase δ; Pol ε, polymerase ε; RNAi, RNA interference.

COMPLIANCE WITH ETHICS GUIDELINES

Haijin He, Marlyn Gonzalez, Fan Zhang, Fei Li declare that they have no conflict of interest.

This article does not contain any studies with human or animal subjects performed by the any of the authors.

REFERENCES

Allshire RC, Karpen GH (2008) Epigenetic regulation of centromeric chromatin: old dogs, new tricks? Nat Rev Genet 9 (12):923–937

Allshire RC, Nimmo ER et al (1995) Mutations derepressing silent centromeric domains in fission yeast disrupt chromosome segregation. Genes Dev 9(2):218–233

Amor DJ, Kalitsis P et al (2004) Building the centromere: from foundation proteins to 3D organization. Trends Cell Biol 14(7):359–368

Bannister AJ, Zegerman P et al (2001) Selective recognition of methylated lysine 9 on histone H3 by the HP1 chromo domain. Nature 410(6824):120–124

Bayne EH, White SA et al (2010) Stc1: a critical link between RNAi and chromatin modification required for heterochromatin integrity. Cell 140(5):666–677

Bell SP, Mitchell J et al (1995) The multidomain structure of Orc1p reveals similarity to regulators of DNA replication and transcriptional silencing. Cell 83(4):563–568

Bernard P, Maure JF et al (2001) Requirement of heterochromatin for cohesion at centromeres. Science 294(5551):2539–2542

Bjerling P, Silverstein RA et al (2002) Functional divergence between histone deacetylases in fission yeast by distinct cellular localization and in vivo specificity. Mol Cell Biol 22(7):2170–2181

Black BE, Cleveland DW (2011) Epigenetic centromere propagation and the nature of CENP-a nucleosomes. Cell 144(4):471–479

Bloom K, Yeh E (2010) Tension management in the kinetochore. Curr Biol 20(23):R1040–R1048

Bui M, Dimitriadis EK et al (2012) Cell-cycle-dependent structural transitions in the human CENP-A nucleosome in vivo. Cell 150 (2):317–326

Burgers PM (2009) Polymerase dynamics at the eukaryotic DNA replication fork. J Biol Chem 284(7):4041–4045

Buscaino A, White SA et al (2012) Raf1 is a DCAF for the Rik1 DDB1-like protein and has separable roles in siRNA generation and chromatin modification. PLoS Genet 8(2):e1002499

Buscaino A, Lejeune E et al (2013) Distinct roles for Sir2 and RNAi in centromeric heterochromatin nucleation, spreading and maintenance. EMBO J 32(9):1250–1264

Camahort R, Li B et al (2007) Scm3 is essential to recruit the histone h3 variant cse4 to centromeres and to maintain a functional kinetochore. Mol Cell 26(6):853–865

Carroll CW, Straight AF (2006) Centromere formation: from epigenetics to self-assembly. Trends Cell Biol 16(2):70–78

Chan SW, Zilberman D et al (2004) RNA silencing genes control de novo DNA methylation. Science 303(5662):1336

Chen ES, Saitoh S et al (2003) A cell cycle-regulated GATA factor promotes centromeric localization of CENP-A in fission yeast. Mol Cell 11(1):175–187

Chen ES, Zhang K et al (2008) Cell cycle control of centromeric repeat transcription and heterochromatin assembly. Nature 451 (7179):734–737

Chikashige Y, Kinoshita N et al (1989) Composite motifs and repeat symmetry in *S. pombe* centromeres: direct analysis by integration of *NotI* restriction sites. Cell 57(5):739–751

Christensen TW, Tye BK (2003) Drosophila MCM10 interacts with members of the prereplication complex and is required for proper chromosome condensation. Mol Biol Cell 14(6):2206–2215

Chuang LS, Ian HI et al (1997) Human DNA-(cytosine-5) methyltransferase–PCNA complex as a target for p21WAF1. Science 277(5334):1996–2000

Cottarel G, Shero JH et al (1989) A 125-base-pair CEN6 DNA fragment is sufficient for complete meiotic and mitotic centromere functions in *Saccharomyces cerevisiae*. Mol Cell Biol 9(8):3342–3349

Couture JF, Trievel RC (2006) Histone-modifying enzymes: encrypting an enigmatic epigenetic code. Curr Opin Struct Biol 16(6):753–760

Cowieson NP, Partridge JF et al (2000) Dimerisation of a chromo shadow domain and distinctions from the chromodomain as revealed by structural analysis. Curr Biol 10(9):517–525

Durso G, Nurse P (1997) Schizosaccharomyces pombe cdc20(+) encodes DNA polymerase epsilon and is required for chromosomal replication but not for the S phase checkpoint. Proc Natl Acad Sci USA 94(23):12491–12496

Dutta A, Bell SP (1997) Initiation of DNA replication in eukaryotic cells. Annu Rev Cell Dev Biol 13:293–332

Dziak R, Leishman D et al (2003) Evidence for a role of MCM (mini-chromosome maintenance)5 in transcriptional repression of sub-telomeric and Ty-proximal genes in *Saccharomyces cerevisiae*. J Biol Chem 278(30):27372–27381

Ehrenhofer-Murray AE, Kamakaka RT et al (1999) A role for the replication proteins PCNA, RF-C, polymerase epsilon and Cdc45 in transcriptional silencing in *Saccharomyces cerevisiae*. Genetics 153(3):1171–1182

Fischer T, Cui B et al (2009) Diverse roles of HP1 proteins in heterochromatin assembly and functions in fission yeast. Proc Natl Acad Sci USA 106(22):8998–9003

Fischle W, Wang Y et al (2003) Binary switches and modification cassettes in histone biology and beyond. Nature 425(6957):475–479

Foltz DR, Jansen LE et al (2009) Centromere-specific assembly of CENP-a nucleosomes is mediated by HJURP. Cell 137(3):472–484

Forsburg SL, Rhind N (2006) Basic methods for fission yeast. Yeast 23(3):173–183

Foss M, McNally FJ et al (1993) Origin recognition complex (ORC) in transcriptional silencing and DNA replication in *S. cerevisiae*. Science 262(5141):1838–1844

Fox CA, Loo S et al (1995) The origin recognition complex has essential functions in transcriptional silencing and chromosomal replication. Genes Dev 9(8):911–924

Gonzalez M, He H et al (2013) Cell cycle-dependent deposition of CENP-A requires the Dos1/2-Cdc20 complex. Proc Natl Acad Sci USA 110(2):606–611

Guerin TM, Palladino F et al (2013) Transgenerational functions of small RNA pathways in controlling gene expression in *C. elegans*. Epigenetics 9(1):37–44

Hall IM, Noma K et al (2003) RNA interference machinery regulates chromosome dynamics during mitosis and meiosis in fission yeast. Proc Natl Acad Sci USA 100(1):193–198

Hatfield MD, Reis AM et al (2006) Identification of MMS19 domains with distinct functions in NER and transcription. DNA Repair 5 (8):914–924

Henderson DS, Banga SS et al (1994) Mutagen sensitivity and suppression of position-effect variegation result from mutations in mus209, the Drosophila gene encoding PCNA. EMBO J 13 (6):1450–1459

Henikoff S, Furuyama T (2010) Epigenetic inheritance of centromeres. Cold Spring Harb Symp Quant Biol 75:51–60

Hong EJ, Villen J et al (2005) A cullin E3 ubiquitin ligase complex associates with Rik1 and the Clr4 histone H3-K9 methyltransferase and is required for RNAi-mediated heterochromatin formation. RNA Biol 2(3):106–111

Horn PJ, Bastie JN et al (2005) A Rik1-associated, cullin-dependent E3 ubiquitin ligase is essential for heterochromatin formation. Genes Dev 19(14):1705–1714

Huen MS, Sy SM et al (2008) Direct interaction between SET8 and proliferating cell nuclear antigen couples H4-K20 methylation with DNA replication. J Biol Chem 283(17):11073–11077

Hyun Y, Yun H et al (2013) The catalytic subunit of Arabidopsis DNA polymerase alpha ensures stable maintenance of histone modification. Development 140(1):156–166

Irvine DV, Zaratiegui M et al (2006) Argonaute slicing is required for heterochromatic silencing and spreading. Science 313(5790):1134–1137

Ito S, Tan LJ et al (2010) MMXD, a TFIIH-independent XPD-MMS19 protein complex involved in chromosome segregation. Mol Cell 39(4):632–640

Jansen LE, Black BE et al (2007) Propagation of centromeric chromatin requires exit from mitosis. J Cell Biol 176(6):795–805

Jia S, Kobayashi R et al (2005) Ubiquitin ligase component Cul4 associates with Clr4 histone methyltransferase to assemble heterochromatin. Nat Cell Biol 7(10):1007–1013

Kagansky A, Folco HD et al (2009) Synthetic heterochromatin bypasses RNAi and centromeric repeats to establish functional centromeres. Science 324(5935):1716–1719

Kloc A, Zaratiegui M et al (2008) RNA interference guides histone modification during the S phase of chromosomal replication. Curr Biol 18(7):490–495

Kuscu C, Zaratiegui M et al (2014) CRL4-like Clr4 complex in *Schizosaccharomyces pombe* depends on an exposed surface of Dos1 for heterochromatin silencing. Proc Natl Acad Sci USA 111 (5):1795–1800

Lee J, Zhou P (2007) DCAFs, the missing link of the CUL4-DDB1 ubiquitin ligase. Mol Cell 26(6):775–780

Li H, Stillman B (2012) The origin recognition complex: a biochemical and structural view. Subcell Biochem 62:37–58

Li F, Goto DB et al (2005) Two novel proteins, Dos1 and Dos2, interact with Rik1 to regulate heterochromatic RNA interference and histone modification. Curr Biol 15(16):1448–1457

Li F, Huarte M et al (2008) Lid2 Is required for coordinating H3K4 and H3K9 methylation of heterochromatin and euchromatin. Cell 135 (2):272–283

Li H, Motamedi MR et al (2009) An alpha motif at Tas3 C terminus mediates RITS cis spreading and promotes heterochromatic gene silencing. Mol Cell 34(2):155–167

Li F, Martienssen R et al (2011a) Coordination of DNA replication and histone modification by the Rik1–Dos2 complex. Nature 475 (7355):244–248

Li PC, Chretien L et al (2011b) *S. pombe* replication protein Cdc18 (Cdc6) interacts with Swi6 (HP1) heterochromatin protein: region specific effects and replication timing in the centromere. Cell Cycle 10(2):323–336

Liachko I, Tye BK (2005) Mcm10 is required for the maintenance of transcriptional silencing in *Saccharomyces cerevisiae*. Genetics 171(2):503–515

Liu J, Ren X et al (2010) Mutation in the catalytic subunit of DNA polymerase alpha influences transcriptional gene silencing and homologous recombination in Arabidopsis. Plant J 61(1):36–45

Luger K (2003) Structure and dynamic behavior of nucleosomes. Curr Opin Genet Dev 13(2):127–135

McIntosh JR, Grishchuk EL et al (2002) Chromosome-microtubule interactions during mitosis. Annu Rev Cell Dev Biol 18:193–219

Min J, Zhang X et al (2002) Structure of the SET domain histone lysine methyltransferase Clr4. Nat Struct Biol 9(11):828–832

Morris CA, Moazed D (2007) Centromere assembly and propagation. Cell 128(4):647–650

Motamedi MR, Verdel A et al (2004) Two RNAi complexes, RITS and RDRC, physically interact and localize to noncoding centromeric RNAs. Cell 119(6):789–802

Nakayama J, Allshire RC et al (2001a) Role for DNA polymerase alpha in epigenetic control of transcriptional silencing in fission yeast. EMBO J 20(11):2857–2866

Nakayama J, Rice JC et al (2001b) Role of histone H3 lysine 9 methylation in epigenetic control of heterochromatin assembly. Science 292(5514):110–113

Natsume T, Tsutsui Y et al (2008) A DNA polymerase alpha accessory protein, Mcl1, is required for propagation of centromere structures in fission yeast. PLoS ONE 3(5):14

Nicolas E, Yamada T et al (2007) Distinct roles of HDAC complexes in promoter silencing, antisense suppression and DNA damage protection. Nat Struct Mol Biol 14(5):372–380

Pak DT, Pflumm M et al (1997) Association of the origin recognition complex with heterochromatin and HP1 in higher eukaryotes. Cell 91(3):311–323

Pal-Bhadra M, Leibovitch BA et al (2004) Heterochromatic silencing and HP1 localization in Drosophila are dependent on the RNAi machinery. Science 303(5658):669–672

Palmer DK, O'Day K et al (1991) Purification of the centromere-specific protein CENP-A and demonstration that it is a distinctive histone. Proc Natl Acad Sci USA 88(9):3734–3738

Papatriantafyllou M (2012) DNA metabolism: MMS19: CIA agent for DNA-linked affairs. Nat Rev Mol Cell Biol 13(9):538

Pluta AF, Mackay AM et al (1995) The centromere: hub of chromosomal activities. Science 270(5242):1591–1594

Poot RA, Bozhenok L et al (2004) The Williams syndrome transcription factor interacts with PCNA to target chromatin remodelling by ISWI to replication foci. Nat Cell Biol 6(12):1236–1244

Prasanth SG, Prasanth KV et al (2004) Human Orc2 localizes to centrosomes, centromeres and heterochromatin during chromosome inheritance. EMBO J 23(13):2651–2663

Prasanth SG, Shen Z et al (2010) Human origin recognition complex is essential for HP1 binding to chromatin and heterochromatin organization. Proc Natl Acad Sci USA 107(34):15093–15098

Probst AV, Dunleavy E et al (2009) Epigenetic inheritance during the cell cycle. Nat Rev Mol Cell Biol 10(3):192–206

Reyes-Turcu FE, Zhang K et al (2011) Defects in RNA quality control factors reveal RNAi-independent nucleation of heterochromatin. Nat Struct Mol Biol 18(10):1132–1138

Rice JC, Allis CD (2001) Histone methylation versus histone acetylation: new insights into epigenetic regulation. Curr Opin Cell Biol 13(3):263–273

Schalch T, Job G et al (2011) The Chp1-Tas3 core is a multifunctional platform critical for gene silencing by RITS. Nat Struct Mol Biol 18(12):1351–1357

Schueler MG, Higgins AW et al (2001) Genomic and genetic definition of a functional human centromere. Science 294 (5540):109–115

Schuh M, Lehner CF et al (2007) Incorporation of Drosophila CID/CENP-A and CENP-C into centromeres during early embryonic anaphase. Curr Biol 17(3):237–243

Shankaranarayana GD, Motamedi MR et al (2003) Sir2 regulates histone H3 lysine 9 methylation and heterochromatin assembly in fission yeast. Curr Biol 13(14):1240–1246

Shanker S, Job G et al (2010) Continuous requirement for the Clr4 complex but not RNAi for centromeric heterochromatin assembly in fission yeast harboring a disrupted RITS complex. PLoS Genet 6(10):e1001174

Shareef MM, King C et al (2001) Drosophila heterochromatin protein 1 (HP1)/origin recognition complex (ORC) protein is associated with HP1 and ORC and functions in heterochromatin-induced silencing. Mol Biol Cell 12(6):1671–1685

Shivaraju M, Unruh JR et al (2012) Cell-cycle-coupled structural oscillation of centromeric nucleosomes in yeast. Cell 150(2):304–316

Strahl BD, Allis CD (2000) The language of covalent histone modifications. Nature 403(6765):41–45

Sugiyama T, Cam HP et al (2007) SHREC, an effector complex for heterochromatic transcriptional silencing. Cell 128(3):491–504

Sun X, Le HD et al (2003) Sequence analysis of a functional Drosophila centromere. Genome Res 13(2):182–194

Taverna SD, Li H et al (2007) How chromatin-binding modules interpret histone modifications: lessons from professional pocket pickers. Nat Struct Mol Biol 14(11):1025–1040

Thon G, Hansen KR et al (2005) The Clr7 and Clr8 directionality factors and the Pcu4 cullin mediate heterochromatin formation in the fission yeast *Schizosaccharomyces pombe*. Genetics 171 (4):1583–1595

Verdel A, Jia S et al (2004) RNAi-mediated targeting of heterochromatin by the RITS complex. Science 303(5658):672–676

Volpe TA, Kidner C et al (2002) Regulation of heterochromatic silencing and histone H3 lysine-9 methylation by RNAi. Science 297(5588):1833–1837

Waga S, Stillman B (1998) The DNA replication fork in eukaryotic cells. Annu Rev Biochem 67:721–751

Weaver BA, Cleveland DW (2007) Aneuploidy: instigator and inhibitor of tumorigenesis. Cancer Res 67(21):10103–10105

Wood V, Gwilliam R et al (2002) The genome sequence of *Schizosaccharomyces pombe*. Nature 415(6874):871–880

Wu XY, Li H et al (2001) The human homologue of the yeast DNA repair and TFIIH regulator MMS19 is an AF-1-specific coactivator of estrogen receptor. J Biol Chem 276(26):23962–23968

Yin H, Zhang X et al (2009) Epigenetic regulation, somatic homologous recombination, and abscisic acid signaling are influenced by DNA polymerase mutation in Arabidopsis. Plant Cell 21(2):386–402

Zaratiegui M, Castel SE et al (2011) RNAi promotes heterochromatic silencing through replication-coupled release of RNA Pol II. Nature 479(7371):135–138

Zhang Z, Shibahara K et al (2000) PCNA connects DNA replication to epigenetic inheritance in yeast. Nature 408(6809):221–225

Zhang K, Mosch K et al (2008) Roles of the Clr4 methyltransferase complex in nucleation, spreading and maintenance of heterochromatin. Nat Struct Mol Biol 15(4):381–388

Zhu B, Reinberg D (2011) Epigenetic inheritance: uncontested? Cell Res 21(3):435–441

3

Competitive virus and host RNAs: the interplay of a hidden virus and host interaction

Changfei Li, Jun Hu, Junli Hao, Bao Zhao, Bo Wu, Lu Sun, Shanxin Peng, George F. Gao, Songdong Meng✉

CAS Key Laboratory of Pathogenic Microbiology and Immunology, Institute of Microbiology, Chinese Academy of Sciences (CAS), Beijing 100101, China

✉ Correspondence: mengsd@im.ac.cn (S. Meng)

ABSTRACT

During virus infection, viral RNAs and mRNAs function as blueprints for viral protein synthesis and possibly as pathogen-associated molecular patterns (PAMPs) in innate immunity. Here, considering recent research progress in microRNAs (miRNAs) and competitive endogenous RNAs (ceRNAs), we speculate that viral RNAs act as sponges and can sequester endogenous miRNAs within infected cells, thus cross-regulating the stability and translational efficiency of host mRNAs with shared miRNA response elements. This cross-talk and these reciprocal interactions between viral RNAs and host mRNAs are termed "competitive viral and host RNAs" (cvhRNAs). We further provide recent experimental evidence for the existence of cvhRNAs networks in hepatitis B virus (HBV), as well as Herpesvirus saimiri (HVS), lytic murine cytomegalovirus (MCMV) and human cytomegalovirus (HCMV) infections. In addition, the cvhRNA hypothesis also predicts possible cross-regulation between host and other viruses, such as hepatitis C virus (HCV), HIV, influenza virus, human papillomaviruses (HPV). Since the interaction between miRNAs and viral RNAs also inevitably leads to repression of viral RNA function, we speculate that virus may evolve either to employ cvhRNA networks or to avoid miRNA targeting for optimal fitness within the host. CvhRNA networks may therefore play a fundamental role in the regulation of viral replication, infection establishment, and viral pathogenesis.

KEYWORDS cvhRNAs, virus RNAs, host mRNAs, miRNA response element, ceRNAs

INTRODUCTION

Currently, viral RNAs are regarded as either templates for translating viral proteins that exert function or as viral pathogen-associated molecular patterns (PAMPs) for detection by host pattern-recognition receptors (PRRs) in innate antiviral immunity (Goubau et al., 2013). However, several recent studies provide intriguing evidence that viral RNAs and host mRNAs with common microRNA (miRNA) binding sequences reciprocally affect each other's levels and activities by directly competing with the targeting miRNAs. In this review, we propose a "competitive viral and host RNAs" (cvhRNAs) hypothesis, including recent studies on miRNAs and noncoding RNAs (ncRNAs) expressed by both the host and virus, as well as the logic of this new and potentially predictable virus-host interaction.

MIRNAS AND COMPETITIVE ENDOGENOUS RNAS (CERNAS)

MiRNAs

MiRNAs are a class of small noncoding double-stranded RNA molecules of approximately 22 nt in length, which regulate (usually inhibit) target gene expression at the post-transcriptional level. Full-length miRNAs (pri-miRNAs) are transcripted by RNA polymerase II, which are processed by Drosha within the nuclear compartment to produce pre-miRNAs of about 65 nucleotides in length. The pre-miRNA is then transported into the cytoplasm and further cleaved by the RNAse III-like endonuclease Dicer to produce their mature form miRNAs (Newman and Hammond, 2010). Upon loading into the RNA-induced silencing complex (RISC), which contains a member of the double-stranded RNA binding protein Argonaute family (Ago), the guide strand of the miRNA duplex recognizes and binds to conserved complementary target sites in target mRNAs (often in the 3′-UTR) through canonical base-pairing between the seed region of approximately 6- to 8- oligonucleotides. MiRNAs regulate target gene expression either by inducing their deadenylation and degradation or by leading to translational inhibition (Miyoshi et al., 2009; Hu et al., 2012). Thus far, >2500 human miRNAs (hsa-mir) and 1900 mouse miRNAs (mmu-mir) have been identified and described at the miRBase website (http://www.mirbase.org, released in Nov. 2013).

CeRNAs

Aside from small regulatory ncRNAs, such as miRNAs, the transcription machinery in mammalian cells also produces long noncoding RNAs (lncRNAs) which are typically >200–300 nt in length. Although currently the function of most of the annotated lncRNAs remains unknown, emerging studies show that certain lncRNAs are involved in various physiological processes, including cell-cycle regulation, apoptosis, and the establishment of cell identity (Ulitsky and Bartel, 2013). Functional lncRNAs may act through several mechanisms, including co-transcriptional regulation, *cis*- or *trans*-regulation of gene expression by bridging proteins and chromatin, titration of RNA-binding factors, and pairing with other RNAs to trigger post-transcriptional regulation (Ulitsky and Bartel, 2013; Augui et al., 2011). Poliseno et al. (2010) further uncovered a unique interaction of the PTEN tumor suppressor gene and its pseudogene (PTENP1), both of which are subject to the same miRNA-mediated post-transcriptional regulation (Poliseno et al., 2010). Pseudogenes harbor premature stop codons, deletions/insertions, and mutations that abrogate their translation into functional proteins. Interestingly, the PTENP1 3′UTR functions as a decoy for PTEN-targeting microRNAs due to its ability to compete for miRNA binding and thus exerts a tumor-suppressive role by modulating the de-repression of cellular levels of PTEN. A subsequent study shows that a putative PTEN competitive endogenous RNA (ceRNA), the ZEB2 transcript, which contains common miRNA recognition elements with PTEN, modulates PTEN protein levels in a protein coding-independent manner (Karreth et al., 2011). These studies indicate a new means of regulatory interaction between mRNAs (including both protein-coding and non-coding mRNAs) that have common microRNA recognition elements.

Further, more recent studies provide compelling evidence that the crosstalk between ceRNAs through competition for their shared miRNAs is involved in signaling pathways and networks in human diseases, such as autoimmune diseases, diabetes and cancer (Ling et al., 2013a, b; Fang et al., 2013; Noorbakhsh et al., 2013; Karreth and Pandolfi, 2013). Together, these newly defined mRNA-mRNA interactions have integral roles in the regulation of gene expression in various physiological and pathological processes and deepen our understanding of a novel aspect of RNA function.

THE CVHRNAS HYPOTHESIS

Somewhat differently from the mechanism mentioned above and unlike cellular endogenous RNAs, viral RNAs and mRNAs are exogenous in virus-infected cells. In addition, viral RNAs are mostly studied as passive targets for host miRNAs (Fig. 1A). Indeed, a large number of impressive experiments have been performed, primarily on binding site validation and functional analyses, which support the notion that the interaction between miRNAs and viral RNAs as their targets is essential to the regulation of viral RNA stability, expression, and translation, as well as viral replication and infection (Jopling et al., 2005; Shimakami et al., 2012; Song et al., 2010; Klase et al., 2012). Hosts employ their miRNAs in defense against viral infection by targeting viral products or inhibiting viral replication. Whereas viruses also encode miRNAs that target specific host genes and pathways to prolong the longevity of infected cells or evade the immune clearance, which may be beneficial for viral infectivity and/or proliferation (Carl et al., 2013; Kincaid and Sullivan, 2012).

Nevertheless, viral RNAs harbor common miRNA-binding sites with host RNAs (Jopling et al., 2005; Shimakami et al., 2012; Song et al., 2010; Klase et al., 2012), indicating that viral RNAs and mRNAs potentially act as bona fide miRNA competitors with host RNAs for the same miRNA pool in infected cells. In addition, a large number of virus-encoded miRNAs and lncRNAs have been uncovered, which play a key role in persistent infections, and importantly, a significant portion of viral miRNAs mimic host miRNAs with similar target sites (Kincaid and Sullivan, 2012). Therefore, we hypothesize that viral RNAs act as sponges that can sequester endogenous miRNAs within infected cells and thus impact the stability and translational efficiency of host mRNAs with shared miRNA response elements (Fig. 1B). We term the reciprocal regulation between virus and host transcripts acting in this manner cvhRNAs.

The basis for cvhRNA crosstalk is dependent on base pairing complementarity and miRNA binding/recognition, and the ability of viral RNAs (or host mRNAs) to sequester and degrade miRNAs and the extent of de-repression in host mRNAs (or viral RNAs) by miRNA down-regulation. Down-regulation of miRNA by interacting viral or host mRNAs could be either via sequestration/sponge or destabilization/degradation, or both (Torres et al., 2011), although the exact mechanisms await further investigation. Thus, the relative abundance of virus vs. host RNAs, levels of common miRNAs, and the number of miRNA response elements may all contribute to cvhRNA interactions, according to a mathematical mass-action model for ceRNA networks (Ala et al., 2013). In addition, the intracellular localization of virus/host mRNAs and miRNAs, the extent of base pairing

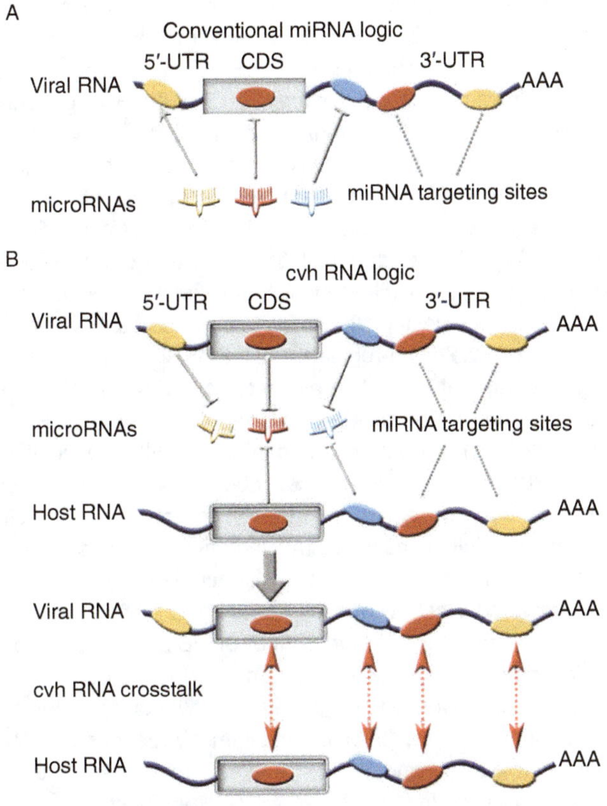

Figure 1. Inference of cvhRNA cross-regulation through miRNA response elements. (A) MiRNAs guide the recognition of target mRNAs through imperfect matches with their targeting sites and thus regulate the expression of target mRNAs at the post-transcriptional level. Each miRNA has multiple (up to tens) miRNA targets. Conversely, each mRNA may harbor multiple miRNA targeting sites (Ala et al., 2013). (B) Viral RNAs and mRNAs harboring miRNA response elements exert their suppression of miRNA as sponges. Host mRNAs may act in a similar way. Thus, viral RNAs and host mRNAs could competitively sequester the same miRNA pool within infected cells. Viral RNAs as miRNA sponges can de-repress the miRNA-mediated inhibition of host mRNAs. Therefore, a cross-regulation may be formed between viral RNAs and host mRNAs that share common miRNA response elements. Stimulation (↑) or inhibition (⊤) is determined following how miRNAs or RNAs impact the activities of RNAs or miRNAs.

complementarity of the competitive RNAs, as well as the time-course of viral RNA expression may also impact ceRNA networks. Given that viruses usually generate highly redundant transcripts upon infection and that multiple highly abundant miRNAs interact with viral RNAs or mRNAs (Shimakami et al., 2012; Kincaid and Sullivan, 2012), we expect that reciprocal interactions between viral RNA-miRNA-host mRNA may orchestrate a permissive molecular environment that enables the generation of cvhRNA networks under virus infection.

THE EXPERIMENTAL EVIDENCE FOR CVHRNA NETWORKS

The existence of cvhRNA cross-talk in viral infections

An early indication that virus-encoded RNAs are involved in regulating host miRNA abundance came from Cazalla et al. (2010), who demonstrated that *Herpesvirus saimiri* (HVS)-expressed noncoding RNAs have certain miRNA binding elements and can interact with host miRNAs in virally transformed T cells. They found that viral ncRNAs down-regulate mature miR-27 in a binding-dependent manner. Transient knockdown of one of the ncRNAs, HSUR 1, resulted in a decreased level of the miR-27 target protein, FOXO1 (Cazalla et al., 2010). Subsequently, several impressive experiments have been performed on the role of coding RNAs and ncRNAs on the degradation and decay of host miRNAs. Lytic murine cytomegalovirus (MCMV) infection leads to rapid degradation of cellular miR-27a and miR-27b via expression of a spliced and highly abundant protein-coding transcript, m169, which facilitates efficient virus replication (Marcinowski et al., 2012). Interestingly, MCMV infection up-regulates expression of genes in mitogen-activated protein kinases (MAPK) pathway, specifically p38, which enhances viral replication (Tang-Feldman et al., 2013). Up-regulation of p38 may attribute to decreased miR-27a as p38 is a potential target of miR-27a. Similarly, a viral intergenic ncRNA element of human cytomegalovirus (HCMV) mediates the selective down-regulation of mature miR-17 and miR-20a through sequence-specific interactions. Viral RNA-induced miR-17 decay is essential for rapid viral production, indicating a potential role for the interaction between the viral ncRNA and cellular miRNAs during viral pathogenesis (Lee et al., 2013a). Indeed, IL-8 as a target of miR-17/20a (Yu et al., 2010), significantly stimulates HCMV replication through cytokine-stimulated transcription of viral late genes and activation of early events in viral replication (Szabó et al., 1999; Redman et al., 2002).

Hepatitis B virus (HBV), a major hepatotropic DNA virus, chronically infects ~350 million people worldwide. Cellular miRNAs have a striking role in mediating HBV replication and infection (Hu et al., 2012; Zhang et al.,2013). MiR-122 is the most abundant liver-specific miRNA, which reaches approximately 70% of the total miRNA population in the adult liver. We observed a significant decrease of miR-122 levels under HBV infection. We further demonstrated that a miR-122 binding element exists in the 3′-UTR of all four HBV mRNAs (i.e., the pre-C/C or pre-genomic RNA (pgRNA), pre-S, S, and X mRNAs), and these viral RNAs act as sponges to bind and sequester endogenous miR-122, which led to decreased miR-122. Furthermore, expression of the HBV 3′-UTR-containing luciferase reporter was sufficient to up-regulate expression of two cellular genes, cyclin G1 and pituitary tumor-transforming gene 1 (PTTG1) binding factor (PBF), whose 3′-UTRs contain a common miR-122 response element as HBV RNAs (Li et al., 2013; Wang et al., 2012). A follow-up study revealed that cyclin G1 significantly enhances HBV expression and replication by suppressing p53-mediated inhibition of HBV enhancers, indicating that the cross-regulation of HBV RNAs and cyclin G1 expression possibly contribute to viral persistence. Further, up-regulation of PBF may contribute to HBV infection-related hepatocellular carcinoma (HCC) because PBF interacts with PTTG1 and increases its transcriptional activity at multiple oncogenes (Li et al., 2013), as well as promotes cell proliferation, invasion, and HCC tumor growth in mice. Under HBV infection, the HBV RNA copy number per cell reaches up to 105 (Li et al., 2013; Wang et al., 2012), and miR-122 is expressed in hepatocytes at 50,000 copies per cell (Hu et al., 2012; Filipowicz and Grosshans, 2011). Conceivably, a positive cross-regulation between HBV RNAs and cyclin G1 or PBF occurs in the presence of the large amounts of viral RNAs which efficiently sequester miR-122, resulting in intermediate miR-122 levels in hepatocytes.

Interestingly, we also identified an IFN-stimulated gene (ISG), NT5C3, which efficiently sequesters miR-122 with its mRNA 3′-UTR through binding sequence-specific interactions (Hao et al., 2013). Conceivably, the possible HBV RNAs-NT5C3 mRNA crosstalk may be a counter mechanism that HBV evolved to antagonize the antiviral activity of IFN.

In addition to the interaction between HBV RNAs and host mRNAs that share a common miR-122 response element, two other groups more recently reported another example of cvhRNA crosstalk in HBV infection (Liu et al., 2013; Wang et al., 2013b). They found that the HBx transcript and HBV mRNAs directly trigger the down-regulation of miR-15a/16 via the miRNA targeting sequences in the viral RNA, and as a consequence, the target of miR-15a/16 (Bcl-2) is up-regulated. Up-regulated Bcl-2 then inhibits the downstream cascade of apoptosis in hepatoma cells. These studies identify a novel HBx transcript and the HBV mRNAs-miR-15a/16-Bcl-2 regulatory pathway that is involved in apoptosis inhibition, which may help facilitate HBV infection and HCC development.

Together, the above studies provide experimental evidence for the existence of cvhRNA networks during HBV infection (Fig. 2). CvhRNAs, including HBV RNAs (e.g., pgRNA, pre-S, S, and X mRNAs) and host mRNAs (e.g., cyclin G1, PBF, NT5C3, and Bcl-2), harbor shared response elements for certain abundant cellular miRNAs (e.g., miR-122 and miR-15a/miR-16). When viral RNAs are highly produced during HBV infection, the repression of host target genes conferred by these common miRNAs can be diluted as viral RNAs down-regulate common miRNAs. Up-regulation of host target genes would contribute to enhanced HBV expression and replication, persistent viral infection, and HCC development.

HCV (hepatitis C virus), another main hepatotropic positive-sense RNA virus, causes chronic infection in about 180 million people worldwide. Extensive studies have demonstrated that miR-122 acts in an unusual manner to stimulate the replication and expression of HCV by binding to two closely spaced target sites in the 5′-UTR of the HCV genome (Jopling et al., 2005; Shimakami et al., 2012; Pang et al., 2012). In this context, interaction of miR-122 and viral RNA stabilizes viral RNA and decreases its decay, rather than induces its degradation as most miRNAs do. Notably, Shan et al. (2007) showed that inhibition of miR-122 up-regulates Heme oxygenase-1 (HO-1) which significantly inhibits HCV replication, suggesting that miR-122 promotes HCV replication partly via down-regulation of HO-1 (Shan et al., 2007). Considering that HCV RNA also interacts with miR-122, it is possible that the positive regulatory pathway of HCV RNA-miR-122-HO-1 may contribute to enhanced HCV replication.

In HBV and HCV infections, robust viral specific T cell responses are observed during resolution of acute and self-limited HBV infections with viral clearance, whereas chronic infection with viral persistence is characterized by only weak and impaired T-cell response. T cells are therefore believed to play a critical role in the control of HBV and HCV replication and infection (Manigold and Racanelli, 2007; Wang et al., 2013a). Conceivably, cvhRNA networks may play an important role in regulation of host immune responses in the liver microenvironment that induce immune tolerance towards viruses. Indeed, the above preliminary studies have shown that the cross-talk and reciprocal interactions between HBV/HCV RNAs and host mRNAs contribute to enhanced and persistent viral expression and replication which inevitably drive T cell tolerance and exhaustion by long-term intensive viral antigenic stimulation. Moreover, it is possible that cvhRNA networks in HBV or HCV infection may directly or indirectly affect expressions of co-inhibitory molecule programmed death (PD)-1 on T cells and its ligand PD-L1 on dendritic cells (DCs), the cytotoxic T lymphocyte antigen-4 (CTLA-4) and proapoptotic protein Bcl2-interacting mediator (Bim) on T cells, or activities of Tregs in the liver microenvironment. Further understanding the regulation of cvhRNA networks in influencing the outcome of viral infections (e.g. viral clearance or persistent infection) will allow the development of novel strategies for therapeutic regimen by targeted reversal of tolerising mechanisms.

Prediction of potential cross-regulation between host and virus by cvhRNA hypothesis

In principle, any viral RNAs sharing common miRNA response elements with host mRNAs could potentially act as cvhRNAs and form regulatory networks across the transcriptome. Besides HSV, MCMV and HCMV, we speculate that other examples of cvhRNA may also exist in viral infections (Table. 1). For instance, HIV-1 may probably up-regulate Notch-3 through cvhRNA crosstalk, as both HIV-1 gag-pol and Notch-3 mRNAs share a common miR-150 response sequence (Huang et al., 2007; Ghisi et al., 2011),

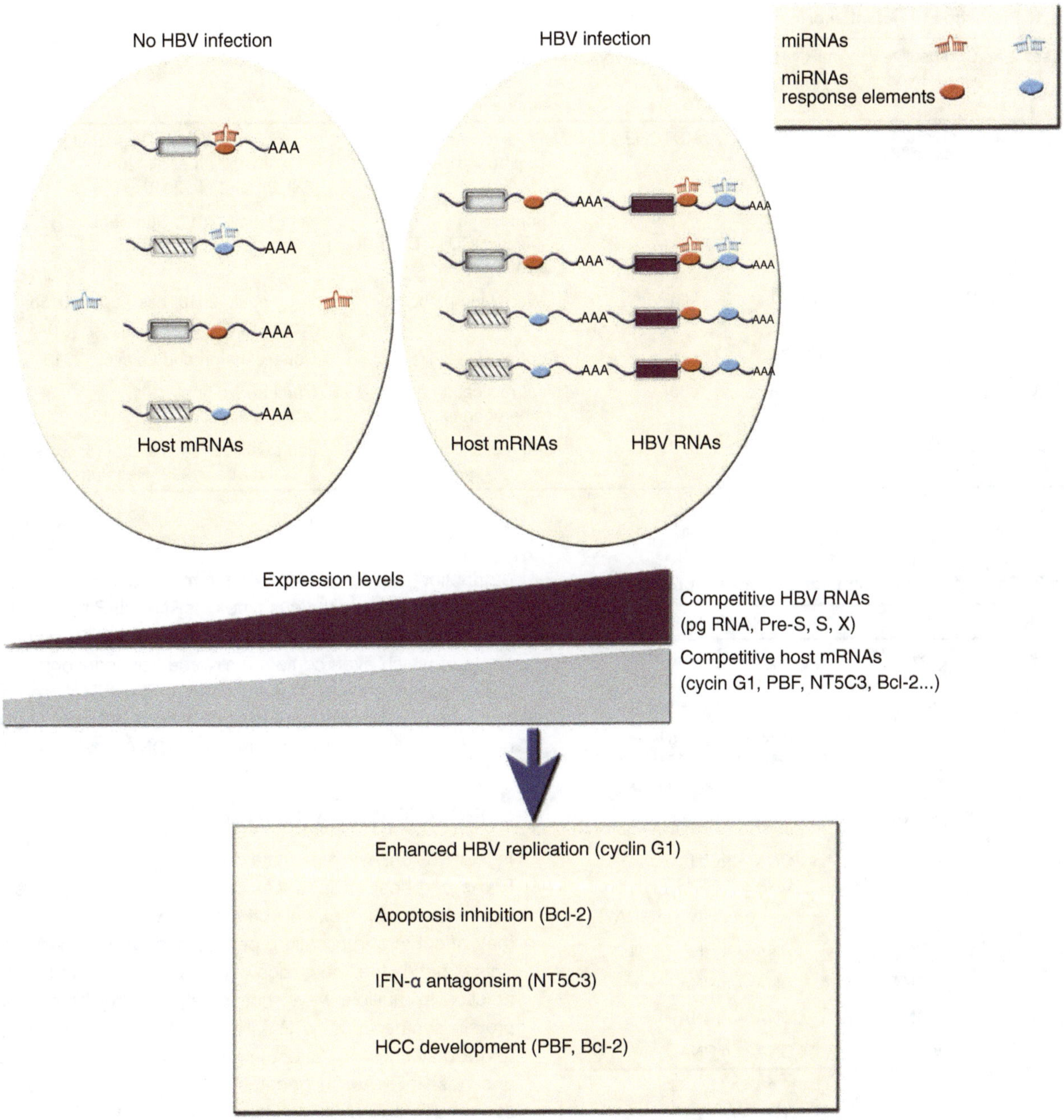

Figure 2. Schematic figure of how cvhRNA networks may mediate viral replication, infection, and the development of HCC during HBV infection. In hepatocytes, miR-122 effectively suppresses the expression of its target genes (cyclin G1, PBF, and NT5C3) at the post-transcriptional level by binding to the miRNA response element within the 3′-UTR of target mRNAs. During HBV infection, high levels of viral RNAs harboring a miR-122 response element in their 3′-UTR competitively sponge and efficiently sequester cellular miR-122, thus blocking the binding of miR-122 to its host target mRNAs. In this manner, viral RNAs de-repress and rescue the expression of host target mRNAs. Similarly, HBV RNAs positively cross-regulate Bcl-2 expression through shared miR-15a/16 response elements. The dose effect between competitive HBV RNAs and host mRNAs is shown. In the networks consisting of four viral RNAs (pgRNA, pre-S, S, and X mRNAs) and four host mRNAs (cyclin G1, PBF, NT5C3, and Bcl-2), elevated expression of de-repressed host genes contributes to enhanced HBV replication, persistent viral infection, and HCC development.

and HIV-1 infection down-regulates miR-150 (Swaminathan et al., 2009). Notch activation in turn is involved in HIV-associated nephropathy (Sharma et al., 2010). The cvhRNA hypothesis also predicts another possible reciprocal interaction between HIV nef and host PTEN mRNAs that have a shared miR-29a response sequence. HIV infection may induce apoptosis of $CD4^+$ T lymphocytes through upregulating PTEN (Kong et al., 2011; Sun et al., 2012;

Table 1. Prediction of potential cross-regulation between host and virus by cvhRNA hypothesis

Viruses	Competitive virus RNAs	miRNAs	Competitive host RNAs	Effects of cvhRNAs on viral infection and pathogenesis	References
HIV-1	gag-pol	miR-150	NOTCH3	HIV-associated nephropathy	Huang et al., 2007; Ghisi et al., 2011; Swaminathan et al., 2009; Sharma et al., 2010
	Nef	miR-29a	PTEN	Apoptosis in infected $CD4^+$ T Lymphocytes	Kong et al., 2011; Sun et al., 2012; Dabrowska et al., 2008
IAV	PB1	miR-491	MMP-9	Severe lung pathology	Song et al., 2010; Lee et al., 2013b; Yuan et al., 2013
HPV	E1,E2	miR-145	KLF-4	Control of viral life cycle	Gunasekharan and Laimins, 2013
MCMV	m169	miR-27a	p38	Viral replication and atherogenesis	Marcinowski et al., 2012; Tang-Feldman et al., 2013
HCMV	UL144-145 RNA	miR-17/20a	IL-8	Viral replication	Lee et al., 2013a; Yu et al., 2010; Szabó et al. 1999; Redman et al., 2002

Table 2. Different mechanisms for viral mRNA production

Viruses	Genome	mRNA production
IAV	(−)ssRNA	(−)vRNA → mRNA (−)vRNA → (+)cRNA → (−)RNA → mRNA
HCV	(+)ssRNA	(+)vRNA → mRNA (+)vRNA → (−)RNA → (+)RNA → mRNA
HIV-1	dsRNA	dsRNA $\overset{RT}{\rightarrow}$ dsDNA → mRNA
HBV	dsDNA	DNA → mRNA DNA → pgRNA $\overset{RT}{\rightarrow}$ cccDNA → mRNA
HVS	dsDNA	DNA → mRNA
HPV	dsDNA	DNA → mRNA
HCMV	dsDNA	DNA → mRNA
MCMV	dsDNA	DNA → mRNA

Dabrowska et al., 2008). Similarly, influenza virus WSN infection induces up-regulation of MMP9 which mRNA shares a common miR-491-5p binding site with viral PB1 mRNA (Song et al., 2010; Lee et al., 2013b; Yuan et al., 2013), indicating the potential cross-regulation between viral mRNA and host mRNA. Furthermore, the notion of cvhRNAs was recently linked to human papillomaviruses (HPVs) (Gunasekharan and Laimins, 2013). HPV-31, whose E1 and E2 open reading frames contain miR-145 binding sequences, could lead to a dramatic decrease miR-145 level and increased the expression of a miR-145 target, the cellular transcription factor KLF-4. HPV-regulated KLF-4 may contribute to the control of viral life cycle during differentiation.

As cvhRNA networks primarily depend on viral mRNA abundance, different mechanisms responsible for mRNA production among different virus families above are listed in Table 2. Among these virus types, ssRNA, dsRNA or dsDNA genomes serve as the template for mRNA transcription, some of which even contain a reverse transcription stage in their replication cycle. This indicates that cvhRNA networks may exist universally among viruses with different types of genomes (e.g. ssRNA, dsRNA and dsDNA).

CONCLUSIONS AND OUTLOOK

In conclusion, we speculate that cross-talk between viral RNAs and host mRNAs via common miRNA response elements forms complex networks during virus infection, which may affect viral replication, persistence, and pathogenesis. This expands the knowledge of how viral RNAs and mRNAs function in multiple ways, not merely as blueprints for viral protein synthesis or as PAMPs in innate immunity. Figs. 1 and 2 only show the direct interactions between viral RNAs and host mRNAs. In general, one miRNA has several or even multiple RNA/mRNA targets, and conversely, one target RNA or mRNA likely harbors several miRNA response elements. Thus, indirect interactions between viral RNAs and host mRNAs may also play an important role in cvhRNA networks, which may be further intimately intertwined by cellular transcription factor networks (Ala et al., 2013). In addition, variations in miRNA response elements (e.g. the seed region of pairing) of viral RNA which inevitably affect base pair complementarity, could thereby influence the interactions between viral and host RNAs. To establish infection, viruses likely evolved to accumulate mutation to increase the number of shared miRNA response elements with supportive host factors and decrease the number shared with inhibitory host factors; this hypothesis deserves further investigation. Global mRNA and miRNA sequencing, Gene Ontology analysis of transcriptomic databases and

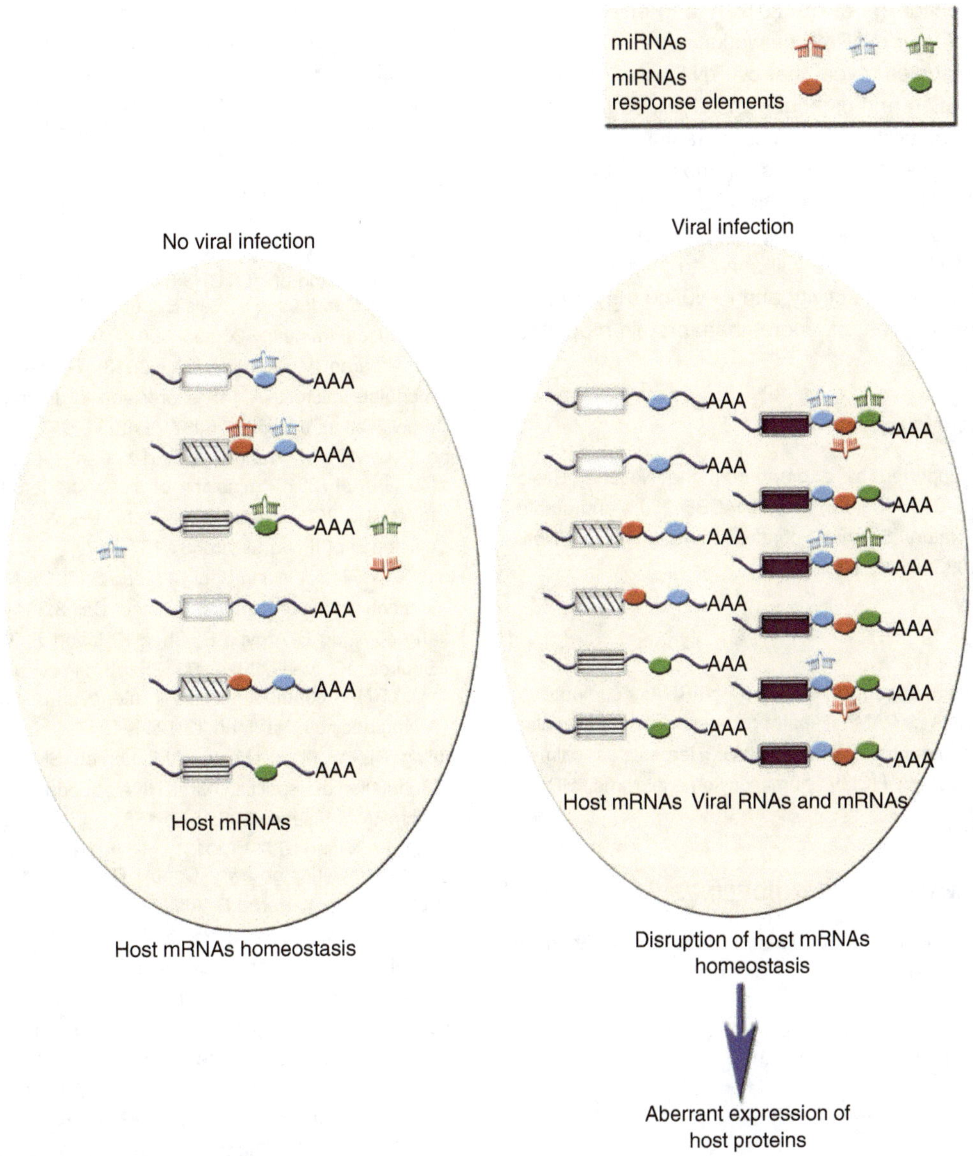

Figure 3. Potential alteration of host mRNA homeostasis by viral RNAs. Many mammalian viruses produce high levels of redundant viral RNAs and mRNAs. According to the cvhRNAs hypothesis, under certain conditions for cvhRNA activity in viral infections, these viral RNAs could modify the balance between cellular miRNAs and host mRNA targets in a miRNA response sequence dependent and coding independent manner. Viral RNAs directly result in the depression of cellular transcripts with shared miRNA response elements and the subsequent aberrant expression of host proteins.

establishment of proper mathematical models are needed to provide further dissection of the impact of cvhRNA networks on virus and host interaction.

A fine-tuned balance between miRNAs and host mRNA targets (including coding and noncoding transcripts) likely exists in cells with no viral infection (Ala et al., 2013). Conceivably, viral infection disturbs mRNA homeostasis through cvhRNA networks, as shown in Fig. 3. Information regarding the balance between miRNA pools and host mRNA homeostasis under viral infection is essential to understand the molecular basis for viral persistence and pathogenesis, which may facilitate the development of new antiviral strategies.

An intriguing question arises from the hypothesis of cvhRNA regulatory networks is why virus evolves to facilitate rather than avoid miRNA binding which inevitably leads to repression of viral RNA function. We speculate that virus may employ either strategy (to usurp cvhRNA networks or to avoid miRNA targeting) for the establishment of persistent infection, depending on which strategy plays a major role in favoring viral infectivity and replication. In particular, some viruses produce redundant RNA/mRNAs during infections as seen in HBV infection (Li et al., 2013; Wang et al., 2012), which facilitate to form cvhRNA networks.

As emerging studies reveal that cvhRNAs play essential roles in viral infection and pathogenesis, it will be important to take into consideration the unique features of direct communication between viral RNAs and host mRNAs when studying the interaction between the virus and host in the long run. We hope that this review of recent advances on ceRNA and cvhRNA networks will serve to stimulate more interest and experimental activity and re-define the concept of virus-host interactions as more than protein-mediated regulation.

ACKNOWLEDGEMENTS

This work was supported by a grant from the National Basic Research program (973 Program) (No. 2014CB542602) and grants from the National Natural Science Foundation of China (Grant Nos. 31230026, 81321063, 81102018).

ABBREVIATIONS

ceRNAs, competitive endogenous RNAs; cvhRNAs, competitive viral and host RNAs; PAMPs, pathogen-associated molecular patterns; ncRNAs, noncoding RNAs; HVS, Herpesvirus saimiri; MCMV, cytomegalovirus; HCMV, human cytomegalovirus; HPVs, human papillomaviruses.

COMPLIANCE WITH ETHICS GUIDELINES

Changfei Li, Jun Hu, Junli Hao, Bao Zhao, Bo Wu, Lu Sun, Shanxin Peng, George F. Gao and Songdong Meng declare that they have no conflict of interest.

This article does not contain any studies with human or animal subjects performed by the any of the authors.

REFERENCES

Ala U, Karreth FA, Bosia C, Pagnani A, Taulli R, Léopold V, Tay Y, Provero P, Zecchina R, Pandolfi PP (2013) Integrated transcriptional and competitive endogenous RNA networks are cross-regulated in permissive molecular environments. Proc Natl Acad Sci USA 110:7154–7159

Augui S, Nora EP, Heard E (2011) Regulation of X-chromosome inactivation by the X-inactivation centre. Nat Rev Genet 12:429–442

Carl JW Jr, Trgovcich J, Hannenhalli S (2013) Widespread evidence of viral miRNAs targeting host pathways. BMC Bioinf 14(Suppl 2):S3

Cazalla D, Yario T, Steitz JA (2010) Down-regulation of a host microRNA by a *Herpesvirus saimiri* noncoding RNA. Science 328:1563–1566

Dabrowska A, Kim N, Aldovini A (2008) Tat-induced FOXO3a is a key mediator of apoptosis in HIV-1-infected human CD4+ T lymphocytes. J Immunol 181:8460–8477

Fang L, Du WW, Yang X, Chen K, Ghanekar A, Levy G, Yang W, Yee AJ, Lu WY, Xuan JW et al (2013) Versican 3′-untranslated region (3′-UTR) functions as a ceRNA in inducing the development of hepatocellular carcinoma by regulating miRNA activity. FASEB J 27:907–919

Filipowicz W, Grosshans H (2011) The liver-specific microRNA miR-122: biology and therapeutic potential. Prog Drug Res 67:221–238

Ghisi M, Corradin A, Basso K, Frasson C, Serafin V, Mukherjee S, Mussolin L, Ruggero K, Bonanno L, Guffanti A et al (2011) Modulation of microRNA expression in human T-cell development: targeting of NOTCH3 by miR-150. Blood 117:7053–7062

Goubau D, Deddouche S, Reis E, Sousa C (2013) Cytosolic sensing of viruses. Immunity 38:855–869

Gunasekharan V, Laimins LA (2013) Human papillomaviruses modulate microRNA 145 expression to directly control genome amplification. J Virol 87:6037–6043

Hao J, Jin W, Li X, Wang S, Zhang X, Fan H, Li C, Chen L, Gao B, Liu G et al (2013) Inhibition of alpha interferon (IFN-α)-induced microRNA-122 negatively affects the anti-hepatitis B virus efficiency of IFN-α. J Virol 87:137–147

Hu J, Xu Y, Hao J, Wang S, Li C, Meng S (2012) MiR-122 in hepatic function and liver diseases. Protein Cell 3:364–371

Huang J, Wang F, Argyris E, Chen K, Liang Z, Tian H, Huang W, Squires K, Verlinghieri G, Zhang H et al (2007) Cellular microRNAs contribute to HIV-1 latency in resting primary CD4+ T lymphocytes. Nat Med 13:1241–1247

Jopling CL, Yi M, Lancaster AM, Lemon SM, Sarnow P (2005) Modulation of hepatitis C virus RNA abundance by a liver-specific MicroRNA. Science 309:1577–1581

Karreth FA, Pandolfi PP (2013) ceRNA cross-talk in cancer: when ce-bling rivalries go awry. Cancer Discov 3:1113–1121

Karreth FA, Tay Y, Perna D, Ala U, Tan SM, Rust AG, DeNicola G, Webster KA, Weiss D, Perez-Mancera PA et al (2011) In vivo identification of tumor-suppressive PTEN ceRNAs in an oncogenic BRAF-induced mouse model of melanoma. Cell 147:382–395

Kincaid RP, Sullivan CS (2012) Virus-encoded microRNAs: an overview and a look to the future. PLoS Pathog 8:e1003018

Klase Z, Houzet L, Jeang KT (2012) MicroRNAs and HIV-1: complex interactions. J Biol Chem 287:40884–40890

Kong G, Zhang J, Zhang S, Shan C, Ye L, Zhang X (2011) Upregulated microRNA-29a by hepatitis B virus X protein enhances hepatoma cell migration by targeting PTEN in cell culture model. PLoS ONE 6:e19518

Lee S, Song J, Kim S, Kim J, Hong Y, Kim Y, Kim D, Baek D, Ahn K (2013a) Selective degradation of host MicroRNAs by an intergenic HCMV noncoding RNA accelerates virus production. Cell Host Microbe 13:678–690

Lee YH, Lai CL, Hsieh SH, Shieh CC, Huang LM, Wu-Hsieh BA (2013b) Influenza A virus induction of oxidative stress and MMP-9 is associated with severe lung pathology in a mouse model. Virus Res 178:411–422

Li C, Wang Y, Wang S, Wu B, Hao J, Fan H, Ju Y, Ding Y, Chen L, Chu X et al (2013) HBV mRNAs-mediated miR-122 inhibition up-regulates PTTG1-binding protein which promotes HCC tumor growth and cell invasion. J Virol 87:2193–2205

Ling S, Birnbaum Y, Nanhwan MK, Thomas B, Bajaj M, Li Y, Li Y, Ye Y (2013a) Dickkopf-1 (DKK1) phosphatase and tensin homolog on chromosome 10 (PTEN) crosstalk via microRNA interference in the diabetic heart. Basic Res Cardiol 108:352

Ling S, Birnbaum Y, Nanhwan MK, Thomas B, Bajaj M, Ye Y (2013b) MicroRNA-dependent cross-talk between VEGF and HIF1α in the diabetic retina. Cell Signal 25:2840–2847

Liu N, Zhang J, Jiao T, Li Z, Peng J, Cuizhu Q, Ye X (2013) Hepatitis B virus inhibits apoptosis of hepatoma cells by sponging miR-15a/16 cluster. J Virol 87:13370–13378

Manigold T, Racanelli V (2007) T-cell regulation by CD4 regulatory T cells during hepatitis B and C virus infections: facts and controversies. Lancet Infect Dis 7:804–813

Marcinowski L, Tanguy M, Krmpotic A, Rädle B, Lisnić VJ, Tuddenham L, Chane-Woon-Ming B, Ruzsics Z, Erhard F, Benkartek C et al (2012) Degradation of cellular mir-27 by a novel, highly abundant viral transcript is important for efficient virus replication in vivo. PLoS Pathog 8:e1002510

Miyoshi K, Okada TN, Siomi H, Siomi MC (2009) Characterization of the miRNA-RISC loading complex and miRNA-RISC formed in the Drosophila miRNA pathway. RNA 15:1282–1291

Newman MA, Hammond SM (2010) Emerging paradigms of regulated microRNA processing. Genes Dev 24:1086–1092

Noorbakhsh J, Lang AH, Mehta P (2013) Intrinsic noise of microRNA-regulated genes andthe ceRNA hypothesis. PLoS ONE 8:e72676

Pang PS, Pham EA, Elazar M, Patel SG, Eckart MR, Glenn JS (2012) Structural map of a Pan PS, microRNA-122: hepatitis C virus complex. J Virol 86:1250–1254

Poliseno L, Salmena L, Zhang J, Carver B, Haveman WJ, Pandolfi PP (2010) A coding-independent function of gene and pseudogene mRNAs regulates tumour biology. Nature 465:1033–1038

Redman TK, Brittm WJ, Wilcox CM, Graham MF, Smith PD (2002) Human cytomegalovirus enhances chemokine production by lipopolysaccharide-stimulated lamina propria macrophages. J Infect Dis 185:584–590

Shan Y, Zheng J, Lambrecht RW, Bonkovsky HL (2007) Reciprocal effects of micro-RNA-122 on expression of heme oxygenase-1 and hepatitis C virus genes in human hepatocytes. Gastroenterology 133:1166–1174

Sharma M, Callen S, Zhang D, Singhal PC, Vanden Heuvel GB, Buch S (2010) Activation of Notch signaling pathway in HIV-associated nephropathy. AIDS 24:2161–2170

Shimakami T, Yamane D, Jangra RK, Kempf BJ, Spaniel C, Barton DJ, Lemon SM (2012) Stabilization of hepatitis C virus RNA by an Ago2-miR-122 complex. Proc Natl Acad Sci USA 109:941–946

Song L, Liu H, Gao S, Jiang W, Huang W (2010) Cellular microRNAs inhibit replication of the H1N1 influenza A virus in infected cells. J Virol 84:8849–8860

Sun G, Li H, Wu X, Covarrubias M, Scherer L, Meinking K, Luk B, Chomchan P, Alluin J, Gombart AF et al (2012) Interplay between HIV-1 infection and host microRNAs. Nucleic Acids Res 40:2181–2196

Swaminathan S, Zaunders J, Wilkinson J, Suzuki K, Kelleher AD (2009) Does the presence of anti-HIV miRNAs in monocytes explain their resistance to HIV-1 infection? Blood 113:5029–5030

Szabó J, Bácsi A, Beck Z, Kiss J, Andirkó I, Tóth FD (1999) Role of interleukin-8 and transforming growth factor-beta1 in enhancement of human cytomegalovirus replication by human T cell leukemia-lymphoma virus type I in macrophages co-infected with both viruses. J Interferon Cytokine Res 19:209–217

Tang-Feldman YJ, Lochhead SR, Lochhead GR, Yu C, George M, Villablanca AC, Pomeroy C (2013) Murine cytomegalovirus (MCMV) infection upregulates P38 MAP kinase in aortas of Apo E KO mice: a molecular mechanism for MCMV-induced acceleration of atherosclerosis. J Cardiovasc Transl Res 6:54–64

Torres AG, Fabani MM, Vigorito E, Gait MJ (2011) MicroRNA fate upon targeting with anti-miRNA oligonucleotides as revealed by an improved Northern-blot-based method for miRNA detection. RNA 17:933–943

Ulitsky I, Bartel DP (2013) lincRNAs: genomics, evolution, and mechanisms. Cell 154:26–46

Wang S, Qiu L, Yan X, Jin W, Wang Y, Chen L, Wu E, Ye X, Gao GF, Wang F et al (2012) Loss of MiR-122 expression in patients with hepatitis B enhances hepatitis B virus replication through cyclin G1 modulated P53 activity. Hepatology 55:730–741

Wang S, Zhang X, Ju Y, Zhao B, Yan X, Hu J, Shi L, Yang L, Ma Z, Chen L et al (2013a) MicroRNA-146a feedback suppresses T cell immune function by targeting Stat1 in patients with chronic hepatitis B. J Immunol 191:293–300

Wang Y, Jiang L, Ji X, Yang B, Zhang Y, Fu XD (2013b) Hepatitis B viral RNA directly mediates down-regulation of the tumor suppressor microRNA miR-15a/miR-16-1 in hepatocytes. J Biol Chem 288:18484–18493

Yu Z, Willmarth NE, Zhou J, Katiyar S, Wang M, Liu Y, McCue PA, Quong AA, Lisanti MP, Pestell RG (2010) microRNA 17/20 inhibits cellular invasion and tumor metastasis in breast cancer by heterotypic signaling. Proc Natl Acad Sci USA 107:8231–8236

Yuan M, Zhan Q, Duan X, Song B, Zeng S, Chen X, Yang Q, Xia J (2013) A functional polymorphism at miR-491-5p binding site in the 3′-UTR of MMP-9 gene confers increased risk for atherosclerotic cerebral infarction in a Chinese population. Atherosclerosis 226:147–152

Zhang X, Hou J, Lu M (2013) Regulation of hepatitis B virus replication by epigenetic mechanisms and microRNAs. Front Genet 4:202

4

MicroRNA-29a modulates axon branching by targeting doublecortin in primary neurons

Hanqin Li[1], Susu Mao[1], Haitao Wang[2], Ke Zen[1✉], Chenyu Zhang[1✉], Liang Li[1✉]

[1] Jiangsu Engineering Research Center for MicroRNA Biology and Biotechnology, State Key Laboratory of Pharmaceutical Biotechnology, Nanjing University School of Life Sciences, Nanjing 210093, China
[2] Department of Pharmacy, 306th Hospital of PLA, Beijing 100101, China
✉ Correspondence: kzen@nju.edu.cn (K. Zen), cyzhang@nju.eud.cn (C. Zhang), lijing@sibs.ac.cn (L. Li)

ABSTRACT

MicroRNAs (miRNAs) are endogenously expressed small, non-coding transcripts that regulate protein expression. Substantial evidences suggest that miRNAs are enriched in central nervous system, where they are hypothesized to play pivotal roles during neural development. In the present study, we analyzed miRNAs expression in mice cerebral cortex and hippocampus at different developmental stages and found miR-29a increased dramatically at postnatal stages. In addition, we provided strong evidences that miR-29a is enriched in mature neurons both *in vitro* and *in vivo*. Further investigation demonstrated that the activation of glutamate receptors induced endogenous miR-29a level in primary neurons. Moreover, we showed that miR-29a directly regulated its target protein Doublecortin (DCX) expression, which further modulated axon branching in primary culture. Together, our results suggested that miR-29a play an important role in neuronal development of mice cerebrum.

KEYWORDS miR-29a, doublecortin, glutamate receptor, mature neurons, axon branching

INTRODUCTION

MiRNAs are approximately 21 nucleotide non-coding transcripts that are derived from hairpin precursors, which regulate protein expression by base pairing to the 3′ untranslated region (UTR) of mRNAs (Chen et al., 2008). Several works show that miRNAs are enriched in central nervous system and emerging evidence have suggested the pivotal roles of miRNAs in neural development (Ji et al., 2013; Shu et al., 2013). Mutations in Drosophila Ago1 cause a severe loss of all types of neurons and glial cells while dicer deficient mice display abnormal neurogenesis and gliogenesis in the developing central nervous system (Kataoka et al., 2001; Kawase-Koga et al., 2009). There are also studies indicating the significant role of specific miRNA in neural development. For example, miR-132 modulates the dopamine neuron differentiation while miR-124 regulates neurite outgrowth during neuronal differentiation (Yang et al., 2012; Yu et al., 2008). Together, these data support the notion that miRNAs modulate cell differentiation, maturation as well as survival in central nervous system.

To investigate the function of miRNAs in the development of mice cerebrum, we screened miRNAs expression in mice cerebrum at different developmental stages. Interestingly, we found miR-29a changed most dramatically among all those miRNAs that increased during cerebrum development. MiR-29 family is consisted of three mature members, miR-29a, miR-29b, and miR-29c, which are encoded by two genome clusters (Wang et al., 2013). Recent studies have demonstrated that miR-29 family is associated with the pathogenesis of sporadic Alzheimer's disease by targeting BACE and NAV3 (Bettens et al., 2009; Shioya et al., 2010). Further investigation indicates that miR-29b has anti-apoptotic effect in mature neuron by targeting BH3-only (Kole et al., 2011). However, the role of miR-29 family during neuronal development remains obscure, which still needs further investigation.

In the present study, we demonstrated that miR-29a, which highly expresses in mature neurons, is up-regulated

Hanqin Li and Susu Mao contribute equally to this work.

Table 1. Copy number of miRNAs that increased significantly during the cerebrum development

	Copy number at different developmental stages				Ratio
	E12.5	E15.5	E18.5	P60	P60/E12.5
miR-22	650.86	753.18	1135	7193.4	11.05
miR-23a	1249.1	1817.3	1482.8	6461.2	5.17
miR-24	3396.8	5393	5354	20931	6.16
miR-29a	61.68	61.58	73.86	5122.3	83.05
miR-31	228.11	245.63	352.49	3118.1	13.67
miR-128	945.99	1333.8	4568.2	7894.6	8.35
miR-132	510.18	901.66	1727.1	16807	32.94
miR-138	1713.5	4395.6	6829.2	23379	13.64
miR-139-5p	359.66	484.26	636.85	4870.7	13.54
miR-143	285	291.6	411.99	2356.9	8.27
miR-145	640.83	638.49	821.15	4777.1	7.45
miR-221	747.36	478.43	901.21	5651.1	7.56
miR-222	463.46	426.43	783.48	9207.9	19.87

both in cortex and hippocampus during cerebrum development especially in postnatal stages. Besides, we provided evidences that activation of glutamate receptors contribute to the increase of miR-29a in primary neurons, suggesting the possible association of neural activity and the increase of miR-29a. Moreover, our data revealed that DCX, a protein that is reported to affect axon branching (Bilimoria et al., 2010), is a direct target of miR-29a. Further functional analysis showed that miR-29a increased axon branching via repressing DCX expression. Taken together, these data indicate that miR-29a plays an important role in neuronal development of mice cerebrum.

RESULTS

Screening miRNAs expression levels in mice cerebral cortex at different developmental stages

To investigate the role of miRNAs in the development of mice cerebrum, we first analyzed the miRNAs levels in the mice cortex at different stages (E12.5, E15.5, E18.5 and P60) by miRNAs array. Here we mainly focused on the increased miRNAs which are selected based on the following two criteria: (1) the minimum copy number of miRNA at P60 is above 2000; (2) the fold change of miRNA between P60 and E12.5 is more than 5. Among all those selected miRNAs, miR-29a increased most significantly in P60 mice cortex (Table 1). Meanwhile, the copy number of miR-29b and miR-29c is less than 300 at the stage of P60 (data not shown). Surprisingly, miR-29a did not change significantly during E12.5–E18.5 stages while it increased drastically (>70 fold) during E18.5–P60 stages. To further understand the correlation between miR-29a and cerebrum development, we analyzed miR-29a expression both in cortex and hippocampus by Q-RTPCR. We found that cortical miR-29a showed a time-dependent increase exclusively in postnatal stages while miR-138 increased gradually during the entire developmental stages analyzed (E12.5–P60), which is consistent with the miRNAs array analysis (Fig. 1A). Similarly, the level of miR-29a in hippocampus began to increase at P7 while the level of miR-138 gradually increased during the whole stages (E18.5–P60) (Fig. 1B). We also compared the expression levels of miR-29 family in cortex and hippocampus of P60 mice by Q-RTPCR and demonstrated that miR-29a concentration is relatively higher than that of the other two family members (Fig. S1).

Expression pattern of miR-29 family in mice cerebrum

To elucidate the expression pattern of miR-29 family in mice cerebrum, we examined three members of miR-29 family including miR-29a, miR-29b and miR-29c by *in situ* hybridization. Our results demonstrated that miR-29a is most intensively expressed in hippocampus neurons while miR-29b and miR-29c showed much weaker signals (Fig. 1C, a–e). Besides, miR-29a is also expressed in cortical neurons, while miR-29b and miR-29c barely showed any positive signals (Fig. 1C, a′–e′). These results are coincident with our Q-RTPCR analysis (Fig. S1). Nevertheless, using lung tissue as positive control, we demonstrated that both miR-29b and miR-29c show relatively strong signals in the mice lung tissue compared with that in cortex and hippocampus (Fig. S2). Importantly, we also analyzed miR-29a expression pattern in dentate gyrus (DG) area at different stages. The signal appeared at P7 and increased at P14 and P60, which is consistent with our Q-RTPCR analysis (Fig. 2A, a–c). Furthermore, the expression of miR-29a in DG area began

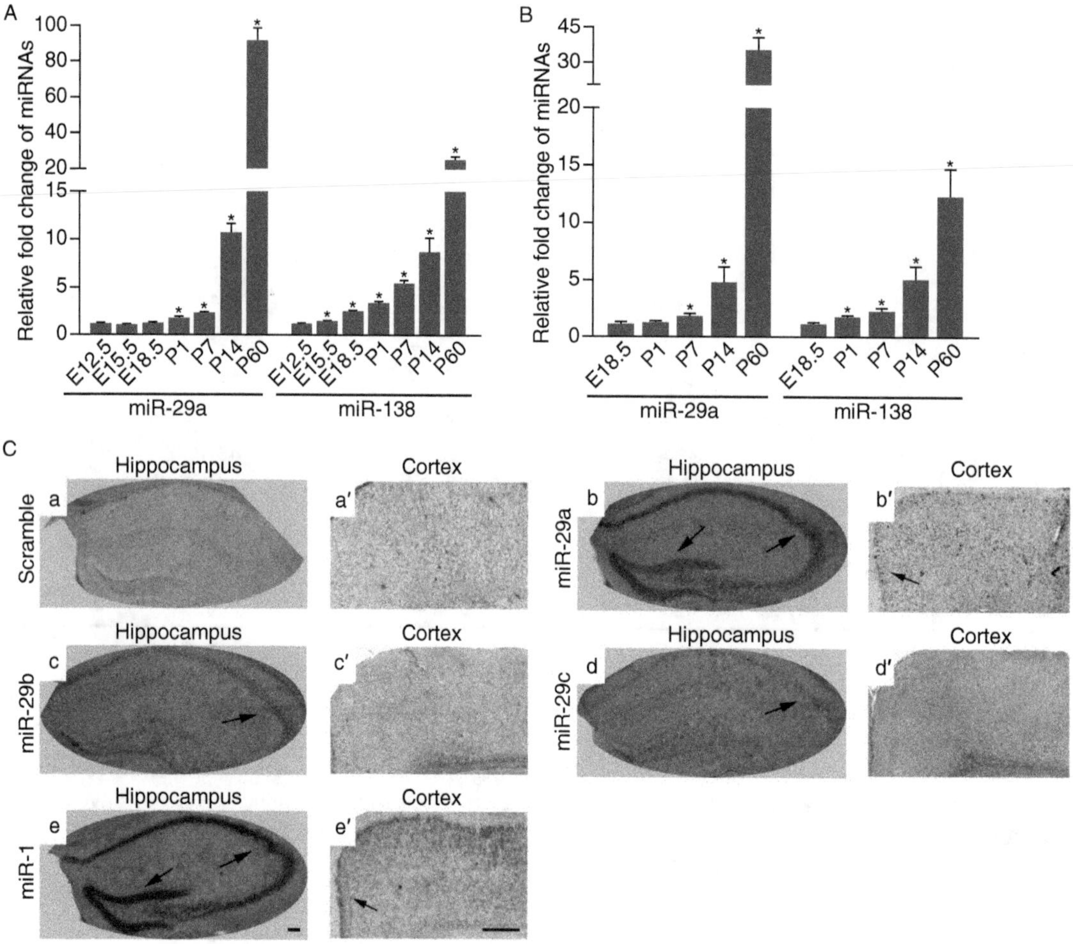

Figure 1. Expression of miR-29a in mice cortex and hippocampus. (A) Relative expression level of miR-29a and miR-138 (U6 is used as internal control) in cortex of different developmental stages, *, $P < 0.01$ compared to the E12.5 stage in each group ($n = 5$); (B) relative expression level of miR-29a and miR-138 in hippocampus of different developmental stages, *, $P < 0.01$ compared to the E18.5 stage in each group ($n = 5$); (C) *in situ* hybridization of different members of miR-29 family in hippocampus (a–e) and cortex (a′–e′) ($n = 3$), miR-124 is used as positive control, arrows indicate neurons with positive signal in cortex or hippocampus, scale bar = 200 μm.

from the outside of granule layer (P7) while gradually spread to the inside (P60), which is similar to the neuron maturation pattern marked by NeuN (mature neuronal marker) (Fig. 2A, d–f) (Snyder et al., 2012). Together, these data provide evidence indicating that miR-29a is highly expressed and functions in the development of cerebrum, especially in mature neurons.

Activation of glutamate receptors increases the level of miR-29a in primary neurons

We have observed the up-regulation of miR-29a during cerebrum development exclusively at postnatal stages (Fig. 1A and 1B). In addition, we also examined the miR-29a level in cultured neuron from cortex or hippocampus at different time points and showed that miR-29a increased as neurons became mature (Fig. 2B). Therefore, it is interesting for us to elucidate the mechanism underlying the regulation of miR-29a expression. We investigated the association between neural activity and miR-29a expression by assessing the level of miR-29a in cortical neurons after glutamate treatment (10 μmol/L, 15 min) and found that miR-29a is significantly increased since 0.5 h after the stimulation. Such increase lasted for 6 h and miR-29a returned to the basal level after 12 h (Fig. 2C). Furthermore, we found that pretreatment of MK-801 (NMDA receptor antagonist) significantly blocked the increase of miR-29a induced by glutamate (Fig. 2D). Similarly, DNQX (AMPA/kainate receptor antagonist) also inhibited such increase and exhibited linear effect when used with MK-801 together (Fig. 2D). These data indicate that the activation of glutamate receptors may contribute to the increase of miR-29a in neurons during cerebrum maturation.

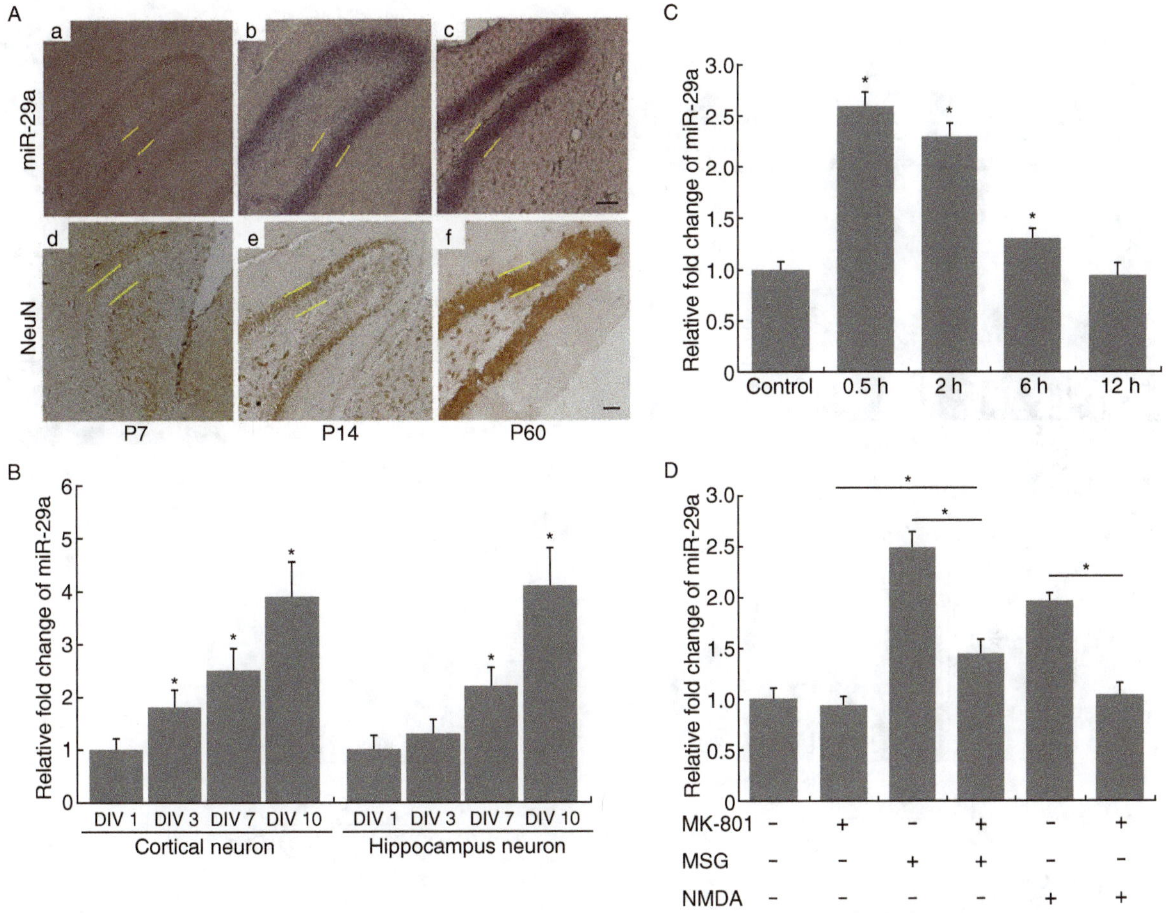

Figure 2. Expression of miR-29a in neurons under different circumstances. (A) Expression pattern of miR-29a (purple) in DG area of hippocampus at different postnatal stages (a–c), expression pattern of NeuN (brown) in DG area of hippocampus at different postnatal stages (d–f), granule cell layers are indicated between two yellow lines (n = 3), scale bar = 50 μm; (B) relative expression level of miR-29a in primary neuron at different cultured stages, *, $P < 0.01$ compared to DIV 1 in each group (n = 5); (C) relative expression level of miR-29a in primary cortical neurons at different time points after the stimulation of 10 μmol/L GSM for 15 min, *, $P < 0.01$ compared to control (n = 3); (D) relative expression level of miR-29a in primary cortical neurons at 2 h after indicated treatments, *, $P < 0.01$ (n = 3).

DCX is a direct target of miR-29a

Through bioinformatics screening, we found that the sequence of miR-29a matched well with 3′ UTR of DCX mRNA (Fig. 3A). DCX is a microtubule-associated protein expressed by neuronal precursor cells and immature neurons, which is reported to modulate axon branching (Bilimoria et al., 2010; Spampanato et al., 2012). In addition, the targeted sequence in DCX mRNA that matched with the seed sequence of miR-29a is highly conservative in many species, indicating the regulation of DCX expression by miR-29a is quite universal. We further evaluate the protein level of DCX in both cortex and hippocampus at different developmental stages. In embryonic stages, DCX level slightly increased in cortex (E18.5) while did not change a lot in hippocampus. However, in postnatal stages, DCX expression significantly decreased both in cortex and hippocampus at P7 (Fig. 3B–D), which is negatively correlated with the change of miR-29a level in cortex and hippocampus (Fig. 1A and 1B). We also demonstrated that the expression patterns of DCX and miR-29a in DG area are mutually complementary (Figs. 3E and 2A). Moreover, we analyzed the expression of DCX during the neuron maturation *in vitro* and found the level of DCX showed a time-course decrease, which is negatively correlated to the level of miR-29a (Fig. 3F and 3G). To prove that DCX is a direct target of miR-29a we conducted luciferase assay and found that miR-29a markedly reduce the translational efficiency of reporter gene tagged with the 3′UTR of DCX mRNA containing the putative miR-29a-binding sites while mutations in these sites significantly

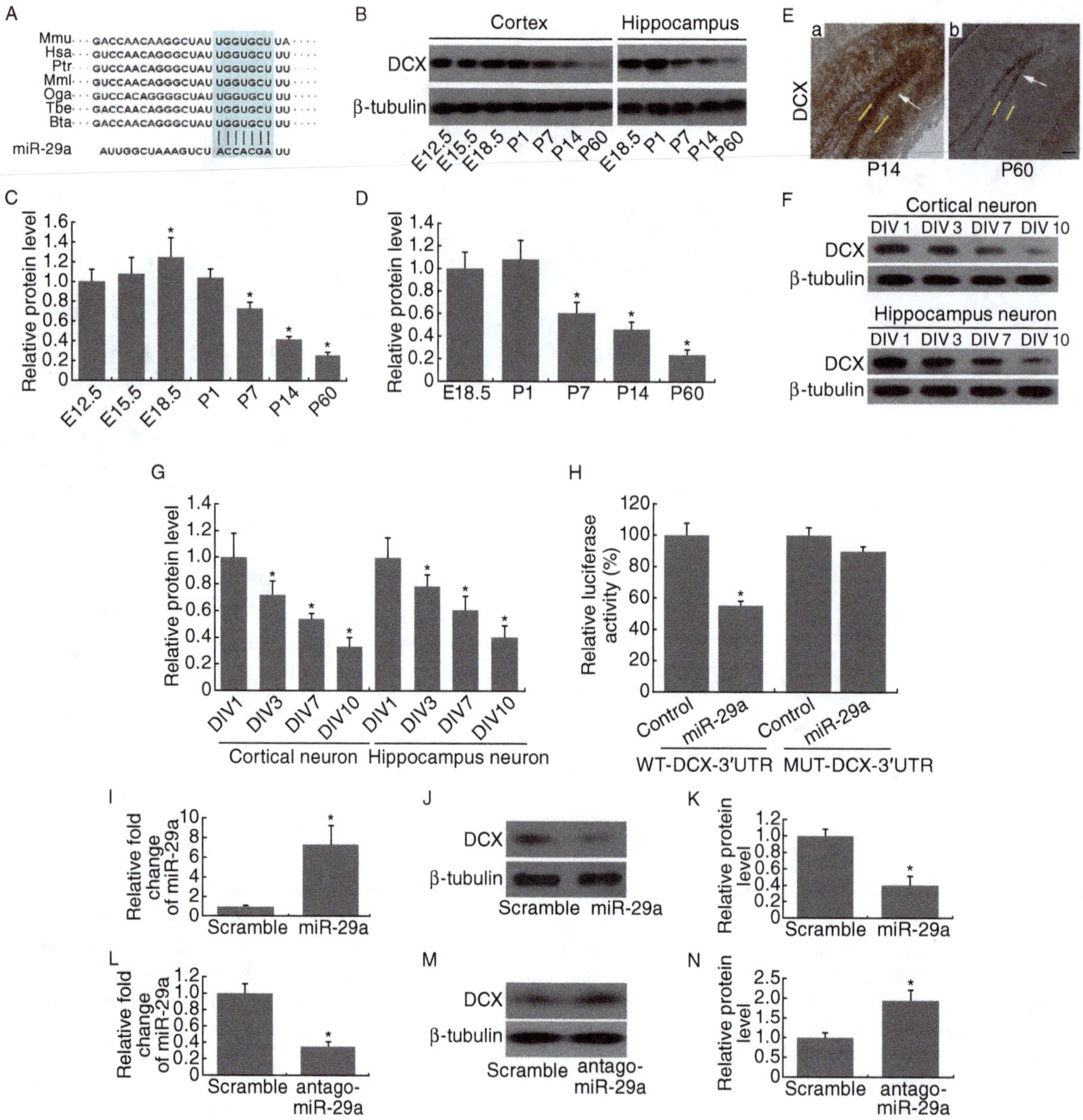

Figure 3. DCX is a direct target of miR-29a. (A) Seed sequence and alignment of the miR-29a binding sites in the 3′UTRs of DCX mRNAs from different species, the predicted base-pairing of miR-29a with target recognition seed sequence is shown in colored; (B) Western-blot of DCX (β-tubulin is used as internal control) in cortex (left panel) or hippocampus (right panel) at different developmental stages indicated (n = 3); (C and D) quantification of DCX protein in cortex (C) or hippocampus (D) at different developmental stages indicated (n = 3); (E) expression pattern of DCX (brown, arrows) in DG area of hippocampus at different postnatal stages, granule cell layers are between two yellow lines (n = 3), scale bar = 50 μm; (F and G) Western-blot and quantification of DCX protein in primary neurons derived from cortex and hippocampus at different time points (n = 3); (H) relative luciferase activity measured in different groups of 293T cells, *, P < 0.01 compared to control (n = 3); (I) relative level of miR-29a in primary cortical neurons transfected with miR-29a mimics, *, P < 0.01 compared to scramble (n = 3); (J and K) Western-blot and quantification of DCX protein in primary cortical neurons treated as in G, *, P < 0.01 compared to scramble (n = 3); (L) relative level of miR-29a in primary cortical neurons transfected with antago-miR-29a, *, P < 0.01 compared to scramble (n = 3); (M, N) Western-blot and quantification of DCX protein in primary cortical neurons treated as in J, *, P < 0.01 compared to scramble (n = 3).

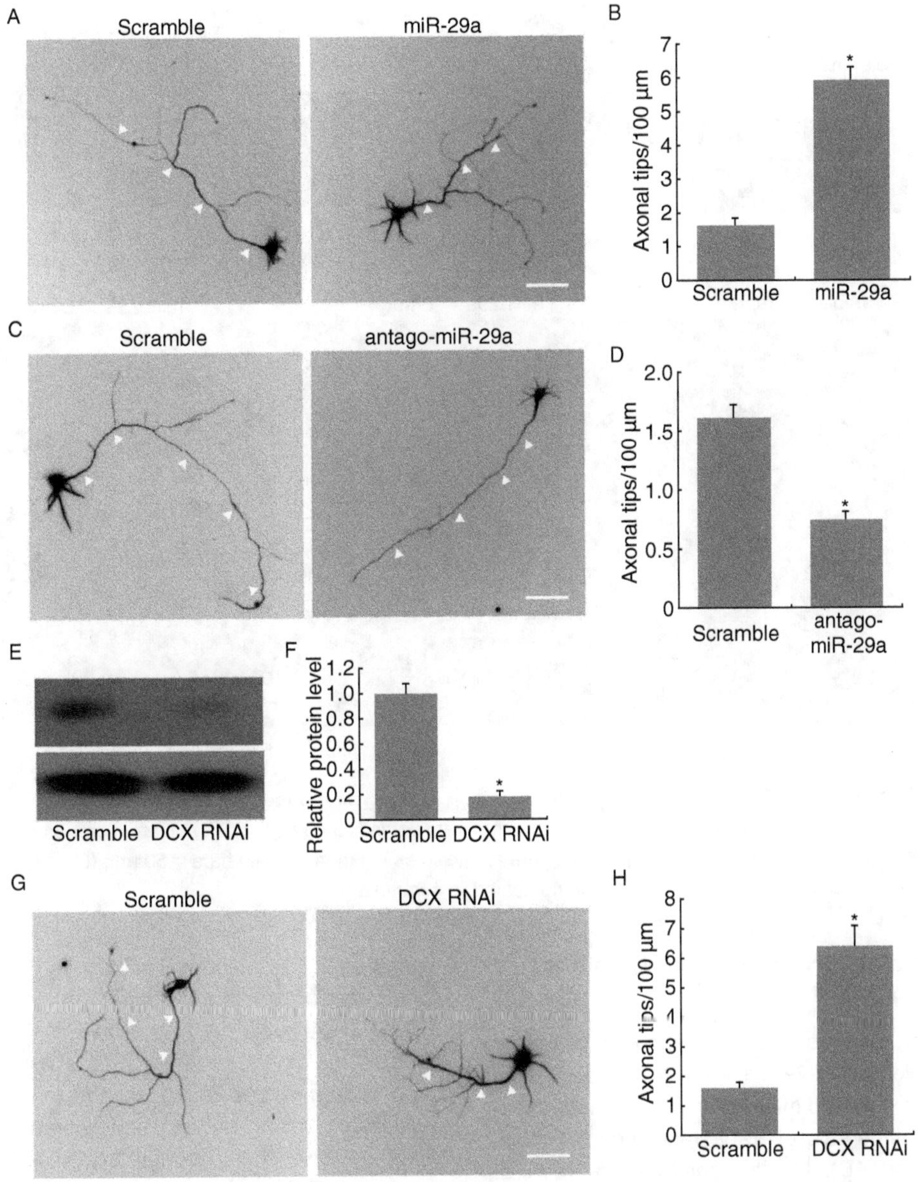

Figure 4. Effects of miR-29a on axon branching of primary cortical neurons. (A) Cortical neurons transfected with scramble or miR-29a mimics together with the eGFP expression plasmid were fixed 4 d after transfection and subjected to immunocytochemistry using the eGFP antibody. In all images of this type, arrowheads indicate primary axons, scale bars = 50 μm; (B) quantification of axon branching of cortical neurons treated in A, *, $P < 0.01$ compared to scramble ($n = 20$); (C) cortical neurons transfected with scramble or antago-miR-29a together with the eGFP expression plasmid were analyzed as in A, scale bars = 50 μm; (B) quantification of axon branching of cortical neurons treated in C, *, $P < 0.01$ compared to scramble ($n = 20$); (E, F) Western-blot and quantification of DCX protein in cortical neurons treated with scrambles or DCX RNAi ($n = 3$); (G) cortical neurons transfected with scrambles or DCX RNAi together with the eGFP expression plasmid were analyzed as in A, scale bars = 50 μm; (H) quantification of axon branching of cortical neurons treated in G, *, $P < 0.01$ compared to scramble ($n = 20$).

blocked the inhibitory effect (Fig. 3H). Next, through gain- and loss-of-function approaches, we evaluated the effect of miR-29a on DCX expression in primary neurons. Over-expression of miR-29a resulted in a remarkable decrease of DCX protein level in primary neurons (Fig. 3I–K). Meanwhile, antago-miR-29a reduced cellular miR-29a level and increased DCX protein expression in cultured neurons (Fig. 3L–N). Together, these evidences support the notion that miR-29a inhibits DCX expression by directly targeting its 3′UTR of mRNA.

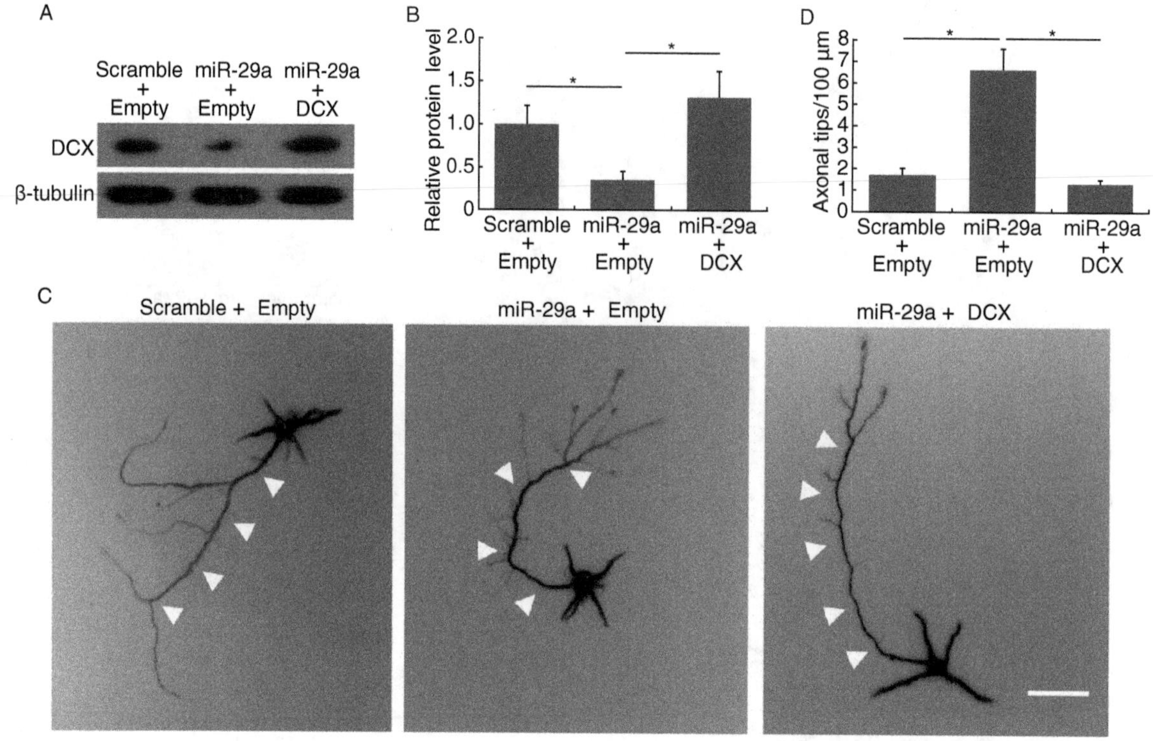

Figure 5. Over-expression of DCX rescues the axon branching phenotype induced by miR-29a. (A and B) Western-blot and quantification of DCX protein in cortical neurons under different treatment indicated (n = 3); (C) different groups of cortical neurons transfected together with the eGFP expression plasmid were analyzed as in Fig. 4A, scale bars = 50 μm; (D) quantification of axon branching of cortical neurons treated in C, *, $P < 0.01$ compared to scramble (n = 20).

MiR-29a increases axon branching in primary cortical neuron via targeting DCX

Since DCX is reported to restrict axon branching we further investigated the role of miR-29a in axon growth of primary neurons. We over-expressed miR-29a in cortical neurons at DIV 3 and found it triggered exuberant branching of axons at DIV 7 (Fig. 4A and 4B). On the contrary, inhibition of endogenous miR-29a by antago-miR resulted in long axons bearing few or no branches of primary neurons (Fig. 4C and 4D). In addition, to elucidate the role of DCX in miR-29a induced axon branching, we specifically knocked down the endogenous DCX protein by RNAi (Fig. 4E and 4F). In the morphology assay, we found that DCX knockdown also induced excessive axonal branches compared with control neurons (Fig. 4G and 4H). Furthermore, we also over-expressed DCX (without miR-29 binding site) and demonstrated an effective inhibition of the axon branching induced by miR-29a over-expression (Fig. 5). These results indicate that miR-29a may increase axon branching in primary neuron by repressing DCX expression.

DISCUSSION

MiRNAs are small, non-coding transcripts that functioned in many tissues by repressing target gene expression (Chen et al., 2012). Circumstantial evidences suggest that miRNAs play a crucial role in the central nervous system (Ji et al., 2013). To elucidate the role of miRNAs in cerebrum development, we analyzed the miRNAs expression pattern in mice cerebrum and found several miRNAs that increase significantly during different developmental stages, among which miR-29a changed most significantly. The expression level of miR-29a showed no significant differences until postnatal stages when it increased dramatically both in cortex and hippocampus, indicating miR-29a may be associated with the process of neural maturation.

Since it is reported that miR-29 expressed more strongly in cultured astrocyte than in neuron (Smirnova et al., 2005), we investigated the expression pattern of miR-29 *in vivo* by *in situ* hybridization. Our results demonstrated that miR-29a mainly expressed in mature neuron in cerebrum while miR-29b and miR-29c showed relatively weak signals. These discrepancies may be the result of differences between *in vitro* and *in vivo* as well as the techniques with different sensitivity used for detecting miRNAs. In addition, we provided evidence that the expression pattern of miR-29a in DG area is similar to that of NeuN (mature neuronal marker), which suggested that miR-29a is increased during neuron maturation. Furthermore, we showed a time-course increase of miR-29a expression in primary neurons. Together, these data indicate that miR-29a may play an important role in mature neurons.

Since neural stimulation helped to fine-tune the process of neuronal maturation (He et al., 2010), we further investigated the association between miR-29a expression and neural activity. Our results demonstrated that activation of glutamate receptors induced the endogenous level of miR-29a in primary neurons indicating that neural activity may modulate the expression of miR-29a.

Through bioinformatics screening and luciferase assay, we searched and verified one target gene of miR-29a—DCX. DCX initially expressed in neuronal precursor cells and immature neurons (E12.5) while it began to down-regulate around P7, which is coincident with the time that miR-29a aroused in cerebrum (Spampanato et al., 2012). Immunostaining results also revealed that DCX expressed in a complementary pattern compared with the expression of miR-29a in DG area. Besides, protein analysis also demonstrated that miR-29a inhibited DCX translation in primary neurons.

As a target protein of miR-29a, DCX is a microtubule associated protein that stabilized microtubules (Cohen et al., 2008). There is evidence that GSK3β/DCX signaling pathway restricts axon branching (Bilimoria et al., 2010). In addition, neural activity that has been proved to affect miR-29a expression is also reported to regulate axon development (Uesaka et al., 2005). Therefore, it is necessary for us to evaluate the effect of miR-29a on axon growth. Our data suggested that miR-29a triggered excessive axon branching while blocking endogenous miR-29a resulted in the opposite effect. Moreover, DCX RNAi also promoted axon branching of cortical neurons, which is similar to the effect of miR-29a over-expression. MiR-29 has multiple targets that may affect neuronal growth (Govek et al., 2005; Wang et al., 2013). To further confirm the role of DCX in the miR-29a regulated axon branching, we over-expressed DCX and found it effectively rescued the axon branching phenotype induced by miR-29a over-expression. Together, these results suggest that miR-29a modulated axon branching via targeting DCX expression.

Taken together, our data demonstrate that activation of glutamate receptors increased the expression of miR-29a, which further promoted axon branching in the maturation of cerebral neurons by targeting DCX. Whether miR-29a-DCX regulation pathway is involved in the neural activity dependent axon growth *in vivo* needs further investigation.

MATERIAL AND METHODS

Primary culture

Animal care and procedures were performed in accordance with the Laboratory Animal Care Guidelines approved by Model Animal Research Center of Nanjing University. Primary neurons were cultured from cortex of E16.5 mice as described previously (David et al., 2007). Neurons were seeded at 3×10^5 per well in 12 well plate for biochemistry analysis, or at 1×10^4 per well for immunofluorescent staining and morphological observation. Half medium was refreshed every 2 d.

Cultured neurons (DIV 3) were transfected with miR-29a mimics (Genepharma), antago-miR-29a (RiboBio Inc) or DCX RNAi (Genepharma, described in (Ramos et al., 2006)) together with eGFP reporter plamid using lipofectamine 2000 (Invitrogen) according to the manufacturer's protocol. Cells were harvested 24 h later for RNA analysis and 48 h later for protein analysis. For morphological study, cells were fixed 4 d after transfection.

Neurons (DIV 3) were subjected to the stimulation of MSG (10 µmol/L) for 15 min. After that RNA samples were collected at the time indicated. For the inhibition of glutamate receptor, cells were pretreated with MK-801 (10 µmol/L, Sigma) or/and DNQX (10 µmol/L, Sigma) for 2 h before subjecting to MSG stimulation.

RNA and protein quantification

Mice cortex and hippocampus were dissected from C57BL/6 J mouse at different stages indicated. Total RNA was purified from the cultures or tissues with TRIZOL (Invitrogen) according to the manufacturer's instructions. The RNA purity was determined by the OD_{260}/OD_{280} ratio, and the concentration was calculated based on OD_{260}. 50 µg of total RNA of cortexes from different stages (E12.5, E15.5, E18.5 and P60) were sent to the Capitalbio Corporation for miRNAs array analysis.

Q-RTPCR was carried out using TaqMan microRNA probes (Applied Biosystems). Briefly, 2 µg of total RNA was reverse-transcribed to cDNA using AMV reverse transcriptase (TaKaRa) and stem-loop RT primers (Applied Biosystems). Real-time PCR was performed using a TaqMan PCR kit on an Applied Biosystems 7300 Sequence Detection System (Applied Biosystems). All reactions, including the no-template controls, were run in triplicate. After the reactions, the *CT* values were determined using fixed-threshold settings. Upon the evaluation of each miRNA's level, the expression level of U6 was used as internal control.

Cultures or tissues were both lysed in 50 mmol/L Tris, pH 7.5, 150 mmol/L NaCl, 2 mmol/L EDTA, and 1% Triton X-100. The protease inhibitors aprotinin, pepstatin, leupeptin, as well as the reducing agent dithiothreitol were added to this buffer before cell lysis. Lysates were cleared of insoluble material by spinning at maximum speed on a tabletop centrifuge, boiled in sample buffer, and examined using standard SDS-PAGE, followed by Western blotting. The antibody used were DCX (Abcam) and β-tubulin (Abcam).

In situ hybridization

This method was adjusted based on the protocol described by Obernosterer et al (Obernosterer et al., 2007). Brains were fixed in 4% paraformaldehyde overnight at 4°C and cryo protected in 15% sucrose followed by 30% sucrose overnight each. Cryostat section (20 µm) were cut and mounted onto Superfrost Gold Plus slide (Fisher). Then sections were post-fixed in 4% paraformaldehyde followed by washed in PBS for three times and 75% alcohol for one time, dried in RT, and incubated in hybridization buffer (DEPC H_2O with 50% de-ionized formamide (*v/v*), 5× SSC, 0.2 mg/mL yeast tRNA, 0.5 mg/mL herring sperm DNA, 10% dextran sulfate (*w/v*), 1/1000 Tween-20) with a concentration of 10 nmol/L for digoxigenin-labeled LNA probes (Exiqon) overnight at 50°C in a humidified chamber. Following hybridization, slides were washed in 0.1× SSC at 55°C for 3 times, 30 min each, then blocked in 10% FBS in TBS (pH = 7.5) for 1 h. AP-conjuncted-anti-digoxigenin antibody (Roche) was incubated with slides overnight at final concentration of 1:1000. NBT/BCIP kit (Invitrogen) was used to do the color reaction.

Plasmid construction and luciferase assay

Part of mmu-DCX 3′UTR containing the predicted target sequence was obtained from cortex of P1 mouse using RT-PCR with primers as indicated. Forward primer 5′-AAGATGATAGGCTAAATCAAAGCC-3′, reverse primer 5′-CAGGATGAATGCTAGACAATACAC-3′. This PCR product was inserted into pMIR-REPORTTM-Luciferase vector (Ambion) after digested with *Hin*dIII and *Sac*I (Takara) to construct WT vector. Mutant vector was generated in Invitrogen by replacing the predicted target region with its reverse sequence (from AAGCAC CAATAGCCTTGTTGGTC to CTGGTTGTTCCGATAACCACGAA). The oligonucleotides used for the RT-PCR cloning of DCX without miR-29 binding site: forward primer 5′-ggttccaccaaaatATGg (ATG initiator), reverse primer 5′-TCAcatggaatcgccaag (TGA terminator codon).

293T cell was cultured in DMEM (Gibco) containing 10% FBS (Gibco) and seeded in 24 well plate. 12 h after plating, 0.2 μg of firefly luciferase reporter plasmid, 0.2 μg of β-galactosidase expression vector (Ambion), and equal amounts (20 pmol) of miR-29a mimics or scrambled negative control RNA were transfected into cells with Lipofectamine 2000 as manufacturer's instructing. The β-galactosidase vector was used as a transfection control. At 24 h post-transfection, cells were analyzed using a luciferase assay kit (Promega).

Immunochemistry

Primary cortical neurons were briefly washed twice with cold PBS, and then fixed in 4% paraformaldehyde for 10 min at room temperature (RT). Neurons after fixation were washed with cold PBS three times, permeabilized and blocked with 2% BSA and 0.05% Triton X-100 in PBS for 1 h (RT). Next, neurons were incubated with primary antibodies (anti-eGFP, Abcam) in 2% BSA/PBS in a humidified chamber overnight at 4°C and rinsed three times in PBS (3 × 5 min). This was followed by incubation with secondary antibodies (Invitrogen) in 2% BSA/PBS in a light-proof container (2 h at RT).

Slides were washed and blocked with PBS containing 0.3% Triton X-100 and 5% BSA (Sigma), followed with the incubation of primary antibodies (NeuN, Millipore; DCX, Abcam) overnight at 4°C. Sections were washed three times in PBS and then incubated with biotinylated secondary antibody (1:500, Vector Laboratories, Burlingame, California, USA) overnight at 4°C. After that signals were further amplified using peroxidase Vectastain ABC reagent (Vector Laboratories). 3,3′-diaminobenzidine (Sigma) was used for the color reaction.

Morphological analysis

Images were captured on an Olympus microscope and analyzed using Necromantic software. Axon tips were defined as terminal points of the primary axon and all of its branches.

Statistical Analysis

Data are presented as sample means and their standard errors. Two-tailed Student's *t* test was used for comparison of the means between any two groups. Differences were considered significant when $P < 0.05$.

ACKNOWLEDGEMENTS

This work was supported by grants from the National Natural Science Foundation of China (Grant Nos. 31000478, 31100777 and 31271378).

COMPLIANCE WITH ETHICS GUIDELINES

Hanqin Li, Susu Mao, Haitao Wang, Ke Zen, Chenyu Zhang and Liang Li declare that they have no conflict of interest.

All institutional and national guidelines for the care and use of laboratory animals were followed.

REFERENCES

Bettens K, Brouwers N, Engelborghs S, Van Miegroet H, De Deyn PP, Theuns J, Sleegers K, Van Broeckhoven C (2009) APP and BACE1 miRNA genetic variability has no major role in risk for Alzheimer disease. Hum Mutat 30:1207–1213

Bilimoria PM, de la Torre-Ubieta L, Ikeuchi Y, Becker EB, Reiner O, Bonni A (2010) A JIP3-regulated GSK3beta/DCX signaling pathway restricts axon branching. J Neurosci 30:16766–16776

Chen X, Ba Y, Ma L, Cai X, Yin Y, Wang K, Guo J, Zhang Y, Chen J, Guo X et al (2008) Characterization of microRNAs in serum: a novel class of biomarkers for diagnosis of cancer and other diseases. Cell Res 18:997–1006

Chen X, Liang H, Zhang J, Zen K, Zhang CY (2012) Horizontal transfer of microRNAs: molecular mechanisms and clinical applications. Protein Cell 3:28–37

Cohen D, Segal M, Reiner O (2008) Doublecortin supports the development of dendritic arbors in primary hippocampal neurons. Dev Neurosci 30:187–199

David A, Tiveron MC, Defays A, Beclin C, Camosseto V, Gatti E, Cremer H, Pierre P (2007) BAD-LAMP defines a subset of early endocytic organelles in subpopulations of cortical projection neurons. J Cell Sci 120:353–365

Govek EE, Newey SE, Van Aelst L (2005) The role of the Rho GTPases in neuronal development. Genes Dev 19:1–49

He S, Ma J, Liu N, Yu X (2010) Early enriched environment promotes neonatal GABAergic neurotransmission and accelerates synapse maturation. J Neurosci 30:7910–7916

Ji F, Lv X, Jiao J (2013) The role of microRNAs in neural stem cells and neurogenesis. J Genet Genomics 40:61–66

Kataoka Y, Takeichi M, Uemura T (2001) Developmental roles and molecular characterization of a Drosophila homologue of Arabidopsis Argonaute1, the founder of a novel gene superfamily. Genes Cells 6:313–325

Kawase-Koga Y, Otaegi G, Sun T (2009) Different timings of Dicer deletion affect neurogenesis and gliogenesis in the developing mouse central nervous system. Dev Dyn 238:2800–2812

Kole AJ, Swahari V, Hammond SM, Deshmukh M (2011) miR-29b is activated during neuronal maturation and targets BH3-only genes to restrict apoptosis. Genes Dev 25:125–130

Obernosterer G, Martinez J, Alenius M (2007) Locked nucleic acid-based in situ detection of microRNAs in mouse tissue sections. Nat Protoc 2:1508–1514

Ramos RL, Bai J, LoTurco JJ (2006) Heterotopia formation in rat but not mouse neocortex after RNA interference knockdown of DCX. Cereb Cortex 16:1323–1331

Shioya M, Obayashi S, Tabunoki H, Arima K, Saito Y, Ishida T, Satoh J (2010) Aberrant microRNA expression in the brains of neurodegenerative diseases: miR-29a decreased in Alzheimer disease brains targets neurone navigator 3. Neuropathol Appl Neurobiol 36:320–330

Shu SY, Qing D, Wang B, Zeng QY, Chen YC, Jin Y, Zeng CC, Bao R (2013) Comparison of microRNA expression in hippocampus and the marginal division (MrD) of the neostriatum in rats. J Biomed Sci 20:9

Smirnova L, Grafe A, Seiler A, Schumacher S, Nitsch R, Wulczyn FG (2005) Regulation of miRNA expression during neural cell specification. Eur J Neurosci 21:1469–1477

Snyder JS, Ferrante SC, Cameron HA (2012) Late maturation of adult-born neurons in the temporal dentate gyrus. PLoS ONE 7: e48757

Spampanato J, Sullivan RK, Turpin FR, Bartlett PF, Sah P (2012) Properties of doublecortin expressing neurons in the adult mouse dentate gyrus. PLoS ONE 7:e41029

Uesaka N, Hirai S, Maruyama T, Ruthazer ES, Yamamoto N (2005) Activity dependence of cortical axon branch formation: a morphological and electrophysiological study using organotypic slice cultures. J Neurosci 25:1–9

Wang Y, Zhang X, Li H, Yu J, Ren X (2013) The role of miRNA-29 family in cancer. Eur J Cell Biol 92:123–128

Yang D, Li T, Wang Y, Tang Y, Cui H, Tang Y, Zhang X, Chen D, Shen N, Le W (2012) miR-132 regulates the differentiation of dopamine neurons by directly targeting Nurr1 expression. J Cell Sci 125:1673–1682

Yu JY, Chung KH, Deo M, Thompson RC, Turner DL (2008) MicroRNA miR-124 regulates neurite outgrowth during neuronal differentiation. Exp Cell Res 314:2618–2633

5

Structural studies on MRG701 chromodomain reveal a novel dimerization interface of MRG proteins in green plants

Yanchao Liu[1,2], Hong Wu[1], Yu Yu[3], Ying Huang[1]✉

[1] State Key Laboratory of Molecular Biology, National Center for Protein Science Shanghai, Shanghai Science Research Center, Shanghai Key Laboratory of Molecular Andrology, Institute of Biochemistry and Cell Biology, Shanghai Institutes for Biological Sciences, Chinese Academy of Sciences, Shanghai 200031, China
[2] University of Chinese Academy of Sciences, Beijing 100049, China
[3] State Key Laboratory of Genetic Engineering, Collaborative Innovation Center of Genetics and Development Institute of Plant Biology, School of Life Science, Fudan University, Shanghai 200433, China
✉ Correspondence: huangy@sibcb.ac.cn (Y. Huang)

ABSTRACT

MRG proteins are conserved during evolution in fungi, flies, mammals and plants, and they can exhibit diversified functions. The animal MRGs were found to form various complexes to activate gene expression. Plant MRG1/2 and MRG702 were reported to be involved in the regulation of flowering time via binding to H3K36me3-marked flowering genes. Herein, we determined the crystal structure of MRG701 chromodomain (MRG701^{CD}). MRG701^{CD} forms a novel dimerization fold both in crystal and in solution. Moreover, we found that the dimerization of MRG chromodomains is conserved in green plants. Our findings may provide new insights into the mechanism of MRGs in regulation of gene expression in green plants.

KEYWORDS MRG701, chromodomain, homodimer

INTRODUCTION

The members of human MORF4 related gene (MRG) family are highly conserved from yeast to human, including human retinoblastoma binding protein 1 (RBP-1) and the MORF4 related gene on chromosome 15 (MRG15), drosophila male specific lethal 3 (Msl3), *S. pombe* altered polarity 13 (Alp13), and *S. cerevisiae* Esa1 associated protein 3 (Eaf3p) (Bertram and Pereira-Smith, 2001). Human mortality factor 4 (MORF4) gene was identified in 1999, which can induce senescence in immortal cell lines (Bertram et al., 1999). There are seven genes with related sequences of MORF4 gene in human, which was referred as MRG family. The most studied instance in this family is MRG15 in humans, which plays crucial roles in various cellular processes such as replicative senescence, cell proliferation, and DNA damage repair, and it also exhibits functions in various complexes that regulate transcription (Leung et al., 2001; Tominaga et al., 2005; Hayakawa et al., 2010). MRG15 was identified in the two distinct nucleoprotein complexes of MAF1 and MAF2 via sucrose gradient analysis (Pardo et al., 2002). The latter is a histone acetyltransferase (HAT)-associated Tip60/NuA4 complex. Furthermore, MRG15 was also found in the Sin3S/Rpd3S complex which is a histone deacetylase (HDAC)-associated complex with antagonizing function (Yochum and Ayer, 2002). Additional studies showed that MRG15 was associated with other complexes beyond transcriptional regulation. For example, MRG15 physically interacted with PALB2 in homologous recombination of DNA repair (Hayakawa et al., 2010). Other MRG proteins have also been reported to be involved in chromatin regulating complexes. *Drosophila* Msl3 is a subunit of the male specific lethal complex (MSL) (Gorman et al., 1995). MSL has HAT activity and is required for the dosage compensation of X-linked genes in male flies (Hilfiker et al., 1997).

Yanchao Liu and Hong Wu have contributed equally to this work.

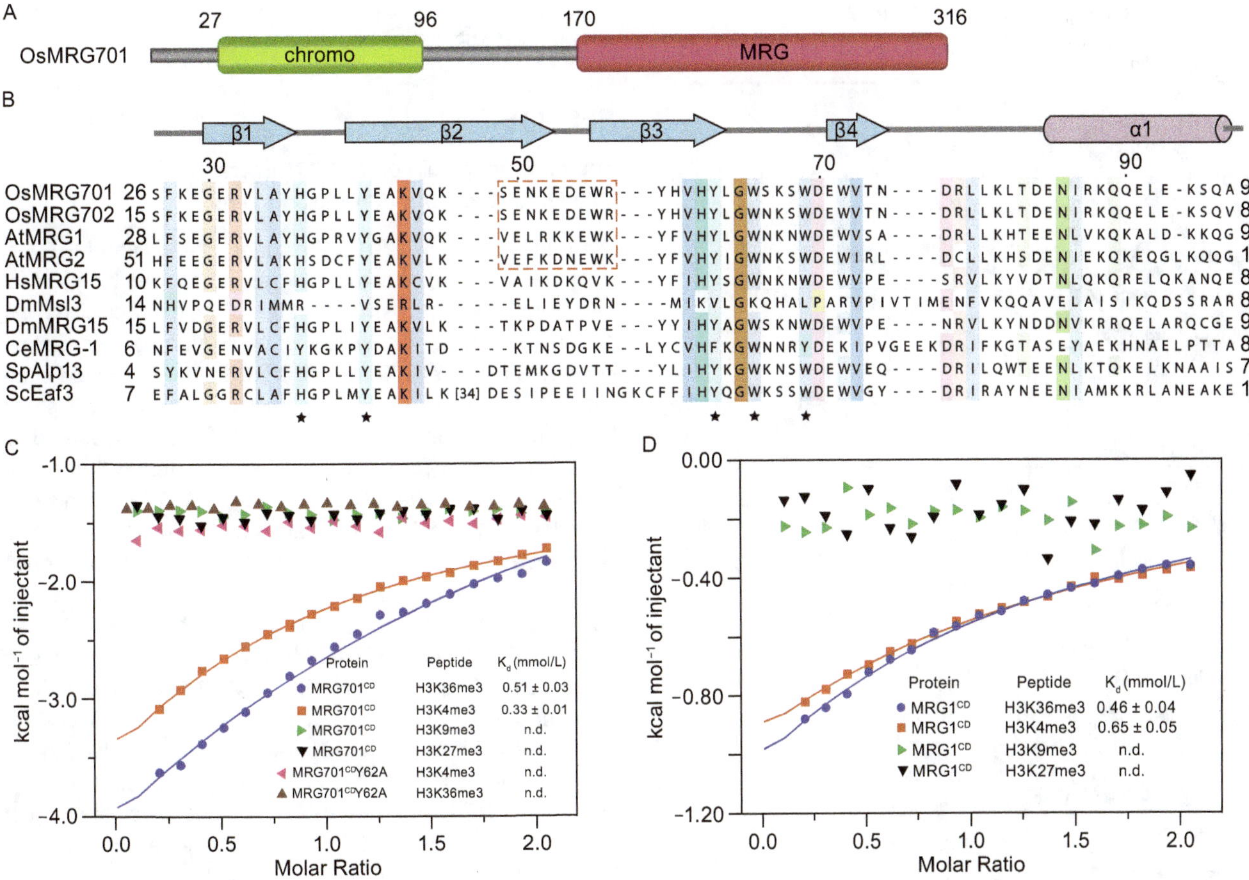

Figure 1. MRG701^{CD} specifically interacts with H3K36me3 and H3K4me3 peptides *in vitro*. (A) Domain architecture of MRG701 in rice. (B) Sequence alignment of the chromodomains of MRG proteins in different species. Amino acids are shaded as a function of conservation with a 30% conservation visibility cutoff that was calculated using Clustal. Secondary structural elements of MRG701^{CD} are displayed above the sequence alignment. The conserved amino acids that represent the aromatic cage for histone peptide binding are marked with asterisks. The dimerization patch is indicated with a red dashed box. A large insertion that is only present in ScEaf3 is omitted for clarity and replaced by the number of amino acids. (C) ITC measurements of binding affinities of wild-type MRG701^{CD} and mutant MRG701^{CD}Y62A to tri-methylated histone peptides. (D) ITC measurements of binding affinities of MRG1^{CD} to tri-methylated histone peptides.

There are two homologues of MRG proteins in green plants, MRG1/2 in *Arabidopsis* and MRG701/702 in rice. MRG2 and MRG702 have been reported to be involved in floral transition, a highly regulated process in plant development between the vegetative stage and reproductive stage (Bu et al., 2014; Jin et al., 2015). In *Arabidopsis*, the two flowering genes of *FLOWERING LOCUS C (FLC)* and *FLOWERING LOCUS T (FT)* function antagonistically to control the flowering time (Searle et al., 2006). MRG2 directly binds to *FT* and increases the expression of *FT* in an H3K36me3-dependent manner. MRG2 also physically interacts with CONSTANS (CO), a transcription factor which functions in the regulation of photoperiodic flowering, and activates the expression of *FT* (Putterill et al., 1995; Samach et al., 2000). Moreover, MRG1/2 activates gene expression by interacting with the HAM1 and HAM2, the histone H4 specific acetyltransferases (Xu et al., 2014).

MRG family proteins contain a chromodomain at the N-terminal and an MRG domain at the C-terminal end. The MRG domains form homodimers and interact with other proteins to form functional complexes (Zhang et al., 2006b; Xie et al., 2015). As a "Royal Family" member, the N-terminal chromodomain functions as a histone reader that specifically recognizes methylated histone markers (Yap and Zhou, 2011). MRG15 mediates the recruitment of functional complexes to target genes with specific histone marks (Luco et al., 2010). Animal MRG chromodomains and fungi MRG chromodomains have been reported to bind to trimethylated histone H3 at lysine 36 (H3K36me3), such as human MRG15, *S. pombe* Eaf3, and the fly homologue Msl3 (Zhang

Table 1. Crystallographic statistics of MRG701^{CD}

Data collection	
Space group	$P2_1$
Cell dimension	
a, b, c (Å)	31.79, 59.80, 40.62
α, β, γ (°)	90.00, 102.58, 90.00
Wavelength (Å)	0.9782
Resolution range (Å)[a]	29.90–1.40 (1.45–1.40)
Completeness (%)[a]	98.6 (98.8)
Rmerge (%)[a]	6.0 (10.0)
I/σI	29.0 (12.3)
Redundancy	5.2 (4.8)
Refinement	
Resolution range (Å)	29.90–1.40
No. of reflections	28,842
Rwork (%)/Rfree (%)	18.22/20.45
R.m.s.deviations	
Bond length (Å)	0.007
Bond angles (°)	1.05
Ramachandran plot	
Most favored region (%)	100
Allowed region (%)	0.0
Outliers (%)	0.0

[a] Values in parentheses are for the highest-resolution shell.

et al., 2006a; Larschan et al., 2007; Sun et al., 2008). H3K36me3 is a histone modification usually located in the middle and at the 3' ends of transcribed coding regions (Pokholok et al., 2005; Kolasinska-Zwierz et al.,2009). However, there have also been reports of MRG chromodomains in plants binding to H3K4me3. H3K4me3 is a histone modification that usually accumulates at the transcription start site (TSS) and associates with actively transcribed genes at the 5' ends (Zhang et al., 2009). For example, the *Arabidopsis* MRG2 chromodomain (MRG2^{CD}) and the rice MRG702 chromodomain (MRG702^{CD}) were reported to bind to H3K4me3 *in vitro* (Bu et al., 2014; Jin et al., 2015).

In order to test the histone binding specificities of other plant MRG chromodomains, we analyzed the binding affinities of two plant MRG chromodomains to various histone peptides via isothermal titration calorimetry. Most green plants have evolved to encode two homologues of the MRG family proteins. Herein, we study the binding specificities of the chromodomains of MRG701 (the homologue of MRG702 in rice) and MRG1 (the homologue of MRG2 in *Arabidopsis*). Both the MRG701 chromodomain (MRG701^{CD}) and the MRG1 chromodomain (MRG1^{CD}) specifically bind to H3K36me3 and H3K4me3. We also investigated the crystal structure of MRG701^{CD}.

RESULTS

The chromodomain of MRG701 interacts with H3K36me3 and H3K4me3 *in vitro*

We purified the MRG701^{CD} (residues 27–96) and MRG1^{CD} (residues 28–110) according to the domain architecture of MRG proteins (Fig. 1A). To test the binding specificities, we performed ITC to measure the equilibrium disassociation constants (K_d). MRG701^{CD} and MRG1^{CD} both showed binding specificities to H3K36me3 (residues 31–41) as well as H3K4me3 (residues 1–9) (Fig. 1C and 1D). The measured K_d is summarized in Table S1, which is consistent with the previously reported values of MRG2^{CD} and MRG702^{CD} to histone peptides (Table S1) (Bu et al., 2014; Jin et al., 2015). Sequence analysis showed that there is high sequence homology between MRG chromodomains from the same species. There are only two amino acids that differ between rice MRG701^{CD} and MRG702^{CD}, and 61% that differ between *Arabidopsis* MRG1^{CD} and MRG2^{CD}. MRG701^{CD} shared 60% sequence identity to MRG2^{CD} and 54% to human MRG15^{CD}, respectively (Fig. 1B). Five aromatic residues involved in the formation of the trimethylated lysine binding pocket are conserved in MRG701^{CD} and MRG1^{CD}. These results suggest that MRG701^{CD} and MRG1^{CD} may bind histone H3K36me3 or H3K4me3 in the same manner as MRG2^{CD}.

Overall structure of MRG701 chromodomain

In order to investigate the molecular mechanism of the histone peptide recognition of MRG701^{CD} and MRG1^{CD}, we co-crystallized them with the histone peptides H3K36me3 (residues 31–41) and H3K4me3 (residues 1–9). Only MRG701^{CD} in complex with H3K36me3 produced crystals that diffracted to 1.4-Å. The structure was determined by molecular replacement using MRG2^{CD} (PDB: 4PLI) as a search model (Bu et al., 2014), and the structure was then refined via Phenix (Adams et al., 2010) (Table 1). There are two MRG701^{CD} molecules (molA and molB) in one asymmetric unit (AU), and the r.m.s.d. between these two molecules is 0.140 Å. The structure of MRG701^{CD} adopts a compact fold with four β-strands shaped like a β-barrel and a C-terminal α-helix (Fig. 2A). The trimethylated lysine binding pocket was formed by the following five aromatic residues: H37, Y42, Y62, W65, and W69 (Fig. 2B and 2C). The backbone Cα alignment of MRG701^{CD}, MRG2^{CD}, and MRG15^{CD} showed that the backbone Cα of MRG701^{CD} matches the backbone Cα of MRG2^{CD} and MRG15^{CD} quite well, with r.m.s.d. of 0.461 and 0.690 Å, respectively (Fig. 2D). The aromatic cage of MRG701^{CD} is also conserved with those of MRG2^{CD} and MRG15^{CD} (Fig. 2E). However, no extra electron density was observed in the trimethylated lysine binding pocket. Comparing the structure with other MRG chromodomain structures, we found that W65 in the aromatic cage was stretched toward the center of the cage. The indole ring of W65 occupied the binding site of the trimethyl group in the lysine side chain. Therefore, there are only MRG701^{CD} molecules in the crystal structure (Fig. S1). In the crystal structure of MRG15^{CD}, the side chain orientation of the corresponding W49 is the same as that of W65 in MRG701^{CD}. W65 in the aromatic cage may function as a "switch." When there is no peptide binding, the side

Figure 3. MRG701^{CD} and MRG2^{CD} showed dimerization in crystal. (A) View highlighting interactions between two molecules of the MRG701^{CD} dimer colored in light green and light blue. Yellow dashed lines indicate the hydrogen bonds. (B) MRG2^{CD} dimer in the P6$_1$ space group is colored in cyan and light pink. (C) MRG2^{CD} dimer in the P6$_5$ space group is colored in cyan and beige. (D) Surface structure of MRG701^{CD} dimer is shown in colored as (A). (E) The Electron density map (2FoFc) of the MRG701^{CD} dimer interface is shown in mesh at 0.5-σ and is colored as in (A). ▶

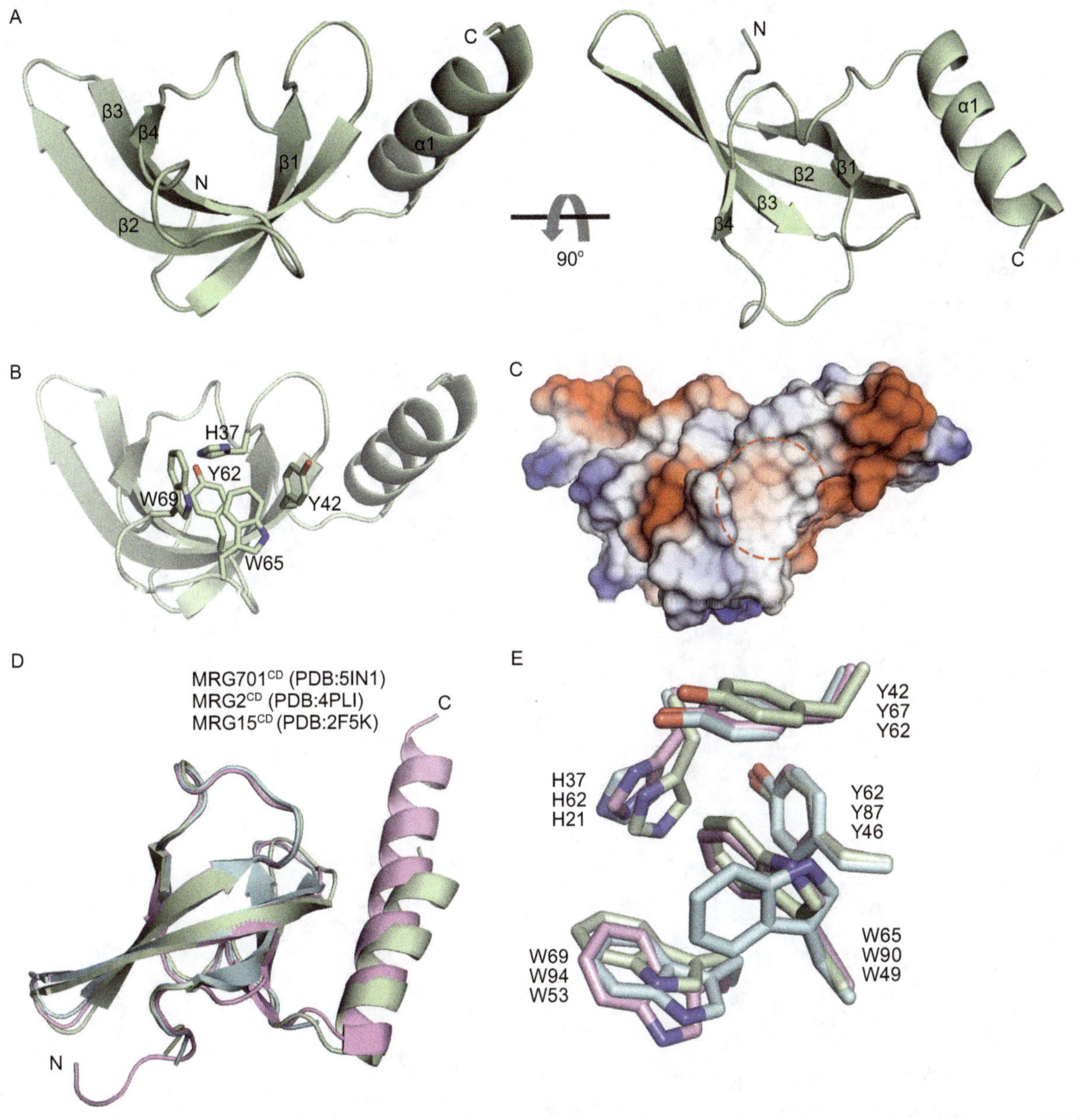

Figure 2. Overall structure of MRG701^{CD}. (A) MRG701^{CD} monomer consisting of four β-strands and one α-helix. View of 90° rotation around the horizontal axis is shown on right. (B) Residues consisting of aromatic cage are shown as sticks. (C) Electrostatic potential surface of MRG701^{CD} with histone binding cage indicated in red dashed circle (scale blue to red; positive to negative). (D) Structures of MRG701^{CD}, MRG2^{CD}, and MRG15^{CD} are superimposed and colored in light green, cyan, and violet, respectively. (E) Comparison of the aromatic cages of MRG701, MRG2, and human MRG15. Residues are shown as colored sticks as in (D).

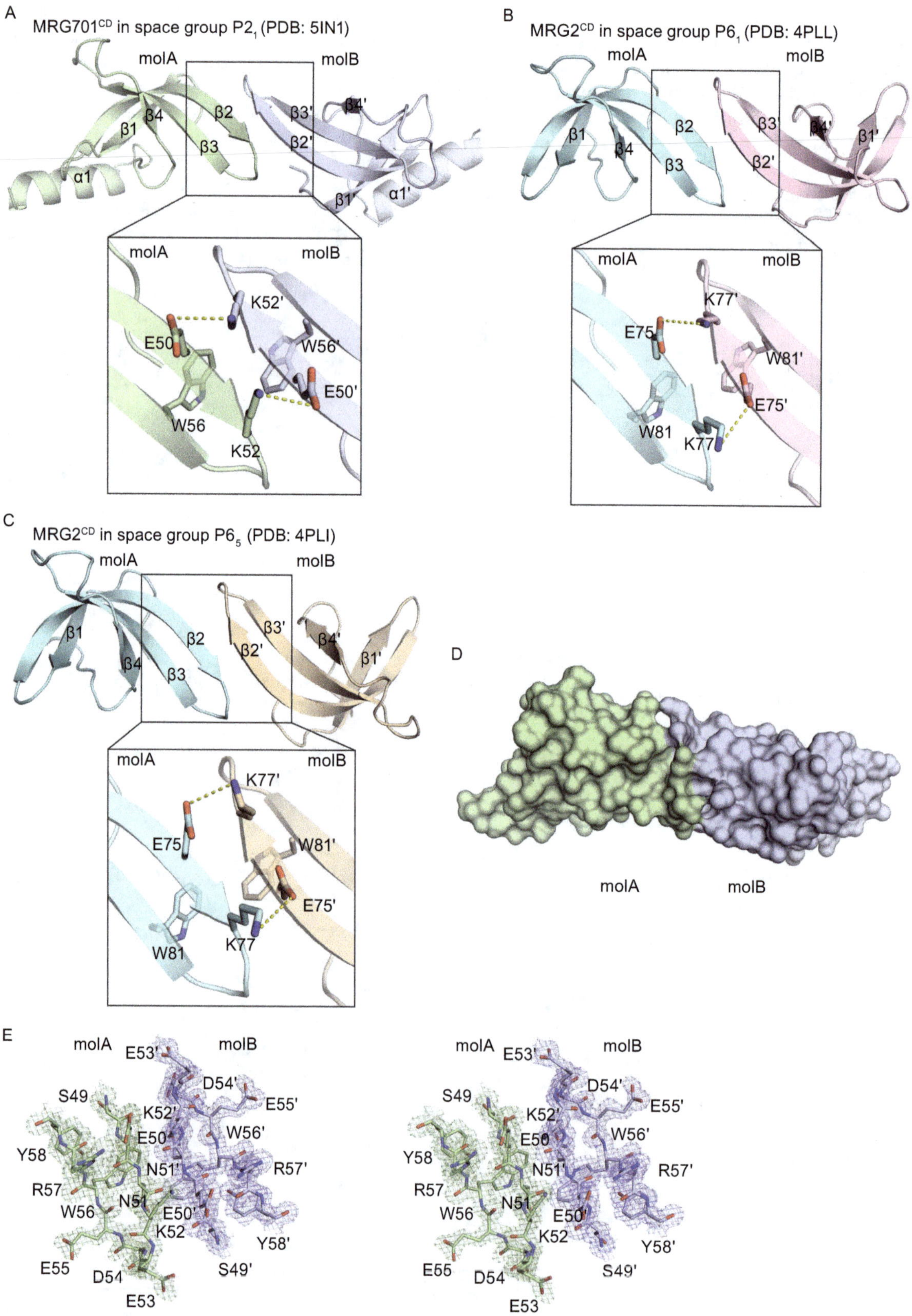
A
MRG701^CD in space group P2_1 (PDB: 5IN1)
molA
molB
β1
β2
β3
β4
α1
β1'
β2'
β3'
β4'
α1'
molA
molB
K52'
E50
W56'
E50'
W56
K52
B
MRG2^CD in space group P6_1 (PDB: 4PLL)
molA
molB
β1
β2
β3
β4
β1'
β2'
β3'
β4'
molA
molB
K77'
E75
W81'
E75'
W81
K77
C
MRG2^CD in space group P6_5 (PDB: 4PLI)
molA
molB
β1
β2
β3
β4
β1'
β2'
β3'
β4'
molA
molB
K77'
E75
W81'
E75'
W81
K77
D
molA
molB
E
molA
molB
E53'
D54'
E55'
S49
K52'
E50'
W56'
Y58
N51'
R57'
R57
W56
N51
E50'
K52
Y58'
E55
D54
S49'
E53

chain of W65 is rotated toward the center of the aromatic cage and block the binding of trimethylated histone peptides. Upon binding, the side chain of W65 is most likely flipped and leaves space for histone binding (Fig. S1).

MRG701^{CD} and MRG2^{CD} showed dimerization in crystal

By comparing MRG701^{CD} with other MRG chromodomains structures, we unexpectedly found dimerization in the crystal structures of plant MRG701^{CD} and MRG2^{CD} (PDB 4PLL, 4PLI, and 4PL6), but we did not find dimerization in the crystal structures of human MRG15^{CD} (PDB 2F5K) or yeast Eaf3^{CD} (PDB 3E9F) (Fig. 3A–C). In the structure of MRG701^{CD}, the β2 of molA stands anti-parallel to the β2' of molB. Specifically, both E50 and K52 on β2 of molA form both hydrogen bonding and electrostatic interactions with K52' and E50' on β2' of molB, respectively. In addition, W56 on β3 of molA contacts W56' on β3' of molB via π-π stacking interaction (Fig. 3A and 3D). The total buried surface area of the dimeric interface in MRG701^{CD} is 645 Å^2, which is approximately 6.9% of the total surface area of the MRG701^{CD} dimer as calculated by CNS (Fig. 3D and 3E) (Brunger et al., 1998; Brunger, 2007). MRG2^{CD} showed two different morphological crystals under the space groups of $P6_1$ and $P6_5$. In both kinds of crystals, the two molecules of MRG2^{CD} in one AU interact with each other in the same way as MRG701^{CD} (Fig. 3B and 3C).

MRG701^{CD}, MRG1^{CD}, and MRG2^{CD} exhibited dimerization in solution

To test whether the formation of dimerization was due to crystal packing, we designed point mutations according to the crystal structures. In each mutant, we either disrupted the π-π stacking interactions between the indole rings (MRG701^{CD}W56A and MRG1^{CD}W58A) or disrupted the hydrogen bonding and electrostatic interactions (MRG2^{CD}E75A). The purified mutant proteins were characterized by circular dichroism spectroscopy. The resulting spectrum showed no obvious change between wild-type and mutant proteins, indicating that the overall folding of the mutants was the same as the wild-type proteins (Fig. S2). We then performed size-exclusion chromatography-coupled with the multi-angle light scattering (SEC-MALS) method to check the oligomerization status of wild-type and mutant proteins in solution. Wild-type MRG701^{CD}, MRG1^{CD}, and MRG2^{CD} were eluted as a single peak on the size-exclusive column with the calculated molecular weights (MW) of 20.1 kDa, 20.7 kDa and 20.7 kDa, respectively (Fig. 4A and 4B). However, the mutant proteins (MRG701^{CD}W56A, MRG1^{CD}W58A, and MRG2^{CD}E75A) all showed an obvious delay on the size-exclusive column exhibiting MWs of 10.3 kDa, 13.4 kDa, and 12.5 kDa, respectively. The observed MWs of the mutants are nearly half the MWs of the wild-types and approximate the theoretical MWs of the corresponding monomers (Table 2). Therefore, MRG701^{CD}, MRG1^{CD}, and MRG2^{CD} form dimers both in crystal and in solution.

Figure 4. MRG701^{CD}, MRG1^{CD}, and MRG2^{CD} showed dimerization in solution. (A–C) The elution profiles of wild-type and mutant proteins on SEC-MALS monitored via absorbance at UV 280 nm. (A) MRG701^{CD} (black) and MRG701^{CD}W56A (green). (B) MRG1^{CD} (black) and MRG1^{CD}W58A (blue). (C) MRG2^{CD} (black) and MRG2^{CD}E75A (red). (D) The theoretical molar mass (black) and observed molar mass (red) of wild-type and mutant proteins via SEC-MALS. (E) ITC measurements for binding affinities of wild-type MRG701^{CD} and mutant MRG701^{CD}W56A to tri-methylated histone peptides. (F) ITC measurements for binding affinities of wild-type MRG1^{CD} and mutant MRG1^{CD}W58A to tri-methylated histone peptides. (G) ITC measurements for binding affinities of wild-type MRG2^{CD} and mutant MRG2^{CD}E75A to tri-methylated histone peptides. ▶

Dimerization of the MRG chromodomains showed no effect on the binding to histone peptides

Dimerization of the MRG chromodomains through the interactions of β-strands leads to a platform with an anti-parallel histone binding site on each side of the platform. In the crystal structure, the histone binding site is embraced by the β-barrel and is spatially proximate to the dimer interface (Fig. 2B). To investigate whether the dimerization of MRG chromodomains affects the binding to histone tails, ITC experiments were performed. The measured K_d values of MRG701^{CD}W56A to H3K36me3 and to H3K4me3 were 0.26 mmol/L and 0.25 mmol/L, respectively (Fig. 4E). These values represent less than a 2-fold change compared to those of MRG701^{CD}, which were respectively 0.51 mmol/L and 0.33 mmol/L. The K_d of MRG2^{CD}E75A and MRG1^{CD}W58A to histone H3K36me3 and H3K4me3 peptides were also measured (Fig. 4F and 4G). Comparison of K_d showed only mild differences between the binding affinities of the wild-type and mutant proteins to histone peptides (Table S1). Thus, dimerization of the MRG chromodomains does not affect the histone binding ability.

The dimerization of MRG chromodomains is conserved among green plants

To investigate whether this MRG chromodomain dimerization may exist in a wider range of species, we performed sequence alignments via Clustal Omega (http://www.ebi.ac.uk/Tools/msa/clustalo/) using MRG701 (residues 49–59) as a reference. The alignment results indicated that residues involved in dimerization (residues E50, K52, and W56 in MRG701) are conserved in most green plants, including dicots and monocots. However, these residues were not conserved in human, fly, worm, or fungi (Fig. 5A). Therefore, this mode of dimerization of MRG chromodomains is unique to green plants.

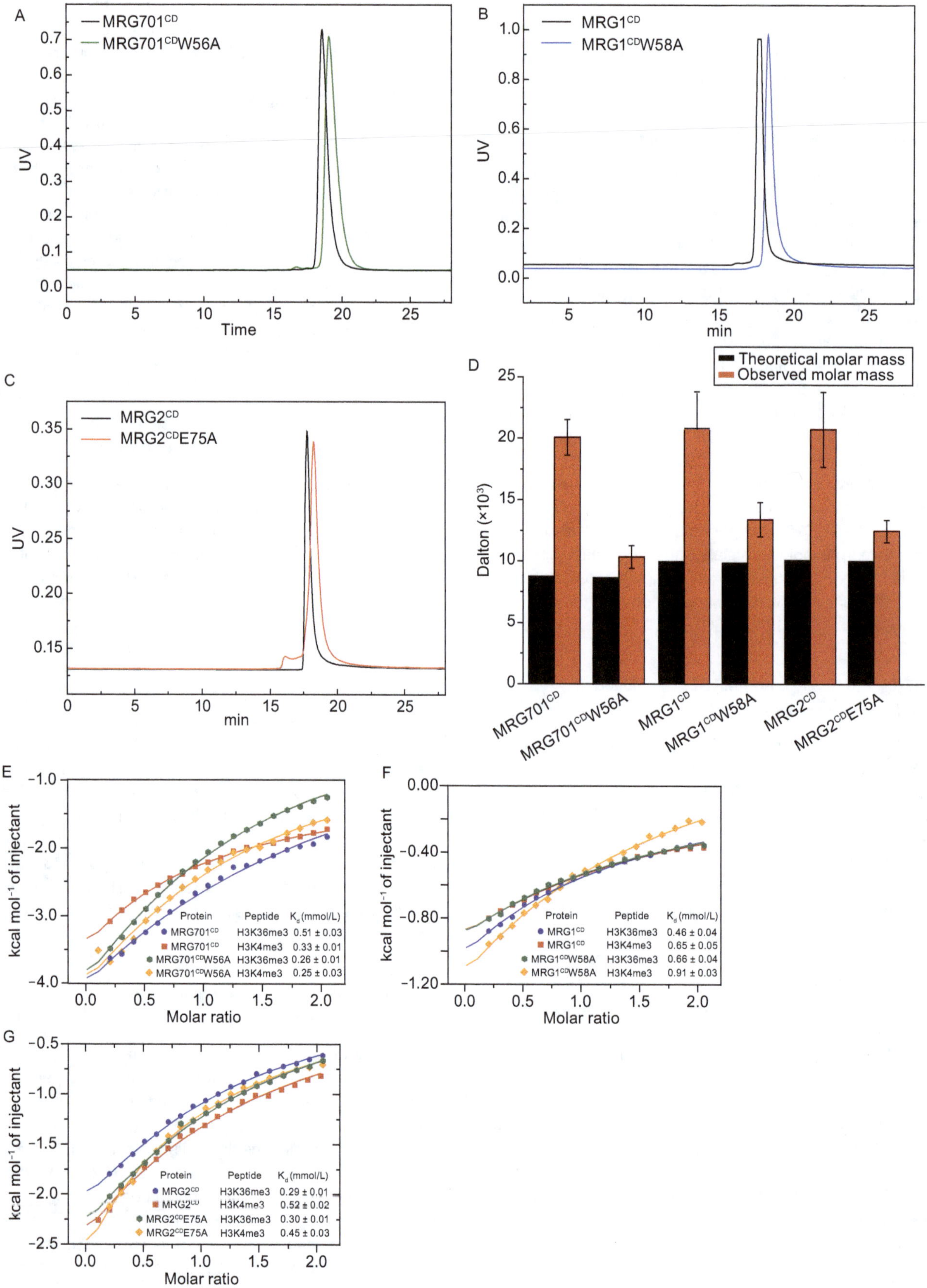

A
MRG701CD
MRG701CDW56A
UV
Time
B
MRG1CD
MRG1CDW58A
UV
min
C
MRG2CD
MRG2CDE75A
UV
min
D
Theoretical molar mass
Observed molar mass
Dalton (×10³)
MRG701CD
MRG701CDW56A
MRG1CD
MRG1CDW58A
MRG2CD
MRG2CDE75A
E
kcal mol⁻¹ of injectant
Molar ratio
Protein Peptide Kd (mmol/L)
MRG701CD H3K36me3 0.51 ± 0.03
MRG701CD H3K4me3 0.33 ± 0.01
MRG701CDW56A H3K36me3 0.26 ± 0.01
MRG701CDW56A H3K4me3 0.25 ± 0.03
F
kcal mol⁻¹ of injectant
Molar ratio
Protein Peptide Kd (mmol/L)
MRG1CD H3K36me3 0.46 ± 0.04
MRG1CD H3K4me3 0.65 ± 0.05
MRG1CDW58A H3K36me3 0.66 ± 0.04
MRG1CDW58A H3K4me3 0.91 ± 0.03
G
kcal mol⁻¹ of injectant
Molar ratio
Protein Peptide Kd (mmol/L)
MRG2CD H3K36me3 0.29 ± 0.01
MRG2CD H3K4me3 0.52 ± 0.02
MRG2CDE75A H3K36me3 0.30 ± 0.01
MRG2CDE75A H3K4me3 0.45 ± 0.03

Table 2. Molar mass (kDa) of wild-type proteins and mutants

	Theoretical molar mass (kDa)	Observed molar mass by SEC-MALS (kDa)
MRG701^{CD}	8.750	20.08 ± 1.46
MRG701^{CD}W64A	8.635	10.34 ± 0.92
MRG2^{CD}	10.071	20.74 ± 3.05
MRG2^{CD}E75A	10.013	12.47 ± 0.92
MRG1^{CD}	9.918	20.67 ± 2.99
MRG1^{CD}W58A	9.803	13.36 ± 1.40

DISCUSSION

In this work, we investigated the binding ability of MRG701^{CD} to the histone peptides H3K36me3 and H3K4me3, and we determined the crystal structure of MRG701^{CD}. In the crystal structure we observed a noncanonical dimerization of MRG701^{CD} formed by three highly conserved residues. Mutation of the interfacial key residues showed that dimerization did not affect the binding to H3K36me3 or H3K4me3. By analyzing the structures of other chromodomains, we found that the residues involved in the formation of the dimer interface are highly conserved among green plants. These findings suggest that the dimerization of MRG chromodomains is a unique phenomenon in green plants, and this process may provide a mechanism for how MRG proteins play different roles in regulating gene expression in different species.

It has long since been established that chromodomains function as monomers, such as the human Cbx family chromodomains (Kaustov et al., 2011). Until recently, several chromodomains have been reported as a dimer. The chromodomain of *Drosophila* Polycomb (dPc^{CD}) forms a dimer through the loop region connecting β4 and the C-terminal α-helix (Fig. 5B) (Min et al., 2003). The chromodomain of the M-phase phosphoprotein 8 (MPP8^{CD}) dimerized through the anti-parallel β-strand paring of two molecules (Fig. 5C) (Chang et al., 2011; Li et al., 2011). Moreover, Rhino, a *Drosophila* HP1 homolog, forms a dimer via the C-terminal α-helix and the loop region between β1 and β2 (Fig. 5D) (Yu et al., 2015). Additionally, dimerization is required for the function of Rhino *in vivo*. Another example is Swi6, a *S. Pombe* HP1 homolog, which binds the H3K9me3-marked nucleosomes and drives the chromatin to switch from an auto-inhibition state to a speading-competition state (Canzio et al., 2013, 2014). In the auto-inhibited state, the chromodomain of Swi6 recognizes a histone-mimic sequence on the loop region of another Swi6 chromodomain.

Dimerization of chromodomains will juxtapose two histone binding sites and enable the chromodomains to bind two histone tails simultaneously. In the dPc^{CD} dimer, the two histone H3K27me3 peptides are proximate to each other. In the crystal structure of Mpp8^{CD}, the two aromatic cages are on the exterior sides of the dimer and the two histone H3K9me3 peptides remain spatially separated. In this paper, we reported that MRG701^{CD} acts as a dimer both in crystal and in solution. Two MRG701^{CD} molecules are related via pseudo-2-fold symmetry in the crystal structure, and they form a dimer interface along the β2-strand of each molecule in an anti-parallel manner. This is a novel dimerization mode of chromodomains and the first case observed in MRG chromodomains. Dimerization of MRG701^{CD} also generates two histone binding sites in an anti-parallel way (Fig. 5E).

Figure 5. The dimerization of MRG chromodomains is conserved among green plants. (A) Amino acid sequence alignment of chromodomain dimerization patch between species. The conserved amino acids essential for dimer formation in plants are marked with asterisks and highlighted in colored shades by sequence identity. (B) Two molecules of the dPc^{CD} dimer colored in cyan and pink. (C) Two molecules of the MPP8^{CD} dimer colored in gray and light yellow. (D) Two molecules of the Rhino^{CD} dimer colored in light pink and light yellow. (E) Two molecules of the MRG701^{CD} dimer colored in light green and light blue. (A–C) All the bound histone H3K27me3 and H3K9me3 peptides are shown in dark orange and green, respectively. (F) Proposed model showing the different binding modes of animal and green plant MRG proteins on euchromatins. ▶

To date, the function of the dimerization of MRG chromodomains remains unclear. One possibility is that dimerization could regulate the recruitment of MRG proteins to their target genes. Histone H3K36me3 markers are known to locate on each side of the nucleosome and near the nucleosome core. In animals and fungi, only the C-terminal MRG domain is dimerized (MRG-MRG dimer). This leaves the N-terminal chromodomain flexible in solution and therefore accessible to the H3K36me3 markers close to the nucleosome core. However, in green plants, both the MRG domain and the chromodomain can be dimerized. If a complete dimer is formed (type AB), MRG molecules are in a rigid conformation and the two chromodomains are not easily accessible to the H3K36me3 markers. If partial dimers are formed (MRG-only or chromodomain-only), MRG molecules will oligomerize in a zig-zag manner (AB via MRG-MRG and BC via CD-CD etc.). This will result in high spatial proximity for the two methylated lysine binding pockets, and therefore the H3K36me3 markers on each side of the nucleosome would be easy to recognize and MRG molecules will spread on the chromatin surface (Fig. 5F).

The above model may explain the different function of MRG proteins between species. In mammals, yeasts, and worms, H3K36me3 marks the exons of transcribed genes and was enriched at the 3'-end of the coding region to regulate alternative splicing and transcriptional activation (Bannister et al., 2005; Pokholok et al., 2005; Kolasinska-Zwierz et al., 2009; Luco et al., 2010). In contrast, H3K36me3 is enriched at the 5'-end of the coding gene in *Arabidopsis*, and this region is also enriched in the active

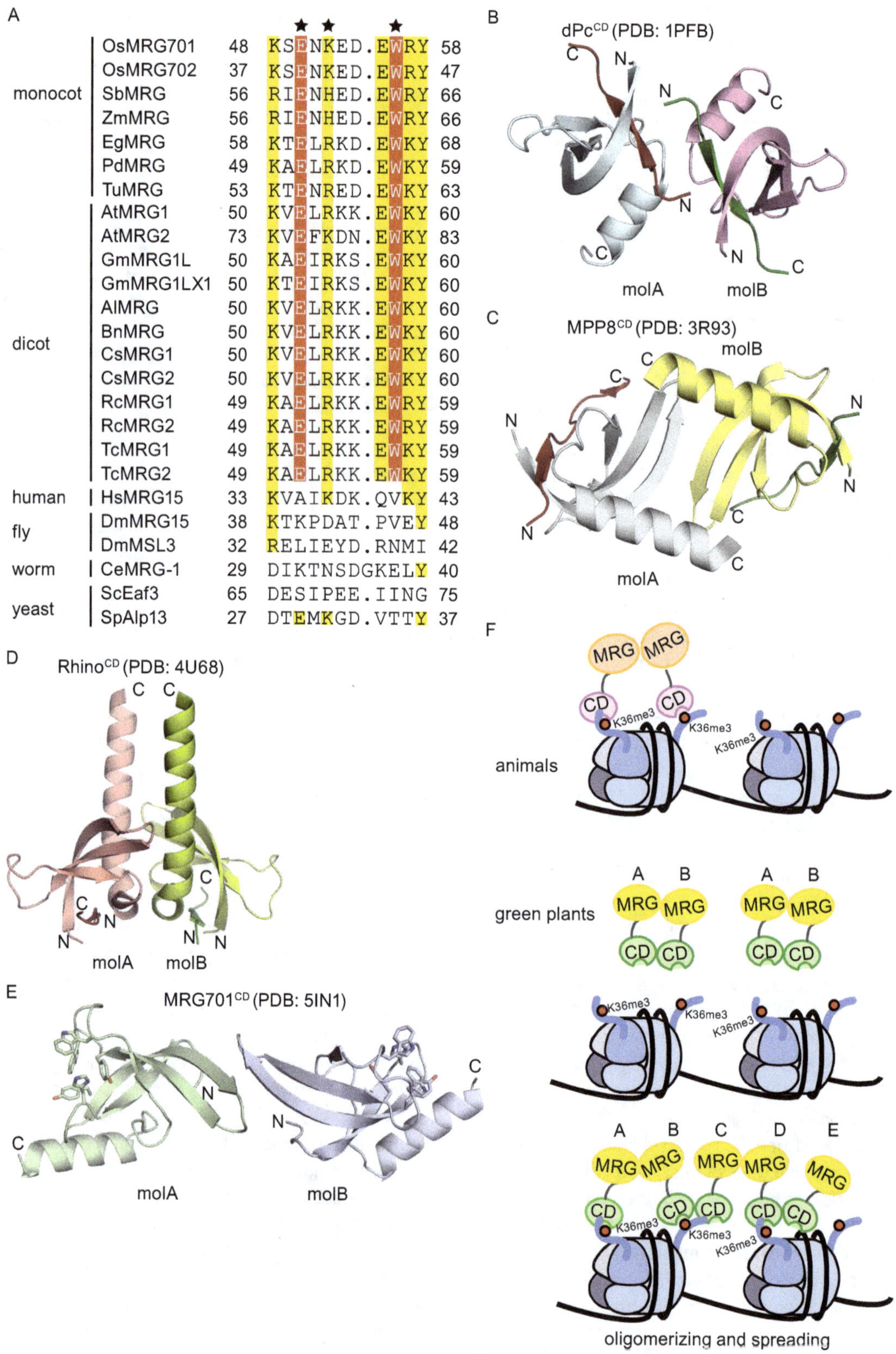

transcriptional marker H3K4me2/3 (Roudier et al., 2011). This suggests a different mechanism of H3K36me3 deposition and recognition (Xu et al., 2014). Our findings may aid in understanding the function of MRG proteins in green plants. Our findings show a novel mode of dimerization of MRG proteins and imply a potential diversity of the recognition mechanism of MRG proteins at the chromatin level.

MATERIALS AND METHODS

Protein expression and purification

The cDNA encoding the chromodomain of *Arabidopsis thaliana* MRG1 (residues 28–110) and *Oryza sativa* MRG701 (residues 27–96) were amplified by PCR and cloned into the modified pET28-SMT3 vector. Point mutants were generated by Site-Directed Mutagenesis kit (NEB) according to the manufacturer's instructions. The plasmid was transformed into *Escherichia coli* strain BL21 (DE3). The cells were cultured in LB medium with 50 μg/mL kanamycin and induced by final concentration of 0.2 mmol/L isopropyl-β-d-thiogalactopyranoside (IPTG) at 18°C for 18 h. Cells were harvested and re-suspended in 20 mmol/L Tris pH 8.0, 500 mmol/L NaCl, 25 mmol/L imidazole and then broken by cell disruptor (JNBIO). The fusion protein was purified first with Ni HiTrap column (GE healthcare) and then removed the tag by ULP enzyme. The target protein was further purified with Superdex G75 Hiload 26/60 column (GE healthcare) and then concentrated to 20 mg/mL for structural and biochemical studies.

Crystallization and diffraction data collection

For crystallization of MRG701 chromodomain (residues 27-96) and peptide complex, a solution of 20 mg/mL protein was incubated for 1 hour with H3K36me3 (residues 31–41) in a 1:2 molar ratio. Crystallization was performed using hanging drop vapor diffusion method. Crystals of the complex were grown at 16°C by mixing 1 μL of the protein-peptide solution with 1μL precipitant solution (0.1 mol/L sodium acetate pH 4.6, 25% PEG 4000, 0.2 mol/L ammonium sulfate). The diffraction data was collected from flash-cooled crystals at 100K at SSRF (Shanghai Synchrotron Radiation Facility).

Structure determination

The data was processed with HKL3000 (Otwinowski and Minor, 1997). The molecular-replacement solution was generated using the program Phaser and the crystal structure of MRG2 (PDB 4PLI) as a search model (McCoy et al., 2007). The initial model was built with COOT and refined using the program Phenix (Adams et al., 2010; Emsley et al.,2010). The final refined structure was shown by Pymol (The PyMOL Molecular Graphics System, Version 1.8 Schrödinger, LLC). Coordinate has been deposited under PDB accession code 5IN1.

Isothermal titration calorimetry

ITC experiments were carried out at 20°C on a MicroCal iTC200. Protein and peptide were kept in an identical buffer of 20 mmol/L Tris pH 8.0, 100 mmol/L NaCl. The sample cell was filled with a 0.1–0.2-mmol/L solution of protein, and the injection syringe with 1–2 mmol/L of the titrating ligand. Each titration consisted of 20 2-μL injections with 2 min interval. Binding isotherms were analyzed by fitting data into the one-site model using the ITC data analysis module Origin 7.0 software.

SEC-MALS analysis

The absolute molar mass of target protein was determined by size exclusion chromatography coupled with multi-angel light scattering. The experiments were performed by an Agilent system connected with a Wyatt DAWN HELEOS-II multi-angle light scattering instrument and a Wyatt OptilabrEX differential refractometer. A Wyatt 015S5 (150 Å) column was used with a 100 μL sample loop at a flow rate of 0.5 mL/min in the running buffer of 10 mmol/L Tris 8.0, 100 mmol/L NaCl. The molar mass was calculated by light scattering intensity and differential refractive index using Wyatt ASTRA 6 software (Wyatt, 1993).

Circular dichroism spectroscopy

Circular dichroism spectroscopy was performed by Jasco J-715 spectropolarimeter. The concentration of sample was 0.2 mg/mL in 5 mmol/L Tris pH 8.0. All measurements were taken in 1mm light path length in 300 μL total volume. Spectra were collected from 180 nm to 340 nm at room temperature, with a scanning speed of 50 nm/min and corrected by the background spectra containing the same buffer.

ACKNOWLEDGEMENTS

This work was supported by grants from the National Basic Research Program of China (973 Program) (No. 2012CB910500), the Strategic Priority Research Program (XDB08010202), and the National Natural Science Foundation of China (Grant No. 31270774). We are extremely grateful to National Center for Protein Science Shanghai for their instrument support and technical assistance. We thank the staff at BL19U of Shanghai Synchrotron Radiation Facility (SSRF) for help with data collection.

ABBREVIATIONS

Alp13, altered polarity 13; AU, asymmetric unit; CD, chromodomain; CO, CONSTANS; dPc, *Drosophila* Polycomb; Eaf3p, Easal associated protein 3; *FLC, FLOWERING LOCUS C*; *FT, FLOWERING LOCUS T*; HADC, histone deacetylase; HAT, histone acetyltransferase; H3K36me3, trimethylated histone H3 at lysine 36, H3K4me3, trimethylated histone H3 at lysine 4; ITC, isothermal titration calorimetry; K^d, equilibrium disassociation constant; MPP8, M-phase phosphoprotein 8; MRG, MORF4 related gene; MRG15, MORF4 related gene on chromosome 15; Msl3, male specific lethal 3; MORF4, mortality factor 4; MSL, male specific lethal complex; MW, molecular weight; RBP-1, retinoblastoma binding protein 1; r.m.s.d, root-mean-square deviation; SEC-MALS, size exclusion chromatography coupled with multi-angle light scattering; TSS, transcription start site.

COMPLIANCE WITH ETHICS GUIDELINES

Yanchao Liu, Hong Wu, Yu Yu and Ying Huang declare that they have no conflict of interest.

This article does not contain any studies with human or animal subjects performed by the any of the authors.

REFERENCES

Adams PD, Afonine PV, Bunkoczi G, Chen VB, Davis IW, Echols N, Headd JJ, Hung LW, Kapral GJ, G'Kunstleve RW et al (2010) PHENIX: a comprehensive python-based system for macromolecular structure solution. Acta Crystallogr Sect D 66:213–221

Bannister AJ, Schneider R, Myers FA, Thorne AW, Crane-Robinson C, Kouzarides T (2005) Spatial distribution of di- and tri-methyl lysine 36 of histone H3 at active genes. J Biol Chem 280:17732–17736

Bertram MJ, P'Smith OM (2001) Conservation of the MORF4 related gene family: identification of a new chromo domain subfamily and novel protein motif. Gene 266:111–121

Bertram MJ, Bérubé NG, Hang-Swanson X, Ran Q, Leung JK, Bryce S, Spurgers K, Bick RJ, Baldini A, Ning Y et al (1999) Identification of a gene that reverses the immortal phenotype of a subset of cells and is a member of a novel family of transcription factor-like genes. Mol Cell Biol 19:1479–1485

Brunger AT (2007) Version 1.2 of the crystallography and NMR system. Nat Protoc 2:2728–2733

Brunger AT, Adams PD, Clore GM, DeLano WL, Gros P, G'Kunstleve RW, Jiang JS, Kuszewski J, Nilges M, Pannu NS et al (1998) Crystallography & NMR system: a new software suite for macromolecular structure determination. Acta Crystallogr D 54:905–921

Bu Z, Yu Y, Li Z, Liu Y, Jiang W, Huang Y, Dong AW (2014) Regulation of Arabidopsis flowering by the histone mark readers MRG1/2 via interaction with CONSTANS to modulate FT expression. PLoS Genet 10:e1004617

Canzio D, Liao M, Naber N, Pate E, Laron A, Wu S, Marina DB, Garcia JF, Madhani HD, Cooke R et al (2013) A conformational switch in HP1 releases auto-inhibition to drive heterochromatin assembly. Nature 496:377–381

Canzio D, Larson A, Narlikar GJ (2014) Mechanisms of functional promiscuity by HP1 proteins. Trends Cell Biol 24:377–386

Chang Y, Sun L, Kokura K, Horton JR, Fukuda M, Espejo A, Izumi V, Koomen JM, Bedford MT, Zhang X et al (2011) MPP8 mediates the interactions between DNA methyltransferase Dnmt3a and H3K9 methyltransferase GLP/G9a. Nat Commun 2:533

Emsley P, Lohkamp B, Scott WG, Cowtan K (2010) Features and development of Coot. Acta Crystallogr D 66:486–501

Gorman M, Franke A, Baker BS (1995) Molecular characterization of the male-specific lethal-3 gene and investigations of the regulation of dosage compensation in *Drosophila*. Development 121:463–475

Hayakawa T, Zhang F, Hayakawa N, Ohtani Y, Shinmyozu K, Nakayama J, Andreassen PR (2010) MRG15 binds directly to PALB2 and stimulates homology-directed repair of chromosomal breaks. J Cell Sci 123:1124–1130

Hilfiker A, Hilfiker-Kleiner D, Pannuti A, Lucchesi JC (1997) mof, a putative acetyl transferase gene related to the Tip60 and MOZ human genes and to the SAS gene of yeast, is required for dosage compensation in *Drosophila*. EMBO J 16:2054–2060

Jin J, Shi J, Liu B, Liu Y, Huang Y, Yu Y, Dong AW (2015) MORF-RELATED GENE702, a reader protein of trimethylated histone H3 lysine 4 and histone H3 lysine 36, is involved in brassinosteroid-regulated growth and flowering time control in rice. Plant Physiol 168:1275–1285

Kaustov L, Ouyang H, Amaya M, Lemak A, Nady N, Duan S, Wasney GA, Li Z, Vedadi M, Schapira M et al (2011) Recognition and specificity determinants of the human cbx chromodomains. J Biol Chem 286:521–529

Kolasinska-Zwierz P, Down T, Latorre I, Liu T, Liu XS, Ahringer J (2009) Differrential chromatin marking of introns and expressed exons by H3K36me3. Nat Genet 41:376–381

Larschan E, Alekseyenko AA, Gortchakov AA, Peng S, Li B, Yang P, Workman JL, Park PJ, Kuroda MI (2007) MSL complex is attracted to genes marked by H3K36 trimethylation using a sequence-independent mechanism. Mol Cell 28:121–133

Leung JK, Berube N, Venable S, Ahmed S, Timchenko N, Pereira-Smith OM (2001) MRG15 activates the B-myb promoter through formation of a nuclear complex with the retinoblastoma protein and the novel protein PAM14. J Biol Chem 276:39171–39178

Li J, Li Z, Ruan J, Xu C, Tong Y, Pan PW, Tempel W, Crombet L, Min J, Zang J (2011) Structural basis for specific binding of human MPP8 chromodomain to histone H3 methylated at lysine 9. PLoS One 6:e25104

Luco RF, Pan Q, Tominaga K, Blencowe BJ, Pereira-Smith OM, Misteli T (2010) Regulation of alternative splicing by histone modifications. Science 327:996–1000

McCoy AJ, G'Kunstleve RW, Adams PD, Winn MD, Storoni LC, Read RJ (2007) Phaser crystallographic software. J Appl Crystallogr 40:658–674

Min J, Zhang Y, Xu RM (2003) Structural basis for specific binding of Polycomb chromodomain to histone H3 methylated at Lys 27. Genes Dev 17:1823–1828

Otwinowski Z, Minor W (1997) Processing of X-ray diffraction data collected in oscillation mode. Methods Enzymol 276:307–326

Pardo PS, Leung JK, Lucchesi JC, P'Smith OM (2002) MRG15, a novel chromodomain protein, is present in two distinct multiprotein complexes involved in transcriptional activation. J Biol Chem 277:50860–50866

Pokholok DK, Harbison CT, Levine S, Cole M, Hannett NM, Lee TI, Bell GW, Walker K, Rolfe PA, Herbolsheimer E et al (2005) Genome-wide map of nucleosome acetylation and methylation in yeast. Cell 122:517–527

Putterill J, Robson F, Lee K, Simon R, Coupland G (1995) The CONSTANS gene of Arabidopsis promotes flowering and encodes a protein showing similarities to zinc finger transcription factor. Cell 80:847–857

Roudier F, Ahmed I, Berard C, Sarazin A, Mary-Huard T, Cortijo S, Bouyer D, Caillieux E, Duvernois-Berthet E, Al-Shikhley L et al (2011) Integrative epigenomic mapping defines four main chromatin states in Arabidopsis. EMBO J 30:1928–1938

Samach A, Onouchi H, Gold SE, Ditta GS, Schwarz-Sommer Z, Yanofsky MF, Coupland G (2000) Distinct roles of CONSTANS target genes in reproductive development of Arabidopsis. Science 288:1613–1616

Searle I, He Y, Turck F, Vincent C, Fornara F, Krober S, Amasino RA, Coupland G (2006) The transcription factor FLC confers a flowering response to vernalization by repressing meristem competence and systemic signaling in Arabidopsis. Genes Dev 20:898–912

Sun B, Hong J, Zhang P, Dong X, Shen X, Lin D, Ding J (2008) Molecular basis of the interaction of Saccharomyces cerevisiae Eaf3 chromo domain with methylated H3K36. J Biol Chem 283:36504–36512

Tominaga K, Kirtane B, Jackson JG, Ikeno Y, Ikeda T, Hawks C, Smith JR, Matzuk MM, P'Smith OM (2005) MRG15 regulates embryonic development and cell proliferation. Mol Cell Biol 25:2924–2937

Wyatt PJ (1993) Light scattering and the absolute characterization of macromolecules. Anal Chim Acta 272:1–40

Xie T, Zmyslowski AM, Zhang Y, Radhakrishnan I (2015) Structural basis for multi-specificity of MRG domains. Structure 23:1049–1057

Xu Y, Gan ES, Zhou J, Wee WY, Zhang X, Ito T (2014) Arabidopsis MRG domain proteins bridge two histone modifications to elevate expression of flowering genes. Nucleic Acids Res 42:10960–10974

Yap KL, Zhou MM (2011) Structure and mechanisms of lysine methylation recognition by the chromodomain in gene transcription. Biochemistry 50:1966–1980

Yochum GS, Ayer DE (2002) Role for the mortality factors MORF4, MRGX, and MRG15 in transcriptional repression via associations with Pf1, mSin3A, and Transducin-Like Enhancer of Split. Mol Cell Biol 22:7868–7876

Yu B, Cassani M, Wang M, Liu M, Ma J, Li G, Zhang Z, Huang Y (2015) Structural insights into Rhino-mediated germline piRNA cluster formation. Cell Res 25:525–528

Zhang P, Du J, Sun B, Dong X, Xu G, Zhou J, Huang Q, Liu Q, Hao Q, Ding J (2006a) Structure of human MRG15 chromo domain and its binding to Lys36-methylated histone H3. Nucleic Acids Res 34:6621–6628

Zhang P, Zhao J, Wang B, Du J, Lu Y, Chen J, Ding J (2006b) The MRG domain of human MRG15 uses a shallow hydrophobic pocket to interact with the N-terminal region of PAM14. Protein Sci 15:2423–2434

Zhang X, Bernatavichute YV, Cokus S, Pellegrini M, Jacobsen SE (2009) Genome-wide analysis of mono-, di- and trimethylation of histone H3 lysine 4 in *Arabidopsis thaliana*. Genome Biol 10:1–14

6

Role of circadian gene Clock during differentiation of mouse pluripotent stem cells

Chao Lu[1,2], Yang Yang[1], Ran Zhao[1], Bingxuan Hua[4], Chen Xu[1], Zuoqin Yan[4], Ning Sun[1,2,3]✉, Ruizhe Qian[1,2]✉

[1] Department of Physiology and Pathophysiology, School of Basic Medical Sciences, Fudan University, Shanghai 200032, China
[2] Research Center on Aging and Medicine, Fudan University, Shanghai 200032, China
[3] State Key Laboratory of Medical Neurobiology, Fudan University, Shanghai 200032, China
[4] Department of Orthopedics, Zhongshan Hospital, Fudan University, Shanghai 200032, China
✉ Correspondence: sunning@fudan.edu.cn (N. Sun), rzqian@shmu.edu.cn (R. Qian)

ABSTRACT

Biological rhythms controlled by the circadian clock are absent in embryonic stem cells (ESCs). However, they start to develop during the differentiation of pluripotent ESCs to downstream cells. Conversely, biological rhythms in adult somatic cells disappear when they are reprogrammed into induced pluripotent stem cells (iPSCs). These studies indicated that the development of biological rhythms in ESCs might be closely associated with the maintenance and differentiation of ESCs. The core circadian gene *Clock* is essential for regulation of biological rhythms. Its role in the development of biological rhythms of ESCs is totally unknown. Here, we used CRISPR/CAS9-mediated genetic editing techniques, to completely knock out the *Clock* expression in mouse ESCs. By AP, teratoma formation, quantitative real-time PCR and Immunofluorescent staining, we did not find any difference between *Clock* knockout mESCs and wild type mESCs in morphology and pluripotent capability under the pluripotent state. In brief, these data indicated *Clock* did not influence the maintaining of pluripotent state. However, they exhibited decreased proliferation and increased apoptosis. Furthermore, the biological rhythms failed to develop in *Clock* knockout mESCs after spontaneous differentiation, which indicated that there was no compensational factor in most peripheral tissues as described in mice models before (DeBruyne et al., 2007b). After spontaneous differentiation, loss of CLOCK protein due to *Clock* gene silencing induced spontaneous differentiation of mESCs, indicating an exit from the pluripotent state, or its differentiating ability. Our findings indicate that the core circadian gene *Clock* may be essential during normal mESCs differentiation by regulating mESCs proliferation, apoptosis and activity.

Chao Lu and Yang Yang have contributed equally to this work.

KEYWORDS Circadian gene *Clock*, mouse embryonic stem cells, gene knockout, pluripotency, cell proliferation, cell apoptosis, cell differentiation

INTRODUCTION

Circadian rhythm is a fundamental biological system in humans, animals and plants that regulates various physiological functions such as sleep–wake cycle, energy metabolism, cell division, post-transcriptional regulation (Koike et al., 2012; Cardinal-Aucoin and Steel, 2016; Lopez-Minguez et al., 2016; McAlpine and Swirski, 2016; Nitschke et al., 2016), and endocrine system (Gachon et al., 2004; Lowrey and Takahashi, 2004). The master circadian clock, located in the suprachiasmatic nucleus (SCN), is an essential activator of downstream molecular events critical to the generation of circadian rhythms (Dunlap, 1999). The master clock is regulated by four major groups of genes: *Clock, Bmal1, Cry* (*Cryptochrome* 1 and 2), and *Per* (*Period* 1 and 2). These core circadian genes are regulated in a transcription-translation feedback loop (TTFL) (Xue et al., 2016).

The core circadian genes regulate critical aspects of cellular processes in many organs. Mutations in the circadian genes affect circadian activities and are hazardous to health (King et al., 1997; Kovanen et al., 2010). For instance, some recent studies have shown associations between clock genes and chronic inflammation, blood pressure and energy intake in humans, supporting the importance of the circadian rhythm in cardiac physiology (Dashti et al., 2016; Johnston and Ordovas, 2016). Mutations in *Per* genes increase cancer incidence (Fu et al., 2002; Wood et al., 2008; Borgs et al., 2009) and cancer cell proliferation (Borgs et al., 2009). Circadian clock dysfunction has been linked to oxidative stress and age-related neurodegeneration *in vivo* (Witting et al., 1990; Hu et al., 2009; Wyse and Coogan, 2010). Adult stem cell functions are also regulated by circadian oscillations (Casanova-Acebes et al., 2013; Janich et al., 2013; Karpowicz et al., 2013).

Circadian rhythms in the suprachiasmatic nucleus (SCN) manifest only a few days before birth (Yagita et al., 2010). Bmal1 is expressed in mice during embryogenesis without oscillation (Johnson et al., 2002). It may have other key roles instead of circadian rhythm control during embryogenesis (Yang et al., 2016). Indeed, circadian rhythms do not occur in embryonic stem cells (ESCs) but develop when ESCs exit from the pluripotent state and differentiate to downstream cells (Yagita et al., 2010). Conversely, biological rhythms in adult somatic cells disappear when they are reprogrammed into induced pluripotent stem cells (iPSCs). To better understand the molecular basis underlying circadian rhythms and ESCs pluripotency and differentiation, we studied the role of the core circadian gene *Clock* in the maintenance, differentiation and development of circadian rhythms in mouse embryonic stem cells (mESCs). Using the CRISPR/CAS9 genome editing tool, we completely knocked out the *Clock* gene in mESCs. We found that *Clock* knockout mESCs still exhibited normal clonal morphology and pluripotency, which indicated that *Clock* gene might not be required for regular maintenance of mESCs. However, *Clock* knockout significantly decreased the proliferation and increased the apoptosis of mESCs. Furthermore, the biological rhythms failed to develop in *Clock* knockout mESCs after spontaneous differentiation. Loss of CLOCK protein due to *Clock* gene silencing in mESCs triggered their spontaneous differentiation, leading to stronger expression of genes in the three embryonic germ layers downstream.

Our findings suggest that the core circadian gene *Clock* influences mESCs differentiation process by regulating mESCs proliferation, apoptosis and activity.

RESULTS

Knock out of *Clock* in mESCs using CRISPR/CAS9 genomic editing

To investigate the role of the core circadian gene *Clock* in the maintenance and differentiation of mammalian pluripotent stem cells, we established the *Clock* knockout mESC line using the CRISPR/CAS9 genomic editing tool. This tool enables the efficient construction of knockout alleles via induction of frameshift mutations. In this study, we designed a single gRNA specifically targeting exon 2 in the mouse *Clock* gene (Fig. 1A) and subcloned it into the PX330 CRISPR/CAS9 vectors. Single wild type mESCs were transfected with the subcloned CRISPR/CAS9 plasmids via electroporation. Single colonies with successful genome editing were selected by puromycin and Sanger's sequencing at the cutting site on exon 2 of the *Clock* gene (Fig. 1B and Fig. S1). We deleted 31 single mESC clones, of which 8 clones (clone 5, clone 13, clone 17, clone 19, clone 20, clone 23, clone 24, and clone 28) showed sequence modifications at the CRISPR/CAS9 cutting site (Table S2). To verify whether the *Clock* gene was completely knocked out, we investigated CLOCK protein expression in the 8 positive clones using Western blot. The results indicated that CLOCK protein expression was totally ablated in 3 clones (clone 17, clone 19, and clone 20) (Fig. 1C). To further confirm whether both alleles of the *Clock* gene were modified in clone 17, clone 19, and clone 20, we performed TA cloning of the PCR products flanking the CRISPR/CAS9 cutting site. The results showed that both alleles were edited at exon 2 of the *Clock* gene in all the 3 homozygous mESC clones (Fig. 1D and Fig. S1). Among the 3 clones, clone 19 was used for further study.

Further, to consider the potential influence of off-target gene modifications (Yee, 2016), we detected the most probable ten off-target modification sites which were designed at the CRISPR website (http://crispr.mit.edu/) (Table S3) and found that there was no off-target gene modification (Fig. S2), which excluded the potential influence of off-target gene modifications.

Clock knockout in mESCs did not affect pluripotency

The *Clock* knockout mESCs exhibited compact clonal morphology without spontaneous differentiation under routine mESC maintenance culture conditions, which was similar to the wild type mESCs (Fig. 2A). To further examine whether silencing of *Clock* gene affected the pluripotency of mESCs, we performed staining for alkaline phosphatase and immunofluorescence of pluripotent markers OCT4, NANOG, and SSEA1. The *Clock* knockout mESCs expressed strong alkaline phosphatase activity as well as OCT4, NANOG, and SSEA1 (Fig. 2B and Fig. 3). Real-time PCR of the mRNA expression levels of pluripotent markers *Oct4, Sox2, Nanog, Klf4, Zfp296, Dax1, Esg1,* and *Eras* revealed no significant

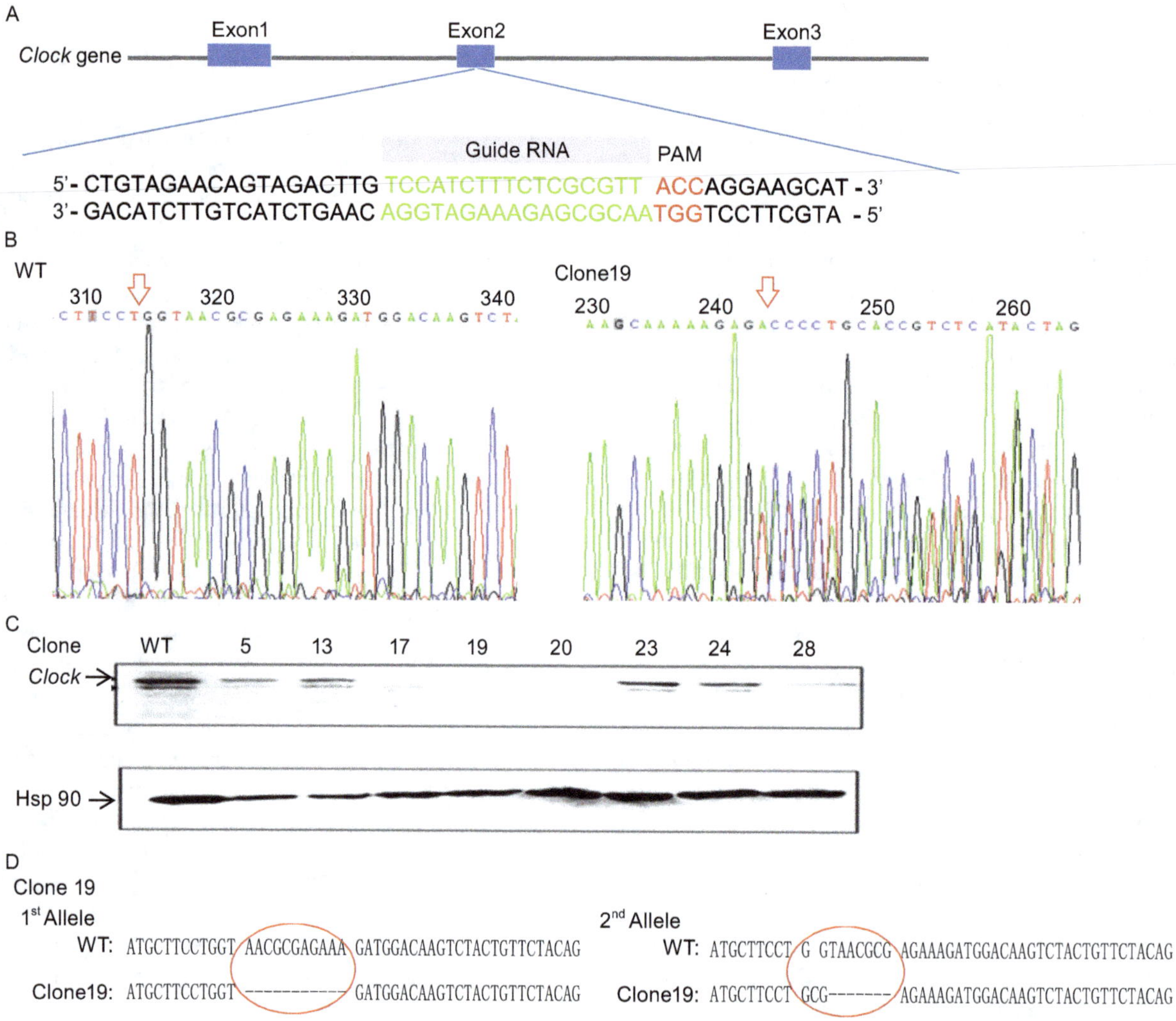

Figure 1. Generation of *Clock* knockout mESC line using CRISPR/CAS9 gene editing. (A) Schematic diagram of the 5′ end of the *Clock* gene (blue bars, exons), the CRISPR/CAS9 gRNA sequence (red bar, the green shaded part), PAM domain (the red shaded part). (B) Representative sequencing showing modifications at the CRISPR/CAS9 cutting site on exon 2 of the mouse *Clock* gene. (C) Western blot revealed CLOCK expression in the positive clones following editing by the CRISPR/CAS9 system. (D) TA cloning results involving a representative homozygous clone (clone 19) show that both alleles of the mouse *Clock* gene were modified on exon 2 by the CRISPR/Cas9 system.

differences ($P > 0.05$) from those of wild type mESCs (Fig. 2C). Furthermore, we used teratoma formation assay as a tool for monitoring pluripotency in stem cell research to study the pluripotency of wild type mESCs and *Clock* knockout mESCs *in vivo* (Fig. 2D) (Nelakanti et al., 2015). As shown in Fig. 2D, after injecting 1×10^7 wild type mESCs or *Clock* knockout mESCs into immunodeficiency mice, we could observe teratoma of 1-2 cm formed after four weeks. Six weeks later, we detected the differential capacity of wild type mESCs and *Clock* knockout mESCs by hematoxylin-eosin staining (HE staining). We found that that both wild type mESCs and *Clock* knockout mESCs could differentiate to endoderm, mesoderm, and ectoderm layers cells, which indicated that wild type mESCs and *Clock* knockout mESCs have similar potential of multilineage differentiation *in vivo*. According to these results we got, silencing of *Clock* gene in mESCs didn't affect their pluripotent nature.

Clock silencing in mESCs decreased cell proliferation rate

During routine culture, we observed that the proliferation rate of *Clock* knockout mESCs was much slower than that of the wild type mESCs. To further confirm our observation, we

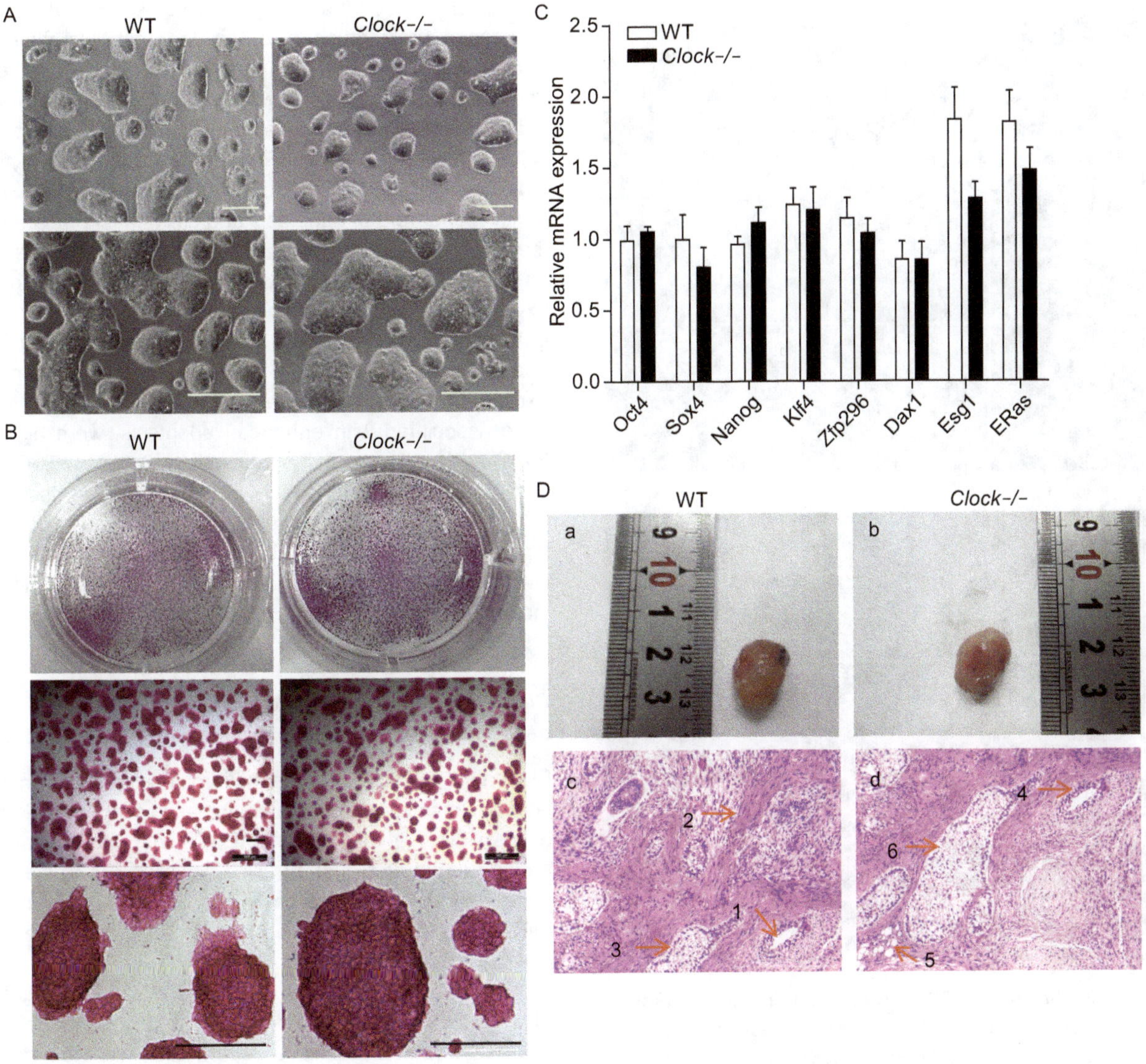

Figure 2. Knockout of *Clock* did not affect pluripotency of mESCs. (A) Phase images of wild type and *Clock-/-* mESCs exhibited compact and round clonal morphology. Bars, 200 μm. (B) Both wild type and *Clock-/-* mESCs stained strongly for alkaline phosphatase. The upper two images represent macroscopic views of a 12-well plate system. Bars, 400 μm. (C) Relative mRNA expression of pluripotent markers *Oct4, Sox2, Nanog, Klf4, Zfp296, Dax1, Esg1,* and *Eras* in wild type and *Clock-/-* mESCs. The mRNA expression levels were normalized to that of the endogenous *GAPDH*. Data represent an average of three independent experiments and are expressed as means ± S.D. Statistical significance was calculated using the Student's *t*-test: * denotes $P < 0.05$, ** denotes $P < 0.01$, *** denotes $P < 0.001$ compared with the Ctrl group, ns, not significant. No statistical significance was found in the expression levels of the above genes between the wild type and *Clock-/-* mESCs ($P > 0.05$). (D) After respectively injecting 10^7 wild type mESCs and *Clock* knockout mESCs into immunodeficiency mice, we could observe teratoma of similar size (1–2 cm) formed after four weeks. Six weeks later, Immunohistochemistry manifested that both wild type mESCs and *Clock* knockout mESCs could differentiate to downstream endoderm, mesoderm, and ectoderm layers cells. Among them, arrow 1 and 4 denote gland cells, which belong to endoderm layer cells; arrow 2 denotes smooth muscle cells, and arrow 5 denotes fat vesicle cells, which belong to mesoderm layer cells; arrow 3 and 6 denote sebaceous gland cells, which belong to ectoderm layer cells.

performed cell proliferation assays of the *Clock* knockout mESCs and wild type mESCs at different time points. As shown in Fig. 4A, the *Clock* knockout slowed the proliferation rate of mESCs significantly compared with that of the wild type mESCs ($P < 0.001$). These results indicated that *Clock* knockout negatively affected the proliferation of mESCs.

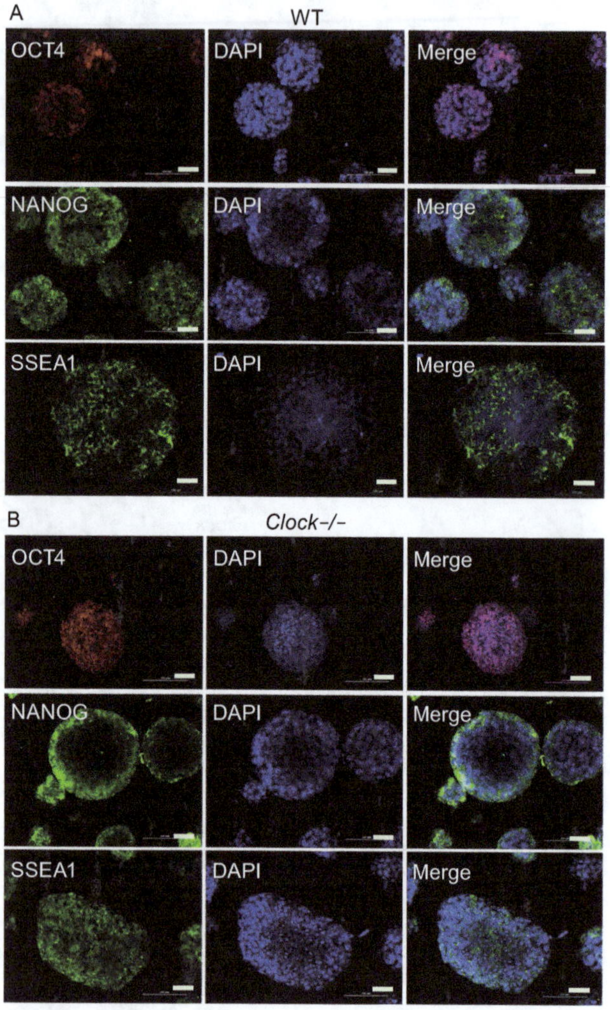

Figure 3. *Clock-/-* mESCs express pluripotent markers OCT4, NANOG, and SSEA1. (A) Immunofluorescence staining of the pluripotent markers OCT4, NANOG, and SSEA1 in wild type mESCs. (B) Immunofluorescence staining of the pluripotency marker OCT4, NANOG, and SSEA1 in *Clock-/-* mESCs. Bars, 100 μm.

Clock knockout in mESCs slowed down the cell cycle and enhanced cell death

The decreased proliferation of the *Clock* knockout mESCs suggested either a slowdown of cell cycle or increased cell death in mESCs. To test our hypotheses, we first examined the expression of cell cycle-related markers *C-Myc, CyclinD1, CDK1, CDK2, PCNA,* and *P27* by real-time PCR in the *Clock* knockout mESCs and compared with those in the wild type mESCs. We found that the expression levels of *C-Myc, CyclinD1, CDK1, CDK2,* and *PCNA*, which facilitated cell proliferation, were lower in *Clock* knockout mESCs than in the wild type mESCs ($P < 0.05$, $P < 0.01$). The P27 mRNA expression, which restrained cell proliferation, was higher in *Clock* knockout mESCs than in the wild type mESCs (Fig. 4B). These results indicated that *Clock* silencing in mESCs decreased the cell cycle. To determine whether the decreased proliferation also resulted from increased cell death, we further investigated cell death in the *Clock* knockout mESCs and the wild type mESCs. Based on six independent experiments using FACS analysis of apoptosis with Annexin V and PI (Propidium Iodide), respectively, indicated that the *Clock* knockout mESCs exhibited significantly higher rate of apoptosis and cell death compared with the wild type mESCs ($P < 0.05$ and $P < 0.01$) (Fig. 5A and B). To further verify this result, we determined the mRNA expression levels of apoptosis-related markers *Bax, Bcl-2, caspase 3,* and *caspase 9* in the *Clock* knockout mESCs and the wild type mESCs. As shown in Fig. 5C, the mRNA expression levels of *Bax, caspase 3,* and *caspase 9*, which were associated with enhanced apoptosis, were higher in the *Clock* knockout mESCs than in the wild type mESCs ($P < 0.05$). The expression level of *Bcl-2*, which was associated with decreased cell apoptosis, was lower than in the wild type mESCs ($P < 0.05$). These results demonstrated that silencing of Clock increased cellular apoptosis, resulting in increased cell death of mESCs.

Clock is required for development of circadian oscillator during mESCs differentiation

Pluripotent mESCs do not exhibit molecular circadian oscillations. However, circadian rhythms developed during the differentiation of pluripotent mESCs into downstream lineages (Yagita et al., 2010). Whether silencing of the *Clock* gene altered the development of circadian oscillation of mESCs during differentiation remains unknown. Therefore, we analyzed the expression of the core clock genes *Bmal1, Clock, Per,* and *Cry* on day 15 after spontaneous differentiation of the *Clock* knockout mESCs and wild type mESCs. Consistent with a previous report, the wild type mESCs developed circadian oscillations on day 15 following spontaneous differentiation. In contrast, the *Clock* knockout mESCs exhibited no circadian oscillation on day 15 after spontaneous differentiation (Fig. 6). These results indicated that silencing of the gene *Clock* disrupted the development of circadian oscillations during the differentiation of pluripotent mESCs.

Clock knockout in mESCs accelerated spontaneous differentiation

We next investigated whether Clock knockout and disruption of the normal circadian oscillations affected mESCs differentiation. 1×10^4 of wild type mESCs or *Clock* knockout mESCs were cultured in differentiation medium. Surprisingly, after day 7, *Clock* knockout mESCs showed more complex cellular structures as well as faster proliferation rate after spontaneous differentiation (Fig. S3A). The mRNA expression levels of the pluripotent markers *Oct4, Sox2, Nanog,* and *Klf4* were gradually decreased in both the *Clock*

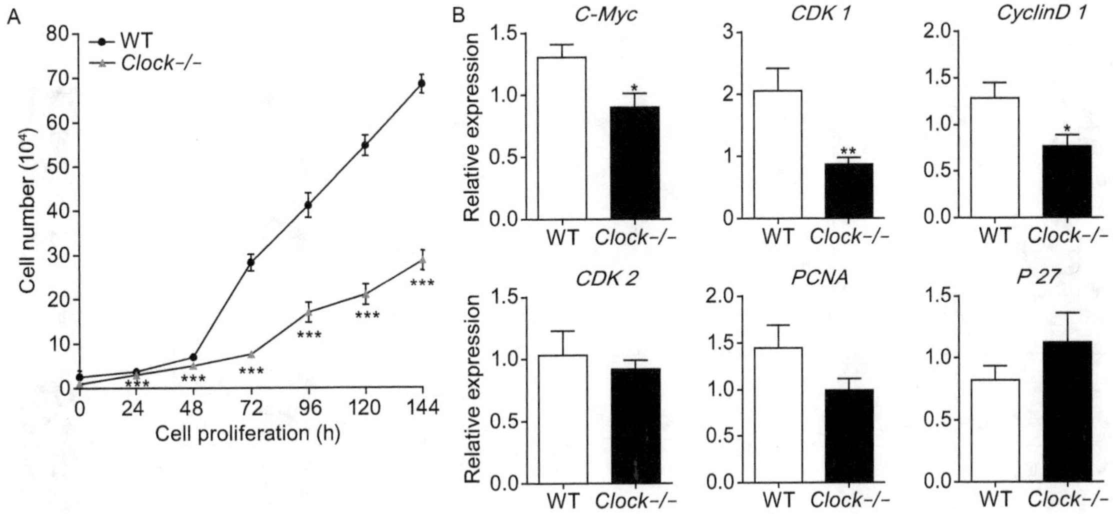

Figure 4. *Clock* silencing inhibited proliferation of mESCs. (A) Cell proliferation curve of the wild type and *Clock-/-* mESCs. *Clock-/-* mESCs exhibited significantly lower proliferation rate than that of the wild type mESCs. (B) Relative mRNA expression levels of cell cycle-related markers *C-Myc, CyclinD1, CDK1, CDK2, PCNA,* and *P27* in wild type and Clock-/- mESCs. The mRNA expression levels were normalized to that of the endogenous *GAPDH*. Data represent an average of three independent experiments and are expressed as means ± S.D. Statistical significance was calculated using the Student's *t*-test: * denotes $P < 0.05$, ** denotes $P < 0.01$, *** denotes $P < 0.001$ compared with the Ctrl group, ns, not significant.

knockout mESCs and wild type mESCs after spontaneous differentiation (Fig. 7A), indicating an exit from the pluripotent state. Spontaneous differentiation also led to a gradual increase in the mRNA expression levels of the endoderm markers *Gata4, Sox17, Foxa2,* and *Lamb1* (Fig. 7B), the mesoderm markers *BMP4, T, Eomes,* and *Actc1* (Fig. 7C), as well as the ectoderm markers *Nestin, Sox1, Neurod1,* and *Otx1* (Fig. 7D). Interestingly, the mRNA expression levels of the above markers in the three embryonic germ layers (ectoderm, mesoderm, and endoderm) in the *Clock* knockout mESCs were relatively higher than in the wild type mESCs. The *Clock* knockout mESCs showed accelerated differentiation compared with wild type mESCs. To further verify these observations, we used immunofluorescence staining of the endoderm marker FOXA2 and the ectoderm marker SOX1 in both the *Clock* knockout mESCs and wild type mESCs on day 9 after spontaneous differentiation. The FOXA2 and SOX1 stained strongly in the *Clock* knockout mESCs than in the wild type mESCs (Fig. 8A and B). These results suggested that knockout the *Clock* triggered spontaneous differentiation of mESCs.

DISCUSSION

In this study, we investigated the role of the core circadian gene *Clock* in the maintenance and differentiation of pluripotent stem cells. We used the CRISPR/CAS9 genome editing tool to induce frameshift mutations in exon 2 of the mouse *Clock* gene and completely ablated its protein expression. This is the first report of a complete knockout of *Clock* expression in pluripotent stem cells. Previous studies mostly used siRNA/shRNA-mediated silencing of *Clock* expression (Mukherjee et al., 2010; Tobback et al., 2012; Tracey et al., 2012; Liang et al., 2013; Li et al., 2015). The complete knockout of *Clock* in our study avoided any potential noise observed in partial knockdown studies involving siRNA/shRNA.

The knockout of the gene *Clock* did not disrupt the clonal morphology or pluripotency of mESCs. However, loss of *Clock* significantly decreased the proliferation and increased the cell death of mESCs. Our findings are consistent with other studies that reported a potential relationship between *Clock* gene and apoptosis as well as cell cycle regulation. One study demonstrated that upregulation of several pro-apoptotic genes in the spleen of *Clock* mutant mouse (Gaddameedhi et al., 2015). Another study also reported that the *Clock* gene controlled the expression of key cell cycle-related regulators, such as *Cdc2, Wee1, P21, PCNA* and *Cdk2* in the intestine (Peyric et al., 2013). Our data also showed that the expression of cell cycle-related proteins in *Clock* knockout mESCs, was affected. The expression of anti-apoptotic protein *Bcl-2* significantly decreased in *Clock* knockout mESCs. The expression of pro-apoptotic protein *Bax*, *cleaved caspase-3,* and *caspase-9* significantly increased in *Clock* knockout mESCs. These data suggested a disruption in the balance of mitochondria and activation of the caspase cascade in *Clock* knockout mESCs (Liang et al., 2013). *Clock* may, thus, contribute to the maintenance of

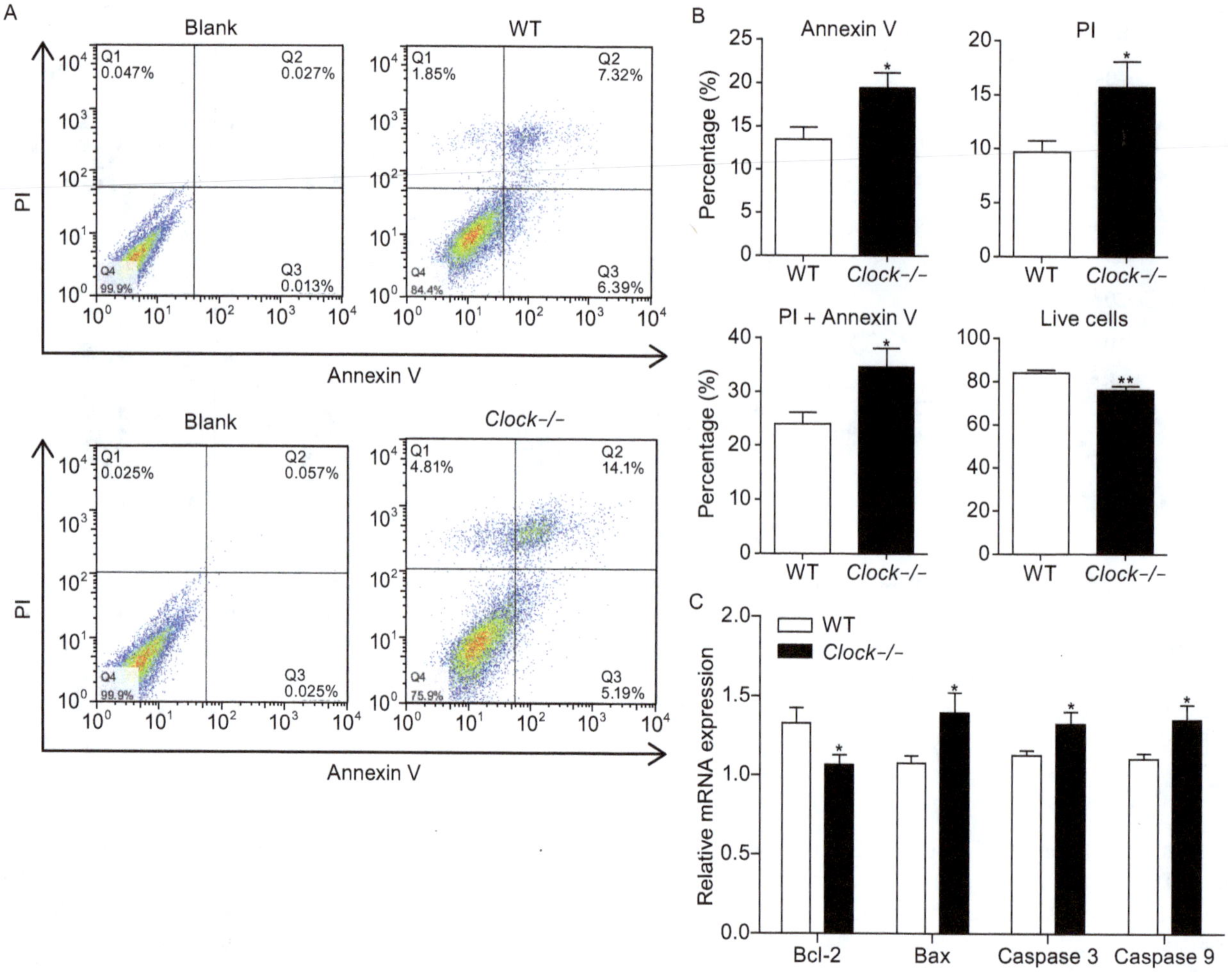

Figure 5. *Clock* silencing accelerated cell death in mESCs. (A) Apoptosis and overall cell death of the wild type and *Clock-/-* mESCs were analyzed by flow cytometry using the Annexin V and PI assay, respectively. *Clock-/-* mESCs exhibited increased apoptosis and overall cell death. (B) Percentage of apoptotic (Annexin V) and dead (PI) cells calculated from six independent experiments. Data were expressed as means ± S.D. (C) Relative mRNA expression levels of apoptosis-related genes *Bax, Bcl-2,* cleaved *caspase-3*, and *caspase-9* in wild type and *Clock-/-* mESCs. The mRNA expression levels were normalized to that of the endogenous *GAPDH*. Data represent the average of three independent experiments and are expressed as means ± S.D. Statistical significance was calculated using the Student's *t*-test: * denotes $P < 0.05$, ** denotes $P < 0.01$, *** denotes $P < 0.001$ compared with the Ctrl group, ns, not significant.

normal proliferation by controlling the balance of cell cycle and apoptosis in mESCs. Overall, our results provide some new insights into the function of the *Clock* gene in the regulation of cell cycle and apoptosis of mESCs.

To consider the potential influence of off-target gene modifications (Yee, 2016), we detected the most probable ten off-target modification sites which were designed at the CRISPR website (http://crispr.mit.edu/) (Supplemental Table. 3) and found that there was no off-target gene modification (Supplemental Fig. 2), which excluded the potential influence of off-target gene modifications. And we further considered the potential influence of alternative cellular roles for *Clock* such as the transcription factor neuronal PAS domain protein 2 *NPAS2* (*MOP4*), which was able to functionally substitute for the loss of *Clock* in the SCN but not in peripheral tissues in mice to regulate circadian rhythmicity (DeBruyne et al., 2007a, b; Debruyne, 2008). In our study, we tested the mRNA expression level of *NPAS2* in the wild type mESCs and *Clock* knockout mESCs after spontaneous differentiation, and we found that both of them barely expressed *NPAS2* after spontaneous differentiation. This indicated that the nerve cells was very few in wild type mESCs and *Clock* knockout mESCs after spontaneous differentiation, so there wasn't compensation of *NPAS2* for the loss of *Clock* (Fig. S3B). All in all, these data indicated that the phenotypes in these *Clock* knockout mESCs should mainly be due to the loss of *Clock* but not the influence

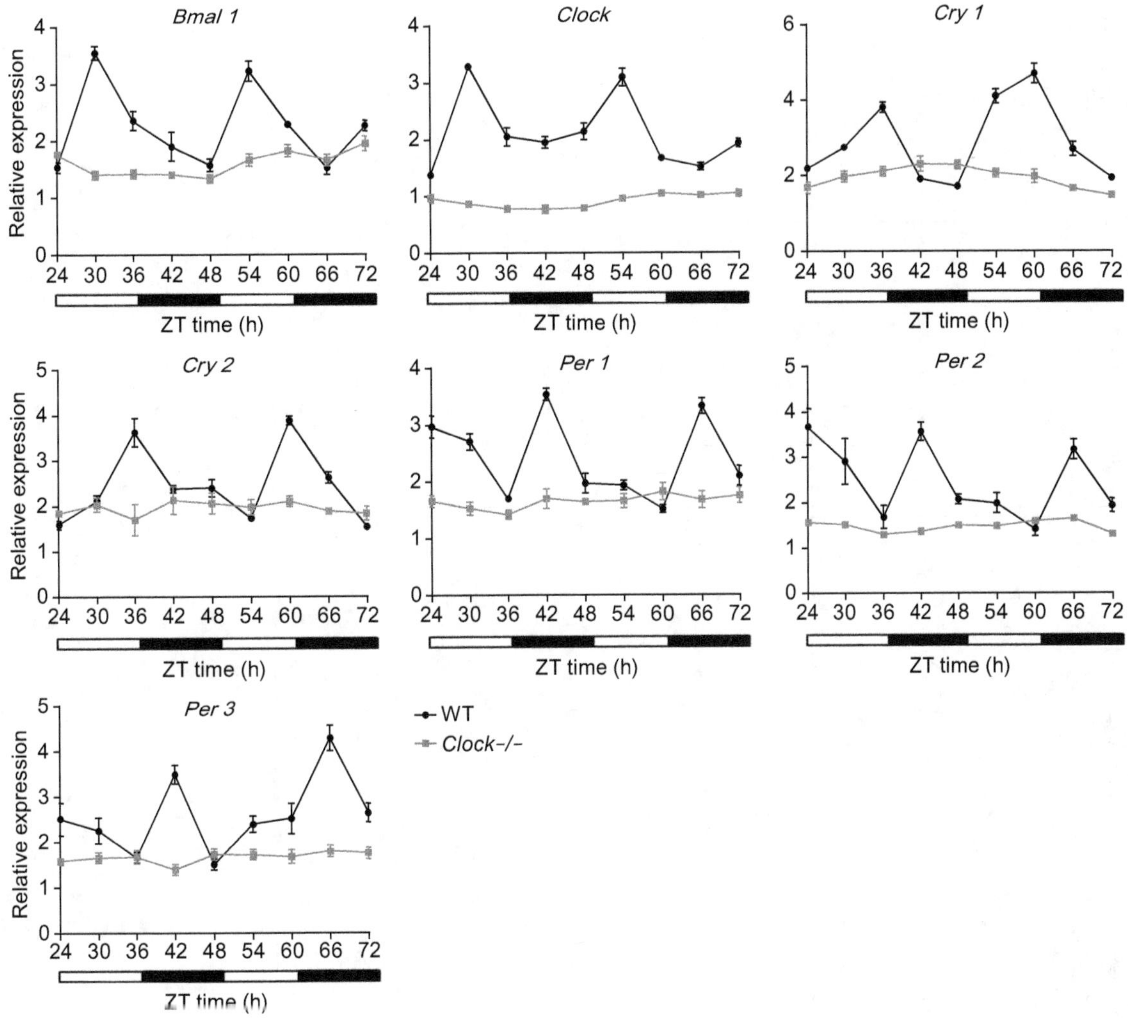

Figure 6. *Clock* silencing disturbed the circadian oscillations during mESC differentiation. Relative mRNA expression of core clock genes *Bmal1, Clock, Cry1, Cry2, Per1, Per2*, and *Per3* in spontaneously differentiated wild type and *Clock-/-* mESCs after serum shock induced biological rhythms. ZT0 to 72 represent 8:00 am to 8:00 am of day 4, respectively. The mRNA expression levels were normalized to that of the endogenous *GAPDH*. Compared with wild type mESCs, *Clock-/-* mESCs showed no circadian oscillation in the expression of core clock genes. Data represent the average of three independent experiments and are presented as means ± S.D.

of off-target gene modifications or loss of other *Clock* functions.

A recent study has shown that the mammalian circadian oscillator in ground state naïve pluripotent stem cells was not developed until after differentiation (Yagita et al., 2010). We also found in this study that naive mESCs did not exhibit circadian oscillations until differentiation, while *Clock* was required for the development of this circadian oscillator in mESCs after differentiation. This finding suggested that *Clock* mediated developmental functions in mammals. Indeed, we found that loss of *Clock* in mESCs triggered spontaneous differentiation. Several critical genes regulating development of the three emryonic germ layers were expressed higher in the *Clock* knockout mESCs, which exhibited accelerated spontaneous differentiation *in vitro*. The disruption of biological rhythms underlying differentiation of *Clock* knockout mESCs may drive these cells toward spontaneous differentiation upon withdrawal of pluripotent signals. Our study indicated that *Clock* medicated the differentiation of mouse pluripotent stem cells. Loss of *Clock* significantly decreased the proliferation and increased the cell death of mESCs indicated that *Clock* accurately regulated the development of mESCs. *Clock* knockout in mESCs accelerated spontaneous differentiation. Some studies have reported that *Clock* knockout in mice is associated with aging and chronic inflammation (Dubrovsky et al., 2010), indicating that this acceleration might be harmful. Our findings indicated that *Clock* knockout mESCs can conduct one

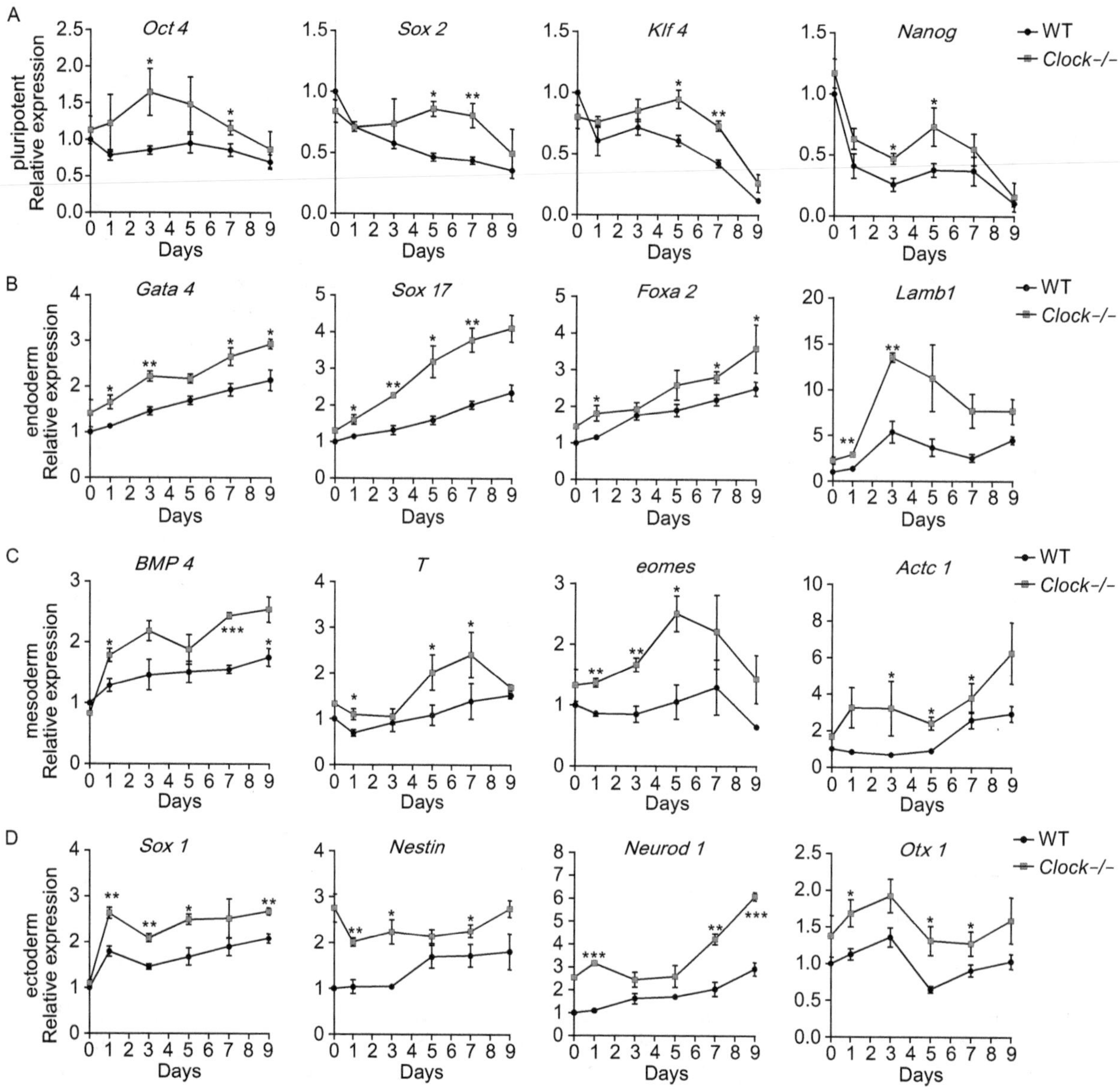

Figure 7. *Clock* silencing promoted the expression of genes controlling the three embryonic germ layers in spontaneously differentiated mESCs. (A) Relative mRNA expression of pluripotent markers *Oct4, Sox2, Nanog,* and *Klf4* after spontaneous differentiation in wild type and *Clock-/-* mESCs. Expression of the pluripotent genes was gradually decreased in wild type and *Clock-/-* mESCs. (B) Relative mRNA expression of endoderm markers *Gata4, Sox17, Foxa2, and Lamb1* after spontaneous differentiation in wild type and *Clock-/-* mESCs. *Clock-/-* mESCs exhibited a higher expression of these endoderm genes. (C) Relative mRNA expression of mesoderm markers *BMP4, T, Eomes,* and *Actc1* after spontaneous differentiation in wild type and *Clock-/-* mESCs. *Clock-/-* mESCs exhibited a higher expression of mesoderm genes. (D) Relative mRNA expression of ectoderm markers *Nestin, Sox1, Neurod1,* and *Otx1* after spontaneous differentiation in wild type and *Clock-/-* mESCs. *Clock-/-* mESCs exhibited a higher expression of these ectoderm genes. The mRNA expression levels were normalized with that of the endogenous *GAPDH*. Data represent the average of three independent experiments and are presented as means ± S.D. Statistical significance was calculated using the Student's *t*-test: *denotes $P < 0.05$, **denotes $P < 0.01$, ***denotes $P < 0.001$ compared with the Ctrl group, ns, not significant.

tool to study these diseases. Yet, additional studies are required to uncover the roles of *Clock* in mammalian development.

In summary, we found that the core circadian gene *Clock* was dispensable for maintaining pluripotency in mouse pluripotent stem cells. However, it was required for

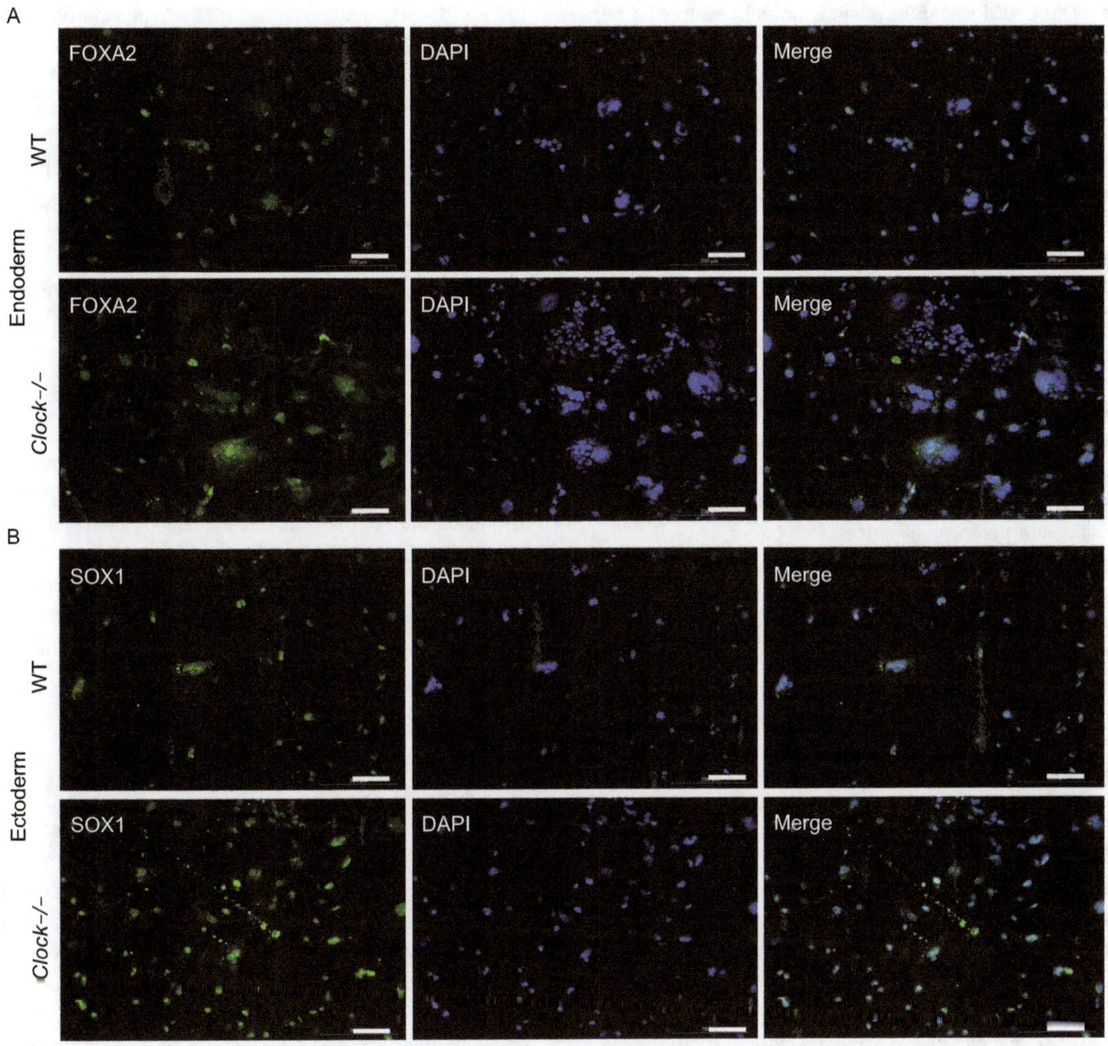

Figure 8. *Clock-/-* mESCs expressed higher endoderm marker FOXA2 and ectoderm marker SOX1 on day 9 after spontaneous differentiation. (A and B) Immunofluorescence staining of the endoderm marker FOXA2 and the ectoderm marker SOX1 in spontaneously differentiated wild type and *Clock-/-* mESCs on day 9. *Clock-/-* mESCs exhibited relatively stronger expression of both FOXA2 and SOX1. Bars, 200 μm.

maintaining regular proliferation and cell death in mESCs. And *Clock* was indispensable for the development of circadian oscillations after differentiation of pluripotent stem cells. Furthermore, our data indicated that *Clock* was essential for the differentiation of mouse pluripotent stem cells. Our findings may provide some new insights into the regulatory mechanisms of *Clock* in pluripotent stem cell development in mammals.

MATERIALS AND METHODS

Culture of mESCs

Mouse 129 ESCs were obtained from ATCC (American Type Culture Collection, Manassas, VA). The wild type and *Clock* knockout mESCs were cultured in the maintenance medium of Dulbecco's Modified Eagle's Media (DMEM, Gibco, USA) supplemented with 3.7 g/L sodium bicarbonate, 1% penicillin and streptomycin, 1.7 mmol/L L-glutamine, non-essential amino acid, 55 mmol/L beta mercaptoethanol, 5 ng/mL mouse leukemia inhibitory factor (LIF), 3 μmol/L EsK3β inhibitor, 1 μmol/L MEK inhibitor and 10% FBS. The mESC maintenance medium was used for spontaneous differentiation without supplementing LIF, the EsK3β inhibitor or the MEK inhibitor.

Construction of the *Clock* knockout mESC line

To construct the *Clock* knockout mESC line, we utilized the clustered, regularly interspaced, short palindromic repeats (CRISPR)/ CRISPR-associated protein 9 (CAS9) (CRISPR/CAS9) genomic editing tool. The CRISPR/CAS9 system is an efficient tool for genome engineering. It induces double-strand breaks (DSBs) and repair using the non-homologous end-joining (NHEJ) mechanism at any specific site. Briefly, one pair of independent oligo primers targeting exon 2 of the *Clock* gene was subcloned into the pX330 vector (Mizuno et al., 2014) (Addgene) to obtain the pX330-Clock plasmid. The primer sequences generating the guide RNA (gRNA) targeting the *Clock* gene were TCCATCTTTCTCGCGTT, then the complementary sequences were AGGTAGAAAGAGCGCAA. As the transcriptional promoter of PX330 plasmid was U6, we replaced base T to base G for the first base was G to be effective transcription. With enzyme sites, the insertion sequences were as follows: oligo1: 5′-CACCGCCATCTTTCTCGCGTTACC-3′, oligo2: 5′-AAACGG-TAACGCGAGAAAGAT GGC-3′. The pX330-Clock plasmid was transfected into single mESCs by electroporation. The correct knockout of *Clock* gene expression was confirmed by Western blot analysis and gene sequencing.

To consider the potential influence of off-target gene modifications (Yee, 2016), we detected the most probable ten off-target modification sites which were designed at the CRISPR website (http://crispr.mit.edu/), whose off-target cleavage site sequences and gene numbers were as follows in Table S3. Specific PCR primers were designed according to the target gene sequences of mouse (Table S3).

Quantitative real-time PCR

The mRNA expression levels of *GAPDH, Oct4, Sox2, Klf4, Nanog, Zfp296, Eras, Dax1, Esg1, C-Myc, PCNA, CDK1, CDK2, CyclinD1, P27, Gata4, Sox17, Foxa2, Lamb1, BMP4, T, Eomes, Actc1, Nestin, Sox1, Neurod1, Otx1, Bcl-2, Bax, cleaved caspase-3,* and *caspase-9* were quantified by RT-PCR. Specific PCR primers were designed according to the target gene sequences of mouse (Supplemental Table. 1). Cells were washed by ice-cold Phosphate buffer solution (PBS, NaCl 137 mmol/L,KCl 2 mmol/L, Na_2HPO_4 10 mmol/L, KH_2PO_4 10 mmol/L pH 7.4). Total RNA was extracted by using Trizol reagent (Invitrogen, USA). Qualities of extracted RNA qualities were measured by OD_{260}/OD_{280} ratios which ranged from 1.9 to 2.1. The cDNA was obtained through reverse transcription kit according to manufacturer's instructions (TOYOBO, Japan). The amplification mixture comprised 1 μL of RT reaction mix, 10 μL of SYBR® Premix Ex Taq TM (2×) (TaKaRa, China), 0.5 μL of 10 μmol/L each of primers and 8.5 μL ddH_2O. Reactions were performed on a fluorescence temperature cycler (Bio-Rad, Hercules, CA, USA). The PCR conditions for *Clock* and *GAPDH* were as follows: one cycle of 3 min at 95°C; 35 cycles of 15 s at 95°C, 30 s at 58°C, 30 s at 72°C. The primer sequences are listed in Supplemental Table 1. The threshold cycle (CT) in RT-PCR was analyzed using the $2^{-\Delta\Delta Ct}$ method

Western blot

Total protein was extracted from the treated cells using Radio-Immunoprecipitation Assay (RIPA) lysis buffer (50 mmol/L Tris/HCl pH 7.4, 150 mmol/L NaCl, 1% Nonidet-P40, 0.5% Sodium deoxycholate, 0.1% SDS). The extraction and isolation of nuclear protein were performed according to the instructions in the Nuclear and Cytoplasmic Protein Extraction Kit (Beyotime, China). Equal amounts (30–50 μg) of protein extracted from cells were boiled at 100°C in 1× SDS loading buffer for 10 min and then were loaded and separated by sodium dodecyl-sulfate (SDS)-polyacrylamide gel electrophoresis (PAGE). Proteins were transferred to a polyvinylidene fluoride (PVDF) membrane (Millipore, Billerica, MA, USA), blocked by 5% nonfat dry milk in Tris-buffered saline containing Tween-20. The membranes were incubated with antibodies against *PKM2* (Cell Signaling), *CK19* (Abcam), and *EpCAM* (Abcam). Signals were detected with horseradish peroxidase (HRP)-conjugated goat anti-mouse or goat anti-rabbit IgG. Immunoreactive bands were visualized by enhanced chemiluminescence (ECL, Amersham Biosciences), which was developed and quantified using an imaging system (Tanon, China).

Alkaline phosphatase staining

The sections were washed with cold 1 × PBS three times, treated with 4% paraform- aldehyde for 1 to 2 min, washed with cold 1 × PBS twice, and with 1 × Tris buffer solution tween (TBST, 50 mmol/L Tris, 150 mmol/L NaCl, 1% Tween-20) once, followed by staining with Alkaline Phosphatase Kit (Millipore, USA) according to the manufacturer's instructions. Images were taken by a Leica DMi8 microscope (Leica, Germany) and analyzed using Image Pro Plus 6.0 software (Media Cybernetics, Rockville, MD).

Teratoma formation assay

To study the pluripotency of wild type mESCs and *Clock* knockout mESCs *in vivo*, we performed teratoma formation assay which could conduct as a tool for monitoring pluripotency in stem cell research (Nelakanti et al., 2015). We injected 1 × 10^7 wild type mESCs and *Clock* knockout mESCs into the oxters of immunodeficiency mice respectively. 1–2 cm teratoma formed after four weeks. Six weeks later, we detected the differential capacity of wild type mESCs and *Clock* knockout mESCs by hematoxylin-eosin staining (HE staining). We found that both wild type mESCs and *Clock* knockout mESCs could differentiate to endoderm, mesoderm, and ectoderm layers cells, which indicated that they have similar potential of multilineage differentiation *in vivo*.

Immunofluorescent staining

The sections were washed with cold 1 × PBS three times, and treated with 4% paraformaldehyde for 15 min. They were again washed with cold 1 × PBS three times, and treated with 5% Triton X-100 in Tris buffer solution (TBS, 50 mmol/L Tris/HCl pH 7.4, 150 mmol/L NaCl) for 20 min at room temperature (RT) (membrane proteins do not need this step). After another wash with cold 1 × PBS three times for 5 min, they were treated with normal goat serum for 30 min at RT, followed by the addition of primary antibodies (1:200) and incubation in a wet box for 2 h at RT or 4 °C overnight. Fluorescent secondary antibodies (1:350) were added and the cells were incubated for 30 min or 1 h at RT, washed three times for 5 min each with PBST (1% Tween in PBS buffer). It was followed by staining with 2-(4-Amidinophenyl)-6-indolecarbamidine dihydrochloride (DAPI) (1:1000) for 5 min at RT, and washed three times for 5 min each with PBS. After treatment with a fluorescence quenching agent, the specimens were sealed and photographed under fluorescence and laser confocal microscope.

The images were analyzed using Image Pro Plus 6.0 software (Media Cybernetics, Rockville, MD).

Cell proliferation curve

Cell viability and number was detected using a cell counting chamber. Wild type mESCs and *Clock* knockout mESCs were transferred to 12-well and 6-well plates (Corning Inc., Corning, New York, USA) at 1×10^4 cells/well and cultured at 6 different time points (24, 48, 72, 96, 120 and 144 h). At the indicated time, cells were digested with 0.05% trypsin, and 20 µL of the resuspended cells were collected in the cell counting chamber (Nexcelom) and counted using the Cellometer Mini software.

Induction of biological rhythms with horse serum shock

Briefly, we induced the biological rhythms of wild type mESCs and *Clock* knockout mESCs after spontaneous differentiation with horse serum shock as previously described (Balsalobre et al., 1998; Xiang et al., 2012). The wild type mESCs and *Clock* knockout mESCs were cultured in the mESC maintenance medium and differentiated spontaneously for 15 days in the differentiation medium. The medium was removed and cells were washed with 1 × PBS, followed by addition of the differentiation medium without FBS for 24 h. The medium was removed and cells were washed with 1 × PBS. A new differentiation medium containing 50% horse serum was added to the cells. After incubation for 2 h, the medium was removed and washed with 1 × PBS, and replaced with the original differentiation medium.

Annexin V and PI assay

The FITC Annexin V and Propidium Iodide (PI) Apoptosis Detection Kit II (Pharmingen, USA) were used to analyze apoptosis and cell death rate using flow cytometry. The wild type mESCs and *Clock* knockout mESCs were collected and washed twice with cold PBS. The resuspended cells in 100 µL of 1 × binding buffer were transferred to a 1.5 mL culture tube, and 5 µL of FITC Annexin V and PI were added to each tube. After gentle vortexing, the cells were incubated for 15 min at room temperature in the dark. Finally, 400 µL of the 1 × binding buffer was added to each tube, and analyzed by flow cytometry in 1 h.

Statistical analysis

All the data were expressed as means ± SD. The statistical significance of differences between the experimental groups was determined using one-way ANOVA by Student's t-test (SPSS 11.0, SPSS Inc., Cary, NC). Probability values (p) of less than 0.05 were considered statistically significant.

ACKNOWLEDGEMENTS

This work was supported by the National Science Foundation Fostering Talents in Basic Research of China No. J1210041 (RZ.Q.); the National Natural Science Foundation of China [No. 81322003 (N.S.), No.31571527 (N.S.), No. 81570771 (RZ.Q.)]; Recruitment Program of Global Experts of the Organization Department of the Central Committee of the CPC (N.S.); Science and Technology Commission of Shanghai Municipality (No. 13JC1401704) (N.S.); and Shanghai Key Laboratory of Clinical Geriatric Medicine No. 13dz2260 700 (C.L).

COMPLIANCE WITH ETHICS GUIDELINES

Chao Lu, Yang Yang, Ran Zhao, Bingxuan Hua, Chen Xu, Zuoqin Yan, Ning Sun and Ruizhe Qian declare that they have no conflict of interest.

All institutional and national guidelines for the care and use of laboratory animals were followed.

REFERENCES

Balsalobre A, Damiola F, Schibler U (1998) A serum shock induces circadian gene expression in mammalian tissue culture cells. Cell 93:929–937

Borgs L, Beukelaers P, Vandenbosch R, Nguyen L, Moonen G, Maquet P, Albrecht U, Belachew S, Malgrange B (2009) Period 2 regulates neural stem/progenitor cell proliferation in the adult hippocampus. BMC Neurosci 10:30

Cardinal-Aucoin M, Steel CG (2016) Circadian control of prothoracicotropic hormone release in an adult insect and the induction of its rhythmicity by light cues. Comp Biochem Physiol A Mol Integr Physiol 201:46–52

Casanova-Acebes M, Pitaval C, Weiss LA, Nombela-Arrieta C, Chevre R, Noelia AG, Kunisaki Y, Zhang D, van Rooijen N, Silberstein LE et al (2013) Rhythmic modulation of the hematopoietic niche through neutrophil clearance. Cell 153:1025–1035

Dashti HS, Aslibekyan S, Scheer FA, Smith CE, Lamon-Fava S, Jacques P, Lai CQ, Tucker KL, Arnett DK, Ordovas JM (2016) Clock genes explain a large proportion of phenotypic variance in systolic blood pressure and this control is not modified by environmental temperature. Am J Hypertens 29:132–140

Debruyne JP (2008) Oscillating perceptions: the ups and downs of the CLOCK protein in the mouse circadian system. J Genet 87:437–446

DeBruyne JP, Weaver DR, Reppert SM (2007a) CLOCK and NPAS2 have overlapping roles in the suprachiasmatic circadian clock. Nat Neurosci 10:543–545

DeBruyne JP, Weaver DR, Reppert SM (2007b) Peripheral circadian oscillators require CLOCK. Curr Biol 17:R538–539

Dubrovsky YV, Samsa WE, Kondratov RV (2010) Deficiency of circadian protein CLOCK reduces lifespan and increases age-related cataract development in mice. Aging (Albany NY) 2:936–944

Dunlap JC (1999) Molecular bases for circadian clocks. Cell 96:271–290

Fu L, Pelicano H, Liu J, Huang P, Lee C (2002) The circadian gene Period2 plays an important role in tumor suppression and DNA damage response in vivo. Cell 111:41–50

Gachon F, Nagoshi E, Brown SA, Ripperger J, Schibler U (2004) The mammalian circadian timing system: from gene expression to physiology. Chromosoma 113:103–112

Gaddameedhi S, Selby CP, Kemp MG, Ye R, Sancar A (2015) The circadian clock controls sunburn apoptosis and erythema in mouse skin. J Invest Dermatol 135:1119–1127

Hu K, Van Someren EJ, Shea SA, Scheer FA (2009) Reduction of scale invariance of activity fluctuations with aging and Alzheimer's disease: Involvement of the circadian pacemaker. Proc Natl Acad Sci USA 106:2490–2494

Janich P, Toufighi K, Solanas G, Luis NM, Minkwitz S, Serrano L, Lehner B, Benitah SA (2013) Human epidermal stem cell function is regulated by circadian oscillations. Cell Stem Cell 13:745–753

Johnson MH, Lim A, Fernando D, Day ML (2002) Circadian clockwork genes are expressed in the reproductive tract and conceptus of the early pregnant mouse. Reprod Biomed Online 4:140–145

Johnston JD, Ordovas JM (2016) Circadian rhythms, metabolism, and chrononutrition in rodents and humans. Adv Nutr 7:399–406

Karpowicz P, Zhang Y, Hogenesch JB, Emery P, Perrimon N (2013) The circadian clock gates the intestinal stem cell regenerative state. Cell Rep 3:996–1004

King DP, Zhao Y, Sangoram AM, Wilsbacher LD, Tanaka M, Antoch MP, Steeves TD, Vitaterna MH, Kornhauser JM, Lowrey PL et al (1997) Positional cloning of the mouse circadian clock gene. Cell 89:641–653

Koike N, Yoo SH, Huang HC, Kumar V, Lee C, Kim TK, Takahashi JS (2012) Transcriptional architecture and chromatin landscape of the core circadian clock in mammals. Science 338:349–354

Kovanen L, Saarikoski ST, Aromaa A, Lonnqvist J, Partonen T (2010) ARNTL (BMAL1) and NPAS2 gene variants contribute to fertility and seasonality. PLoS One 5:e10007

Li R, Cheng S, Wang Z (2015) Circadian clock gene plays a key role on ovarian cycle and spontaneous abortion. Cell Physiol Biochem 37:911–920

Liang X, Cheng S, Jiang X, He X, Wang Y, Jiang Z, Hou W, Li S, Liu Y, Wang Z (2013) The noncircadian function of the circadian Clock gene in the regulation of male fertility. J Biol Rhythms 28:208–217

Lopez-Minguez J, Gomez-Abellan P, Garaulet M (2016) Circadian rhythms, food timing and obesity. Proc Nutr Soc. doi:10.1017/S0029665116000628

Lowrey PL, Takahashi JS (2004) Mammalian circadian biology: elucidating genome-wide levels of temporal organization. Annu Rev Genomics Hum Genet 5:407–441

McAlpine CS, Swirski FK (2016) Circadian influence on metabolism and inflammation in atherosclerosis. Circ Res 119:131–141

Mizuno S, Dinh TT, Kato K, Mizuno-Iijima S, Tanimoto Y, Daitoku Y, Hoshino Y, Ikawa M, Takahashi S, Sugiyama F et al (2014) Simple generation of albino C57BL/6J mice with G291T mutation in the tyrosinase gene by the CRISPR/Cas9 system. Mamm Genome 25:327–334

Mukherjee S, Coque L, Cao JL, Kumar J, Chakravarty S, Asaithamby A, Graham A, Gordon E, Enwright JF III, DiLeone RJ et al (2010) Knockdown of Clock in the ventral tegmental area through RNA interference results in a mixed state of mania and depression-like behavior. Biol Psychiatry 68:503–511

Nelakanti RV, Kooreman NG, Wu JC (2015) Teratoma formation: a tool for monitoring pluripotency in stem cell research. Curr Protoc Stem Cell Biol 32(4a.8):1–17

Nitschke S, Cortleven A, Iven T, Feussner I, Havaux M, Riefler M, Schmulling T (2016) Circadian stress regimes affect the circadian clock and cause jasmonic acid-dependent cell death in cytokinin-deficient arabidopsis plants. Plant Cell. doi:10.1105/tpc.16.00016

Peyric E, Moore HA, Whitmore D (2013) Circadian clock regulation of the cell cycle in the zebrafish intestine. PLoS One 8:e73209

Tobback J, Vuerinckx K, Boerjan B, Huybrechts R (2012) RNA interference mortality points to noncircadian functions for the clock gene in the desert locust Schistocerca gregaria. Insect Mol Biol 21:369–381

Tracey CJ, Pan X, Catterson JH, Harmar AJ, Hussain MM, Hartley PS (2012) Diurnal expression of the thrombopoietin gene is regulated by CLOCK. J Thromb Haemost 10:662–669

Witting W, Kwa IH, Eikelenboom P, Mirmiran M, Swaab DF (1990) Alterations in the circadian rest-activity rhythm in aging and Alzheimer's disease. Biol Psychiatry 27:563–572

Wood PA, Yang X, Taber A, Oh EY, Ansell C, Ayers SE, Al-Assaad Z, Carnevale K, Berger FG, Pena MM et al (2008) Period 2 mutation accelerates ApcMin/+ tumorigenesis. Mol Cancer Res 6:1786–1793

Wyse CA, Coogan AN (2010) Impact of aging on diurnal expression patterns of CLOCK and BMAL1 in the mouse brain. Brain Res 1337:21–31

Xiang S, Mao L, Duplessis T, Yuan L, Dauchy R, Dauchy E, Blask DE, Frasch T, Hill SM (2012) Oscillation of clock and clock controlled genes induced by serum shock in human breast epithelial and breast cancer cells: regulation by melatonin. Breast Cancer (Auckl) 6:137–150

Xue T, Song C, Wang Q, Wang Y, Chen G (2016) Investigations of the CLOCK and BMAL1 proteins binding to DNA: a molecular dynamics simulation study. PLoS One 11:e0155105

Yagita K, Horie K, Koinuma S, Nakamura W, Yamanaka I, Urasaki A, Shigeyoshi Y, Kawakami K, Shimada S, Takeda J et al (2010) Development of the circadian oscillator during differentiation of mouse embryonic stem cells in vitro. Proc Natl Acad Sci USA 107:3846–3851

Yang G, Chen L, Grant GR, Paschos G, Song WL, Musiek ES, Lee V, McLoughlin SC, Grosser T, Cotsarelis G et al (2016) Timing of expression of the core clock gene Bmal1 influences its effects on aging and survival. Sci Transl Med 8:324316

Yee JK (2016) Off-target effects of engineered nucleases. FEBS J. doi:10.1111/febs.13760

7

Molecules and mechanisms controlling the active DNA demethylation of the mammalian zygotic genome

Jun-Yu Ma[1], Teng Zhang[1,2], Wei Shen[1], Heide Schatten[3], Qing Yuan Sun[1,2]✉

[1] College of Animal Science and Technology, Qingdao Agricultural University, Qingdao 266109, China
[2] State Key Laboratory of Reproductive Biology, Institute of Zoology, Chinese Academy of Sciences, Beijing 100101, China
[3] Department of Veterinary Pathobiology, University of Missouri, Columbia, MO 65211, USA
✉ Correspondence: sunqy@ioz.ac.cn (Q. Y. Sun)

ABSTRACT

The active DNA demethylation in early embryos is essential for subsequent development. Although the zygotic genome is globally demethylated, the DNA methylation of imprinted regions, part of repeat sequences and some gamete-specific regions are maintained. Recent evidence has shown that multiple proteins and biological pathways participate in the regulation of active DNA demethylation, such as TET proteins, DNA repair pathways and DNA methyltransferases. Here we review the recent understanding regarding proteins associated with active DNA demethylation and the regulatory networks controlling the active DNA demethylation in early embryos.

KEYWORDS active DNA demethylation, zygote, 5-hmC, 5-mC, preimplantation embryo, TET proteins

INTRODUCTION

Prior to fertilization, both the oocyte and spermatozoon have formed a gamete-specific DNA methylation pattern and specific genomic imprinting (Kobayashi et al., 2012; Smith et al., 2012). For the spermatozoon, 80%–90% of low density CpG sites are methylated whereas only 50% are methylated in the oocyte (Kobayashi et al., 2012). For the moderate or high density CpG sites, both oocyte and sperm genomes are hypomethylated. At the transposable element DNA regions, the sperm genome is highly methylated but the oocyte genome is only moderately methylated (Kobayashi et al., 2012). Reduced representation bisulfite sequencing result showed that the number of oocyte-specific methylated DNA regions was 376 and that of the spermatozoon was 4894 (Smith et al., 2012). In addition to the imprinting control regions (ICRs) and gamete-specific methylated CpG sites, whole-genome shotgun bisulfite sequencing data of the oocyte and spermatozoon also showed that more than 1600 CpG islands (CGIs) were differentially methylated between oocytes and spermatozoa (Kobayashi et al., 2012).

After a spermatozoon fertilizes the oocyte, the protamines on the sperm chromatin are replaced by histones (van Meel and Pearson, 1979) before paternal pronucleus (PPN) and maternal pronucleus (MPN) formation. During subsequent early embryo development, the zygotic genome is firstly demethylated from the zygote to morula stage and the genomic DNA is then *de novo* methylated at the blastocyst stage (Ma et al., 2012; Santos et al., 2002; Smith et al., 2012) (Fig. 1). During demethylation of the zygotic genome, different demethylation mechanisms are utilized in maternal or paternal genomes. The paternal genome is actively demethylated (DNA replication independent and mediated by enzymes) whereas the maternal genome is mainly passively demethylated (diluted through DNA replication) (Inoue and Zhang, 2011; Rougier et al., 1998). For the active DNA demethylation, 5-methylcytosines (5-mCs) can be oxidized by the tet methylcytosine dioxygenase 3 (TET3) into 5-hydroxymethylcytosines (5-hmCs) (Iqbal et al., 2011; Ruzov et al., 2011). Although the maternal and paternal genomic CGIs undergo global DNA demethylation, by comparing the DNA methylation profiles of gametes with those of the blastocyst or inner cell mass (ICM) cells, evidence has shown that about half of the gamete-specific DNA methylation patterns is partially maintained during preimplantation embryo development (Borgel et al., 2010; Kobayashi et al., 2012; Smallwood et al., 2011).

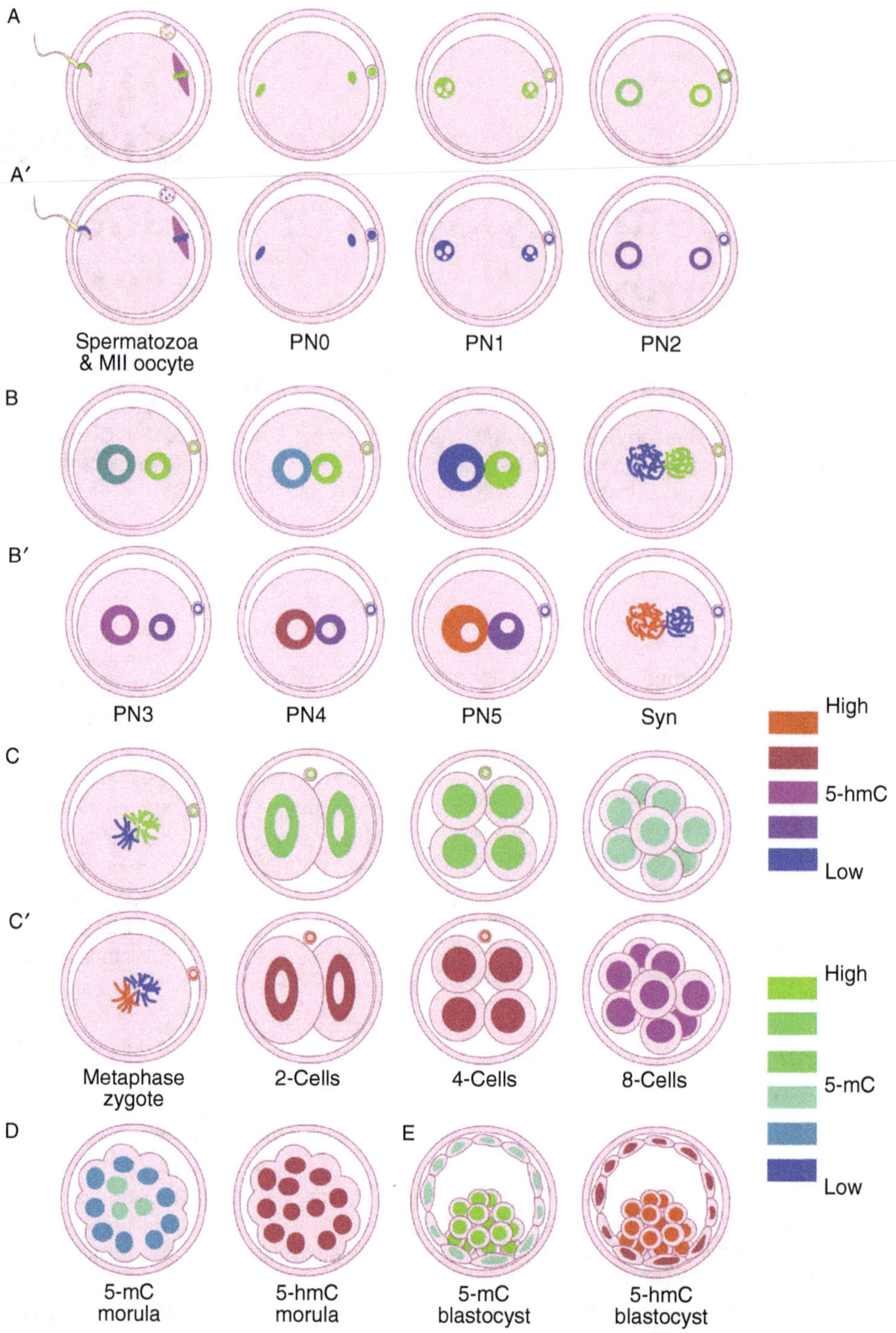

Figure 1. DNA methylation dynamics in mouse preimplantation embryos. (A, B, C, D left and E left), dynamics of 5-mCs in oocyte and preimplantation embryos; (A′, B′, C′, D right and E right), dynamics of 5-hmCs in oocyte and preimplantation embryos.

Recent studies have revealed more details about the mechanisms underlying active DNA demethylation and the essential roles of zygotic genome reprogramming. Here we mainly review details of zygotic genome DNA demethylation and the molecules controlling active DNA demethylation during mammalian early embryo development.

TET PROTEINS MEDIATE ACTIVE DNA DEMETHYLATION

There are three TET proteins exist in mouse, Tet1, Tet2 and Tet3. All the TET proteins contain a cysteine-rich region and a double-stranded β-helix fold at their C-terminals taking functions as the catalytic domain. The Tet1 and Tet3 also contain a CXXC region, which can bind to the genomic CpG islands (Tan and Shi, 2012; Zhao and Chen, 2013). It was firstly discovered in Purkinje neurons and the brain that TET proteins mediate active DNA demethylation (Kriaucionis and Heintz, 2009; Tahiliani et al., 2009). TET proteins were shown to be partners of myeloid/lymphoid leukemia protein (MLL) protein, and oxidized 5-mC to 5-hmC depending on 2-oxoglutarate and Fe(II) and Vitamin C (Blaschke et al., 2013; Tahiliani et al., 2009). Further studies showed that the 5-hmCs could be further transformed into 5-formylcytosines (5-fCs) and 5-carboxylcytosines (5-caCs) by the TET proteins, and that the 5-fCs and 5-caCs could be replaced by cytosines through the base excision repair (BER) pathway (He et al., 2011; Ito et al., 2011; Song et al., 2013). The 5-hmC concentration in mammalian genomic nucleotide bases was less than 1% (Kriaucionis and Heintz, 2009), and traditional bisulfite sequencing methods could not distinguish 5-hmCs from 5-mCs (Huang et al., 2010), which increased the difficulty of investigating the biological functions of 5-hmCs. By using the biotin modified 5-hmC DNA fragments enrichment sequencing and immunofluorescence labeling, it was shown that the 5-hmCs were widely distributed in tissues and cultured cells (Ruzov et al., 2011; Song et al., 2011). The 5-hmC genome localization data of mouse embryonic stem cells (ESCs) showed that the 5-hmCs were mainly enriched at the gene body regions and formed peaks in the vicinity of transcription start sites (TSSs) (Williams et al., 2011). Data showed that 58% of 5-hmCs were identified at the gene body regions, and 6% at the promoter regions (Wu et al., 2011). Referring to the genome-wide ChIP-Seq data of chromatin code proteins, most 5-hmCs highly enriched promoters showed intensive histone H3 trimethyl Lys4 (H3K4me3) or H3K4me3/histone H3 trimethyl Lys27 (H3K27me3) signals, which indicated the 5-hmC highly enriched promoters were mostly corresponding to the active or the poised genes (Ficz et al., 2011; Pastor et al., 2011).

ACTIVE DNA DEMETHYLATION IN EARLY EMBRYOS

Active DNA demethylation in preimplantation embryos was firstly discovered in the mouse zygotic PPN (Mayer et al., 2000; Oswald et al., 2000). Immunofluorescence labeling results showed that the signals of PPN 5-mCs decreased firstly at the PN2 stage, and reached the lowest level at the PN4-PN5 stage. By subsequent dilution of the 5-mCs of paternal and maternal genomes by DNA replication-dependent passive DNA demethylation, the total genomic DNA 5-mCs reached the lowest level at the morula stage. At the blastocyst stage, the genomic DNA was remethylated (Santos et al., 2002).

The 5-hmC was firstly detected by immunofluorescence staining at both paternal and maternal pronuclei in the PN1 zygote. With zygote development, the 5-hmCs increased mainly in the PPN, which coincided with the decrease of PPN 5-mCs (Iqbal et al., 2011; Wang et al., 2014). Immunofluorescence labeling data showed that the 5-hmC also existed in the blastomeres of all embryo stages (Ruzov et al., 2011) (Fig. 1). Using chromosome spreading and 5-hmC labeling of blastomeres from the zygote to 8-cell embryo, Inoue and Zhang observed that 5-hmCs mainly localized at the paternal chromosomes and most 5-hmC positive genomic regions were demethylated in a DNA replication-dependent manner (Inoue and Zhang, 2011). However, very recent evidence showed that the 5-hmCs, the 5-fCs and the 5-caCs in early embryos were not just demethylated passively but that most of the oxidized derivatives of 5-mCs were demethylated actively (Wang et al., 2014).

ACTIVE DNA DEMETHYLATION-ASSOCIATED MOLECULES IN EARLY EMBRYOS

The DNA methylation modification proteins could be classified into three categories: the DNA methyltransferase enzymes like DNA *de novo* methylation enzymes Dnmt3a, Dnmt3b and Dnmt3l, and the DNA methylation maintaining enzyme Dnmt1; the proteins that function in the active DNA demethylation pathways, and the unclassified factors such as the metabolite folic acid (Pufulete et al., 2005) and environment chemical biphenol A (Singh and Li, 2012). These molecules regulate the genomic DNA demethylation by coordination or by taking actions alone. However, the mechanisms by which they target the genomic-specific regions, regulate specific gene expression and participate in different developmental events are still unresolved. Here we summarize the functions of these proteins in preimplantation embryos. The expression levels and cellular localization of the main DNA demethylation associated proteins are displayed in Fig. 2.

Roles of TET proteins in zygotic genome reprogramming

During early embryo development, all three TET proteins are expressed. QRT-PCR results showed that *Tet1* was highly expressed at the morula stage, *Tet2* mRNAs existed from the zygote stage through the blastocyst stage, and *Tet3* mRNAs mainly existed at the zygote stage. In mouse embryonic stem cells (ESCs), both Tet1 and Tet2 could bind to the *Nanog* gene promoter to catalyze 5-mC oxidation (Costa et al., 2013). Meanwhile, Gao et al. proved that *Tet1* could be used to replace the *Oct4* of Yamanaka factors (*Oct4*, *Sox2*, *Klf4* and *cMyc*, OSKM) to induce pluripotency in somatic cells (Gao et al., 2013). Recent data also showed that the OySyNyK [highly efficient modified factors of *Oct4*, *Sox2* and *Nanog*, which could induce the expression of Oct4-GFP of mouse embryonic fibroblast (MEF) as early as 24 h] increased *Tet1* expression at an early stage of somatic cell reprogramming (Zhu et al., 2014). A previous study

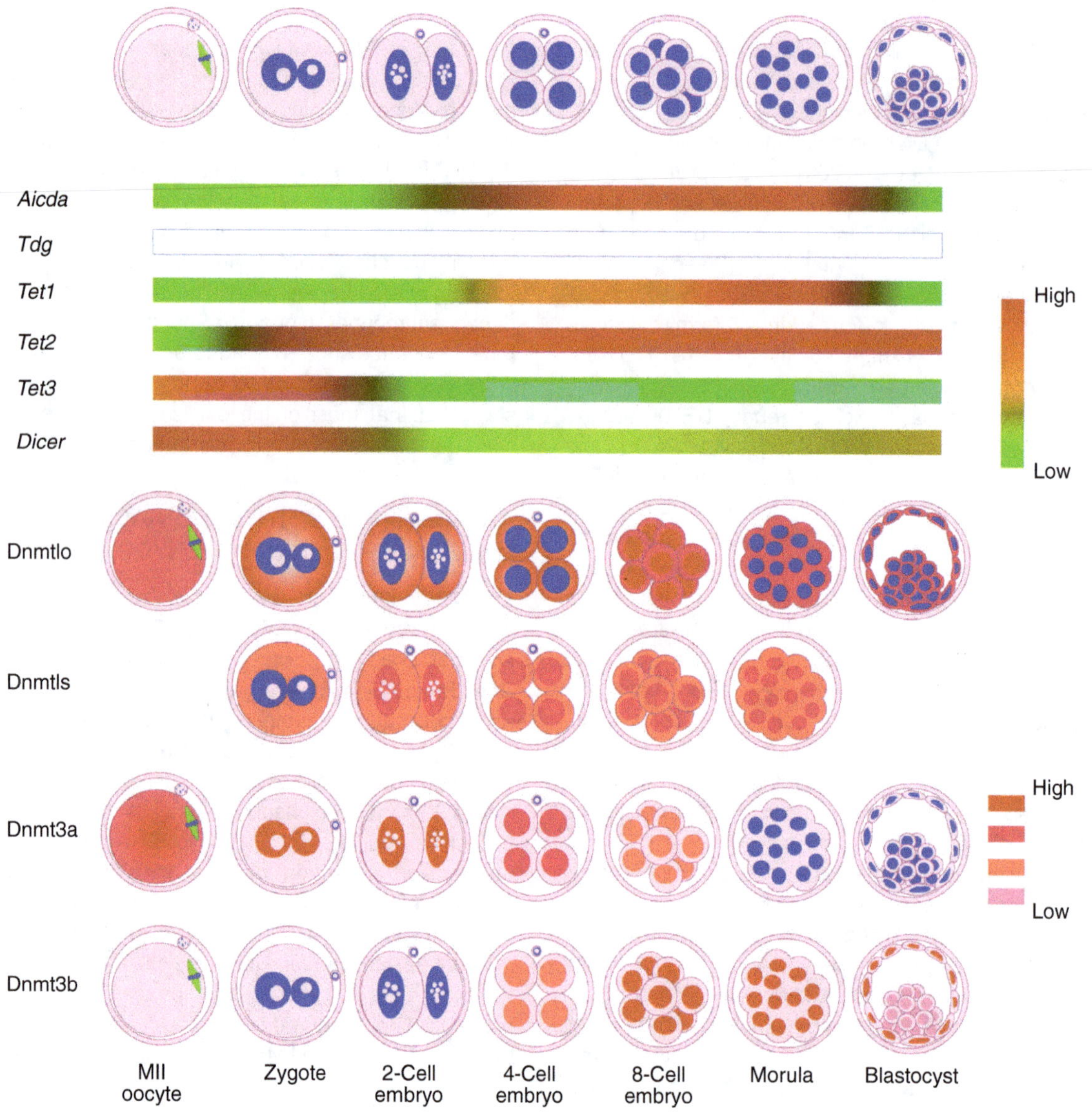

Figure 2. Expression patterns of DNA demethylation-associated genes in mouse preimplantation embryos. Aicda, (Kang et al., 2014); Tet1, Tet2 and Tet3, (Yu et al., 2013); Dicer1, (Murchison et al., 2007); Dnmt1o and Dnmt1 s, (Cirio et al., 2008; Hirasawa et al., 2008; Howell et al., 2001; Kurihara et al., 2008); Dnmt3a and Dnmt3b, (Hirasawa et al., 2008). Gene symbol: italic, mRNAs; bold regular, protein localization.

showed that the expression of *Tet2* but not *Tet1* or *Tet3* was induced at the early stage of MEF reprogramming using OSKM and that the Tet2 protein could target the genomic sites of *Nanog* and *Esrrb* (Doege et al., 2012). For the human, evidence showed that TET2 but not TET1 or TET3 was essential for somatic cell reprogramming (Wang et al., 2013). By depletion of Tet1 and Tet2 in ESCs, it was shown that Tet1 and Tet2 indeed played distinct roles in DNA demethylation. Tet1 mainly targeted the transcription start sites whereas Tet2 mainly targeted the gene bodies (Huang et al., 2014). All the results from ESCs and iPSCs indicate that both Tet1 and Tet2 are associated with maintaining pluripotency of early embryonic cells.

Tet3 was mainly expressed in the zygote and located at the PPN (Nakamura et al., 2012). After knocking down Tet3 expression in mouse zygotes using siRNAs, the demethylation of PPN was diminished (Wossidlo et al., 2011), but the transcription activity showed no significant difference between Tet3 knock-down zygotes and normal zygotes (Inoue et al., 2012). In addition, transcript levels of transposable elements like *LINE1*, *ERVL* as well as the major satellite showed no obvious changes in zygotes or 2-cell embryos after Tet3 knock-down (Inoue et al., 2012), indicating that the global transcription of the zygote might not be affected by PPN active demethylation. In the Tet3-deficient zygotes, the activation of paternal *Oct4* gene was delayed, the reprogramming ability of zygotes was decreased, and the developmental failure of offspring increased (Gu et al., 2011), indicating that the epigenetic reprogramming induced by Tet3 at the zygote stage was important for further embryo development.

The Dppa3 (developmental pluripotency associated 3, also named PGC7 or Stella), is essential for the transition from the Non-Surrounded Nucleolus (NSN) type oocytes to the Surrounded Nucleolus (SN) type oocytes (Liu et al., 2012; Ma et al., 2013). When Dppa3 was depleted in the oocytes, the fertilized embryos could not develop beyond the 4-cell embryo stage (Bortvin et al., 2004). Evidence showed that the Dppa3 was mainly recruited to the MPN in zygote by the histone H3K9me2 to prevent the demethylation of MPN 5-mCs, when the zygotes were microinjected with the mRNAs of lysine (K)-specific demethylase 3A (Kdm3a, also known as Jhdm2a) which specially demethylates the H3K9me1 and H3K9me2, the levels of MPN H3K9me2 and Dppa3 decreased and the 5-hmCs level in MPN were increased (Nakamura et al., 2012). When Dppa3 was depleted, the 5-mCs of MPN could be oxidized to 5-hmCs by the TET proteins (Wossidlo et al., 2011). It was also shown that the DNA methylation levels of maternal imprinted genes like *Peg1*, *Peg3* and *Peg10*, paternal imprinted genes like *H19* and *Rasgrf1* and the transposable element *IAP* could not be maintained in the Dppa3-depleted zygote, whereas the DNA methylation levels of maternal imprinted genes *Snrpn* and *Peg5*, paternal imprinted genes *Meg3* and *LINE1* showed no obvious change when Dppa3 was depleted (Nakamura et al., 2007).

In addition to Dppa3, Tet3 protein localization is also controlled by the O-linked β-GlcNAc (O-GlcNAc) transferase (Ogt) and activation-induced cytidine deaminase (Aicda, also known as Aid). Tet3, but not Tet1 or Tet2, would be transferred to the cytoplasm when O-GlcNAcylated by Ogt (Zhang et al., 2014). When co-expressed with Aicda, the nuclear TET proteins would be translocated to the cytoplasm (Arioka et al., 2012).

DNA methyltransferases

In preimplantation embryos, there are two isoforms of Dnmt1. The oocyte-specific Dnmt1o is mainly located at the cytoplasm of preimplantation embryonic blastomeres and enters the nucleus only at the 8-cell embryo stage (Howell et al., 2001). The oocyte-inherited somatic form Dnmt1s mainly existed at the zygote and 2-cell embryo stage, and at the 2-cell embryo stage the zygotic-originated Dnmt1s were expressed (Cirio et al., 2008). Dnmt1s located to the cytoplasm at the zygote stage and entered the nucleus in subsequent preimplantation embryo stages (Cirio et al., 2008) (Fig. 2). It was shown that maternal Dnmt1 was essential for the maintenance of genomic imprinting and the DNA methylation levels of transposable element *IAP* (Gaudet et al., 2004; Howell et al., 2001), *H19, Snrpn* and *Peg3* (Howell et al., 2001).

Both Dnmt3a and Dnmt3b are responsible for DNA *de novo* methylation, however, the target regions of Dnmt3a and Dnmt3b are not fully overlapping. The methylation of pericentric satellite DNA is controlled by Dnmt3b (Okano et al., 1999). The DNA methylation of genomic CGIs are mainly controlled by Dnmt3a with its parter Dnmt3l (Smallwood et al., 2011). The establishment of DNA methylation and its maintenance are regulated by the histone modifications (Kelsey and Feil, 2013). For example, the chromatin binding pattern of Dnmt3a showed negative correlation with that of the H3K4me3 (Smallwood et al., 2011). In addition to DNA *de novo* methylation activity, new evidence also showed that Dnmt3a and Dnmt3b could transform the 5-hmC to cytosines without the BER pathway (Chen et al., 2012), indicating the possible dual functions of Dnmt3a and Dnmt3b in preimplantation embryos.

DNA damage repair associated proteins

During preimplantation embryo development, DNA double-strand breaks (DSB) occurred spontaneously at the zygotic PPN and blastomeres of 4-cell embryo, 8-cell embryo, morula and blastocyst (Ziegler-Birling et al., 2009). At the zygote stage, DNA DSBs coincided with the DNA replication of PPN and MPN. The DNA DSB marker, γH2A.X foci, appeared earlier and more abundant in PPN than in MPN (Derijck et al., 2008; Wossidlo et al., 2010). Although there was still no evidence proving the correlation between DNA DSBs and active DNA demethylation, chromosome spread results showed that the DNA DSB-induced sister chromatid exchange occurred in early embryos which might dilute the 5-mC on a single chromatid (Inoue and Zhang, 2011).

In addition to γH2A.X-marked DNA DSBs, proteins associated with BER and DNA single strand breaks (SSBs), such as the Parp1 [poly (ADP-ribose) polymerase family, member 1] and Xrcc1 (X-ray repair complementing defective repair in Chinese hamster cells 1), were also recruited to the paternal nucleus DNA (Hajkova et al., 2010). When the BER pathway was inhibited by apurinic/apyrimidinic endonuclease 1 (APE1) inhibitor (CRT0044876) or poly (ADP-ribose) polymerase family protein (PARP) inhibitor (3-aminobenzamide or ABT-888), the DNA demethylation of zygotic PPN became reduced (Hajkova et al., 2010).

In the BER pathway, the cytidine deaminase Aicda and thymine DNA glycosylase (Tdg) can recognize and excise the damaged bases. The Tdg could excise the 5-caC produced by TET proteins and the 5-hmUs which could be produced by deamination of 5-hmC mediated by Aicda (Ma et al., 2012). Tdg and Aicda play important roles in regulating DNA demethylation, however, both the Tdg and Aicda proteins are not detected in the mouse zygote (Hajkova et al., 2010). Tdg null embryos died at about embryonic day

10.5–12.5. Data from Tdg null MEFs showed that the DNA methylation levels increased in CpG islands. In addition, the histone modification H3K4me2 decreased whereas H3K27me3 and H3K9me3 increased in Tdg null MEFs (Cortazar et al., 2011).

The damage-specific DNA binding protein 1 (Ddb1) is the subunit of UV-DNA damaged DNA-binding protein complex and also the component of CUL4 complex which ubiquitinates the histones at UV-DNA damaged sites (Chen et al., 2001; Lan et al., 2012). When Ddb1 was depleted in the zygote, the Tet3-mediated 5-mC hydroxylation was blocked and the 5-mC level at PPN was maintained (Yu et al., 2013). All the above results indicate that the DNA repair pathway not only exerts functions directly on the active DNA demethylation but also maintains the progression of active DNA demethylation and repairs the active DNA demethylation-induced DNA damages.

Other DNA demethylation-associated proteins or molecules

Dicer1 is a critical enzyme controlling the synthesis of miRNAs. In the Dicer1 null cells, the decrease of *miR-290* cluster miRNAs increased the expression of retinoblastoma-like 2 protein (Rbl2) which suppressed the expression of DNMTs (Benetti et al., 2008). RT-PCR data showed that the Dicer1 mRNAs mainly existed in fully grown oocytes and sharply decreased at the 2-cell embryo stage, indicating that Dicer1 exerted functions as a maternal effector. When conditionally knocking out of Dicer1 in growing oocytes, the Dicer1 null oocytes were mainly arrested at metaphase of the first meiosis (MI) stage (Murchison et al., 2007). Knocking out of Dicer1 induced the high expression of mouse transposon (*MT*) and *SINE* elements *B1* and *B2* (Murchison et al., 2007), which may be induced by abnormal DNA methylation pattern in Dicer1 null oocytes (Jeong and Lee, 2005). There is still no information about the effect of Dicer1 on early embryo DNA methylation dynamics, however, knocking down of Dicer1 could reduce the protein level of pluripotency factor Oct4 (Cui et al., 2007), indicating the pivotal roles of Dicer1 in preimplantation embryo development.

Other active DNA demethylation-associated factors include Elp3 (Okada et al., 2010), Gcm1 and Gcm2 (Hitoshi et al., 2011) and the mechanism details still need to be further analyzed.

ACTIVE DNA DEMETHYLATION IS ESSENTIAL FOR EMBRYO DEVELOPMENT

From the experiments of knockdown or knockout of genes like *Tet3*, it had been shown that the PPN DNA active demethylation was essential for mouse embryo development. As is well known, parthenogenetically activated oocyte-derived embryos mostly develop poorly after implantation and could not develop to term (Surani et al., 1984). However, Kono et al. had produced a full-term parthenogenetic mouse by activation of the MII oocyte which was composed of enucleated MII oocyte cytoplasm and the nucleus from another donor MII oocyte. The donor oocyte was produced by transferring the H19-depleted diplotene oocyte (from one day old mouse) nucleus into a fully grown oocyte, and if the nucleus was from the wild type, parthenogenetic embryos could only live for 14 days (Kono et al., 2004; Kono et al., 1996). In addition, the parthenogenetically derived ESCs could be used to produce live parthenote pups by tetraploid embryo complementation (Chen et al., 2009). These results may indicate that: firstly, paternal DNA demethylation was not essential for embryo development and the active DNA demethylation was critical for the development of mammalian placenta; and secondly, if the genomic imprinting was adaptive, embryos could develop to term without PPN active DNA methylation.

The 5-mC oxidation-mediated active DNA demethylation also occurred in somatic cell nuclear transfer (SCNT) mouse 1-cell embryos (Wossidlo et al., 2011). However, the quality of SCNT-produced embryos was obviously lower than that of *in vitro* fertilization (IVF) or intracytoplasmic sperm injection (ICSI)-induced zygotes, which indicated that genomic reprogramming of the somatic cell nucleus by the MII oocyte might not be complete (Thuan et al., 2010; Yang et al., 2007). Recent evidence showed that the nucleus of ESCs, fetal fibroblast or cumulus cells could be reprogrammed by the cytoplasm of interphase 2-cell embryo blastomeres after cell cycle synchronization, however, the interphase 2-cell embryo cytoplasm only could support the cell cycle synchronized ESC nuclear transfer embryo development to term (Kang et al., 2014). These results showed that the epigenetic code established during the preimplantation stage is essential for embryo development to term, in which active DNA demethylation is critically essential.

SPECIES-SPECIFIC ZYGOTIC DNA DEMETHYLATION

Immunofluorescence data showed that the DNA methylation dynamic patterns of different species are not conserved. For human, monkey, rat and mouse, the 5-mC signals were mostly lost in the PPNs whereas for the sheep, there was no obvious 5-mC signal lost in the PPN. The PPN 5-mC dynamic data of pig, cow, goat and rabbit is still not validated (Ma et al., 2012). Compared with the PPN 5-mC, the data of 5-hmC was not enough to analyze the conservation among mammalian species. Evidence showed that the 5-hmC signals mainly increased in PPNs of cow and rabbit (Wossidlo et al., 2011), whereas the 5-hmC signals existed in both the PPN and MPN of pig zygotes (Lee et al., 2014). TET3-induced DNA demethylation was critical for the maternal-zygote transition (MZT) and the expression level of NANOG in blastocysts of pig, which suggested that the 5-mCs hydroxydation was important for early porcine embryo

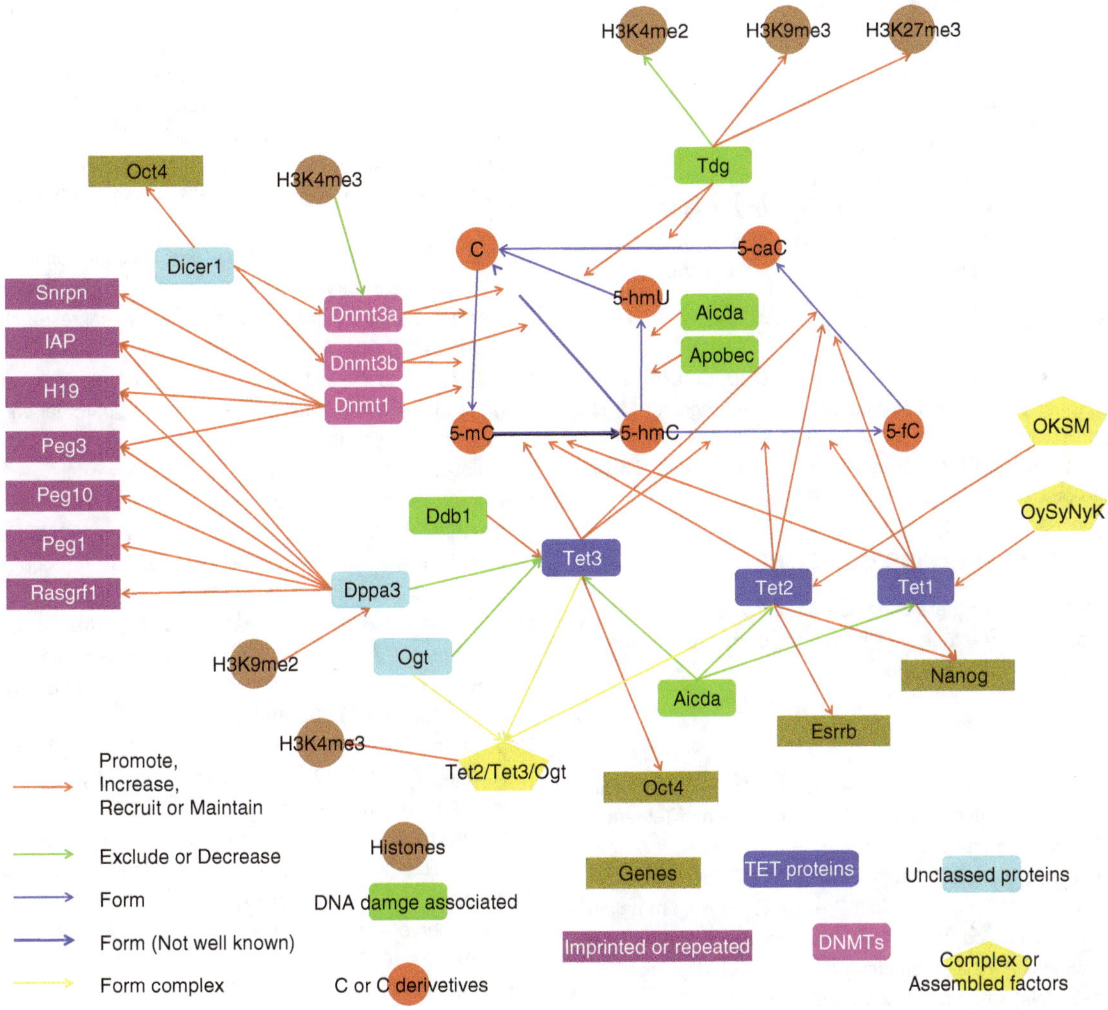

Figure 3. Molecular network regulating DNA demethylation in preimplantation embryos. The 5-mCs are mainly oxidated by TET proteins into 5-hmCs, 5-fCs or 5-caCs. The 5-hmCs could be targeted by Aicda, Apobec, Dnmt3a and Dnmt3b. The 5-caCs could be repaired into Cs (cytosines) by Tdg mediated base excision repair pathway. Most proteins participating directly in the DNA demethylation process could be taken as targets by other factors to regulate the DNA methylation pattern of specific genome regions, such as the localization of TET proteins could be affected by Aicda, Ogt and Dapp3, and the expression of *Tet* genes could be regulated by OSKM factors. In the preimplantation embryos, the DNA methylation imprinted genes and transposable elements *IAP* were mainly maintained by Dnmt1 and the Dapp3, however, the target sites of Dnmt1 and Dapp3 are different (see manuscript). In addition, the histone modifications and miRNAs can also affect or be affected by the active DNA demethylation process.

development (Lee et al., 2014). Although the information about active DNA demethylation among mammals is still not sufficient, due to the conservation of TET proteins in these species, we suspect that the zygotic active DNA demethylation may be ubiquitous in higher mammals.

PERSPECTIVES

The DNA demethylation in early embryos is complex and the DNA methylation patterns of CGIs, non CGI CpG sites, transposable elements, macro and minor satellites, and imprinted sequences are regulated by different proteins (Fig. 3). Although more and more details about active DNA demethylation have been discovered, many questions remain to be answered, including: Which distinct functions of the three TET proteins are critical for preimplantation embryo development? By which mechanisms is Dnmt1 specifially targeted to the imprinted regions and some repeat sequences and by which mechanisms are TET proteins targeted to specific 5-mCs? Whether or not the order of the gene promoter active DNA demethylation is important for embryo development? Are the DNA methylation pattern differences in different blastomeres affecting their developmental fates? What are the key DNA regions for SCNT embryos in reprogramming? How to change DNA methylation pattern of specific region artificially? And so on.

ACKNOWLEDGEMENTS

We apologize to authors whose work could not be cited due to the scope of this review. Q.Y.S. was supported by the National Basic Research Program (973 Program) (Nos. 2012CB944404 and 2011CB944501). We would like to thank the members of the Shen laboratory for their kind and fruitful discussion of the manuscript.

ABBREVIATIONS

BER, base excision repair; CGI, CpG island; DSB, double-strand break; ICM, inner cell mass; ICR, imprinting control regions; MPN, maternal pronucleus; MZT, maternal-zygote transition; PPN, paternal pronucleus; SCNT, somatic cell nuclear transfer; SSB, single-strand break.

COMPLIANCE WITH ETHICS GUIDELINES

The authors declare that there is no conflict of interest. This article does not contain any studies with human or animal subjects performed by the any of the authors.

REFERENCES

Arioka Y, Watanabe A, Saito K, Yamada Y (2012) Activation-induced cytidine deaminase alters the subcellular localization of Tet family proteins. PloS One 7:e45031

Benetti R, Gonzalo S, Jaco I, Munoz P, Gonzalez S, Schoeftner S, Murchison E, Andl T, Chen T, Klatt P et al (2008) A mammalian microRNA cluster controls DNA methylation and telomere recombination via Rbl2-dependent regulation of DNA methyltransferases. Nat struct mol biol 15:998

Blaschke K, Ebata KT, Karimi MM, Zepeda-Martinez JA, Goyal P, Mahapatra S, Tam A, Laird DJ, Hirst M, Rao A et al (2013) Vitamin C induces Tet-dependent DNA demethylation and a blastocyst-like state in ES cells. Nature 500:222–226

Borgel J, Guibert S, Li Y, Chiba H, Schubeler D, Sasaki H, Forne T, Weber M (2010) Targets and dynamics of promoter DNA methylation during early mouse development. Nature genetics 42:1093–1100

Bortvin A, Goodheart M, Liao M, Page DC (2004) Dppa3/Pgc7/stella is a maternal factor and is not required for germ cell specification in mice. BMC dev biol 4:2

Chen X, Zhang Y, Douglas L, Zhou P (2001) UV-damaged DNA-binding proteins are targets of CUL-4A-mediated ubiquitination and degradation. J biol chem 276:48175–48182

Chen Z, Liu Z, Huang J, Amano T, Li C, Cao S, Wu C, Liu B, Zhou L, Carter MG et al (2009) Birth of parthenote mice directly from parthenogenetic embryonic stem cells. Stem Cells 27:2136–2145

Chen CC, Wang KY, Shen CK (2012) The mammalian de novo DNA methyltransferases DNMT3A and DNMT3B are also DNA 5-hydroxymethylcytosine dehydroxymethylases. J biol chem 287:33116–33121

Cirio MC, Ratnam S, Ding F, Reinhart B, Navara C, Chaillet JR (2008) Preimplantation expression of the somatic form of Dnmt1 suggests a role in the inheritance of genomic imprints. BMC dev biol 8:9

Cortazar D, Kunz C, Selfridge J, Lettieri T, Saito Y, MacDougall E, Wirz A, Schuermann D, Jacobs AL, Siegrist F et al (2011) Embryonic lethal phenotype reveals a function of TDG in maintaining epigenetic stability. Nature 470:419–423

Costa Y, Ding J, Theunissen TW, Faiola F, Hore TA, Shliaha PV, Fidalgo M, Saunders A, Lawrence M, Dietmann S et al (2013) NANOG-dependent function of TET1 and TET2 in establishment of pluripotency. Nature 495:370–374

Cui XS, Shen XH, Kim NH (2007) *Dicer1* expression in preimplantation mouse embryos: Involvement of Oct3/4 transcription at the blastocyst stage. Biochemical and biophysical research communications 352:231–236

Derijck A, van der Heijden G, Giele M, Philippens M, de Boer P (2008) DNA double-strand break repair in parental chromatin of mouse zygotes, the first cell cycle as an origin of de novo mutation. Hum mol genet 17:1922–1937

Doege CA, Inoue K, Yamashita T, Rhee DB, Travis S, Fujita R, Guarnieri P, Bhagat G, Vanti WB, Shih A et al (2012) Early-stage epigenetic modification during somatic cell reprogramming by Parp1 and Tet2. Nature 488:652–655

Ficz G, Branco MR, Seisenberger S, Santos F, Krueger F, Hore TA, Marques CJ, Andrews S, Reik W (2011) Dynamic regulation of 5-hydroxymethylcytosine in mouse ES cells and during differentiation. Nature 473:398–402

Gao Y, Chen J, Li K, Wu T, Huang B, Liu W, Kou X, Zhang Y, Huang H, Jiang Y et al (2013) Replacement of Oct4 by Tet1 during iPSC induction reveals an important role of DNA methylation and hydroxymethylation in reprogramming. Cell stem cell 12:453–469

Gaudet F, Rideout WM 3rd, Meissner A, Dausman J, Leonhardt H, Jaenisch R (2004) Dnmt1 expression in pre- and postimplantation embryogenesis and the maintenance of IAP silencing. Mol cell biol 24:1640–1648

Gu TP, Guo F, Yang H, Wu HP, Xu GF, Liu W, Xie ZG, Shi L, He X, Jin SG et al (2011) The role of Tet3 DNA dioxygenase in epigenetic reprogramming by oocytes. Nature 477:606–610

Hajkova P, Jeffries SJ, Lee C, Miller N, Jackson SP, Surani MA (2010) Genome-wide reprogramming in the mouse germ line entails the base excision repair pathway. Science 329:78–82

He YF, Li BZ, Li Z, Liu P, Wang Y, Tang Q, Ding J, Jia Y, Chen Z, Li L et al (2011) Tet-mediated formation of 5-carboxylcytosine and its excision by TDG in mammalian DNA. Science 333:1303–1307

Hirasawa R, Chiba H, Kaneda M, Tajima S, Li E, Jaenisch R, Sasaki H (2008) Maternal and zygotic Dnmt1 are necessary and sufficient for the maintenance of DNA methylation imprints during preimplantation development. Genes dev 22:1607–1616

Hitoshi S, Ishino Y, Kumar A, Jasmine S, Tanaka KF, Kondo T, Kato S, Hosoya T, Hotta Y, Ikenaka K (2011) Mammalian Gcm genes induce Hes5 expression by active DNA demethylation and induce neural stem cells. Nat neurosci 14:957–964

Howell CY, Bestor TH, Ding F, Latham KE, Mertineit C, Trasler JM, Chaillet JR (2001) Genomic imprinting disrupted by a maternal effect mutation in the *Dnmt1* gene. Cell 104:829–838

Huang Y, Pastor WA, Shen Y, Tahiliani M, Liu DR, Rao A (2010) The behaviour of 5-hydroxymethylcytosine in bisulfite sequencing. PloS One 5:e8888

Huang Y, Chavez L, Chang X, Wang X, Pastor WA, Kang J, Zepeda-Martinez JA, Pape UJ, Jacobsen SE, Peters B et al (2014) Distinct roles of the methylcytosine oxidases Tet1 and Tet2 in mouse embryonic stem cells. Proc Natl Acad Sci USA 111:1361–1366

Inoue A, Zhang Y (2011) Replication-dependent loss of 5-hydroxymethylcytosine in mouse preimplantation embryos. Science 334:194

Inoue A, Matoba S, Zhang Y (2012) Transcriptional activation of transposable elements in mouse zygotes is independent of Tet3-mediated 5-methylcytosine oxidation. Cell res 22:1640–1649

Iqbal K, Jin SG, Pfeifer GP, Szabo PE (2011) Reprogramming of the paternal genome upon fertilization involves genome-wide oxidation of 5-methylcytosine. Proc Natl Acad Sci USA 108:3642–3647

Ito S, Shen L, Dai Q, Wu SC, Collins LB, Swenberg JA, He C, Zhang Y (2011) Tet proteins can convert 5-methylcytosine to 5-formylcytosine and 5-carboxylcytosine. Science 333:1300–1303

Jeong KS, Lee S (2005) Estimating the total mouse DNA methylation according to the B1 repetitive elements. Biochem biophys res commun 335:1211–1216

Kang E, Wu G, Ma H, Li Y, Tippner-Hedges R, Tachibana M, Sparman M, Wolf DP, Scholer HR, Mitalipov S (2014) Nuclear reprogramming by interphase cytoplasm of two-cell mouse embryos. Nature 509:101–104

Kelsey G, Feil R (2013) New insights into establishment and maintenance of DNA methylation imprints in mammals. Philos Trans R Soc Lond Ser B, Biol sci 368:20110336

Kobayashi H, Sakurai T, Imai M, Takahashi N, Fukuda A, Yayoi O, Sato S, Nakabayashi K, Hata K, Sotomaru Y et al (2012) Contribution of intragenic DNA methylation in mouse gametic DNA methylomes to establish oocyte-specific heritable marks. PLoS genet 8:e1002440

Kono T, Obata Y, Yoshimzu T, Nakahara T, Carroll J (1996) Epigenetic modifications during oocyte growth correlates with extended parthenogenetic development in the mouse. Nat genet 13:91–94

Kono T, Obata Y, Wu Q, Niwa K, Ono Y, Yamamoto Y, Park ES, Seo JS, Ogawa H (2004) Birth of parthenogenetic mice that can develop to adulthood. Nature 428:860–864

Kriaucionis S, Heintz N (2009) The nuclear DNA base 5-hydroxymethylcytosine is present in Purkinje neurons and the brain. Science 324:929–930

Kurihara Y, Kawamura Y, Uchijima Y, Amamo T, Kobayashi H, Asano T, Kurihara H (2008) Maintenance of genomic methylation patterns during preimplantation development requires the somatic form of DNA methyltransferase 1. Dev biol 313:335–346

Lan L, Nakajima S, Kapetanaki MG, Hsieh CL, Fagerburg M, Thickman K, Rodriguez-Collazo P, Leuba SH, Levine AS, Rapic-Otrin V (2012) Monoubiquitinated histone H2A destabilizes photolesion-containing nucleosomes with concomitant release of UV-damaged DNA-binding protein E3 ligase. J biol chem 287:12036–12049

Lee K, Hamm J, Whitworth K, Spate L, Park KW, Murphy CN, Prather RS (2014) Dynamics of TET family expression in porcine preimplantation embryos is related to zygotic genome activation and required for the maintenance of NANOG. Dev biol 386:86–95

Liu YJ, Nakamura T, Nakano T (2012) Essential role of DPPA3 for chromatin condensation in mouse oocytogenesis. Biol reprod 86:40

Ma JY, Liang XW, Schatten H, Sun QY (2012) Active DNA demethylation in mammalian preimplantation embryos: new insights and new perspectives. Mol hum reprod 18:333–340

Ma JY, Li M, Luo YB, Song S, Tian D, Yang J, Zhang B, Hou Y, Schatten H, Liu Z et al (2013) Maternal factors required for oocyte developmental competence in mice: transcriptome analysis of non-surrounded nucleolus (NSN) and surrounded nucleolus (SN) oocytes. Cell Cycle 12:1928–1938

Mayer W, Niveleau A, Walter J, Fundele R, Haaf T (2000) Demethylation of the zygotic paternal genome. Nature 403:501–502

Murchison EP, Stein P, Xuan Z, Pan H, Zhang MQ, Schultz RM, Hannon GJ (2007) Critical roles for Dicer in the female germline. Genes dev 21:682–693

Nakamura T, Arai Y, Umehara H, Masuhara M, Kimura T, Taniguchi H, Sekimoto T, Ikawa M, Yoneda Y, Okabe M et al (2007) PGC7/Stella protects against DNA demethylation in early embryogenesis. Nat cell biol 9:64–71

Nakamura T, Liu YJ, Nakashima H, Umehara H, Inoue K, Matoba S, Tachibana M, Ogura A, Shinkai Y, Nakano T (2012) PGC7 binds histone H3K9me2 to protect against conversion of 5mC to 5hmC in early embryos. Nature 486:415–419

Okada Y, Yamagata K, Hong K, Wakayama T, Zhang Y (2010) A role for the elongator complex in zygotic paternal genome demethylation. Nature 463:554–558

Okano M, Bell DW, Haber DA, Li E (1999) DNA methyltransferases Dnmt3a and Dnmt3b are essential for de novo methylation and mammalian development. Cell 99:247–257

Oswald J, Engemann S, Lane N, Mayer W, Olek A, Fundele R, Dean W, Reik W, Walter J (2000) Active demethylation of the paternal genome in the mouse zygote. Curr biol 10:475–478

Pastor WA, Pape UJ, Huang Y, Henderson HR, Lister R, Ko M, McLoughlin EM, Brudno Y, Mahapatra S, Kapranov P et al (2011) Genome-wide mapping of 5-hydroxymethylcytosine in embryonic stem cells. Nature 473:394–397

Pufulete M, Al-Ghnaniem R, Khushal A, Appleby P, Harris N, Gout S, Emery PW, Sanders TA (2005) Effect of folic acid supplementation on genomic DNA methylation in patients with colorectal adenoma. Gut 54:648–653

Rougier N, Bourc'his D, Gomes DM, Niveleau A, Plachot M, Paldi A, Viegas-Pequignot E (1998) Chromosome methylation patterns during mammalian preimplantation development. Genes dev 12:2108–2113

Ruzov A, Tsenkina Y, Serio A, Dudnakova T, Fletcher J, Bai Y, Chebotareva T, Pells S, Hannoun Z, Sullivan G et al (2011) Lineage-specific distribution of high levels of genomic 5-hydroxymethylcytosine in mammalian development. Cell res 21:1332–1342

Santos F, Hendrich B, Reik W, Dean W (2002) Dynamic reprogramming of DNA methylation in the early mouse embryo. Dev biol 241:172–182

Singh S, Li SS (2012) Epigenetic effects of environmental chemicals bisphenol a and phthalates. Int j mol sci 13:10143–10153

Smallwood SA, Tomizawa S, Krueger F, Ruf N, Carli N, Segonds-Pichon A, Sato S, Hata K, Andrews SR, Kelsey G (2011) Dynamic CpG island methylation landscape in oocytes and preimplantation embryos. Nat genet 43:811–814

Smith ZD, Chan MM, Mikkelsen TS, Gu H, Gnirke A, Regev A, Meissner A (2012) A unique regulatory phase of DNA methylation in the early mammalian embryo. Nature 484:339–344

Song CX, Szulwach KE, Fu Y, Dai Q, Yi C, Li X, Li Y, Chen CH, Zhang W, Jian X et al (2011) Selective chemical labeling reveals the genome-wide distribution of 5-hydroxymethylcytosine. Nat biotechnol 29:68–72

Song CX, Szulwach KE, Dai Q, Fu Y, Mao SQ, Lin L, Street C, Li Y, Poidevin M, Wu H et al (2013) Genome-wide profiling of 5-formylcytosine reveals its roles in epigenetic priming. Cell 153:678–691

Surani MA, Barton SC, Norris ML (1984) Development of reconstituted mouse eggs suggests imprinting of the genome during gametogenesis. Nature 308:548–550

Tahiliani M, Koh KP, Shen Y, Pastor WA, Bandukwala H, Brudno Y, Agarwal S, Iyer LM, Liu DR, Aravind L et al (2009) Conversion of 5-methylcytosine to 5-hydroxymethylcytosine in mammalian DNA by MLL partner TET1. Science 324:930–935

Tan L, Shi YG (2012) Tet family proteins and 5-hydroxymethylcytosine in development and disease. Development 139:1895–1902

Thuan NV, Kishigami S, Wakayama T (2010) How to improve the success rate of mouse cloning technology. J reprod Dev 56:20–30

van Meel FC, Pearson PL (1979) Replacement of protamine by F1 histone during reactivation of fused human sperm nuclei. Histochemistry 63:329–339

Wang T, Wu H, Li Y, Szulwach KE, Lin L, Li X, Chen IP, Goldlust IS, Chamberlain SJ, Dodd A et al (2013) Subtelomeric hotspots of aberrant 5-hydroxymethylcytosine-mediated epigenetic modifications during reprogramming to pluripotency. Nat cell biol 15:700–711

Wang L, Zhang J, Duan J, Gao X, Zhu W, Lu X, Yang L, Li G, Ci W, Li W et al (2014) Programming and inheritance of parental DNA methylomes in mammals. Cell 157:979–991

Williams K, Christensen J, Pedersen MT, Johansen JV, Cloos PA, Rappsilber J, Helin K (2011) TET1 and hydroxymethylcytosine in transcription and DNA methylation fidelity. Nature 473:343–348

Wossidlo M, Arand J, Sebastiano V, Lepikhov K, Boiani M, Reinhardt R, Scholer H, Walter J (2010) Dynamic link of DNA demethylation, DNA strand breaks and repair in mouse zygotes. EMBO j 29:1877–1888

Wossidlo M, Nakamura T, Lepikhov K, Marques CJ, Zakhartchenko V, Boiani M, Arand J, Nakano T, Reik W, Walter J (2011) 5-Hydroxymethylcytosine in the mammalian zygote is linked with epigenetic reprogramming. Nat commun 2:241

Wu H, D'Alessio AC, Ito S, Wang Z, Cui K, Zhao K, Sun YE, Zhang Y (2011) Genome-wide analysis of 5-hydroxymethylcytosine distribution reveals its dual function in transcriptional regulation in mouse embryonic stem cells. Genes dev 25:679–684

Yang X, Smith SL, Tian XC, Lewin HA, Renard JP, Wakayama T (2007) Nuclear reprogramming of cloned embryos and its implications for therapeutic cloning. Nat genet 39:295–302

Yu C, Zhang YL, Pan WW, Li XM, Wang ZW, Ge ZJ, Zhou JJ, Cang Y, Tong C, Sun QY et al (2013) CRL4 complex regulates mammalian oocyte survival and reprogramming by activation of TET proteins. Science 342:1518–1521

Zhang Q, Liu X, Gao W, Li P, Hou J, Li J, Wong J (2014) Differential regulation of the ten-eleven translocation (TET) family of dioxygenases by O-linked beta-N-acetylglucosamine transferase (OGT). J biol chem 289:5986–5996

Zhao H, Chen T (2013) Tet family of 5-methylcytosine dioxygenases in mammalian development. J Hum genet 58:421–427

Zhu G, Li Y, Zhu F, Wang T, Jin W, Mu W, Lin W, Tan W, Li W, Street RC et al (2014) Coordination of Engineered Factors with TET1/2

Williams K, Christensen J, Pedersen MT, Johansen JV, Cloos PA, Rappsilber J, Helin K (2011) TET1 and hydroxymethylcytosine in transcription and DNA methylation fidelity. Nature 473:343–348

Wossidlo M, Arand J, Sebastiano V, Lepikhov K, Boiani M, Reinhardt R, Scholer H, Walter J (2010) Dynamic link of DNA demethylation, DNA strand breaks and repair in mouse zygotes. EMBO j 29:1877–1888

Wossidlo M, Nakamura T, Lepikhov K, Marques CJ, Zakhartchenko V, Boiani M, Arand J, Nakano T, Reik W, Walter J (2011) 5-Hydroxymethylcytosine in the mammalian zygote is linked with epigenetic reprogramming. Nat commun 2:241

Wu H, D'Alessio AC, Ito S, Wang Z, Cui K, Zhao K, Sun YE, Zhang Y (2011) Genome-wide analysis of 5-hydroxymethylcytosine distribution reveals its dual function in transcriptional regulation in mouse embryonic stem cells. Genes dev 25:679–684

Yang X, Smith SL, Tian XC, Lewin HA, Renard JP, Wakayama T (2007) Nuclear reprogramming of cloned embryos and its implications for therapeutic cloning. Nat genet 39:295–302

Yu C, Zhang YL, Pan WW, Li XM, Wang ZW, Ge ZJ, Zhou JJ, Cang Y, Tong C, Sun QY et al (2013) CRL4 complex regulates mammalian oocyte survival and reprogramming by activation of TET proteins. Science 342:1518–1521

Zhang Q, Liu X, Gao W, Li P, Hou J, Li J, Wong J (2014) Differential regulation of the ten-eleven translocation (TET) family of dioxygenases by O-linked beta-N-acetylglucosamine transferase (OGT). J biol chem 289:5986–5996

Zhao H, Chen T (2013) Tet family of 5-methylcytosine dioxygenases in mammalian development. J Hum genet 58:421–427

Zhu G, Li Y, Zhu F, Wang T, Jin W, Mu W, Lin W, Tan W, Li W, Street RC et al (2014) Coordination of Engineered Factors with TET1/2

8

Secreted miR-34a in astrocytic shedding vesicles enhanced the vulnerability of dopaminergic neurons to neurotoxins by targeting Bcl-2

Susu Mao[1,2], Qi Sun[1,2,3], Hui Xiao[1,2], Chenyu Zhang[1,2]✉, Liang Li[1,2]✉

[1] State Key Laboratory of Pharmaceutical Biotechnology, Nanjing Advanced Institute for Life Sciences (NAILS), Nanjing University School of Life Sciences, Nanjing 210093, China
[2] Jiangsu Engineering Research Center for microRNA Biology and Biotechnology, Nanjing 210093, China
[3] Center of Kidney Disease, 2nd Affiliated Hospital, Nanjing Medical University, Nanjing 210003, China
✉ Correspondence: cyzhang@nju.edu.cn (C. Zhang), lijing@sibs.ac.cn (L. Li)

ABSTRACT

MicroRNAs (miRNAs) are a class of noncoding RNAs that regulates target gene expression at posttranscriptional level, leading to further biological functions. We have demonstrated that microvesicles (MVs) can deliver miRNAs into target cells as a novel way of intercellular communication. It is reported that in central nervous system, glial cells release MVs, which modulate neuronal function in normal condition. To elucidate the potential role of glial MVs in disease, we evaluated the effects of secreted astrocytic MVs on stress condition. Our results demonstrated that after Lipopolysaccharide (LPS) stimulation, astrocytes released shedding vesicles (SVs) that enhanced vulnerability of dopaminergic neurons to neurotoxin. Further investigation showed that increased astrocytic miR-34a in SVs was involved in this progress via targeting anti-apoptotic protein Bcl-2 in dopaminergic neurons. We also found that inhibition of astrocytic miR-34a after LPS stimulation can postpone dopaminergic neuron loss under neurotoxin stress. These data revealed a novel mechanism underlying astrocyte-neuron interaction in disease.

Susu Mao, Qi Sun contributed equally to this work.

KEYWORDS astrocyte, shedding vesicles, miR-34a, dopaminergic neurons, Bcl-2

INTRODUCTION

Parkinson's disease (PD) is characterized by the progressive degeneration of dopamine (DA) neurons in the substantia nigra (SN). Circumstantial evidence has shown that the neuron-glia interaction plays an important role in PD pathogenesis (Furman and Norris, 2014; Lee et al., 2013; Oeckl et al., 2012). It has been suggested that reactive astrocytes may amplify the devastating effects of PD by accelerating the dopaminergic (DA) neuron loss (Saijo et al., 2009).

Microvesicles (MVs) were originally described as cellular debris with limited biological function (Morel et al., 2004). However, increasing amounts of evidence suggest that MV shedding is a constitutive mode of intercellular communication (Al-Nedawi et al., 2009; Gupta and Pulliam, 2014; Hu et al., 2012; Keller et al., 2006; Shantsila et al., 2010). MVs can be divided into two groups—shedding vesicles (SVs) and exosomes—with different discharging processes (Bianco et al., 2009). The release of shedding vesicles is a result of direct budding from the plasma membrane, while exosomes are produced by exocytosis of multivesicular bodies (Zomer et al., 2010). Bianco F et al. showed that in the central nervous system, glial cells can release both type of MVs (Bianco et al., 2009). Further investigation produced evidence that MVs released from microglia stimulate synaptic activity via enhanced sphingolipid metabolism, indicating that microglia MVs may have significant biological functions under normal conditions (Antonucci et al., 2012). Our previous work showed that exosomes can deliver microRNAs (miRNAs) into endothelial cells and then modulate the migration of the cell by repressing the target protein expression (Zhang et al., 2010). In recent years, accumulating evidence also showed that secreted miRNAs in MVs have significant biological functions including effects on cell proliferation, development, differentiation as well as cell death and cancer progression (Hu et al., 2012; Janowska-Wieczorek et al., 2005; Jung and Suh, 2014; Liu et al., 2013; Vlassov et al., 2012; Zhou et al., 2014; Zhu and Fan, 2011). However, there are few studies regarding the function of secreted miRNAs in neuron-glia interaction of the central nervous system either in physiological or disease conditions. Whether secreted miRNAs in astrocytic MVs are involved in PD pathogenesis is still unknown; therefore it is important to investigate the potential role of MVs in astrocyte-neuron interactions in PD.

In the present study, we used both *in vitro* and *in vivo* PD models to evaluate the effects of secreted astrocytic MVs under different conditions on DA neurons survival. We demonstrated that astrocytes under conditions of LPS stress released SVs, which enhanced the vulnerability of DA neurons to neurotoxin. We also found evidence that miR-34a was increased in astrocytic SVs after LPS stimulation and that miR-34a then entered DA cells and repressed the anti-apoptotic protein Bcl-2 (Hockenbery et al., 1993; Wang et al., 2009; Yang et al., 1997; Zhou et al., 2014), thus compromising cellular resistance to neurotoxins. In addition, we showed that blocking the astrocytic miR-34a can rescue the anti-apoptotic function of DA neurons *in vitro* and alleviate DA neuron loss as well as abnormal behavior induced by apomorphine under 6-OHDA stress *in vivo*. These data suggest a new mechanism by which astrocytes influence neuronal survival under disease conditions.

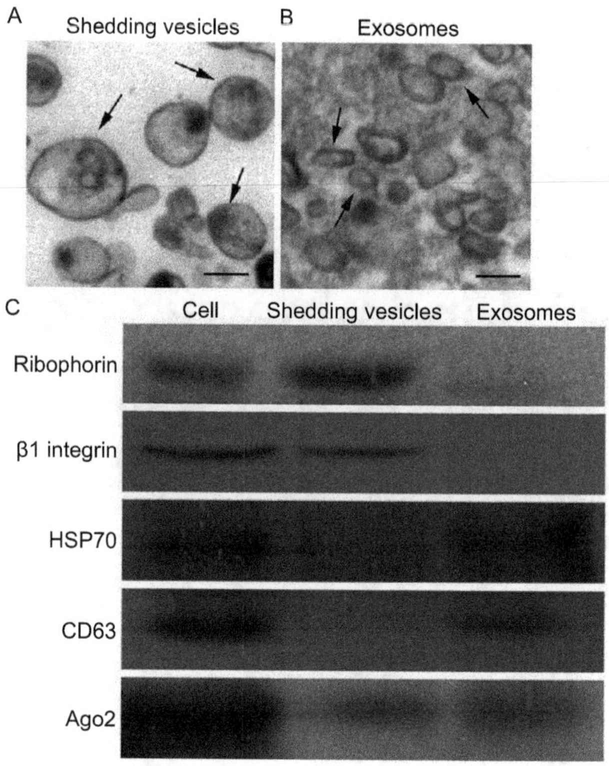

Figure 1. Morphology and characterization of purified microvesicles of U-87 MG cells. (A) EM imaging of purified shedding vesicles secreted by U-87 MG cells (arrows, scale bar = 100 nm); (B) EM imaging of purified exosomes secreted by U-87 MG cells (arrows, scale bar = 100 nm); (C) Western blot of the molecular markers of shedding vesicles (Ribophorin and β1 integrin) and exosomes (CD63 and HSP70); both types of microvesicles contained the miRNA-associated protein Ago2.

RESULTS

Biochemical and morphological characterization of distinct types of microvesicles released from astrocytic cells

We collected the shedding vesicles (SVs) and exosomes from the culture medium by differential ultracentrifugation. EM imaging showed the different sizes of the distinct types of microvesicles. SVs consisted of cup-shaped vesicles in the range of 100–200 nm, while exosomes contained relatively small vesicles of approximately 30–80 nm (Fig. 1A and 1B).

We also analyzed the different types of vesicles by Western blotting of some SVs and exosome markers. Typical SV markers, such as β1 integrin and ribophorin, were enriched in our 10,000 ×*g* pellet, while exosome markers, such as CD63 and heat shock protein HSP70, were found mainly in the 110,000 ×*g* pellet. Furthermore, we also found Ago2 protein expression in both types of the releasing microvesicles (Fig. 1C). Ago2 is reported to be associated with functional miRNAs, indicating that the released microvesicles might have carried miRNAs with Ago2 proteins.

Astrocyte-derived shedding vesicles under stress conditions enhanced the vulnerability of DA neurons to neurotoxins

Next, we investigated the impact of astrocyte-derived microvesicles under stress conditions on cell survival. We found that neither SVs nor exosomes derived from the LPS-stimulated U87-MG astroglial cell line had any effects on SH-SY5Y cell viability under normal conditions (Fig. S1). However, further investigation showed that pretreatment of SVs derived from LPS-stimulated U87-MG cells increased the vulnerability of the SH-SY5Y cells to threshold concentrations of neurotoxins, such as 0.2 mmol/L MPP+ or 10 μmol/L 6-OHDA (Figs. 2A, 2B and S1).

MiRNA profiling of SVs derived from LPS-stimulated U-87 MG cells

We collected the SVs from LPS (LPS SVs) and PBS (control SV) treated U-87 MG cells, respectively. In the present study,

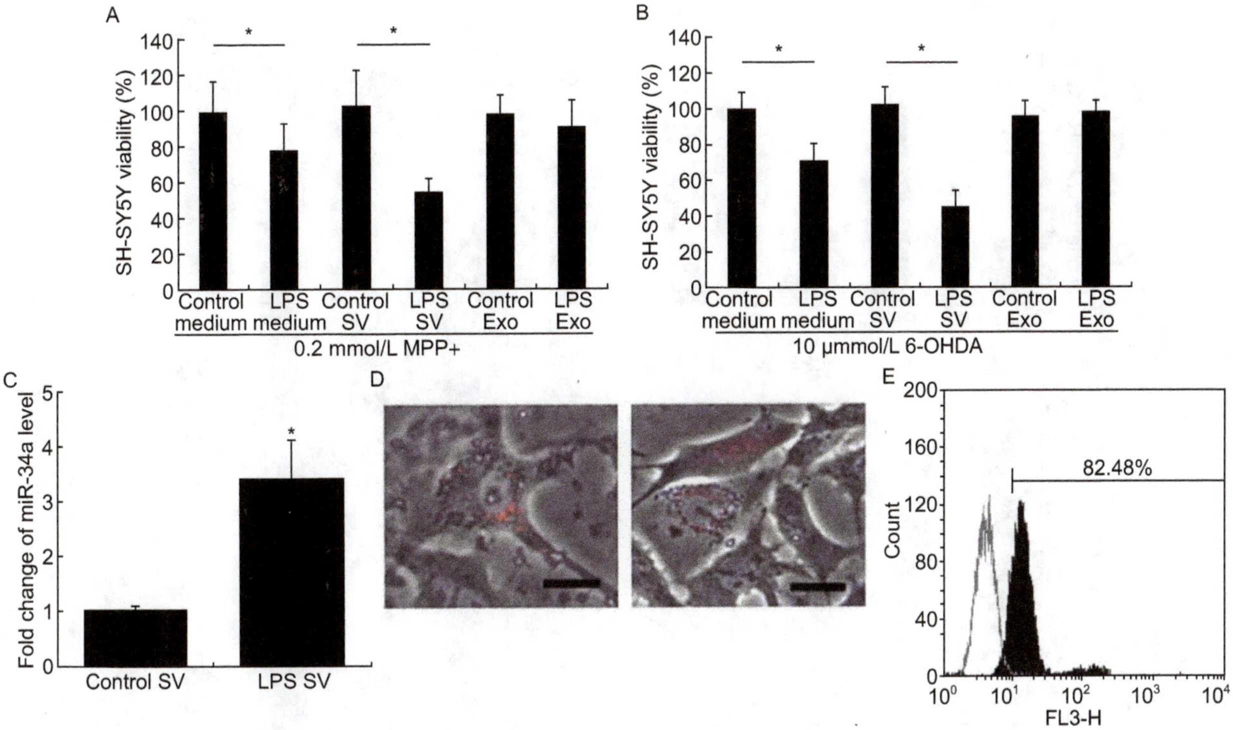

Figure 2. SVs derived from LPS-stimulated U-87 MG cells contain increased miR-34a and reduce SH-SY5Y cell viability to neurotoxins. (A) Cell viability assessed by CCK-8 showed that pretreatment with LPS SVs increased the vulnerability of SH-SY5Y cells to 0.2 mmol/L MPP+ treatment, $^*P < 0.05$; (B) Cell viability assessed by CCK-8 showed that pretreatment with LPS SVs increased the vulnerability of SH-SY5Y cells to 10 μmol/L 6-OHDA treatment, $^*P < 0.05$. Control SV: SVs derived from PBS-stimulated U-87 MG cells; LPS SVs: SVs derived from LPS-stimulated U-87 MG cells; Control Exo: Exosomes derived from control U-87 MG cells; LPS Exo: Exosomes derived from LPS-stimulated U-87 MG cells; (C) Relative miRNA levels between control SV and LPS SVs, $^*P < 0.05$; (D) Fluorescent images showed that labeled SVs derived from U-87 MG cells can enter SH-SY5Y cells after co-incubation; red dots are labeled shedding vesicles, Scale bar = 10 μm; (E) FACS analysis showed that more than 80% of SH-SY5Y cells carry red fluorescence after co-incubation with labeled SVs; the left histogram represents the control group; the right histogram represents SH-SY5Y cells co-cultured with the fluorescence-labeled SVs.

we first analyzed the miRNA levels in LPS SV and control SV using miRNA array. We focused on those miRNAs with a fold-change (LPS SV/control SV) >2 or <0.5, respectively. In each group, 10 miRNAs were listed by their signal intensity in a descending manner (Table S1). Among these altered miRNAs profiles, miR-34a was of particular interest, as it was upregulated 2.86-fold in LPS SV, and more so, as one of its predicted genes, Bcl-2, has been demonstrated to regulate cell apoptosis (Hockenbery et al., 1993; Yang et al., 1997). We further assessed the expression level of miR-34a in LPS SV by qPCR. Consistent with the result of miRNA array, we found that miR-34a was significantly increased, with a fold change of 3.4 ± 0.7 (Fig. 2C). Because it has been reported that miR-34a targets Bcl-2 protein, which regulates cell apoptosis, we proposed that up-regulation of miR-34a in LPS SVs may contribute to the increased vulnerability of SH-SY5Y cells to neurotoxins via repressing Bcl-2 expression.

Secreted miR-34a in LPS SVs enhanced the vulnerability of SH-SY5Y cells to neurotoxins by repressing Bcl-2 protein expression

We co-cultured the fluorescence-labeled SVs of U-87 MG cells with SH-SY5Y cells and found labeled SVs in cultured SH-SY5Y cells by fluorescence microscopy (Fig. 2D). FACS analysis also showed that the 82.48% of cultured SH-SY5Y cells were positive for fluorescence-labeled SVs (Fig. 2E). These results indicated the possibility that SVs may be able to deliver miRNAs into target cells by co-culture.

As it has been reported that miR-34a may target Bcl-2 protein, we first performed the luciferase assay and Western blot in SH-SY5Y cells which showed that miR-34a can indeed target Bcl-2 mRNA directly and repress Bcl-2 protein levels in SH-SY5Y cells (Fig. S2). Next, we further investigated the role of secreted miR-34a in neuronal loss under

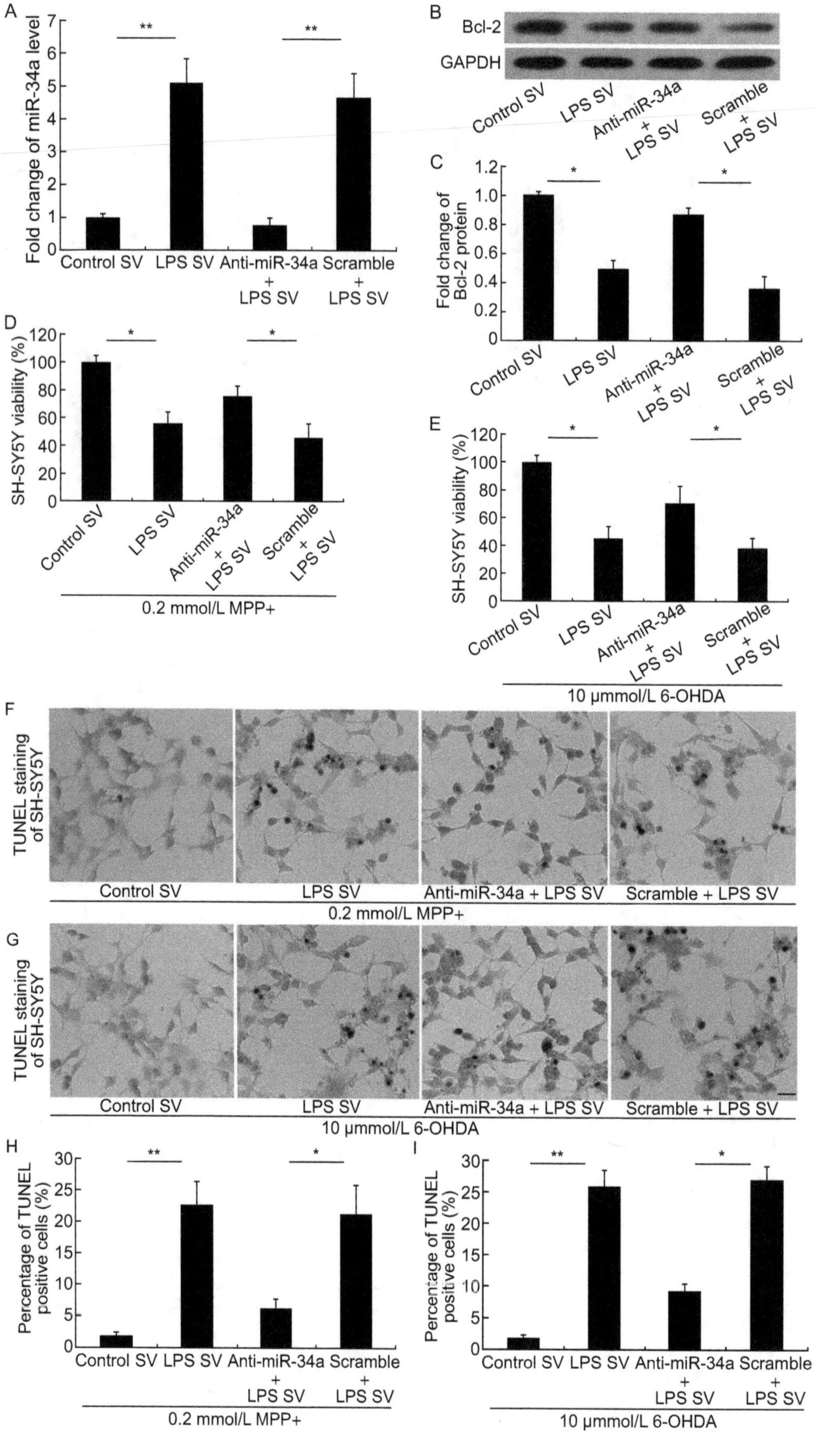
A
Fold change of miR-34a level
Control SV
LPS SV
Anti-miR-34a + LPS SV
Scramble + LPS SV
B
Bcl-2
GAPDH
C
Fold change of Bcl-2 protein
D
SH-SY5Y viability (%)
0.2 mmol/L MPP+
E
10 μmmol/L 6-OHDA
F
TUNEL staining of SH-SY5Y
G
H
Percentage of TUNEL positive cells (%)
I

◀ **Figure 3. Co-incubation with LPS SVs increases miR-34a level while down-regulates Bcl-2 level that further compromises neurotoxin resistance of SH-SY5Y cells.** (A) QPCR analysis of miR-34a levels in SH-SY5Y cells after co-incubation with different groups of SVs, $^{**}P < 0.01$; (B and C) Western-blot and quantification of Bcl-2 protein levels in SH-SY5Y cells after co-incubation with different groups of SVs, $^{*}P < 0.05$; (D and E) Viability of SH-SY5Y cells pre-treated with different groups of SVs following 0.2 mmol/L MPP+ or 10 μmol/L 6-OHDA stress, $^{*}P < 0.05$; (F and G) TUNEL staining of SH-SY5Y cells pre-treated with different groups of SVs after 0.2 mmol/L MPP+ or 10 μmol/L 6-OHDA stress, scale bar = 50 μm; (H and I) Percentage of TUNEL positive cells among SH-SY5Y cells pre-treated with different types of SVs after 0.2 mmol/L MPP+ or 10 μmol/L 6-OHDA stress, $^{*}P < 0.05$, $^{**}P < 0.01$; Control SV: SVs derived from control U-87 MG cells; LPS SV: SVs derived from LPS-stimulated U-87 MG cells; anti miR-34a + LPS SV: SVs derived from LPS-stimulated U-87 MG cells transfected with miR-34a inhibitor; Scramble + LPS SV: SVs derived from LPS-stimulated U-87 MG cells transfected with scramble RNA.

neurotoxin treatment. We found that the level of mature miR-34a was significantly increased in SH-SY5Y cells after the treatment with LPS SVs (Fig. 3A). As a result, the protein level of Bcl-2 was repressed in these cells (Fig. 3B and 3C). However, pretreatment with anti-miR-34a in U-87 MG cells blocked the increase of miR-34a and negated the repression of Bcl-2 caused by LPS SVs (Fig. 3A–C). Further investigation showed that cellular viability was reduced and apoptotic ratio was increased in LPS SV-treated SH-SY5Y cells after 0.2 mmol/L MPP+ exposure. On the other hand, pretreating U-87 MG cells with anti-miR-34a significantly lessened the effect of LPS SVs on SH-SY5Y cell viability and apoptotic ratio (Fig. 3D, 3F and 3H). Similar results were achieved when 6-OHDA was used (Fig. 3E, 3G and 3I).

In addition, we transfected miR-34a into U-87 MG cells and then collected the SVs (miR-34a SVs) without LPS stimulation. In these SVs, miR-34a level was increased by approximately 22.1 ± 2.7 fold (Fig. S3). Pretreatment with miR-34a SVs increased the levels of mature miR-34a in SH-SY5Y cells, leading to a decrease in Bcl-2 protein levels (Fig. 4A–C). Moreover, we also observed that miR-34a SVs reduced the cellular viability and increased the percentage of apoptotic SH-SY5Y cells after 0.2 mmol/L MPP+ or 10 μmol/L 6-OHDA treatment (Fig. 4D–I).

Together, these data provide strong evidence that secreted miR-34a in LPS SVs plays a major role in regulating apoptosis of SH-SY5Y cells exposed to neurotoxins by repressing Bcl-2 protein expression.

SVs derived from LPS-stimulated astrocytes increased primary DA neuron loss after neurotoxin exposure

Here, we further used primary culture to verify the results demonstrated in the cell lines. We collected LPS SVs from primary astrocytes and found that the miR-34a levels were also increased (Fig. S4). Pretreatment with primary astrocytic LPS SVs caused significant TH-positive cell loss after low concentration MPP+ (0.1 μmol/L) or 6-OHDA (2 μmol/L) treatment, without significant effect on the total neuron number showed by DAPI (Fig. 5A–D). This effect was strongly attenuated after blocking the up-regulation of miR-34a in astrocytic LPS SVs by anti-miR-34a pretreatment in astrocytes. To further verify that miR-34a is necessary and sufficient to induce apoptosis, we also used antagomiR oligonucleotides directed against miR-34a (antagomiR-34a) to block the level of miR-34a in primary neurons. Similar results were obtained (Fig. 5A–D). These results indicated that in an *ex vivo* situation, stressed astrocytes released SVs that increased TH neuron vulnerability, in which an increased level of secreted miR-34a may be involved.

Secreted miR-34a in LPS SVs enhanced the DA neuron vulnerability to 6-OHDA *in vivo*

To further understand the role of LPS SVs *in vivo*, we evaluated the effects of different groups of SVs in the 6-OHDA rat model. We injected 6-OHDA unilaterally into the striatum while simultaneously injecting SVs into the SN in the same side (Fig. 6A). We found that the number of DA neurons was significantly decreased in the LPS SVs-treated rats after 1 week, while blocking the up-regulation of miR-34a in LPS SVs can reduce the DA neuron loss at the same time point. However, when we examined DA neurons 3 weeks after the injection, we found that all groups of rats showed severe DA neuron loss with no significant differences (Figs. 6B and 2C). We further assessed apomorphine-induced contralateral rotation of these rat groups at different time points (1 week, 2 weeks and 3 weeks respectively) after the surgery. Among 15 rats for each group that exhibit abnormal apomorphine-induced rotation at the third week (>6/min), we found that the probability of disease onset of the anti-miR-34a + LPS SV group is 3/15 on the first week and 5/15 on the second week while the numbers in scramble + LPS SV group are 8/15 and 11/15 respectively (Fig. 6D). These results showed that blocking the up-regulation of miR-34a in LPS SVs can significantly delay the disease onset but not able to reverse the disease progression after the injection.

Taken together, our data suggest that LPS-stimulated astrocyte SVs can enhance DA neuron vulnerability, which accelerates the progression of DA neuron loss, and that miR-34a plays a crucial role in this process.

DISCUSSION

Neuron-glia interaction is thought to play an important role in the pathogenesis of many neurodegenerative diseases (Malavolta et al., 2013; Saijo et al., 2009). In the present study, we revealed a novel mechanism by which astrocytes can influence neuronal survival under PD conditions. Our results demonstrated that reactive astrocytes release SVs (not exosomes), which increase the percentage of apoptotic SH-SY5Y cells after treatment with low concentrations of neurotoxins. Thus, these SVs may increase DA neuron vulnerability. Furthermore, we demonstrated that there are differences in miRNA expression patterns between SVs produced by reactive and resting astrocytes and that miR-34a was significantly up-regulated. Our previous work demonstrated that MVs can deliver miRNAs into target cells and shut down target gene expression (Zhang et al., 2010). In addition, several studies revealed that MVs mediated cell interaction exists in central nervous system (Antonucci et al., 2012; Hu et al., 2012). Moreover, there is evidence that miR-34a targets the anti-apoptotic protein, Bcl-2, which can further induce tubular cell apoptosis (Liu et al., 2011; Sun et al., 2012; Wang et al., 2009; Zhou et al., 2014). Here we also performed luciferase assay which ensure that miR-34a can directly bind and regulate Bcl-2 in SH-SY5Y cells. These data together indicate the possibility that astrocytic miR-34a may be transferred into neurons via SVs, where it then represses Bcl-2 protein levels and decreases the anti-apoptotic function of the targeted neurons. Notably, several other miRNAs listed in Table S1 also have important functions in modulating cell apoptosis. Among up-regulated miRNAs, miR-18a was validated to regulate cell apoptosis by targeting Bcl-2 directly in mice and humans (Scherr et al., 2014), and miR-296-3p was found to increase apoptosis levels of an islet cell line αTC1-6 cells (Barbagallo et al., 2013). On the other hand, in those down-regulated miRNAs, inhibition of miR-708 promoted the apoptosis of SW480 cells (Lei et al., 2014) and T24 cells (Song et al., 2013). These data suggest that in addition to miR-34a, those altered miRNAs summarized in Table S1 may also affect cell apoptosis. Nevertheless, our data demonstrated that the level and variation of miR-34a in astrocytic SVs is one of the highest among these miRNAs upon LPS stimulation, which indicates that miR-34a plays a major role in modulating cell apoptosis.

Furthermore, we provided direct evidence that fluorescence-marked astrocytic SVs can enter target cells after 24 h of incubation, increase miR-34a levels and decrease Bcl-2 protein levels in SH-SY5Y cells. Blocking the increase in the miR-34a levels in astrocytic SVs by a miR-34a inhibitor alleviated the DA cell loss caused by LPS-stressed astrocytic SVs. This evidence indicates that increased miR-34a levels may contribute to the disruptive effect of reactive-astrocytic SVs on DA cell survival. Since our miRNAs array data indicated that the level of several other miRNAs also changed in astrocytic SVs after LPS stimulation, we further testify the effect of miR-34a by over-expression experiment. When we increased the miR-34a levels in astrocytic SVs without LPS stimulation by transfecting astrocytes with pre-miR-34a, we found that those SVs still enhanced the vulnerability of DA neurons to low concentrations of neurotoxins, leading to an increased percentage of apoptotic cells. These results strongly indicated that the upregulation of miR-34a in astrocytic SVs is sufficient to increase neuronal vulnerability under disease conditions.

Figure 4. Overexpression of miR-34a in SVs is sufficient to increase the vulnerability of SH-SY5Y cells to neurotoxins. (A) QPCR analysis of miR-34a levels in SH-SY5Y cells after co-incubation with miR-34a SV compared with cells co-incubated with control SV, $*P < 0.05$; (B and C) Western-blot and quantification of Bcl-2 protein levels in SH-SY5Y cells after co-incubation with miR-34a SV compared with cells co-incubated with control SV, $*P < 0.05$; (D and E) Viability of SH-SY5Y cells pre-treated with different groups of SVs after 0.2 mmol/L MPP+ or 10 μmol/L 6-OHDA stress, $*P < 0.05$; (F and H) TUNEL staining of SH-SY5Y cells pre-treated with different groups of SVs after 0.2 mmol/L MPP+ or 10 μmol/L 6-OHDA stress, scale bar = 50 μm; (G and I) Percentage of TUNEL positive cells of SH-SY5Y cells pre-treated with different types of SVs after 0.2 mmol/L MPP+ or 10 μmol/L 6-OHDA stress, $*P < 0.05$, $**P < 0.01$. Control SV: SVs derived from control U-87 MG cells; miR-34a SV: SVs derived from U-87 MG cells that overexpressed miR-34a.

In primary culture, we found that LPS-stimulated rat astrocytes also generated SVs with higher levels of miR-34a. In addition, SVs derived from LPS-stressed primary astrocytes enhanced primary DA neuron vulnerability to low concentrations of both MPP+ and 6-OHDA. This disruptive effect was alleviated when we blocked the increase of miR-34a using a miR-34a inhibitor. All of these results were consistent with the data obtained from the cell lines, which suggests that a similar mechanism exists for both primary and cell line culture systems.

Next, we investigated the role of secreted miR-34a in astrocytic SVs in regulating neuron survival *in vivo*. Using the 6-OHDA model, we demonstrated that SVs derived from LPS-stimulated primary astrocytes significantly accelerated DA neuron loss following 6-OHDA treatment. Most of the rats lost more than 30% of their DA neurons in the first week and showed apomorphine-induced rotation. When we blocked the increase of miR-34a in astrocytic SVs by transfecting miR-34a inhibitor into primary astrocytes, we found that the acceleration of DA neuron loss was reduced and that apomorphine-induced rotation was delayed compared with scramble group. These data indicate that the increased levels of secreted miR-34a in LPS-stimulated-astrocytic SVs play a crucial role in the acceleration of DA neuron loss. Recently, it is reported that miR-34a is shuttled in exosomes which induces cell apoptosis in fibrotic kidney (Zhou et al., 2014). This is similar with our findings in nervous system except that in our model we demonstrated that SVs contain the increase of miR-34a and display robust effect rather than exosomes in regulating cell apoptosis. When we examined those PD rats three weeks after 6-OHDA injection, we found all groups showed severe DA neuron loss. Combined with behavioral tests, it is indicated that blocking the function of secreted miR-34a in LPS-stimulated astrocytic SVs can postpone but not reverse the progress of disease.

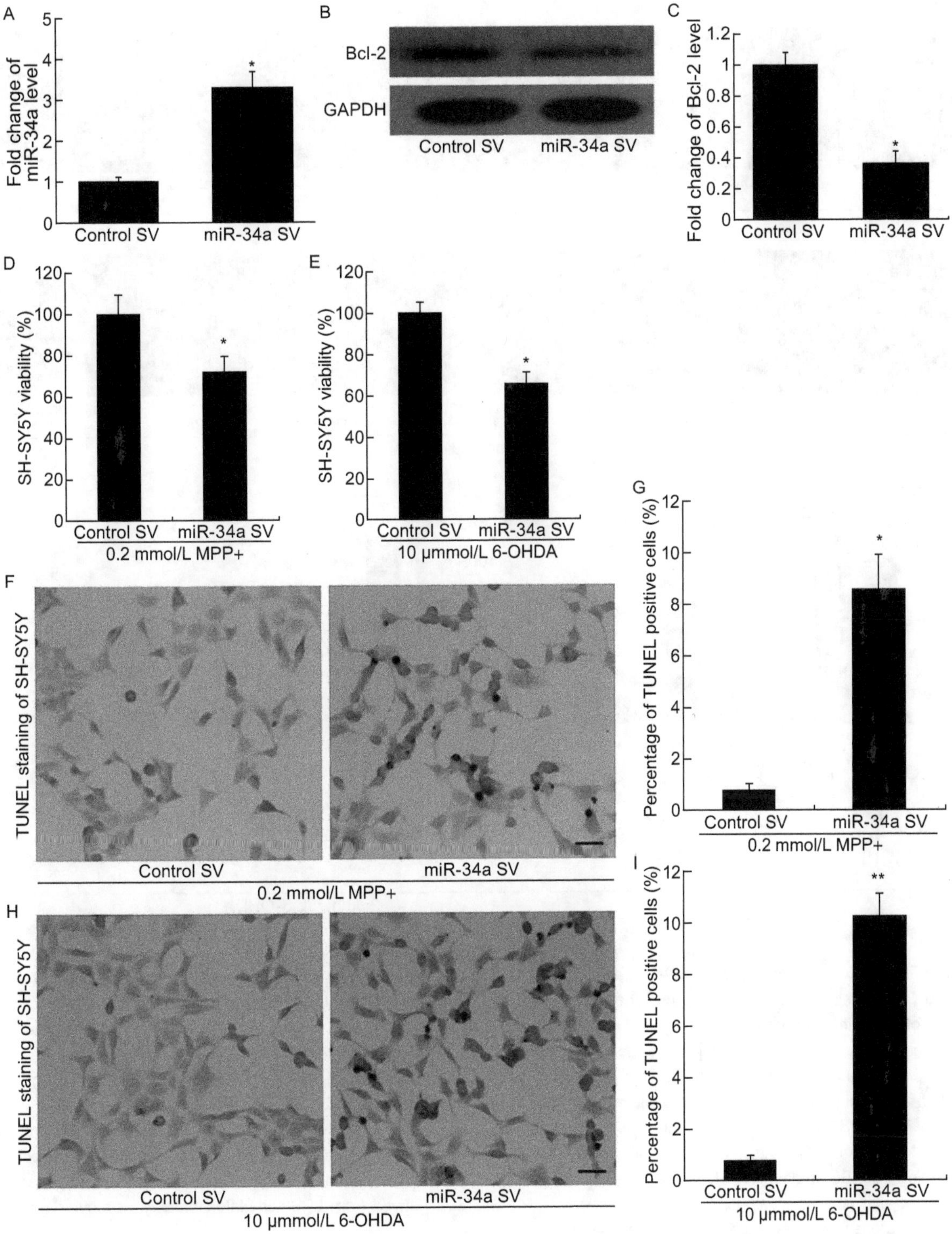
A
Fold change of miR-34a level
Control SV
miR-34a SV
B
Bcl-2
GAPDH
Control SV
miR-34a SV
C
Fold change of Bcl-2 level
Control SV
miR-34a SV
D
SH-SY5Y viability (%)
Control SV
miR-34a SV
0.2 mmol/L MPP+
E
SH-SY5Y viability (%)
Control SV
miR-34a SV
10 μmmol/L 6-OHDA
F
TUNEL staining of SH-SY5Y
Control SV
miR-34a SV
0.2 mmol/L MPP+
G
Percentage of TUNEL positive cells (%)
Control SV
miR-34a SV
0.2 mmol/L MPP+
H
TUNEL staining of SH-SY5Y
Control SV
miR-34a SV
10 μmmol/L 6-OHDA
I
Percentage of TUNEL positive cells (%)
Control SV
miR-34a SV
10 μmmol/L 6-OHDA

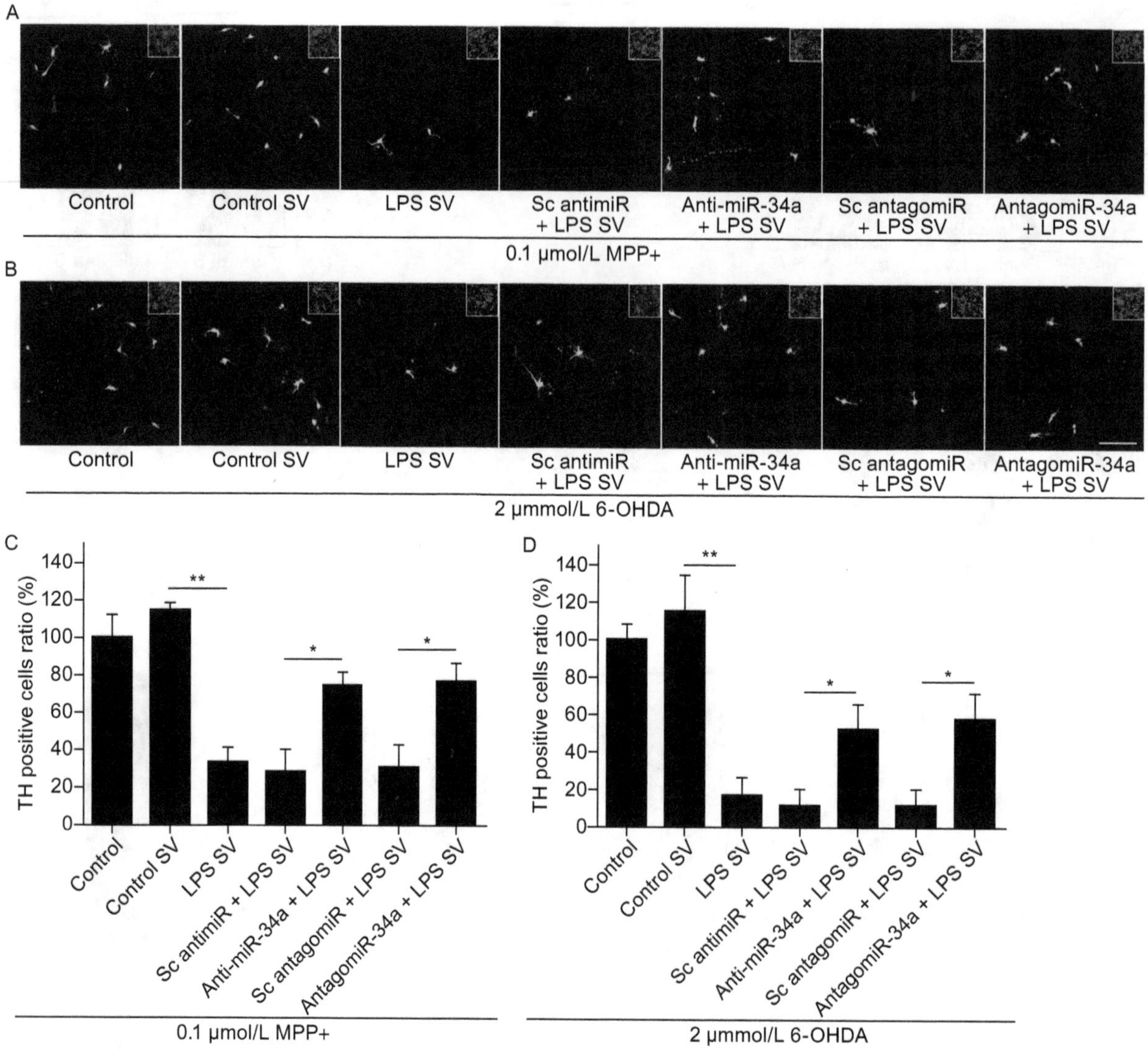

Figure 5. SVs derived from LPS-stimulated primary astrocytes enhance DA neurons loss after treatment with neurotoxins. (A and C) TH staining and quantification of the percentage of TH-positive cells among primary neurons pre-treated with different groups of SVs after 0.1 μmol/L MPP+ stress, *$P < 0.05$, **$P < 0.01$. Inset: DAPI staining of the cell nuclei in the field of view. (B and D) TH staining and quantification of the percentage of TH-positive cells among primary neurons pre-treated with different groups of SVs after 2 μmol/L 6-OHDA stress, *$P < 0.05$, **$P < 0.01$, scale bar = 200 μm. Inset: DAPI staining of the cell nuclei in the field of view. Control: a control group of neuron culture without SVs incubation; Control SV: SVs derived from primary astrocytes; LPS SV: SVs derived from LPS-stimulated primary astrocytes; anti-miR-34a + LPS SV: SVs derived from LPS-stimulated primary astrocytes transfected with miR-34a inhibitor; Sc antimiR + LPS SV: SVs derived from LPS-stimulated primary astrocytes transfected with scramble RNA; Sc antagomiR + LPS SV: SVs derived from LPS-stimulated primary astrocytes, and the primary neurons were pretreated with scramble antagomiR; antagomiR-34a + LPS SV: SVs derived from LPS-stimulated primary astrocytes, and the primary neurons were pretreated with antagomiR-34a.

In many but not all cases of Parkinson's disease, an increase in the number of reactive astrocytes has been observed (Forno et al., 1992), which suggest that astrocytes play vital roles in PD. In the 6-OHDA animal model of PD, Stromberg found an increased amount of reactive astrocytes in the striatum (Stromberg et al., 1986). Activation of astrocytes is a prevalent response to neuronal damage in neurodegenerative diseases, with potential neurotoxic and neuroprotective consequences. Several findings suggest a neuroprotective role for astrocytes in PD as well as animal models of PD. Saura et al. found infusion of IL-1β led to an increased activation of astrocytes in the the substantia nigra and the dopaminergic neurons were protected

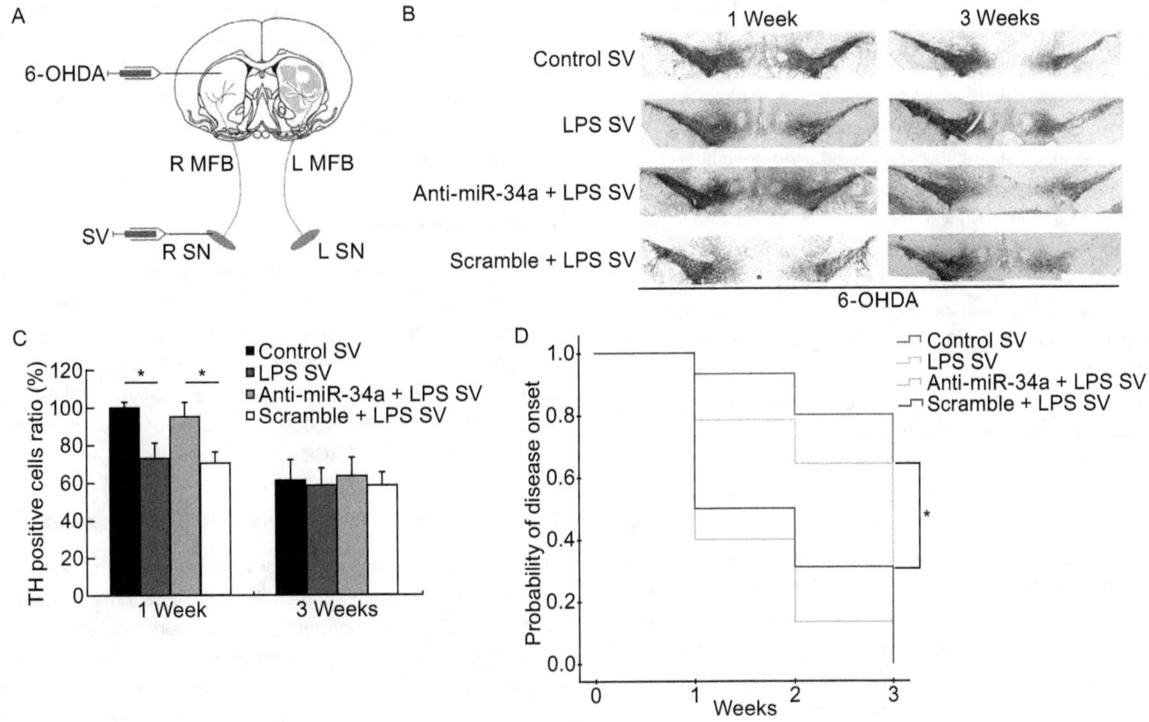

Figure 6. SVs derived from LPS-stimulated primary astrocytes accelerates DA neurons loss as well as disease onset in 6-OHDA rat model. (A) Schematic diagram of the experiment process; (B) TH-immunoreactive cell bodies in SN of different groups of rats; (C) Quantification of DA neurons loss among different groups of rats, $^*P < 0.05$ (n = 10 in each group); (D) The Kaplan-Meier curves demonstrate the probability of disease onset, $^*P < 0.05$. (n = 15 in each group).

against 6-OHDA (Saura et al., 2003). It also has been shown that astrocytes can protect neurons from NO toxicity by a glutathione dependent mechanism (Chen et al., 2001). Additionally, astrocytes express monoamine oxidase-B and catechol-O-methyl-transferaseare, thus reducing the oxidative stress by metabolizing dopamine (Hirsch et al., 1999). On the other hand, Li et al. found that suitable activation of astrocytes increased their protection effect on dopaminergic neurons stimulated by LPS, while excessive activation attenuated it, which suggested that astrocytes played a double-faced effect on dopaminergic neurons (Li et al., 2009). Also, astrocytes can release chemokines and cytokines that are deleterious to neurons in various models of PD, such as reactive oxygen species, nitrite, S100B and so on (Niranjan et al., 2010). The present study suggested that astrocytes had a destructive effect through SV-miR-34a-Bcl2 pathway under LPS stress condition. These results suggest that in stress condition, activated astrocytes may act as a double-edged sword in regulating DA neurons survival. Such functional balance of reactive astrocytes needs to be further explored. Altogether, our results suggested that under stress conditions, such as LPS stimulation, increased levels of miR-34a in astrocytic SVs can be transported into DA neurons, where they enhance the vulnerability of the neurons to neurotoxins by targeting the anti-apoptotic protein Bcl-2. The increase of astrocytic miR-34a in disease conditions may be a potential target for alleviation of PD progress.

MATERIALS AND METHODS

Animals and cell culture

All of the animal care and experimental procedures were performed in accordance with the Laboratory Animal Care Guidelines approved by the Model Animal Research Center of Nanjing University. U-87 MG and SH-SY5Y cell lines were maintained in Dulbecco's Modified Eagle Medium: Nutrient Mixture F-12 (DMEM/F-12, Life Technologies, Grand Island, NY, USA) with 10% Fetal Bovine Serum (FBS, Life Technologies). Primary cultures of rat brain astrocytes were prepared as follows: briefly, newborn (P0) rat cerebral cortices were separated in Hank's Balanced Salt Solution (HBSS, Life Technologies) and digested into cell suspensions. Digestion was terminated by the addition of 10% FBS. Cells were seeded in 75 cm^2 flask and maintained in DMEM/F-12 medium containing 10% FBS. After 14 days of culture, the astrocytes were separated from the microglia and oligodendrocytes by shaking for 12 h in an orbital shaker at 240 rpm. Cell samples were checked regularly for the expression of glial fibrillary acid protein, a marker for astrocytes. Only those cultures containing >95% astrocytes were used. Primary cultures of rat DA neurons were prepared as follows: briefly, embryonic (E14.5) rat ventral mesencephalon (VM) was removed from the brains in HBSS and digested into cell suspensions. Digestion was terminated by the addition of 10% FBS. Cells were seeded in 24-well plates that were coated with 0.05% poly-D-lysine (Sigma-Aldrich, St Louis, MO) and maintained in neurobasal medium (Life Technologies) containing 2% B27 (Life Technologies) and 0.2 mmol/L L-glutamine (Life Technologies) for 4 more days *in vitro* before use. U-87 MG cells and primary astrocytes were treated with different concentrations of LPS (Sigma-Aldrich) for 6 h and washed three times with PBS. The conditioned medium was collected after another 48 h culture. SH-SY5Y cells and primary DA neurons were treated with different groups of astrocytic SVs in an astrocyte-to-neuron ratio of 9:1 for 24 h before exposure to neurotoxin. SH-SY5Y cells were exposed to different concentrations of 1-Methyl-4-phenylpyridinium iodide (MPP+, Sigma-Aldrich) or 6-Hydroxydopamine hydrobromide (6-OHDA, Sigma-Aldrich) for 6 h and cultured for another 24 h before subsequent assays. Primary DA neurons were exposed to different concentrations of MPP+ or 6-OHDA for 1 h and cultured for another 24 h before subsequent assays.

MVs isolation and electronic microscopy

MVs were isolated from culture medium by differential centrifugation as in previous publications (Bianco et al., 2009). After removing cells and other debris by centrifugation at 300 ×*g* and 1200 ×*g*, the supernatant was centrifuged at 10,000 ×*g* for 1 h to collect the shedding vesicles (SVs) and at 110,000 ×*g* for 2 h to collect the exosomes; all steps were performed at 4°C. MVs were collected from the pellet and resuspended in FBS-free medium. In preparation for Electron Microscope (EM) imaging, MVs were precipitated, fixed in a 0.1 mol/L phosphate buffer containing 2.5% glutaraldehyde (Sigma-Aldrich), and cut into 50 μm-thick sections using a vibratome. The sections were postfixed with 1% OsO_4, dehydrated and embedded in Durcupan (ACM; Fluka, Buchs, Switzerland) on a microscope slide and covered with a coverslip. The sections were cut again using a Reichert ultramicrotome into 70-nm-thick sections. The ultrathin sections were then stained with uranyl acetate and lead citrate and evaluated with an electron microscope.

RNA isolation, microRNA array and quantitative real-time PCR of mature miRNAs

The total RNA of the MVs derived from 10^8 cells was extracted using TRIzol Reagent (Life Technologies). Samples of total RNA were extracted for miRNAs Affymetrix GeneChip miRNA 3.0 array analysis (Life Technologies). Quantitative real-time PCR (qPCR) was performed using TaqMan microRNA probes (Applied Biosystems, Foster City, CA, USA) according to the manufacturer's instructions. Briefly, 1 μg of the total RNA was reverse-transcribed to produce cDNA using Avian Myeloblastosis Virus reverse transcriptase (Takara, Dalian, China) and stem-loop RT primers (Applied Biosystems). Real-time PCR was performed using a TaqMan PCR kit and an Applied Biosystems 7300 Sequence Detection System (Applied Biosystems). All of the reactions, including the no-template controls, were run in triplicate. After the reactions, the CT values were determined using fixed-threshold settings.

Fluorescence labeling of SVs

U-87 MG cells were treated with SynaptoGreen (Sigma-Aldrich) for 6 h and washed five times with PBS; after the cells were cultured for another 48 h, the medium was collected and the SVs were isolated. The collected SVs were added into the SH-SY5Y cells in an astrocyte-to-neuron ratio of 9:1 for 24 h. Subsequently, the SH-SY5Y cells were collected and analyzed by fluoromicroscopy or Fluorescence Activated Cell Sorter (FACS).

Cell transfection with ncRNA, anti-miR-34a or pre-miR-34a

U-87 MG cells, primary astrocytes and primary neurons were transfected using Lipofectamine 2000 (Life Technologies) according to the manufacturer's instructions. For overexpression of miR-34a, 50 nmol/L of pre-miR-34a or scrambled negative control pre-miRNA (pre-ncRNA) was used. For knockdown of miR-34a, 20 nmol/L of anti-miR-34a, 50 nmol/L of antagomiR oligonucleotides directed against miR-34a or scrambled negative control anti-miRNA (anti-ncRNA) was used. Cell media were collected 48 h after transfection.

Luciferase assay

The 3′UTR of Bcl-2 containing the predicted target sequence was cloned and inserted into pMIR-REPORT^{TM}-Luciferase vector (Ambion, Austin, TX, USA). Forward primer: 5′-CATGCCTGCCCCAAACAAATA-3′; reverse primer: 5′-AGGGCATTTTTCCCATCGCT-3′. Mutated vector was generated in Invitrogen by replacing the predicted target region with its reverse sequence (from ACUGCC to TGACGG). SH-SY5Y cells were seeded in 24 well plates for 12 h. Afterwards, 0.2 μg of firefly luciferase reporter plasmid, 0.2 μg of β-galactosidase (β-gal) expression vector (Ambion), and 50 nmol/L of miR-34a mimics or scrambled control were transfected into cells. Cells were harvested for luciferase assay (Promega, Madison, WI, USA) 24 h later, and luciferase activities were normalized to β-gal activities.

CCK-8 Assay

Cytotoxicity was also assayed using CCK-8 (Dojindo, Kumamoto, Japan) assay. Briefly, SH-SY5Y cells were cultured in a 96-well plate with different treatments. The cells were then treated with 10 μL CCK-8 solution per well and incubated for 3 h at 37°C. The amount of formazan dye generated by cellular dehydrogenase activity was measured by absorbance at 450 nm with a microplate reader.

TUNEL assay

For the TUNEL assay, the cells were harvested and washed with PBS, fixed and permeabilized, and then TUNEL-labeled using a One Step TUNEL Apoptosis Assay Kit (Beyotime, China) as instructed. The total percentage of apoptotic cells was estimated by determining the percentage of cells with positive TUNEL staining in five randomly selected fields from each slide under a microscope.

TH staining

After treatment, primary DA cultures were washed with PBS and fixed with 4% paraformaldehyde (PFA) for 15 min. Then, the cells were blocked in PBS with 5% horse serum and 0.03% Triton X-100 for 1 h. Immunocytochemistry was performed using mouse anti-tyrosine hydroxylase monoclonal antibody (Sigma). The percentage of cells with positive TH staining was estimated in five randomly selected fields under a fluorescent microscope.

Experimental animals were anesthetized and perfused transcardially with 0.9% saline followed by 4% PFA. Brain samples were postfixed with 4% PFA overnight and equilibrated in 15% and 30% sucrose. Coronal sections of 40 μm were prepared with a sliding microtome. Immunohistochemistry was performed using mouse anti-tyrosine hydroxylase monoclonal antibody. Unbiased quantification of TH-immunoreactive neurons in the SN was performed according to the optical fractionator principle (Gundersen et al., 1988).

Intracerebral injection

Animals received two 2.5-μL stereotaxic injections of 2.0 μg/μL 6-OHDA (total dose = 10 μg 6-OHDA) delivered at a rate of 0.5 μL/min and the syringe was left in place for an extra 5 min and then withdrawn gently, and the skin was sutured. Striatum injection coordinates were as follows: site 1: AP +1.3, ML −2.8, DV −4.5; site 2: AP −0.6, ML −4.0, DV −5.5; and tooth bar set at −3.2. The lesion was allowed to progress for 1–3 weeks after that animals were killed for further analyses.

The vesicles (1 μg/μL, protein concentration) were injected at two sites (site 1: AP −4.8, ML −2.0, DV −7.1; site 2: AP −5.5, ML −1.9, DV −7.0) stereotaxically over the SN (2.5 μL /site). Injection rate was 0.5 μL/min and the needle was left in place for 10 min before it was slowly retracted from the brain.

Behavioral analysis

The animals were tested for apomorphine-induced (0.2 mg/kg, i.p.) turning 1, 2 and 3 weeks after intracerebral injection. Only the rats that exhibited a mean rotation toward the healthy side at least 6.0 full body turns per min were considered as disease onset.

Statistical analysis

All experiments were performed in triplicate. Data were presented as the mean ± SEM; statistical significance of difference (P value) was assessed using Student's t-test by GraphPad Prism 5 software (GraphPad, San Diego, CA, USA), and $P < 0.05$ was considered significant.

ACKNOWLEDGEMENTS

This work was supported by grants from the National Natural Science Foundation of China (Grant Nos. 31000478, 81170309 and 31471019) and the National Basic Research Program (973 Program) (No. 2011CB504803).

ABBREVIATIONS

6-OHDA, 6-Hydroxydopamine hydrobromide; DA, dopamine; DMEM/F-12, Dulbecco's Modified Eagle Medium/Nutrient Mixture F-12; EM, electron microscope; FACS, Fluorescence Activated Cell Sorter; FBS, Fetal Bovine Serum; HBSS, Hank's Balanced Salt Solution; LPS, Lipopolysaccharide; MiRNAs, microRNAs; MPP+, 1-Methyl-4-phenylpyridinium iodide; MVs, microvesicles; PD, Parkinson's disease; PFA, paraformaldehyde; qPCR: quantitative real-time PCR; SN, substantia nigra; SVs, shedding vesicles; TH, tyrosine hydroxylase; VM, ventral mesencephalon.

COMPLIANCE WITH ETHICS GUIDELINES

Susu Mao, Qi Sun, Hui Xiao, Chenyu Zhang and Liang Li declare that they have no conflict of interest. All institutional and national guidelines for the care and use of laboratory animals were followed.

REFERENCES

Al-Nedawi K, Meehan B, Rak J (2009) Microvesicles: messengers and mediators of tumor progression. Cell cycle 8:2014–2018

Antonucci F, Turola E, Riganti L, Caleo M, Gabrielli M, Perrotta C, Novellino L, Clementi E, Giussani P, Viani P et al (2012) Microvesicles released from microglia stimulate synaptic activity via enhanced sphingolipid metabolism. EMBO J 31:1231–1240

Barbagallo D, Piro S, Condorelli AG, Mascali LG, Urbano F, Parrinello N, Monello A, Statello L, Ragusa M, Rabuazzo AM et al (2013) miR-296-3p, miR-298-5p and their downstream networks are causally involved in the higher resistance of mammalian pancreatic alpha cells to cytokine-induced apoptosis as compared to beta cells. Bmc Genomics 14:62

Bianco F, Perrotta C, Novellino L, Francolini M, Riganti L, Menna E, Saglietti L, Schuchman EH, Furlan R, Clementi E et al (2009) Acid sphingomyelinase activity triggers microparticle release from glial cells. EMBO J 28:1043–1054

Chen Y, Vartiainen NE, Ying W, Chan PH, Koistinaho J, Swanson RA (2001) Astrocytes protect neurons from nitric oxide toxicity by a glutathione-dependent mechanism. J Neurochem 77:1601–1610

Forno LS, DeLanney LE, Irwin I, Di Monte D, Langston JW (1992) Astrocytes and Parkinson's disease. Prog Brain Res 94:429–436

Furman JL, Norris CM (2014) Calcineurin and glial signaling: neuroinflammation and beyond. J Neuroinflamm 11:158

Gundersen HJ, Bagger P, Bendtsen TF, Evans SM, Korbo L, Marcussen N, Moller A, Nielsen K, Nyengaard JR, Pakkenberg B et al (1988) The new stereological tools: disector, fractionator, nucleator and point sampled intercepts and their use in pathological research and diagnosis. APMIS 96:857–881

Gupta A, Pulliam L (2014) Exosomes as mediators of neuroinflammation. J Neuroinflamm 11:68

Hirsch EC, Hunot S, Damier P, Brugg B, Faucheux BA, Michel PP, Ruberg M, Muriel MP, Mouatt-Prigent A, Agid Y (1999) Glial cell participation in the degeneration of dopaminergic neurons in Parkinson's disease. Adv Neurol 80:9–18

Hockenbery DM, Oltvai ZN, Yin XM, Milliman CL, Korsmeyer SJ (1993) Bcl-2 functions in an antioxidant pathway to prevent apoptosis. Cell 75:241–251

Hu G, Yao H, Chaudhuri AD, Duan M, Yelamanchili SV, Wen H, Cheney PD, Fox HS, Buch S (2012) Exosome-mediated shuttling of microRNA-29 regulates HIV Tat and morphine-mediated neuronal dysfunction. Cell death Dis 3:e381

Janowska-Wieczorek A, Wysoczynski M, Kijowski J, Marquez-Curtis L, Machalinski B, Ratajczak J, Ratajczak MZ (2005) Microvesicles derived from activated platelets induce metastasis and angiogenesis in lung cancer. Int J Cancer 113:752–760

Jung HJ, Suh Y (2014) Circulating miRNAs in ageing and ageing-related diseases. J Genet Genomics 41:465–472

Keller S, Sanderson MP, Stoeck A, Altevogt P (2006) Exosomes: from biogenesis and secretion to biological function. Immunol Lett 107:102–108

Lee KW, Im JY, Woo JM, Grosso H, Kim YS, Cristovao AC, Sonsalla PK, Schuster DS, Jalbut MM, Fernandez JR et al (2013) Neuroprotective and anti-inflammatory properties of a coffee component in the MPTP model of Parkinson's disease. Neurotherapeutics 10:143–153

Lei SL, Zhao H, Yao HL, Chen Y, Lei ZD, Liu KJ, Yang Q (2014) Regulatory roles of microRNA-708 and microRNA-31 in proliferation, apoptosis and invasion of colorectal cancer cells. Oncol Lett 8:1768–1774

Li XZ, Bai LM, Yang YP, Luo WF, Hu WD, Chen JP, Mao CJ, Liu CF (2009) Effects of IL-6 secreted from astrocytes on the survival of dopaminergic neurons in lipopolysaccharide-induced inflammation. Neurosci Res 65:252–258

Liu C, Zhou CF, Gao F, Cai SY, Zhang C, Zhao LQ, Zhao F, Cao F, Lin J, Yang YY et al (2011) MiR-34a in Age and Tissue Related Radio-Sensitivity and Serum miR-34a as a Novel Indicator of Radiation Injury. Int J Biol Sci 7:221–233

Liu Y, Zhao L, Li D, Yin Y, Zhang CY, Li J, Zhang Y (2013) Microvesicle-delivery miR-150 promotes tumorigenesis by up-regulating VEGF, and the neutralization of miR-150 attenuate tumor development. Protein Cell 4:932–941

Malavolta M, Basso A, Piacenza F, Costarelli L, Giacconi R, Mocchegiani E (2013) Peripheral mononuclear cell rejuvenation for senescence surveillance in Alzheimer disease. Curr Pharm Des 19:1720–1726

Morel O, Toti F, Hugel B, Freyssinet JM (2004) Cellular microparticles: a disseminated storage pool of bioactive vascular effectors. Curr Opin Hematol 11:156–164

Niranjan R, Nath C, Shukla R (2010) The mechanism of action of MPTP-induced neuroinflammation and its modulation by melatonin in rat astrocytoma cells, C6. Free Radical Res 44:1304–1316

Oeckl P, Lattke M, Wirth T, Baumann B, Ferger B (2012) Astrocyte-specific IKK2 activation in mice is sufficient to induce neuroinflammation but does not increase susceptibility to MPTP. Neurobiol Dis 48:481–487

Saijo K, Winner B, Carson CT, Collier JG, Boyer L, Rosenfeld MG, Gage FH, Glass CK (2009) A Nurr1/CoREST pathway in microglia and astrocytes protects dopaminergic neurons from inflammation-induced death. Cell 137:47–59

Saura J, Pares M, Bove J, Pezzi S, Alberch J, Marin C, Tolosa E, Marti MJ (2003) Intranigral infusion of interleukin-1beta activates astrocytes and protects from subsequent 6-hydroxydopamine neurotoxicity. J Neurochem 85:651–661

Scherr M, Elder A, Battmer K, Barzan D, Bomken S, Ricke-Hoch M, Schroder A, Venturini L, Blair HJ, Vormoor J et al (2014) Differential expression of miR-17~92 identifies BCL2 as a therapeutic target in BCR-ABL-positive B-lineage acute lymphoblastic leukemia. Leukemia 28:554–565

Shantsila E, Kamphuisen PW, Lip GY (2010) Circulating microparticles in cardiovascular disease: implications for atherogenesis and atherothrombosis. J Thromb Haemost 8:2358–2368

Song T, Zhang X, Zhang L, Dong J, Cai W, Gao JP, Hong BF (2013) miR-708 promotes the development of bladder carcinoma via direct repression of Caspase-2. J Cancer Res Clin 139:1189–1198

Stromberg I, Bjorklund H, Dahl D, Jonsson G, Sundstrom E, Olson L (1986) Astrocyte responses to dopaminergic denervations by 6-hydroxydopamine and 1-methyl-4-phenyl-1,2,3,6-tetrahydropyridine as evidenced by glial fibrillary acidic protein immunohistochemistry. Brain Res Bull 17:225–236

Sun L, Wu Z, Shao Y, Pu Y, Miu W, Yao J, Wu Y, Yang Z (2012) MicroRNA-34a suppresses cell proliferation and induces apoptosis in U87 glioma stem cells. Technol Cancer Res Treat 11:483–490

Vlassov AV, Magdaleno S, Setterquist R, Conrad R (2012) Exosomes: current knowledge of their composition, biological functions, and diagnostic and therapeutic potentials. Biochim Biophysica Acta 1820:940–948

Wang X, Liu P, Zhu H, Xu Y, Ma C, Dai X, Huang L, Liu Y, Zhang L, Qin C (2009) miR-34a, a microRNA up-regulated in a double transgenic mouse model of Alzheimer's disease, inhibits bcl2 translation. Brain Res Bull 80:268–273

Yang J, Liu X, Bhalla K, Kim CN, Ibrado AM, Cai J, Peng TI, Jones DP, Wang X (1997) Prevention of apoptosis by Bcl-2: release of cytochrome c from mitochondria blocked. Science 275:1129–1132

Zhang Y, Liu D, Chen X, Li J, Li L, Bian Z, Sun F, Lu J, Yin Y, Cai X et al (2010) Secreted monocytic miR-150 enhances targeted endothelial cell migration. Molecular cell 39:133–144

Zhou Y, Xiong M, Niu J, Sun Q, Su W, Zen K, Dai C, Yang J (2014) Secreted fibroblast-derived miR-34a induces tubular cell apoptosis in fibrotic kidney. J Cell Science 127:4494–4506

Zhu H, Fan GC (2011) Extracellular/circulating microRNAs and their potential role in cardiovascular disease. Am J Cardiovasc Dis 1:138–149

Zomer A, Vendrig T, Hopmans ES, van Eijndhoven M, Middeldorp JM, Pegtel DM (2010) Exosomes: Fit to deliver small RNA. Communicative Integr Biol 3:447–450

9

Single-cell transcriptomics reveals gene signatures and alterations associated with aging in distinct neural stem/progenitor cell subpopulations

Zhanping Shi[1], Yanan Geng[1], Jiping Liu[1], Huina Zhang[1], Liqiang Zhou[1], Quan Lin[1,2], Juehua Yu[1], Kunshan Zhang[1], Jie Liu[1], Xinpei Gao[1], Chunxue Zhang[1], Yinan Yao[1], Chong Zhang[1], Yi E. Sun[1,2,3]✉

[1] Stem Cell Translational Research Center, Tongji Hospital, Tongji University School of Medicine, Shanghai 200065, China
[2] Department of Psychiatry and Biobehavioral Sciences and Intellectual Development and Disabilities Research Center, David Geffen School of Medicine, University of California, Los Angeles, CA 90095, USA
[3] Collaborative Innovation Center for Brain Science, Tongji University, Shanghai 200092, China
✉ Correspondence: yi.eve.sun@gmail.com (Y. E. Sun)

ABSTRACT

Aging associated cognitive decline has been linked to dampened neural stem/progenitor cells (NSC/NPCs) activities manifested by decreased proliferation, reduced propensity to produce neurons, and increased differentiation into astrocytes. While gene transcription changes objectively reveal molecular alterations of cells undergoing various biological processes, the search for molecular mechanisms underlying aging of NSC/NPCs has been confronted by the enormous heterogeneity in cellular compositions of the brain and the complex cellular microenvironment where NSC/NPCs reside. Moreover, brain NSC/NPCs themselves are not a homogenous population, making it even more difficult to uncover NSC/NPC sub-type specific aging mechanisms. Here, using both population-based and single cell transcriptome analyses of young and aged mouse forebrain ependymal and subependymal regions and comprehensive "big-data" processing, we report that NSC/NPCs reside in a rather inflammatory environment in aged brain, which likely contributes to the differentiation bias towards astrocytes versus neurons. Moreover, single cell transcriptome analyses revealed that different aged NSC/NPC subpopulations, while all have reduced cell proliferation, use different gene transcription programs to regulate age-dependent decline in cell cycle. Interestingly, changes in cell proliferation capacity are not influenced by inflammatory cytokines, but likely result from cell intrinsic mechanisms. The Erk/Mapk pathway appears to be critically involved in regulating age-dependent changes in the capacity for NSC/NPCs to undergo clonal expansion. Together this study is the first example of using population and single cell based transcriptome analyses to unveil the molecular interplay between different NSC/NPCs and their microenvironment in the context of the aging brain.

KEYWORDS NSC/NPCs, SEZ/SVZ, single cell transcriptome, aging, cell cycle, Erk1/2

INTRODUCTION

Adult neural stem/progenitor cells (NSC/NPCs) exist in the subependyma/subventricular zone (SEZ/SVZ) of the forebrain as well as the subgranular zone (SGZ) of hippocampus (Zhao et al., 2008). These NSC/NPCs support the on-going adult neurogenesis throughout life to replenish olfactory bulb inter-neurons as well as granule neurons in hippocampal dentate gyrus (Ming and Song, 2011). Recently, our single cell transcriptome analyses revealed that CD133 positive ependymal cells throughout the whole ventricular surface of the central nervous system (CNS) harbored quiescent neural stem cell properties, which upon VEGF stimulation can be mitotically activated, and with subsequent bFGF treatment, enter a transit amplifying stage and then differentiate into neurons and glia in CNS regions that were not known to harbor adult neurogenesis (Luo et al., 2015). Since VEGF is abundant after CNS injury, we propose that CD133 positive ependymal cells are mitotically activated after injury, and with proper cellular environment, these cells are capable for neural repair (Yang et al., 2015; Duan et al., 2015).

Adult NSC/NPCs reside in highly complex cellular environment and themselves are also highly heterogeneous cell populations (Beckervordersandforth et al., 2010; Coskun et al., 2008; Morrison and Spradling, 2008). When it comes to investigating alterations in molecular characteristics associated with aging in distinct subpopulations of adult NSC/NPCs, conventional RNA sequencing becomes insufficient, because it only reflects the sum and/or average molecular features of all cells (Shapiro et al., 2013; Nolan et al., 2013). To resolve such a problem, single-cell-based transcriptome analyses are essential. Single-cell RNA-seq can precisely reflect the molecular characteristics of individual cells, which might be hidden from total-cell RNA-seq, if these features belong to rare cells in heterogeneous populations (Shalek et al., 2013). However, these features may be of particular importance for critical biological functions. Because of the high resolution of single cell transcriptome analyses, new subtypes of adult NSC/NPCs were gradually discovered in mouse SEZ/SVZ (Llorens-Bobadilla et al., 2015; Dulken et al., 2017). Single-cell transcriptome analysis, being able to reveal the molecular heterogeneity among cells, has been employed to demonstrate dynamic alterations in various biological processes and innate signals (Luo et al., 2015; Kim DH et al., 2015).

Aging related cognitive decline has recently been attributed, at least in part, to reduced NSC/NPC activities. With progression of aging, adult neurogenesis gradually declines and the number of NSC/NPCs continues to decrease in mouse SEZ/SVZ (Enwere et al., 2004; Maslov et al., 2004). Our previous work and others' have shown that there are at least three quiescent and three active subpopulations of NSCs in mouse ependymal zone (EZ) or SEZ/SVZ (Llorens-Bobadilla et al., 2015; Luo et al., 2015; Dulken et al., 2017). Whether similar or different molecular mechanisms underlying distinct NSC/NPC subtypes during aging remains unclear.

In current study, by both population-based and single-cell transcriptome analysis, we identified three subpopulations of NSC/NPCs that expressed distinct combination of NSC/NPC markers in young and aged mouse SEZ/SVZ. It became clear that a handful of cell type specific markers were insufficient to define a homogenous cell population. Instead, the whole transcriptome or a large cluster of genes or a gene module, may be utilized to better define a distinct cell population. Moreover, we report here age-dependent transcriptional alterations in distinct NSC/NPC subpopulations as well as features of the microenvironment where these cells reside. Both cell extrinsic and intrinsic mechanisms underlying NSC/NPC aging were uncovered.

RESULTS

NSC/NPCs from aged mice showed reduced proliferation capacity and neurogenic potentials

To determine the potential proliferation and differentiation capacities of aged NSC/NPCs in the SEZ/SVZ, we microdissected walls (EZ together with SEZ/SVZ) of lateral ventricles covering striatum and dissociated them into single cells (Fig. 1A). When plated on uncoated surfaces at a clonal density in the presence of fibroblast growth factor (bFGF), NSCs isolated from aged mouse (29 mths) SEZ/SVZ formed significantly less and smaller neurospheres than that from young mice (2 mths) (Fig. 1B and 1C). The number of Ki67$^+$ proliferating cells was dramatically decreased in NSC/NPCs isolated and cultured from aged SEZ/SVZ (Fig. 1D and 1E), suggesting reduced cell cycling in aged NSC/NPCs. Upon differentiation, aged NSC/NPCs differentiated into less neurons, indicative of reduced neurogenic potentials (Fig. 1F and 1G).

Consistent with *in vitro* NSC/NPCs culture results, numbers of Ki67 and Sox2 positive NSC/NPCs as well as doublecortin (Dcx)-positive neuroblasts, decreased in SEZ/SVZ of aged mice (Fig. 2A–D). Moreover, mRNA expression levels of an ependymal NSC marker, CD133, neuroblast markers, Dlx2 and Dcx, significantly decreased in SEZ/SVZ regions of aged mice (Fig. 2E), while Gfap, an astrocyte or NSC marker, increased expression at mRNA levels. Taken together, these observations suggest that aged SEZ/SVZ NSC/NPCs do decrease proliferation and neurogenic potential both *in vitro* and *in vivo*.

The microenvironment of NSC/NPCs in aged brain is pro-inflammation

To explore the molecular mechanisms underlying NSC/NPCs aging, we first performed transcriptome analyses of brain tissues surrounding EZ-SEZ/SVZ lining lateral ventricles covering striatum from young and aged mice (Fig. 3A). Differential gene expression clearly demonstrated that comparing to young mice, cell cycle and cell proliferation-related genes were downregulated in aged brain, consistent with reduced Ki67 immuno-labeling (Figs. 2A and 3B–D). In contrast, old mice had many immune response, inflammation-related genes elevated expression (Fig. 3B, 3E, and 3F), suggesting that aged brains possess a quite inflammatory microenvironment surrounding the EZ-SEZ/SVZ where adult neural stem cells reside. To confirm the gene expression result, we performed immuno-histochemical analyses to measure the amount of microglia, which are residential immune cells in the brain. It is obvious that within brain regions near EZ-SEZ/SVZ many IBA1 positive

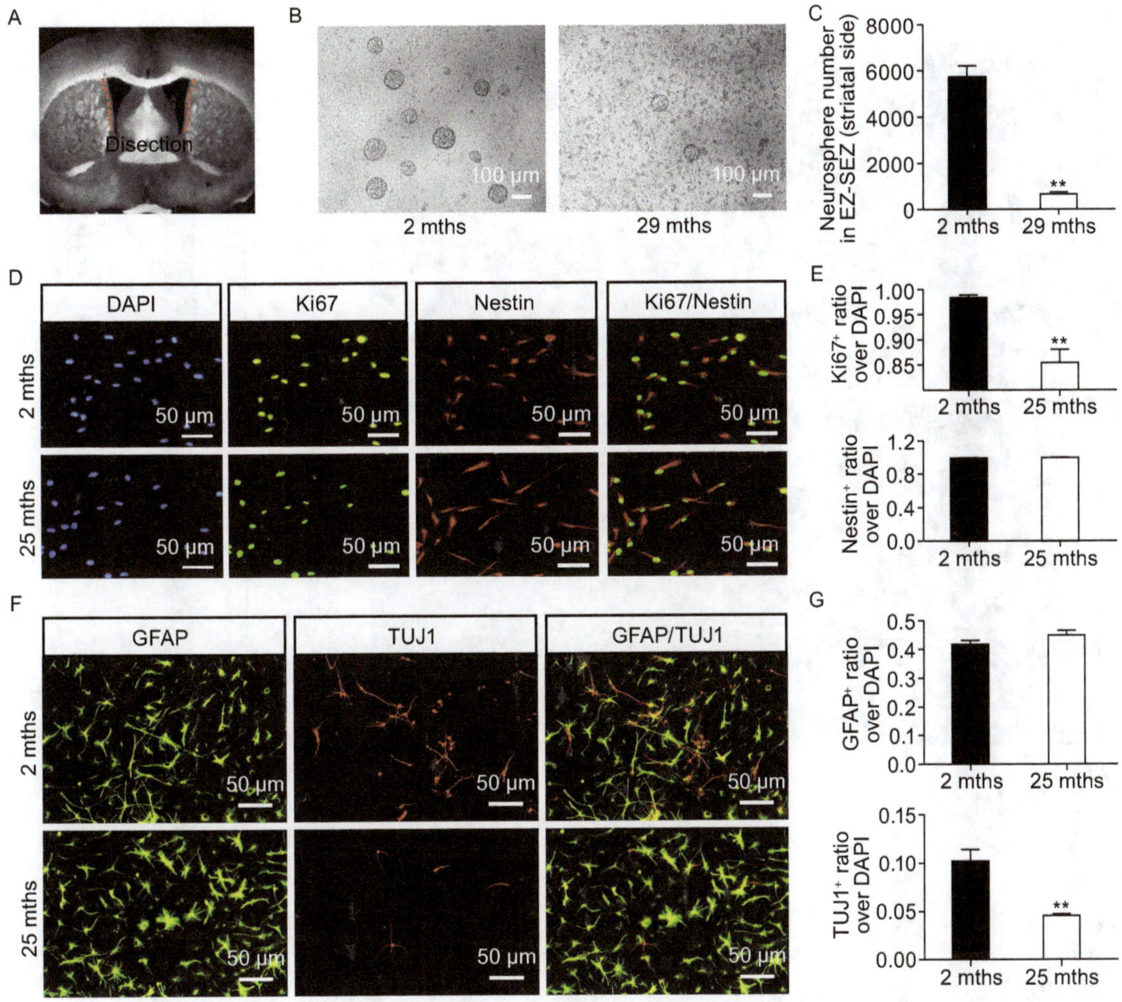

Figure 1. Decreased proliferation and neurogenic potential of cultured NSC/NPCs derived from aged mouse SEZ/SVZ. (A) Schematic showing SVZ dissection site. (B) Neurospheres clonally derived from NSC/NPCs from young (2 mths) and old (29 mths) mouse SEZ/SVZs. (C) Quantification of neurospheres in (B) (n = 3). (D) Immunofluorescence staining of Ki67 and Nestin in cultured NSC/NPCs from young (2 mths) and old (25 mths) mouse SEZ/SVZ. (E) Quantification of (D). (F) Representative fields of GFAP and Tuj1 immunofluorescence staining of cultured NSC/NPCs from young (2 mths) and old (25 mths) mice, after 7 days of spontaneous differentiation. (G) Quantification of (F). (**P < 0.01, $n \geq 3$).

microglial cells were present, indicative of elevated inflammation (Fig. 3G and 3H). Our previous studies have demonstrated that inflammatory cytokines IL-6, IL-1β, LIF, CNTF, potently induce astrocyte differentiation or reactive gliosis through activation of the JAK-STAT pathway, while at the same time inhibit neuronal differentiation (Bonni et al., 1997; Sun et al., 2001). We speculate that the highly inflammatory microenvironment in aged brain contributes to the reduced neurogenic potentials of aged NSC/NPCs.

Single cell RNA-seq of a discrete population of cells isolated from SEZ/SVZ adjacent to striatum in young and old mice

To further investigate molecular characteristics of various NSC/NPCs during aging, we performed single cell RNA-seq of a discrete NSC/NPCs population from EZ and SEZ/SVZ from young and old mice. The EZ-SEZ/SVZ adjacent to striatum was micro-dissected. Single cell suspension was subjected to flow cytometry analysis. According to the size and complexity of cells, we identified two cell populations, of which one was bigger and more discrete (population 1, P1), while the other was relatively small and aggregated (population 2, P2) (Fig. 4A). In order to determine the population in which NSC/NPCs are located, single cells from EZ-SEZ/SVZ were sorted into 96-well plates as single cells by flow cytometry (Fig. 4A). We found the clonal formation rate of neurospheres from P1 was significantly higher than that of P2 (Fig. 4B). In addition, a cell surface marker CD133 (*Prom1*) was used for flow cytometry analysis of P1 and P2. Consistent with neurosphere formation potentials, P1 contains more CD133⁺ NSC/NPCs than P2 (Fig. S1). Based on

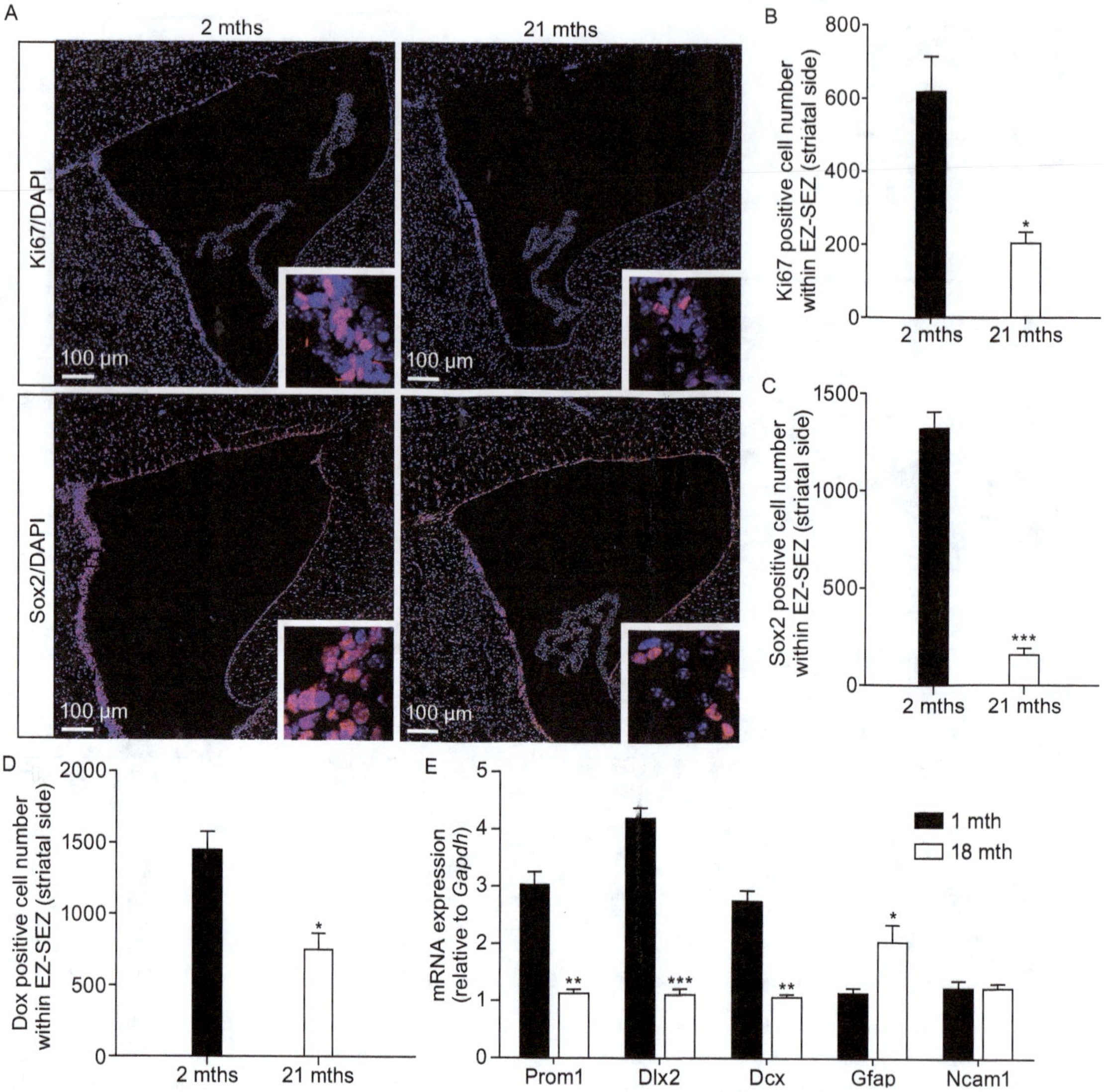

Figure 2. Reduced cell proliferation and neurogenesis in aged mouse forebrain neurogenic zone. (A) Micrographs of sagittal sections from young (2 mths) and old (21 mths) mouse EZ-SEZ/SVZ region immunostained with a proliferation marker Ki67 and a neural stem cell marker Sox2. The insets respectively indicate Ki67 and Sox2 positive cells. (B–D) Quantification of Ki67, Sox2, Dcx positive cells in young and old mouse EZ-SEZ/SVZ region lining the striatal side ($n \geq 3$ per age group, see MATERIALS AND METHODS). (E) The mRNA expression of several NPC marker genes (*Prom1*, *Dlx2*, *Dcx*, *Gfap*, and *Ncam1*) were measured by qPCR of total RNA extracted from young (1 mth) and old (18 mths) mouse EZ-SEZ/SVZ tissues ($n = 3$ per group). Error bars indicate the SEM from three independent experiments. $^*P < 0.05$; $^{**}P < 0.01$; $^{***}P < 0.001$. CC, corpus callosum; Cpu, caudate putamen; LV, lateral ventricle; Fi, fimbria of the hippocampus.

the distribution pattern of P1, it is obvious that P1 still represented rather heterogeneous NSC/NPC populations (Fig. 4A). To probe the heterogeneous cellular composition of the P1 cell fraction, we randomly harvested single-cells from the P1 fraction and carried out reverse transcription using the Smart-seq2 method, followed by qPCR validation with several NSC/NPC markers, such as *Prom1*, *Gfap*, *Dlx2*, and *Dcx*. In this way, we identified 22 cells expressing these markers from both young and old brains.

Libraries of 22 single cells were constructed using Smart-seq2 method and sequenced. Sequencing samples have an average depth of 9.75 million reads, with an average transcriptome alignment rate of 49% (Table S1). We set technical replicates of different single cell cDNA libraries and sequenced different batches using the same single-cell cDNA sample, to ensure the stability of library construction and sequencing platforms. High Pearson correlation coefficient (0.96–0.97) between the replicates demonstrated our RNA-seq system was quite stable (Fig. 4C).

We used t-distributed stochastic neighbor embedding (t-SNE) on transcriptomes of 22 cells with gene filtration. Two criteria were used, i.e., FPKM has to be bigger than 1,

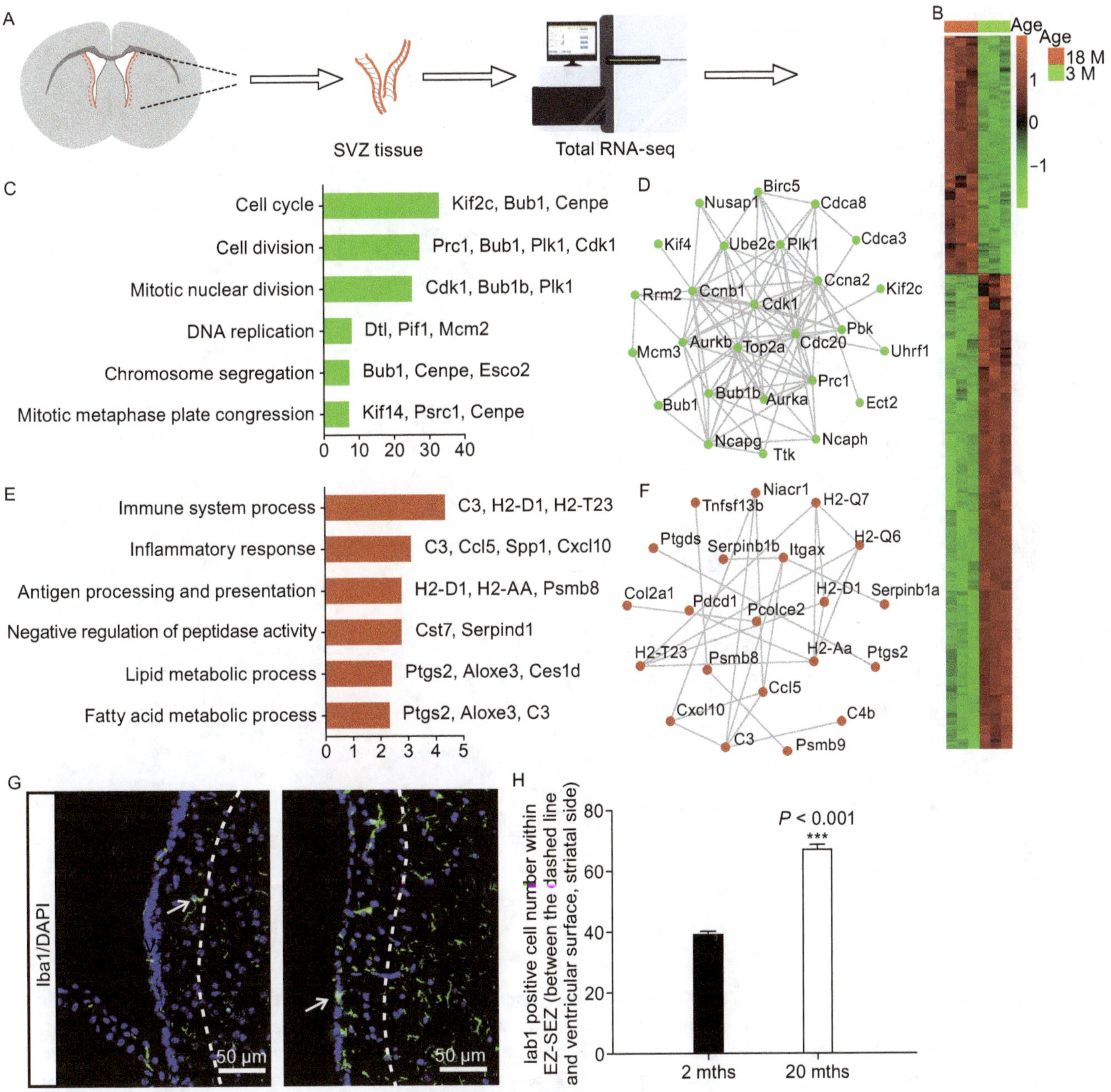

Figure 3. Total RNA-seq of brain tissue surrounding EZ-SEZ/SVZ revealed NSC/NPCs reside in a rather inflammatory environment in aged brain. (A) Schematic showing EZ-SEZ/SVZ tissue dissection and subsequent total RNA-seq. (B) Heatmap of differentially expressed genes in EZ-SEZ/SVZ tissue respectively derived from young (3 weeks) and old (18 mths) mice ($n = 3$ per age group). (C) GO analyses and (D) STRING protein-protein interaction network of down-related genes in aged group. Length of bars indicated the significance (−log10 transferred *P*-value, Fisher exact test). (E) GO analyses and (F) STRING protein-protein interaction network of up-related genes in old group. Length of bars indicated the significance (−log10 transferred *P*-value, Fisher exact test). (G) Representative micrograph of sagittal sections from young (2 mths) and old (20 mths) mouse SEZ/SVZ region immune-stained with Iba1 (green). (H) Quantification of Iba1 positive cells in young (2 mths, $n \geq 3$) and old (20 mths, $n \geq 3$) mouse SEZ/SVZ region. Error bars indicate the SEM from at least three independent experiments. ***$P < 0.001$. Cpu, caudate putamen; LV, lateral ventricle.

and the gene had to be expressed in at least 3 cells out of 22. Through this filtration 10,181 genes were obtained for t-SNE and other analyses. This unbiased analysis clearly divided 22 cells into three main clusters (Fig. 4D). The clusters were assigned to three distinct cell types based on their expression of known NSC/NPC signature genes (Fig. 4E). The Prom1^{+} cluster expressed active neural stem cell (aNSC) markers, including *Hes5*, Tlx (*Nr2e1*), *Egfr*, Glast

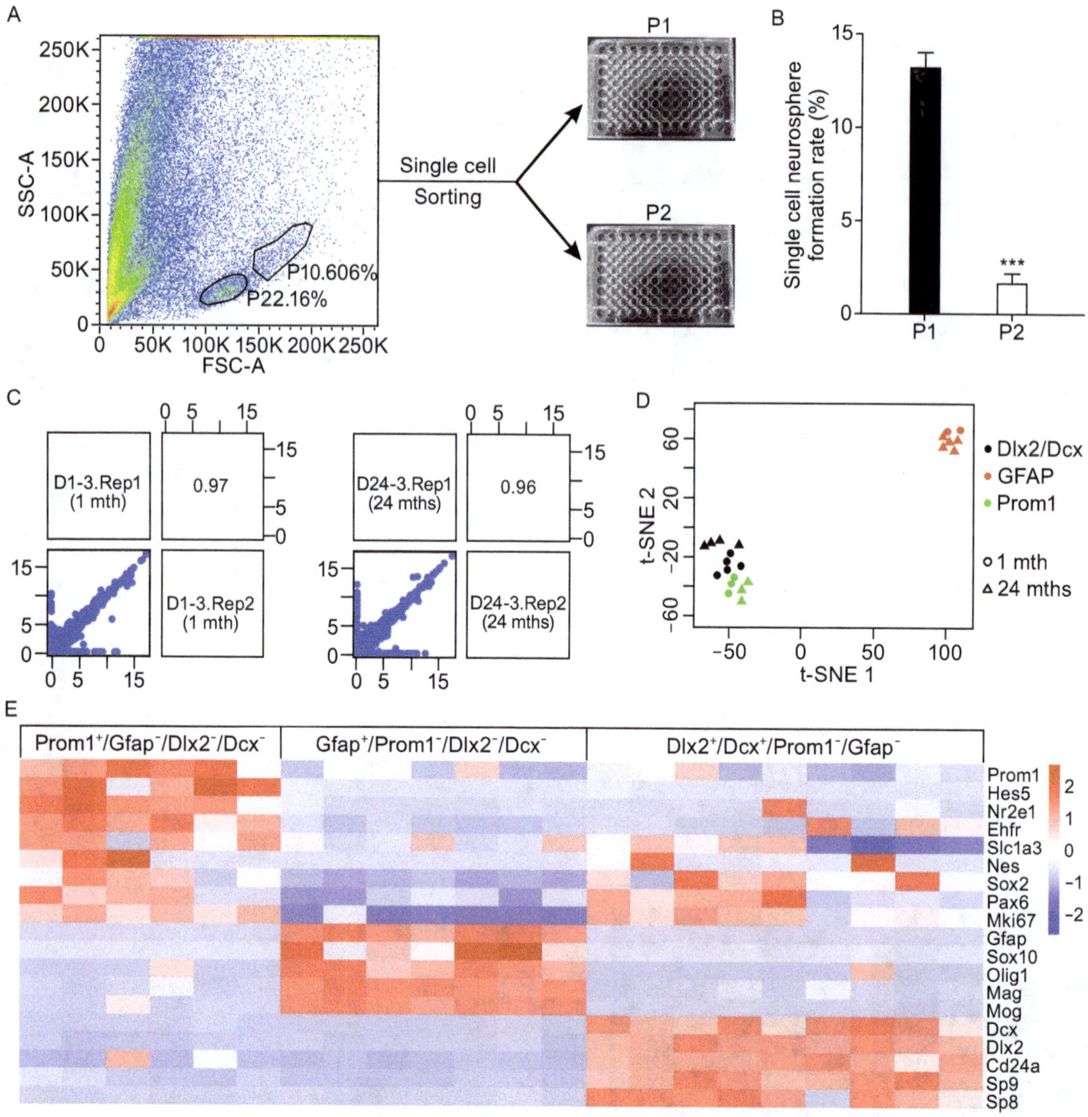

Figure 4. Transcriptome analysis of single cells derived from a discrete SEZ/SVZ population of cells. (A) P1 population rather than P2 was identified as enriched by NPCs, by flow cytometry and single cell neurosphere formation analysis was performed. (B) Quantification of single cell neurosphere formation rate in P1 and P2 populations. At least three independent experiments were quantified (***$P < 0.001$, $n \geq 3$). (C) cDNA samples from two single cells were subjected to two independent library construction and sequencing in two batches. The respective Pearson correlation coefficient of two replicates was 0.97 and 0.96, indicatively minimum sequencing batch effect and reliable library construction and sequencing technology. (D) The t-SNE plot of 22 single cell transcriptome after gene filtration. Three different types of NSC/NPCs were clustered together, i.e., red (GFAP$^+$/Prom1$^-$/Dlx2$^-$/Dcx$^-$), green (Prom1$^+$/GFAP$^-$/Dlx2$^-$/Dcx$^-$), and black (Dlx2$^+$/Dcx$^+$/Prom1$^-$/GFAP$^-$). (E) Three different types of NSC/NPCs express discrete set of marker genes.

(*Slc1a3*) and cell-cycle-related gene, *Mki67*. The Gfap$^+$ cluster expressed genes, such as *Sox10*, *Olig1*, *Mag* and *Mog,* indicative of oligodendroglial lineage. The Dlx2$^+$/Dcx$^+$ cluster expressed *Cd24a*, *Sp9*, and *Sp8*, which is consistent of them being neuroblasts (NBs) in forebrain SEZ/SVZ. Both Prom1$^+$ and NB subpopulations also expressed the well-known NSC/NPC markers, Sox2, Nestin, and Pax6. To further confirm the identity of these three cell clusters, we performed weighted gene co-expression network analyses (WGCNA) and identified three gene clusters (brown, turquoise, and red modules) that correlated well with Prom1$^+$, Gfap$^+$, and Dlx2$^+$/Dcx$^+$ cell clusters, respectively, regardless of the age of the cells (Fig. 5A and 5B). Cross reference of our data with those published (Llorens-Bobadilla et al., 2015), demonstrated that our Prom1$^+$ cell-specific "red" module of genes were enriched in published aNSCs (Fig. 5C and 5D), while our Gfap+ cell-specific "turquoise" gene module was highly enriched in oligodendrocytes (Oligo), and lastly, our Dlx2$^+$/Dcx$^+$ cell-specific "brown" module had enriched expression in NB as well as transit

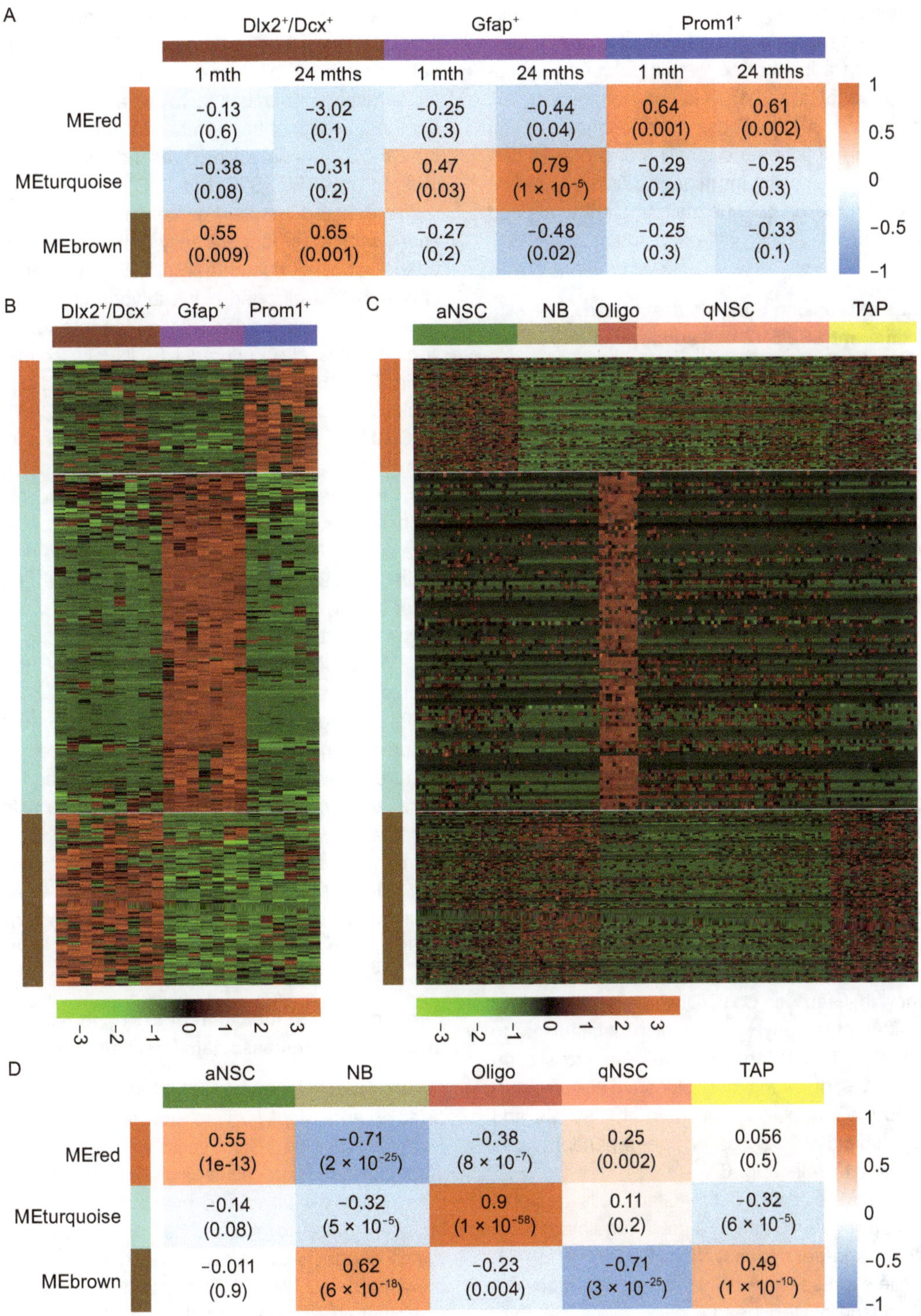

Figure 5. NSC/NPC subtype identification of 22 sequenced single cells based on transcriptome analyses. (A) WGCNA revealed three specific gene modules (red, turquoise, and brown), demonstrating enrichment in Prom1$^+$, Gfap$^+$, and Dlx2$^+$/Dcx$^+$ cells populations. (B) Heatmap of genes of red, turquoise, and brown modules in 22 sequenced single cells. (C) Cross-referencing gene expression of red, turquoise, and brown identified from our 22 single cell transcriptome, in published data set by Llorens-Bobadilla et al. in *Cell Stem Cell* (2015). (D) Correlations between the three modules with 5 published cell types by Llorens-Bobadilla et al. (2015, *Cell Stem Cell*).

amplifying progenitors (TAP) (Fig. 5C and 5D). All these cells have previously been reported to exist in the NSC/NPC pool from EZ-SEZ/SVZ in adult mouse via single cell transcriptome analyses (Luo et al., 2015; Llorens-Bobadilla et al., 2015; Dulken et al., 2017). These data suggested that although single cell transcriptome profiling is prone to carrying lots of noises, when using large gene clusters to define cell type identity, the result could be quite unambiguous and can be cross-referenced among different studies.

Differential gene expression analysis associated with aging in different SEZ/SVZ cell subpopulations

From t-SNE analyses as well as WGCNA, it is obvious that within each of the three cell clusters (cell types), young and aged samples (cells) were segregated, which indicated distinct gene expression between old and young cells. To further understand molecular mechanisms underlying aging of each of the three clusters of cells and evaluate whether they use same or different aging regulatory gene network, we extrapolated differential gene expression in young and aged cells for each cell type, and cross referenced expression of those genes in the other cell types. We omitted genes with reads less than 5, and the gene had to be expressed in at least 3 cells out of 22. Using DEseq2 analyzing package, by comparing transcriptome of Prom1^{+}/Egfr^{+} cells from 1-mth-old mouse EZ-SEZ/SVZ to that of in 24-mths-old mice (Fig. 6A), we found a total of 141 age-dependent genes. Gene ontology (GO) analyses of these genes indicated that aged cells had increased "cell cycle arrest", increased "cellular response to stress", increased "ER overload response", and decreased "cell cycle" (e.g., Melk, Mapk3, etc.) as well as decreased "mitochondrial organization" genes in addition to others (Fig. 6A–C). Interestingly, these age-dependent gene clusters in Prom1^{+}/Egfr^{+} cells did not show consistent age-dependency in Gfap^{+} or Dlx2^{+}/Dcx^{+} cells, suggesting that they use different underlying molecular mechanisms to regulate aging (Fig. 6A). In addition, we also cross-compared transcriptomes of young and old brain tissues surrounding EZ-SEZ/SVZ as well as cultured NSC/NPCs and found that they also did not have same age-dependent gene expression as Prom1^{+} cells (Fig. 6A).

Similar analyses were carried out for Gfap^{+} cells, a total of 578 genes (408 up-regulated and 170 down-regulated) were obtained showing age-dependency (Fig. 6A, 6D, and 6E). These cells, while express Gfap mRNA, were not astrocytes, but rather oligodendrocyte lineage cells, because they expressed a large series of oligodendrocyte lineage genes but not enough astrocyte genes (Fig. S3). Aged Gfap^{+} oligo cell cluster expressed increased "autophagy", "DNA repair" genes, while decreased expression of "cell cycle" genes, such as *Mki67* (Fig. 6A, 6D, and 6E). Aged Dlx2^{+}/Dcx^{+} neuroblasts, on the other hand, increased express of "calcium, magnesium responsive genes" among others, but reduced expression of "telomere maintenance" genes as well as a lot of "cell cycle" genes including *Mki67*, *Melk*, *Mapk1* (Figs. 6A, 6F, 6G, and S2A). An NB age-depended cell cycle checkpoint gene, *Bub1*b, was chosen for immune-histochemical validation in young and aged mice, and result demonstrated that the number of Bub1b expressing cells in EZ-SEZ/SVZ (striatal side) was indeed significantly reduced in aged mice (Fig. S2B and S2C).

While age-dependent genes for each cell subtype did not share similar age-dependency in the other two cell types or brain tissues or cultured NSC/NPCs, GO analysis strongly indicated that all aged cell subtypes showed reduced "cell cycle" gene expression. This is also true for aged brain tissue samples and cultured NSC/NPCs isolated from aged brain (Figs. 6 and S4), suggesting reduced cell proliferation could be a cardinal feature for aged NSC/NPCs. Cultured NSC/NPCs increased genes in "mitochondrial electron transport" and "redox processes", while decreased expression of genes involved in "cell cycle" and "neurogenesis" such as *Myc*, *Ascl1*, *Sox1*, *Jun*, *Tgfb2*, *Axin2*. From cross-data comparison, it is clear that aging gene network of aged brain tissue samples is relatively similar to that of Dlx2^{+}/Dcx^{+} cells, and to a lesser extent, cultured NSC/NPCs (Fig. S4), suggesting that in brain tissues surrounding EZ-SEZ/SVZ, Dlx2^{+}/Dcx^{+} NB might represent one of the major cell populations.

Taken together, relatively large scale, unbiased transcriptome analyses of NSC/NPC single cells, cultured cell mixture, as well as brain tissues surrounding the neurogenic zone, all demonstrated reduction of cell proliferation being associated with aged NSC/NPCs. In addition, microenvironment of aged NSC/NPCs is quite inflammatory. Given that inflammatory cytokines did not appear to slow down cell proliferation *in vitro*, it is possible that the reduction in cell cycling is an intrinsic feature of aged NSC/NPCs.

Phosphorylation inhibition of Erk1/2 in NSCs may be involved in aging-associated decline in NSC/NPCs

GO analysis has revealed that Mapk cascade related genes (*Mapk3* and *Mapk1*) were down-regulated respectively in Prom1^{+} and Dlx2^{+}/Dcx^{+} cells, with aging (Fig. 7A). Mapk, also known as extracellular signal-regulated kinases (ERKs), often plays a vital role in mediating intracellular signaling, including those to regulate cell proliferation. Emerging evidence suggests that Erk1/2 activity is essential to maintain proliferation of NSC/NPCs during normal neurogenesis (Satoh et al., 2011; Vithayathil et al., 2015). Down-regulation of Erk1/2 in NSC/NPCs and neuroblasts during aging may result in a decrease in overall Erk1/2 activity, leading to a decline in NSC/NPC numbers and consequently reduced production of new neurons. Quantitative RT-PCR and validated sequencing result on Erk1/2 (Fig. 7A and 7B). Moreover, quantification of cells immune-labeled with both CD133 and p-Erk1/2 (phosphorylated Erk1/2) in the EZ-SEZ/SVZ suggested that the classic MAPKase pathway was reduced

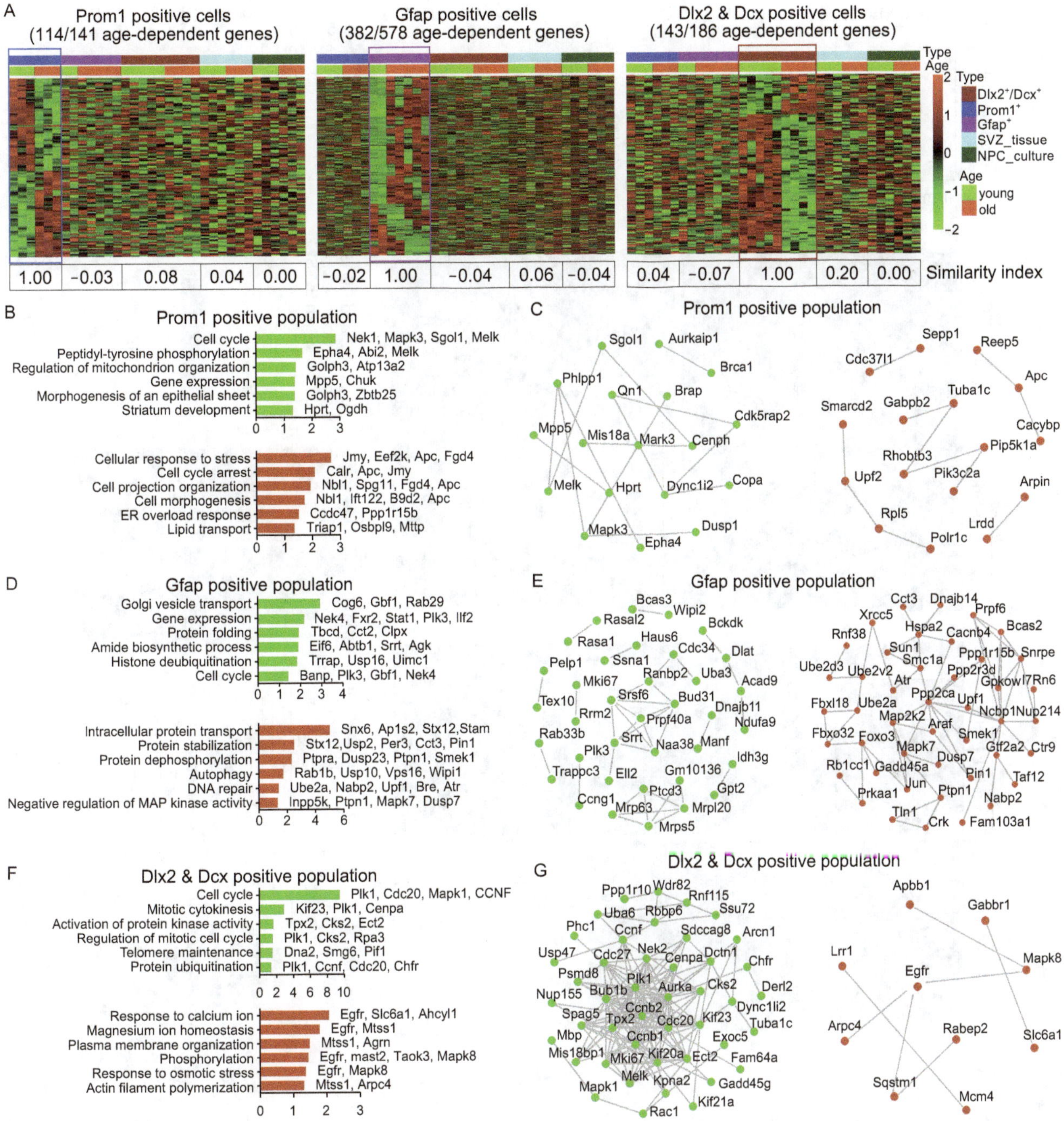

Figure 6. Differential gene expression analysis revealed age-dependent molecular alterations in distinct SEZ/SVZ cell subpopulations. (A) Heatmap of age-dependent differential genes from three distinct SEZ/SZV subpopulations, including Prom1^{+} cells, Gfap^{+} cells, and Dlx2^{+}/Dcx^{+} cells respectively. Cross-data comparisons of expressions of each of these three cell type-specific age-dependent gene panels (excluding those not detected in any one of the 5 types of samples) in the rest of the 5 samples including SEZ/SVZ tissue samples and cultured NSC/NPCs. Similarity index was calculated for each of the three gene panels across 5 types of samples. (B and C) GO analyses and STRING protein-protein interaction network of age-dependent 141 genes (up- and down-related genes) in Prom1^{+} subpopulation. (D and E) GO analyses and STRING protein-protein interaction network of age-dependent 578 genes (up- and down-related genes) in Gfap^{+} subpopulation. (F and G) GO analyses and STRING protein-protein interaction network of age-dependent 186 genes (up- and down-related genes) in NBs subpopulation. Length of bars indicated the significance (−log10 transferred P-value, Fisher exact test).

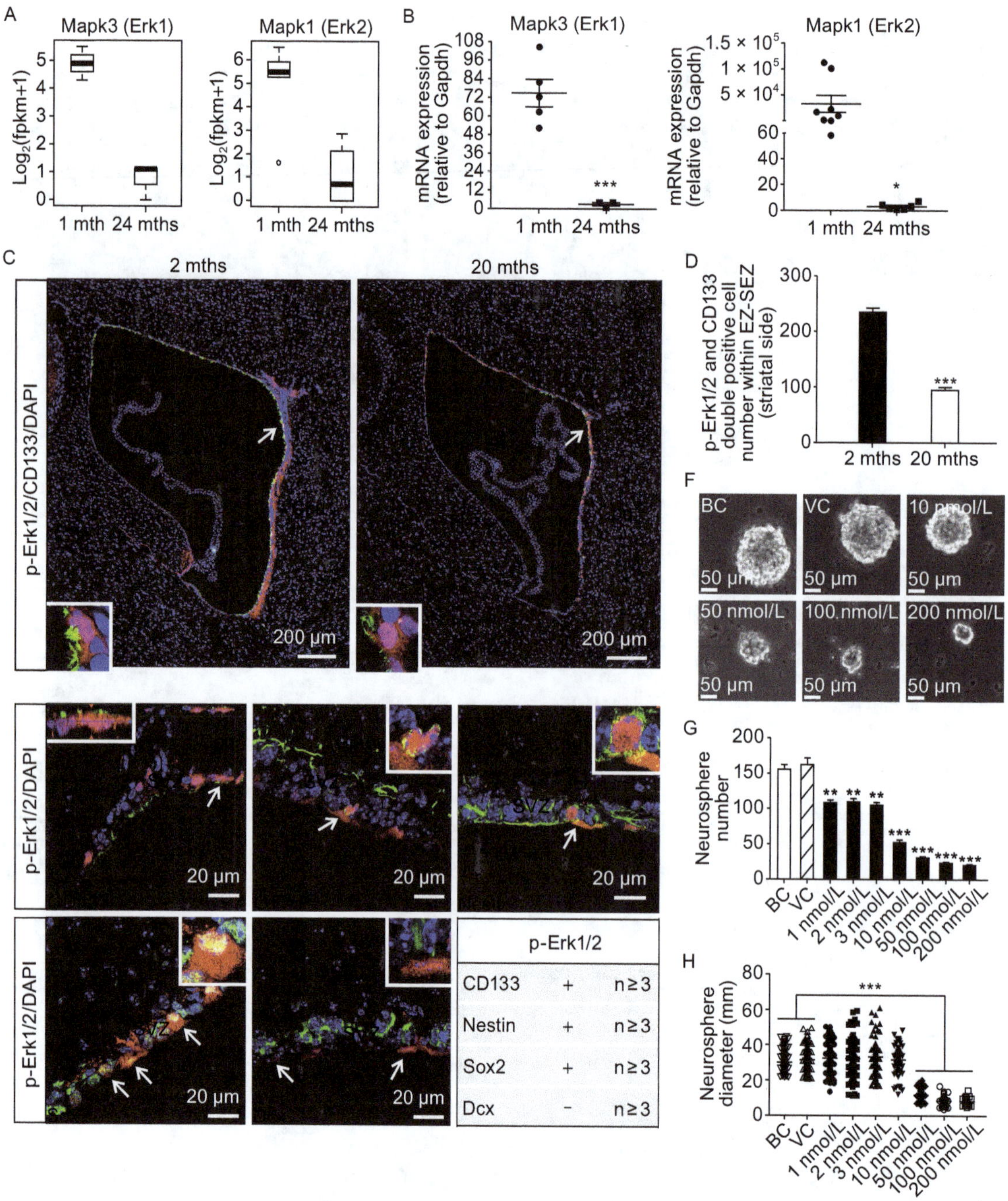

Figure 7. Dampened pErk1/2-Mapkase pathway occurred in aged NSC/NPCs and suppressed clonal expansion capacity of NSC/NPCs. (A) Box plot of Mapk3 and Mapk1, in Prom1^{+} cells and NBs, respectively, demonstrate reduction with aging based on single cell RNA-seq data. (B) qRT-PCR validated sequencing result in (A). (C) Representative micrograph of sagittal sections of mouse SEZ/SVZ from young (2 mths) and old (20 mths), immunostained with CD133 (green) and p-Erk1/2 (red). Arrows and insets indicate CD133 and p-Erk1/2 double positive cells within SEZ/SVZ. (D) Quantification of CD133^{+}/p-Erk1/2^{+} cells in young (2 mths, $n \geq 3$) and old (20 mths, $n \geq 3$) mouse SEZ/SVZ (see MATERIALS AND METHODS). (E) Immunostaining micrographs of sagittal sections of the adult mouse SEZ/SVZ demonstrate p-Erk1/2 (red) could be co-located with the NSC markers (green, including CD133, Gfap, Nestin and Sox2), but not neuroblast marker Dcx (green). Arrows and insets indicate p-Erk1/2^{+}/CD133^{+}, p-Erk1/2^{+}/Gfap^{+}, p-Erk1/2^{+}/Nestin^{+}, p-Erk1/2^{+}/Sox2^{+} NCS/NPCs, as well as p-Erk1/2^{-}/Dcx^{+} neuroblasts within SEZ/SVZ. (F) Representative images of decreased neurosphere formation after exposure to 10 nmol/L, 50 nmol/L, 100 nmol/L, 200 nmol/L specific inhibitor of Erk1/2 phosphorylation (SCH772984), as compared to blank controls (BC) and vehicle controls (VC). (G and H) Quantification of decreased neurosphere numbers and size after exposure to SCH772984. Error bars indicate the SEM from three independent experiments. $^{**}P < 0.01$; $^{***}P < 0.001$.

in aged NSC/NPCs (Fig. 7C and 7D). Interestingly, we found pERK1/2 could be co-localized with CD133, GFAP, Sox2, Nestin single positive cells, while was never present in Dcx^+ NBs (Fig. 7E). Given that $CD24^+$ NBs cannot form clonally grown neurospheres (Coskun et al., 2008), we speculated that a lack of Erk1/2 phosphorylation might eradicate neural sphere formation capacity of NSC/NPCs. To test this, we used SCH772984 to inhibit pErk1/2 pathways and found that the capacity for NSC/NPCs to undergo clonal neurosphere forming was greatly reduced with drug treatment in a dose-dependent manner (Fig. 7F–H). Taken together, it is clear that dampened pErk1/2 pathway likely contributes to declined clonal expansion of aged NSC/NPCs.

DISCUSSION

Aging of the brain is a rather complex and progressive biological process. Aging of NSC/NPCs in the brain, though only represent one perspective of brain aging, is still complicated. Stem/progenitor cell aging involves not only alterations in the molecular characteristics of distinct subtypes of NSC/NPCs, but also changes in the cellular microenvironment, including blood supplies, content of cerebrospinal fluid, and interplays amongst not only NSC/NPCs but also surrounding cells. In-depth understanding of these alterations, and comprehensive analysis of linkages between them may shed light on revealing the molecular mechanisms underlying NSC/NPCs aging.

The pools of adult NSC/NPCs are extremely heterogeneous. Total RNA-seq of EZ-SEZ/SVZ tissue can only represent alterations in the overall transcription profile, while ignoring changes in individual subsets of cells. It remains uncertain that which subtypes of cells are responsible for those alterations. This is a major obstacle for studying the underlying molecular mechanisms. With single cell sequencing applied to adult neural stem cell research, some novel cell subtypes have been discovered (Luo et al., 2015). The investigation on gene signatures and alterations of distinct cell subpopulation in heterogeneous pools of adult NSCs in relation to aging may bring a new strategy for revealing novel mechanisms for stem cell aging.

As a proof of principle study, we carried out both tissue-based, cultured NSC/NPC-based, and single cell transcriptome analyses of NSC-NPCs isolated from young and aged mouse forebrain neurogenic zone. We incidentally sequenced three distinct NSC/NPC populations, but there are a lot more NSC/NPC subtypes. For example, in our published paper, we identified $CD133^+$ ($Prom1^+$) but Egfr negative quiescent NSCs in the ependymal region, which rely on VEGF but not bFGF as mitogens. The $Prom1^+Egfr^+$ active NSCs share some well-known NSC genes including *Nestin*, *Sox2*, *Pax6*, with neuroblasts. These features do not exist in quiescent $CD133^+$ NSC populations. Moreover, our $Gfap^+$ oligodendroglial lineage cells are also drastically different from the $Gfap^+$ SVZ type B neural stem cells (Doetsch et al., 1999). Our study pointed out a future direction of using large scale massive parallel single cell RNA sequencing, e.g., the 10x Genomics platform (Zheng et al., 2017), to further dissect out the cellular composition and molecular characterization of the extremely heterogeneous NSC/NPC populations in the brain.

It is known that single cell RNA sequencing is prone to generating noises in the sequencing result. From this study we realized that using large gene clusters to define cell subtypes is a valid approach as compared to using a handful of biomarkers, which, based on single cell RNA sequencing, often are not that specific and still mark heterogeneous populations. It is comforting that when using large number of gene sets, it is very easy to cross-reference single cell sequencing result for cell subtype identification. Biologically, our comprehensive sequencing data sets had uniformly pointed towards one theme, i.e., aged NSC/NPCs had reduced cell proliferation. On the other hand, each of these three distinct cell populations use different molecular mechanisms to reduce cell cycling, because differential gene expression profile between young and old NSC/NPCs of a distinct subpopulation is not shared by the other subpopulations. Moreover, pErk1/2 signaling appeared to be important for the reduced capacity for clonal expansion of aged NSCs. Brain tissue-based sequencing indicated that in addition to reduction in cell cycle, aged brain expresses many immune response/inflammation related genes. It is likely that neuro-inflammation is an important element for brain aging.

Taken together, through this proof of principle study we demonstrated the power of using single cell transcriptome analyses to reveal potential molecular mechanisms underlying NSC/NPCs aging. Such fundamental studies will help us better understand brain aging and potentially reverse aging.

MATERIALS AND METHODS

Subjects

The mice used in the experiment were all C57BL/6, respectively purchased from Shanghai SLAC Laboratory Animal Co., Ltd. and Beijing Vital River Laboratory Animal Technology Co., Ltd. Mice were housed four per cage, maintained on a 12 h light/dark schedule, and allowed free access to food and water, following protocols approved by the Animal Research Committee of Tongji University School of Medicine, China.

Neurosphere culture *in vitro*

All mice were anesthetized and euthanized in accordance with institutional guidelines. Mice brains were quickly removed from the skull and put into cold HBSS. After twice washes with cold HBSS, the brains were dissected under stereomicroscope. The lateral ventricle wall near the striatum was obtained and enzymatically digested with papain (Worthington LS003127) at 37°C water bath for 30 min. Digested cell suspension was filtered with 40 μm cell strainer. After centrifugation, the cell pellet was resuspended with DMEM-F12 (Thermo Fisher Scientific) supplemented with 1× B27 (Thermo Fisher Scientific) and plated on uncoated 96-well plate for neurosphere culture *in vitro*. Recombinant murine FGF-basic (bFGF) (10 ng/mL; Peprotech) was added every day. Five days later, the neurospheres were counted and measured.

NPC spontaneous differentiation *in vitro*

To examine the spontaneous differentiation status of primary NPCs, we plated the single cell suspension of NPCs onto PDL (10 μg/mL in sterile distilled water; Sigma)/Laminin (5 μg/mL in DMEM-F12) coated coverslips at the density of 1 × 10^4 cells/cm^2. When the cell confluence reached approximately 80%, bFGF was withdrawn. After 7 days of continuous culture without bFGF, the cells were fixed with 4% PFA and performed immunostaining with GFAP and Tuj1 antibody.

Single cell Isolation from the SVZ

As mentioned above, NPCs from the lateral ventricle wall near the striatum were obtained. After twice PBS washes, the cells were sorted using BD FACS Aria II. According to size and complexity, cells were divided into two populations, P1 and P2 population. Single cell neurosphere formation experiment showed that P1 population contained more NSCs than P2. Consequently, P1 population was sorted into dishes followed by manual picking of single cell under microscope.

Fragment library construction and sequencing

After the generation of cDNA from a single cell through Smart-seq2 method (Picelli et al., 2014; Picelli et al., 2013), 8 ng of cDNA was used for fragment library preparation. Using the Covaris^{TM} S2 System (Covaris, Inc.), cDNA was sheared into 150 bp short fragments according to the manufacturer's instructions. The sheared fragments were constructed into fragment library with DNA Library Prep Kit for Illumina (New England Biolabs, Inc.) according to the manufacturer's instructions. A brief overview is as follows: first, short fragments were repaired using the end repair enzyme, then adaptors were added to the end of the fragments by T/A ligase. Afterwards, the adaptor-ligated DNA fragments were purified with AMpure XP beads and amplified for 10 cycles. Last, the libraries were used for Illumina deep sequencing. Reads were aligning to GRCm38 with HISAT2 (Kim D et al., 2015). Gene expression levels were estimated by StringTie (Pertea et al., 2015). RNA-Seq data was deposited at GSE100389.

t-SNE

We used t-distributed stochastic neighbor embedding (t-SNE) (Laurens and Hinton, 2008) to visualize single cell RNA-Seq data. Briefly, t-SNE was calculated on all expressed genes. We used 50 principal components; perplexity = 5 and correlation as a distance measure.

WGCNA

Genes with FPKM ≥ 1 in at least 3 samples were identified as *bona fide* expressed genes. Highly variable genes were detected by ANOVA (FDR < 0.05 for any 3 cell types and ages). 2,203 highly variable genes were supplied to weighted gene co-expression network analysis (WGCNA) as described before (Luo et al., 2015; Zhang et al., 2014). Specifically, soft-power of 7 was chosen to construct a topological overlap matrix from gene correlation network. Modules were detected by dynamic hybrid cut. Highly correlated modules (Pearson correlation of module eigengene >0.9) were merged as one module.

Differential gene expression analysis

Differential expression between the putative groups was conducted using the R package DESeq2 (Love et al., 2014), genes which were expressed at least 5 read counts in 3 samples would take into consideration. To identify significant genes, we select genes with criteria of *P*-values < 0.05 and absolute values of the logarithm (to basis 2) of the fold change (LogFC) > 1. Functional enrichment analysis based on gene ontology (GO) database was performed by using David 6.7 program (http://david.abcc.ncifcrf.gov/) and protein-protein interactions (PPI) network of DEGs were constructed using STRING database (http://www.string-db.org/). Similarity index was calculated by Pearson correlation coefficient. Briefly, we transform subtype-specific differential expressed genes into 3 levels, 1 and −1 for significantly up- or down-regulated genes respectively and 0 for insignificant genes. Pearson correlation coefficient was calculated using transformed gene alteration degree.

Immunostaining

The tissue slices or cultured samples were treated with methanol or 4% PFA for 10 min at −20°C. After three washes, all slides were incubated in blocking buffer (10% BSA and 5% normal donkey serum and 0.2% triton X-100 in PBS) for 1 h at room temperature. Slides or coverslips in 24-well plate were immersed in primary antibody buffer (200 μL per slide or 300 μL per well) for incubation overnight at 4°C. The primary antibodies that were used in this study are: anti-GFAP (Dako) 1:1000, anti-Tuj1 (R&D) 1:500, anti-doublecortin (Cell Signaling Technology) 1:400 (Santa Cruz) 1:1000, anti-Ki67 (Abcam) 1:200, anti-sox2 (Abcam) 1:200, anti-Bub1b (Abcam) 1:100, anti-p-Erk1/2 (Cell Signaling Technology) 1:200, anti-CD133 (Ebioscience) 1:500, anti-Nestin (Millipore) 1:1000, anti-Iba1 (Abcam) 1:200. The following day, the coverslips and slides were washed three times. Then they were incubated in appropriate Alexa 488-, Alexa 568- or Cy3-conjugated secondary antibodies (Thermo Fisher Scientific) for 1 h at room temperature at 1:1000 dilution. DAPI staining was used to label the nuclei. After mounting overnight at room temperature, the slides were examined via Nikon fluorescent microscopes or a confocal system (Leica; TCS SP5 II). Quantification of Tuj1$^+$, GFAP$^+$, Ki67$^+$, Sox2$^+$, Bub1b$^+$, Bub1b$^+$/Dcx$^+$, Iba1$^+$ and CD133$^+$/p-Erk1/2$^+$ cells were performed by counting the labeled cells within the SVZ from three independent experiments. Statistical analysis was performed using unpaired *t* test.

Quantification of immunohistochemical staining

Each experimental group contained at least 3 mice, 12 serial sections (sagittal section, 10 μm) were chosen for subsequent immunostaining per mouse, according to similar anatomical locations among each mouse.

Quantitative real-time PCR

Quantitative real-time PCR (qRT-PCR) was performed using TaqMan real-time PCR system (Thermo Fisher Scientific) and SYB-Green real-time PCR system (TaKaRa). 1 µL of cDNAs were used as template to run 20 µL real-time PCR reactions to check the expression of house-keeping gene GAPDH and other genes to be detected, including *CD133*, *GFAP*, *Dlx2*, *Dcx*, *Ncam1*, *Mapk1*, and *Mapk3*. All reactions were triplicated. The PCR was done as follows in an AB7500 thermocycler (Applied Biosystems) with 96-well plates: 95°C for 10 min; then 40 cycles of 95°C for 15 s; and 60°C; for 1 min. The primers of quantitative real-time PCR are shown in Table S2.

The inhibitor of Erk1/2 administration neurosphere formation *in vitro*

The single cell suspension derived from two 2-mths-old C57BL/6 was plated on 9 wells of two 6-well plate. The culture medium is DMEM-F12 (Thermo Fisher Scientific) supplemented with 1× B27 (Thermo Fisher Scientific). And bFGF (10 ng/mL; Peprotech) was daily added. Erk1/2 inhibitors were respectively added into 7 wells at the final concentration of 1 nmol/L, 2 nmol/L, 3 nmol/L, 10 nmol/L, 50 nmol/L, 100 nmol/L, and 200 nmol/L. Vehicle control was added the same amount of inhibitor solvent while blank control was added nothing but culture medium. Five days later, the neurospheres were counted and measured. The experiments were triplicated.

Graphics

Unless otherwise specified, plots were generated in R or GraphPad Prism 5.

ACKNOWLEDGEMENTS

This study was supported by China National Key Research and Development Program (2016YFA0100801 YS), and the National Natural Science Foundation of China (Grant Nos. 8133030 YS and 31620103904 YS), and grants: 2016YFC102705 YS; 2014BAI04B07 WZL; 81470715 YS; TJ1504219036 WZL.

ABBREVIATIONS

Dcx, doublecortin; Egfr, epidermal growth factor receptor; ERKs, extracellular signal-regulated kinases; FACS, fluorescence-activated cell sorting; Glast, glutamate aspartate transporter; GO, gene ontology; NSC, neural stem/progenitor cell; SVZ, subventricular zone; t-SNE, t-distributed stochastic neighbor embedding; WGCNA, weighted gene co-expression network analysis

COMPLIANCE WITH ETHICS GUIDELINES

Zhanping Shi, Yanan Geng, Jiping Liu, Huina Zhang, Liqiang Zhou, Chong Zhang, Quan Lin, Juehua Yu, Kunshan Zhang, Jie Liu, Xinpei Gao, Chunxue Zhang, Yinan Yao, and Yi E. Sun declare that they have no conflict of interest.

AUTHOR CONTRIBUTIONS

Y.E.S. designed the research; Z.P.S., Y.N.G., H.N.Z., C.Z., J. L. X.P. G., C.X.Z., Y.N.Y., Q.L. performed research; J.P.L., L.Q.Z., J.H.Y., K. S.Z. analyzed data; and Y.E.S., Z.P.S. wrote the paper.

REFERENCES

Beckervordersandforth R et al (2010) In vivo fate mapping and expression analysis reveals molecular hallmarks of prospectively isolated adult neural stem cells. Cell Stem Cell 7:744–758

Bonni A et al (1997) Regulation of gliogenesis in the central nervous system by the JAK-STAT signaling pathway. Science 278:477–483

Coskun V et al (2008) CD133+ neural stem cells in the ependyma of mammalian postnatal forebrain. PNAS 105(3):1026–1031

Doetsch F et al (1999) Subventricular zone astrocytes are neural stem cells in the adult mammalian brain. Cell 97:703–716

Duan H et al (2015) Transcriptome analyses reveal molecular mechanisms underlying functional recovery after spinal cord injury. PNAS 112(43):13360–13365

Dulken BW et al (2017) Single-cell transcriptomic analysis defines heterogeneity and transcriptional dynamics in the adult neural stem cell lineage. Cell Rep 18:777–790

Enwere E et al (2004) Aging results in reduced epidermal growth factor receptor signaling, diminished olfactory neurogenesis, and deficits in fine olfactory discrimination. J Neurosci 24:8354–8365

Kim D et al (2015) HISAT: a fast spliced aligner with low memory requirements. Nat Methods 12:357–360

Kim DH et al (2015) Single-cell transcriptome analysis reveals dynamic changes in lncRNA expression during reprogramming. Cell Stem Cell 16:88–101

Laurens VDM, Hinton G (2008) Visualizing data using t-SNE. J Mach Learn Res 9(2605):2579–2605

Llorens-Bobadilla E et al (2015) Single-cell transcriptomics reveals a population of dormant neural stem cells that become activated upon brain injury. Cell Stem Cell 17:329–340

Love MI et al (2014) Moderated estimation of fold change and dispersion for RNA-seq data with DESeq2. Genome Biol 15:550

Luo Y et al (2015) Single-cell transcriptome analyses reveal signals to activate dormant neural stem cells. Cell 161:1175–1186

Maslov AY et al (2004) Neural stem cell detection, characterization, and age-related changes in the subventricular zone of mice. J Neurosci 24:1726–1733

Ming GL, Song H (2011) Adult neurogenesis in the mammalian brain: significant answers and significant questions. Neuron 70:687–702

Morrison SJ, Spradling AC (2008) Stem cells and niches: mechanisms that promote stem cell maintenance throughout life. Cell 132:598–611

Nolan DJ et al (2013) Molecular signatures of tissue-specific microvascular endothelial cell heterogeneity in organ maintenance and regeneration. Dev Cell 26:204 219

Pertea M et al (2015) StringTie enables improved reconstruction of a transcriptome from RNA-seq reads. Nat Biotechnol 3:290–295

Picelli S et al (2013) Smart-seq2 for sensitive full-length transcriptome profiling in single cells. Nat Methods 10:1096–1098

Picelli S et al (2014) Full-length RNA-seq from single cells using Smart-seq2. Nat Protoc 9:171–181

Satoh Y et al (2011) Deletion of ERK1 and ERK2 in the CNS causes cortical abnormalities and neonatal lethality: Erk1 deficiency enhances the impairment of neurogenesis in Erk2-deficient mice. J Neurosci 31:1149–1155

Shalek AK et al (2013) Single-cell transcriptomics reveals bimodality in expression and splicing in immune cells. Nature 498:236–240

Shapiro E, Biezuner T, Linnarsson S (2013) Single-cell sequencing-based technologies will revolutionize whole-organism science. Nat Rev Genet 14:618–630

Sun Y et al (2001) Neurogenin promotes neurogenesis and inhibits glial differentiation by independent mechanisms. Cell 104:365–376

Vithayathil J et al (2015) Dentate gyrus development requires ERK activity to maintain progenitor population and MAPK pathway feedback regulation. J Neurosci 35:6836–6848

Yang et al (2015) NT3-chitosan elicits robust endogenous neurogenesis to enable functional recovery after spinal cord injury. PNAS 112(43):13354–13359

Zhang K et al (2014) Identification and functional analysis of long non-coding RNAs in mouse cleavage stage embryonic development based on single cell transcriptome data. BMC Genom 15:845

Zhao C et al (2008) Mechanisms and functional implications of adult neurogenesis. Cell 132:645–660

Zheng et al (2017) Massively parallel digital transcriptional profiling of single cells. Nat Commun 8:14049

10

Epigenetic reprogramming, gene expression and *in vitro* development of porcine SCNT embryos are significantly improved by a histone deacetylase inhibitor—*m*-carboxycinnamic acid bishydroxamide (CBHA)

Yuran Song[1,2], Tang Hai[1], Ying Wang[1], Runfa Guo[1], Wei Li[1], Liu Wang[1], Qi Zhou[1]✉

[1] State Key Laboratory of Reproductive Biology, Institute of Zoology, Chinese Academy of Sciences, Beijing 100101, China
[2] Graduate University of the Chinese Academy of Sciences, Beijing 100049, China
✉ Correspondence: zhouqi@ioz.ac.cn (Q. Zhou)

ABSTRACT

Insufficient epigenetic reprogramming of donor nuclei is believed to be one of the most important causes of low development efficiency of mammalian somatic cell nuclear transfer (SCNT). Previous studies have shown that both the *in vitro* and *in vivo* development of mouse SCNT embryos could be increased significantly by treatment with various histone deacetylase inhibitors (HDACi), including *Trichostatin A*, *Scriptaid*, and *m*-carboxycinnamic acid bishydroxamide (CBHA), in which only the effect of CBHA has not yet been tested in other species. In this paper we examine the effect of CBHA treatment on the development of porcine SCNT embryos. We have discovered the optimum dosage and time for CBHA treatment: incubating SCNT embryos with 2 μmol/L CBHA for 24 h after activation could increase the blastocyst rate from 12.7% to 26.5%. Immunofluorescence results showed that the level of acetylation at histone 3 lysine 9 (AcH3K9), acetylation at histone 3 lysine 18 (AcH3K18), and acetylation at histone 4 lysine 16 (AcH4K16) was raised after CBHA treatment. Meanwhile, CBHA treatment improved the expression of development relating genes such as *pou5f1*, *cdx2*, and the imprinted genes like *igf2*. Despite these promising *in vitro* results and histone reprogramming, the full term development was not significantly increased after treatment. In conclusion, CBHA improves the *in vitro* development of pig SCNT embryos, increases the global histone acetylation and corrects the expression of some developmentally important genes at early stages. As in mouse SCNT, we have shown that nuclear epigenetic reprogramming in pig early SCNT embryos can be modified by CBHA treatment.

KEYWORDS swine, nuclear transfer, epigenetic reprogramming, histone deacetylase inhibitor

INTRODUCTION

Somatic cell nuclear transfer (SCNT) has been a successful technology to derive cloned animals in many species (Wilmut et al., 1997; Kato et al., 1998; Wakayama et al., 1998; Baguisi et al., 1999; Onishi et al., 2000; Chesne et al., 2002; Shin et al., 2002; Galli et al., 2003; Woods et al., 2003; Zhou et al., 2003; Lee et al., 2005; Li et al, 2006). It is a valuable tool in fundamental research, biomedical and reproduction fields (Verma et al., 2011; Bendixen et al., 2013). As pigs share much similar physiological characteristics with human beings, it is a potential animal model for human disease and xenotransplantation (Gock et al., 2011; Ekser et al., 2012; Luo et al., 2012). SCNT is a practical way to produce targeted genetic modification in pigs.

However, like in other animals, SCNT in pigs is inefficient, due to the low cloning efficiency, fetus abnormality, and placenta deficiency (Polejaeva et al., 2000; Walker et al.,

Table 1. Pre-implantation development of pig SCNT embryos under treatment with different CBHA concentration

CBHA Treatment (μmol/h)	No. of reconstructed embryos	No. of 2-cell (% reconstructed)	No. of blastocysts (% 2-cell)	No. of blastocysts (% reconstructed)
0	212	132 (62.3)	27 (20.5)	27 (12.7 ± 5.7)[a]
0.02/24	58	40 (69.0)	8 (20.0)	8 (13.8 ± 1.4)[a]
0.2/24	123	75 (61.0)	25 (33.3)	25 (20.3 ± 9.0)[b]
2/24	211	117 (55.4)	56 (47.9)	56 (26.5 ± 12.3)[c]
10/24	90	60 (66.7)	17 (28.3)	17 (18.9 ± 17.9)[b]
20/24	213	110 (51.6)	35 (31.8)	35 (16.4 ± 10.7)[b]
50/24	90	35 (38.9)	12 (34.3)	12 (13.3 ± 3.8)[a]
200/24	65	1 (1.5)	0 (–)	0 (–)
400/24	65	0 (–)	0 (–)	0 (–)

Significant χ^2 comparisons: a versus c, $P < 0.05$; a versus b, $P > 0.05$; b versus c, $P > 0.05$.

2002). It is widely believed that the crucial cause of poor developmental capacity in nuclear transfer embryos is the aberrant nuclear reprogramming. There is a close relationship between embryo reprogramming and epigenetic modification. DNA methylation and histone acetylation are the main modification patterns in epigenetic reprogramming. Compared with normal fertilized zygotes, in SCNT embryos nuclear, reprogramming has to occur within a short time span, so a more "open" chromatin configuration may be beneficial to SCNT embryos (William et al., 2001; Whitworth et al., 2010; Ogura et al., 2013). Histone deacetylase inhibitor (HDACi), has been proved to increase the acetylation level in somatic cells or in nuclear transfer embryos, as it makes the chromatin more flexible and allowing the combination of transcriptional factors (Kishigami et al., 2006; Lager et al., 2008; Dai et al., 2010).

In pigs, some histone deacetylase inhibitors, such as Trichostatin (TSA), *Scriptaid*, and Valproic acid (VPA), have been proved to increase the *in vitro* development of SCNT embryos (Li et al., 2008; Miyoshi et al., 2010; Zhao et al., 2010; Kim et al., 2011). Our previous study has shown that *m*-carboxycinnamic acid bishydroxamide (CBHA), a member of type II HDACi, could improve the *in vitro* and full term development of mouse SCNT embryos and improve the cell line establishment efficiency of embryonic stem cells (ES) derived from SCNT embryos (SCNT-ESCs) (Dai et al., 2010). However, CBHA has not yet been studied in domestic animals. In this study we use this HDACi to treat porcine nuclear transfer embryos to investigate whether it affects the developmental efficiency in pigs.

We have examined the optimum treatment program of CBHA on pig nuclear transfer embryos and observed the development rates in various treated and non-treated groups. To further explore how it improves the development of porcine nuclear transfer embryos, we detected the global acetylation levels of Histone 3 at Lysine 9 (AcH3K9), Histone 3 at Lysine 18 (AcH3K18), and Histone 4 at Lysine 16 (AcH4K16) using immunofluorescence in the CBHA treated embryos and control embryos. Then we analyzed the effect of CBHA treatment on the transcription level of some development-related genes, histone deacetylase genes, and imprinted genes using quantitative PCR. Finally, we detected the full term developmental efficiency of CBHA treated and non-treated embryos.

RESULTS

In vitro development of porcine SCNT embryos are improved by CBHA treatment

To test whether CBHA could improve the early development of porcine SCNT embryos, we treated the SCNT embryos with different concentrations of CBHA and calculated the developmental rates from 2-cell to blastocyst stage (Table 1). We found that proper CBHA treatment had no effect on the cleavage rates of porcine SCNT embryos, but could significantly increase the *in vitro* development rate to blastocyst stage. The optimum treatment concentration of CBHA was 2 μmol/L, while 200 μmol/L or higher concentrations were detrimental to embryos. After treatment with 2 μmol/L CBHA for 24 h, the blastocyst ratio of CBHA-SCNT embryos was two folds (26.5%) of the embryos in control group (12.7%).

Next, to optimize the protocol for CBHA treatment, we examined the effect of different CBHA incubation times on the blastocyst ratios of SCNT embryos. The SCNT embryos were treated with 2 μmol/L or 10 μmol/L CBHA for 16 h, 20 h, and 24 h, respectively. There was no significant difference among the three groups, but the treatment of 2 μmol/L, 24 h gave the best blastocyst ratio (Table 2).

Blastocyst quality was improved after CBHA treatment

To unravel the improvement effects of CBHA treatment on SCNT embryo development, we examined the blastocyst cell number in two groups (Fig. 1). Though the cell numbers in the treated group were higher than the non-treated group, there was no significant difference between the two groups

Table 2. Effect of time duration of CBHA treatment on the pre-implantation development of SCNT embryos

Concentration of CBHA (μmol /L)	Time duration (h)	No. of reconstructed embryos	No. of 2-cell (% reconstructed)	No. of blastocysts (% 2-cell)	No. of blastocysts (% reconstructed)
2	16	77	40 (52.0)	11 (27.5)	11 (14.3 ± 12.5)
	20	78	44 (56.4)	17 (38.6)	17 (21.8 ± 3.8)
	24	77	34 (44.2)	16 (47.1)	16 (20.8 ± 7.2)
20	16	73	34 (46.6)	13 (38.2)	13 (17.8 ± 13.2)
	20	80	38 (47.5)	13 (34.2)	13 (16.3 ± 13.9)
	24	78	36 (45.2)	10 (27.8)	10 (12.8 ± 4.0)
0	–	80	33 (41.3)	14 (42.4)	14 (17.5 ± 6.7)

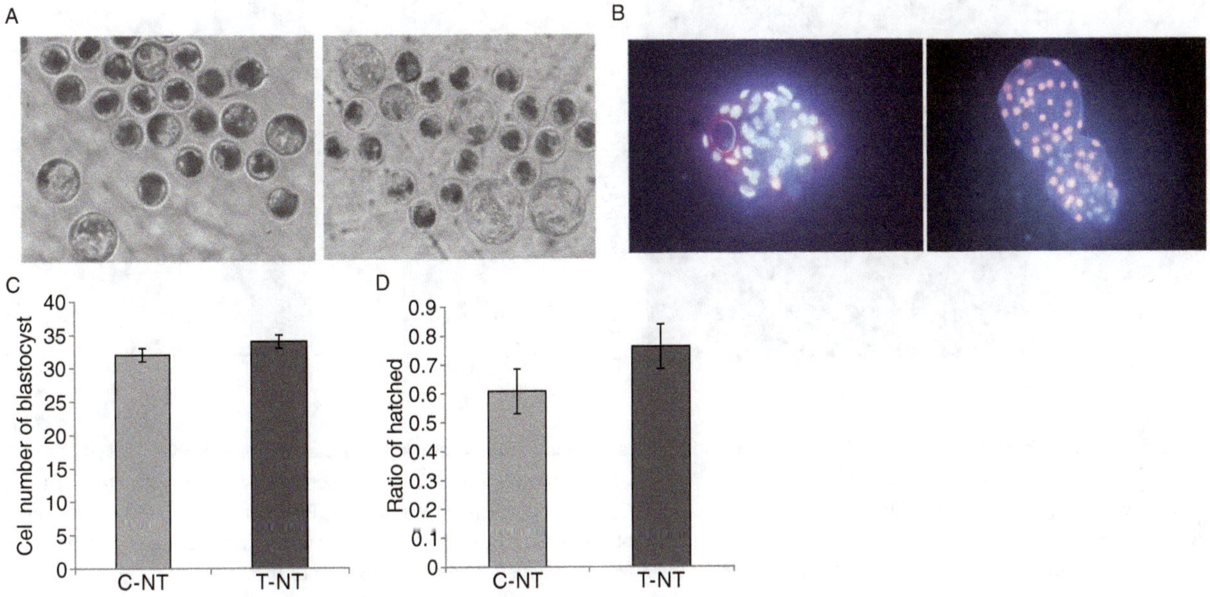

Figure 1. Effect of CBHA on hatching rate and cell number of SCNT blastocysts. (A) Representative photographs of pig blastocysts. Day 7 blastocysts developed from 0 μmol/L CBHA treated SCNT embryos (C-NT group) and 2 μmol/L CBHA treated SCNTembryos (T-NT group). Original magnification was 100×. (B) Fluorescent photographs of pig blastocysts stained with Hoechst. Original magnification was 200×. (C) Statistics of the number of cells in blastocyst from 0 μmol/L CBHA treated SCNT embryos (C-NT group) and 2 μmol/L treated embryos (T-NT group). (D) Ratio of hatched blastocysts from 0 μmol/L CBHA treated SCNT embryos (C-NT group) and 2 μmol/L treated embryos (T-NT group).

(34 vs. 32). Cell numbers from both groups were lower than their *in vivo* and *in vitro* fertilized counterparts.

Embryo implantation and further development are largely related to the blastocyst-hatching rate and the communication between embryo and the uterus. Hence we examined the blastocyst-hatching rate in the two groups. The hatched blastocysts were observed under a stereomicroscope at 168 h after activation (Fig. 1). Although the hatching rate in the CBHA treated group (76.25%) was much higher than those in the control group (67%), there was no significant difference between these two groups. Nevertheless, we found that embryos treated with CBHA could reach the blastocyst stage much earlier than those in the control groups, indicating that at the developmental phase they are more like the *in vivo* embryos, who reach the blastocyst stage on about day 5.

CBHA treatment increases global histone acetylation levels of pig SCNT embryos

As shown in embryos, the global histone acetylation level is closely related to the reprogramming state of chromatin. Meanwhile, we showed that an open, decondensed chromatin state was relevant to the pluripotency of iPSCs

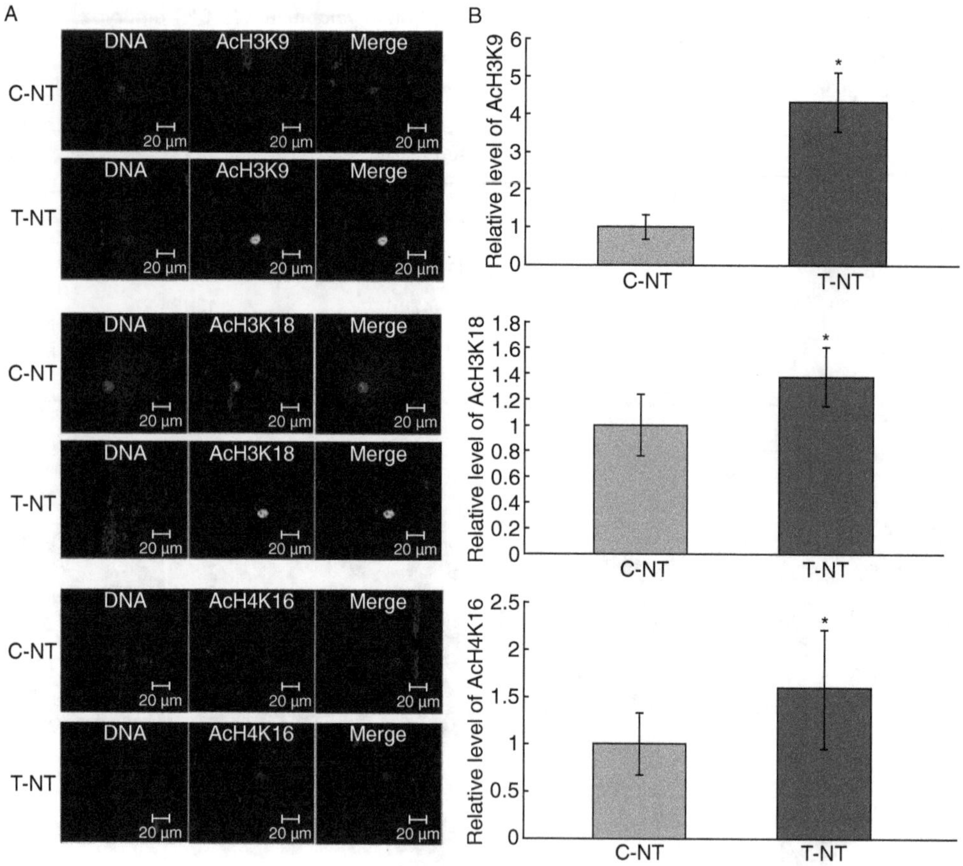

Figure 2. The global AcH3K9, AcH3K18, and AcH4K16 levels of 1-cell stage embryos. (A) Staining of AcH3K9, AcH3K18, and AcH4K16 (green) in non-treated control SCNT embryos (C-NT), and 2 µmol/L CBHA treated embryos (T-NT) at 1-cell stage. Each sample was counterstained with DAPI to visualize DNA (red). Original magnification was 200×. (B) Quantification of AcH3K9 and AcH3K18 signal intensities in C-NT (gray bars) and T-NT (black bars) embryos. Labeling intensity was expressed relative to that of the C-NT embryos (set as 100%). $^{*}P < 0.05$. The experiments were replicated 3 times. $n = 8–10$ per group.

(induced pluripotent stem cells) and ESCs (embryonic stem cells). We further examined the embryo histone acetylation levels by assaying three epigenetic markers: AcH3K9, AcH3K18, and AcH4K16 at 1-cell, 2-cell, and blastocyst stages. We found that CBHA treatment increased the level of AcH3K9, AcH3K18, and AcH4K16 at 1-cell (Fig. 2) and 2-cell stages (Fig. 3). However, apart from AcH3K18, at the blastocyst stage there were no differences between the control and experiment groups (Fig. 4).

Relative transcription level of development related genes, imprinted genes, and histone deacetylase genes

In mammalian early embryonic development, *pou5f1* (*oct4*), an indicator for inner cell mass, is widely accepted to be a key factor in embryonic quality and pluripotency maintenance. Meanwhile, *cdx2*, a pivotal trophectoderm (TE) marker, is important for placenta formation. The expression level of imprinted genes and histone deacetylase genes is also vital for nuclear reprogramming. So, the transcription levels of the five genes (*pou5f1*, *cdx2*, *igf2*, *igf2r*, and *hdac2)* whose functions are mentioned above were detected by quantitive PCR (Fig. 5) at the blastocyst stage. The expression of two key developmental genes, *pou5f1* and *cdx2,* was significantly increased by CBHA treatment. However, the expression of these two genes in both groups was significantly lower than the IVF (*in vitro*, fertilized) group. The expression of the imprinted gene *IGF2* was increased after the CBHA treatment, which was similar to the IVF ones and significantly higher than the control group. However, the expression of another imprinted gene, *igf2r*, and the histone deacetylase gene, *hdac2*, was not affected by CBHA treatment, and both of them were significantly lower than the IVF ones.

Full term development of porcine SCNT embryos following CBHA treatment

To determine whether CBHA could improve the full term development of porcine SCNT embryos, 2-cell stage embryos with and without CBHA treatment were transferred to the oviducts of surrogates on the day of or 1 day after the onset of estrus. In the first batch of experiments, we

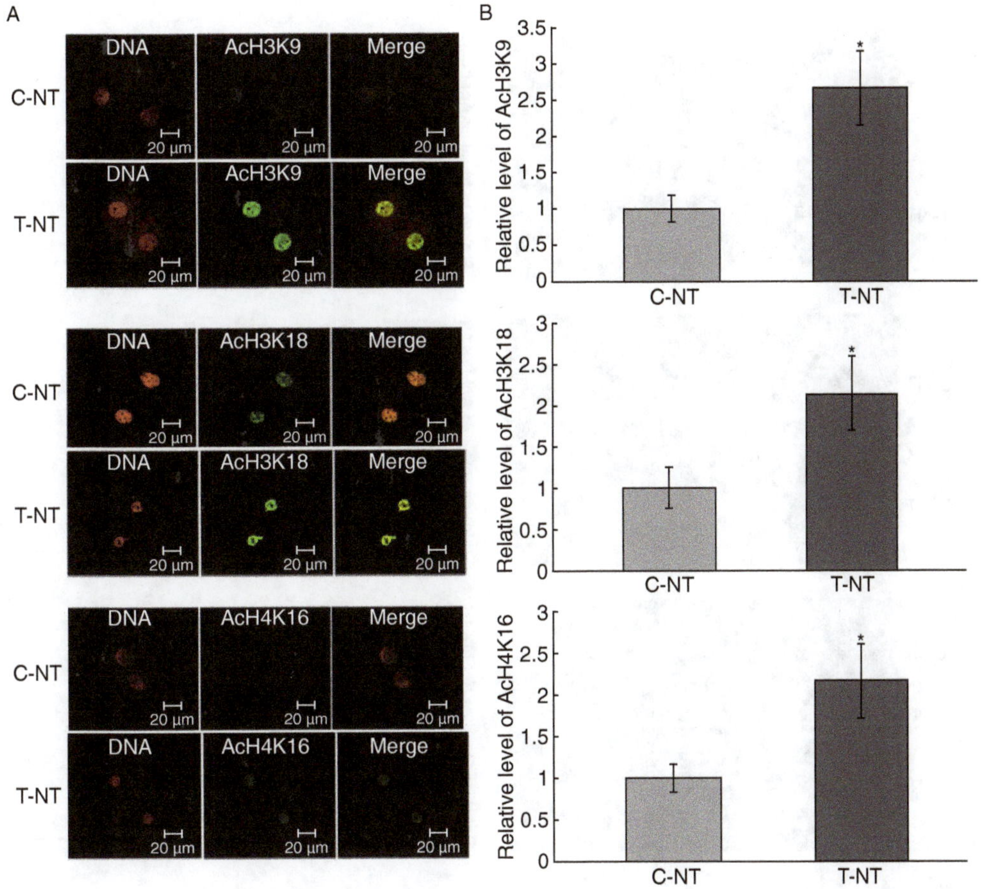

Figure 3. The global AcH3K9, AcH3K18, and AcH4K16 levels of 2-cell stage embryos. (A) Staining of AcH3K9, AcH3K18, and AcH4K16 (green) in non-treated control SCNT embryos (C-NT), and 2 µmol/L CBHA treated embryos (T-NT) at 2-cell stage. Each sample was counterstained with DAPI to visualize DNA (red). Original magnification was 200×. (B) Quantification of AcH3K9 and AcH3K18 signal intensities in C-NT (gray bars) and T-NT (black bars) embryos. Labeling intensity was expressed relative to that of the C-NT embryos (set as 100%). **P* < 0.05, †*P* < 0.01. The experiments were replicated 3 times. *n* = 8–10 per group.

transferred more than 200 embryos to each surrogate. We found that there was no significant difference between the two groups (Table 3), which is inconsistent with the beneficial effect of CBHA in the *in vitro* development. To further exclude the noise caused by the excessive number of embryos transferred, we designed a second batch of experiments, which transferred only 100 embryos to each recipient (Table 4). We obtained a total of 15 live, 6 dead, and 1 mummy from the CBHA treated group and 17 live, and 7 dead from the control group. The full term development ratio of CBHA-treated embryos was 4.2%, which is not significantly different to the control group (4.4%), suggesting CBHA treatment does not improve the full term development of porcine SCNT embryos.

DISCUSSION

Nowadays, somatic cells can be reprogrammed in many ways, the most common being somatic cell nuclear transfer (SCNT) and induced pluripotent stem cells (iPSCs) (Wilmut et al., 1997; Kato et al., 1998; Wakayama et al., 1998; Baguisi et al., 1999; Onishi et al., 2000; Chesne et al., 2002; Shin et al., 2002; Han et al., 2003; Galli et al., 2003; Woods et al., 2003; Zhou et al., 2003; Lee et al., 2005; Li et al, 2006; Takahashi et al., 2006; Yu et al., 2007). After being transfected with pluripotent genes or small molecules, a fraction of somatic cells could be reprogrammed to a pluripotent state like embryonic stem cells (ESCs). However, it has been reported that there are differences in the epigenetic modification and expression of genes between iPSCs and ESCs and the expression of exogenous pluripotent factors is likely to cause tumors (Okita et al., 2007; Bock et al., 2011; Menendez et al., 2011). Although several iPS cell lines have been reported, the silencing of exogenous genes and germline transmission remain unsolved. Thus nuclear transfer is still an indispensable method for somatic cell reprogramming. Meanwhile for pigs, besides therapeutic cloning, SCNT is a useful tool in transgenic research and especially in xenotransplantation and disease models. However, the

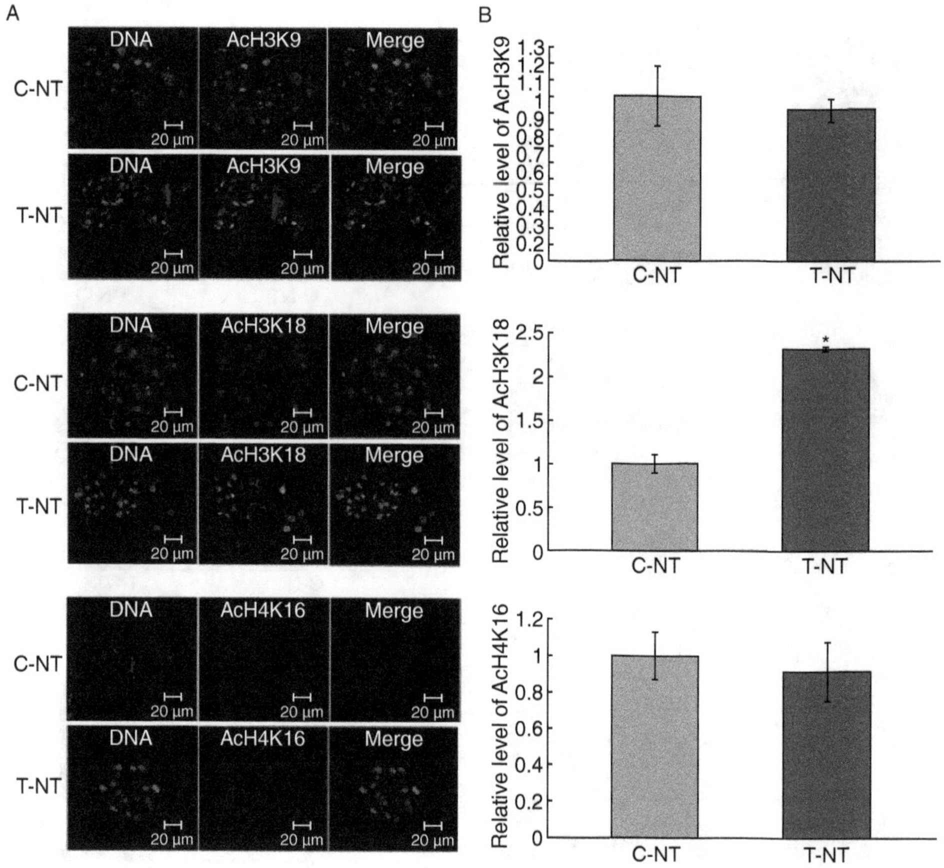

Figure 4. The global AcH3K9, AcH3K18, and AcH4K16 levels of blastocyst stage embryos. (A) Staining of AcH3K9, AcH3K18, and AcH4K16 (green) in non-treated control SCNT embryos (C-NT), and 2 μmol/L CBHA treated embryos (T-NT) at blastocyst stage. Each sample was counterstained with DAPI to visualize DNA (red). Original magnification was 200×. (B) Quantification of AcH3K9 and AcH3K18 signal intensities in C-NT (gray bars) and T-NT (black bars) embryos. Labeling intensity was expressed relative to that of the C-NT embryos (set as 100%). $^{*}P < 0.05$. The experiments were replicated 3 times. n = 8–10 per group.

efficiency of SCNT is still low and the mechanism of nuclear reprogramming is poorly understood.

To improve the efficiency of reprogramming, numerous factors or molecules have been used to alter the epigenetic modification in somatic cells or reconstructed embryos. HDACi is a widely used epigenetic modifier in cancer treatment and somatic cell nuclear reprogramming such as SCNT and iPSCs derivation. However, not all HDACi are beneficial to SCNT embryos, with some of them being unprofitable or detrimental. Generally, type II HDACi is used in epigenetic reprogramming (Monteriro et al., 2011).

Here we used CBHA, a hybrid polar compound, to improve the development of pig SCNT embryos. We found that the efficiency of SCNT and the quality of SCNT embryos are improved after CBHA treatment. Compared with non-treated embryos, CBHA treated SCNT embryos showed a higher global histone acetylation level. Expression of the development-related genes and the imprinted genes at blastocyst stage was also improved with the CBHA treatment.

In mouse SCNT, the optimum CBHA treatment protocol is 20 μmol/L incubation for 10 h post activation (Dai et al., 2010). However, in pig SCNT, we found that the best protocol was 2–10 μmol/L treated for 20–24 h after activation. This difference may be due to the differing developmental time spans between mice and pigs.

Histone acetylation is a significant epigenetic modification in embryo development and nuclear reprogramming. In previous studies, it has been shown that global hyperacetylation associated with a more permissive chromatin state which is an indicator of more fully reprogramming state in nuclear transfer and iPSCs (Li et al., 2002; Rybouchkin et al., 2006). In iPS cells, compared to partially reprogrammed ones, the fully reprogrammed iPSCs exhibit a more open and decondensed chromatin state and are more identical to ES cells (Mattout et al., 2011). We tested the histone acetylation level of reconstructed embryos after the CBHA treatment to examine the effect of this HDACi. Compared with the control SCNT embryos, the CBHA SCNT embryos showed a higher level of global histone acetylation. AcH3K9,

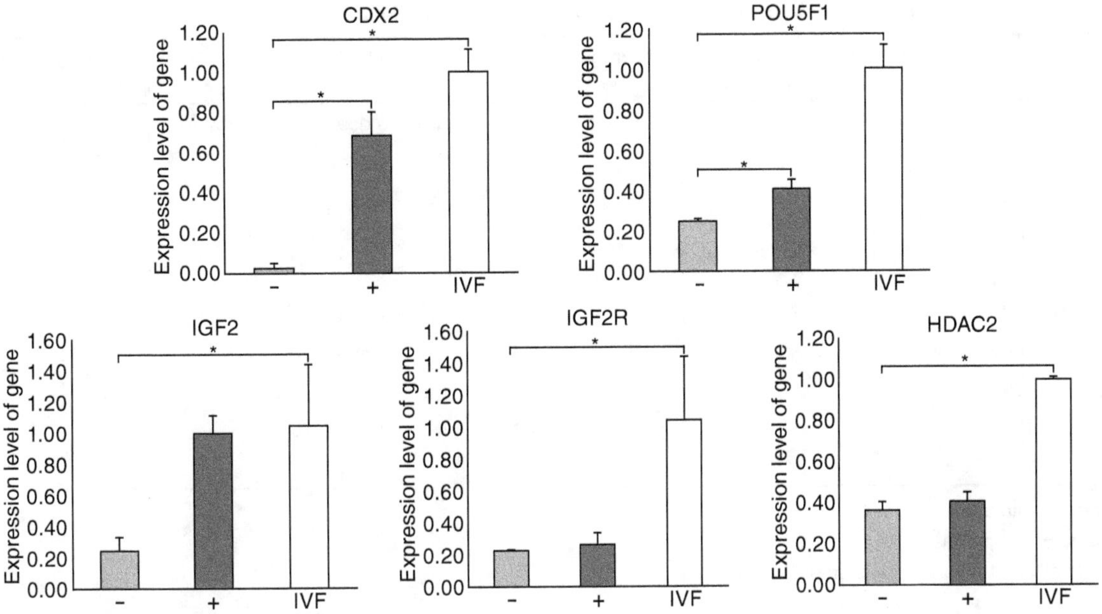

Figure 5. Relative expression at the transcriptional level of histone deacetylase genes, development-related, and imprinted genes. Blastocysts, IVF (open bars), non-treated control SCNT embryos (C-NT) (gray bars), and 2 μmol/L CBHA treated embryos (T-NT) (black bars) were collected on day 7. $^{*}P < 0.05$; $n = 15$.

Table 3. Full-term development of CBHA treated SCNT embryos

Cell type	No. of transferred embryos	CBHA treatment[a]	Pregnancy check[b]	Piglets born	Cloning efficiency (%)[c]
PEFCs	231	+	–	0	0.94
PEFCs	187	+	+	2 dead + 1 mummy	
PEFCs	270	+	+	1 live + 1 dead + 2 mummy	
PEFCs	263	+	+	1 live + 1 dead	
PEFCs	262	–	–	0	
PEFCs	280	–	+	2 live + 5 dead	1.12
PEFCs	259	–	+	1 live + 3 dead	
PEFCs	269	–	+	2 live + 1 dead	
PEFCs	272	–	+	1 dead	

[a] +, Treated; –, untreated.

[b] +, pregnant; –, not pregnant.

[c] No. of piglets/No. of embryos transferred.

AcH3K18, and AcH4K16 were all increased at the 1-cell and 2-cell stages, but there were no significant differences at the blastocyst stage except in AcH3K18. We speculated that CBHA depresses the expression of certain histone deacetylases, thus limiting the modification of deacetylation and increasing the global acetylation level. The hyperacetylation of histone may affect the chromatin remodeling and provide a more permissible and accessible configuration for transcription factors or chaperones combining (Vignon et al., 2002; Lee et al., 2004; Deshmukh et al., 2012; Mason et al., 2012). Embryonic gene activation (EGA) in pigs happened at the 4-cell stage (Maalouf et al., 2009), and the increase of global histone acetylation was mainly discovered at the 1-cell and 2-cell stages, which suggested that a more "open" chromatin structure might be beneficial for EGA.

To further explore the effect of CBHA on SCNT embryo development, we tested the expression of five important genes at the transcription level in the blastocysts of SCNT embryos, CBHA-SCNT embryos, and IVF embryos. This included two development-related genes: *cdx2* and *pou5f1*,

Table 4. Full-term development of CBHA treated SCNT embryos with minimum transfer volume

Cell type	No. of transferred embryos	CBHA treatment[a]	Pregnancy check[b]	Piglets born	Cloning efficiency (%)[c]
PEFCs	100	+	+	6 live + 3 dead + 1 mummy	4.2
PEFCs	100	+	+	6 live + 3 dead	
PEFCs	100	+	+	3 live	
PEFCs	100	+	−	0	
PEFCs	100	+	−	0	
PEFCs	100	−	+	5 live + 1 dead	4.4
PEFCs	100	−	+	5 live + 1 dead	
PEFCs	100	−	+	2 live	
PEFCs	100	−	+	5 live + 3 dead	
PEFCs	100	−	−	0	

[a] +, Treated; −, untreated.

[b] +, pregnant; −, not pregnant.

[c] No. of piglets/No. of embryos transferred.

two imprinted genes: *igf2* and *igf2r*, and histone deacetylase 2 (*hdac2*). In IVF embryos, the expression of *cdx2* and *pou5f1* are significantly higher than the control SCNT and CBHA treated SCNT embryos. However, CBHA treatment improved the expression of the two genes to a large extent compared with the control SCNT embryos. *Cdx2* plays a key role in embryo placental development, and in mouse early embryo development, interaction between *oct3/4* and *cdx2* determines the level of trophectoderm differentiation (Tomanck et al., 1989; Chawengsaksophac et al., 2004; Strumpf et al., 2005). The extremely low expression of *cdx2* in the control SCNT embryos may explain the abnormal placental development in some cloned embryos. While *pou5f1* (*oct4*) is a well-known gene in embryo development, it interacts with regulatory enzymes and epigenetic histone markers in mice. As a central regulator of pluripotency and cell differentiation, it positively modulates the transcription of other recognized stem cell regulators, and it plays a vital role in inner cell mass (ICM) maintenance and ES cell self-renewal (Niwa et al., 2005; Loh et al., 2006). A previous study has shown that in porcine SCNT embryos, the expression of *oct4* and *oct4*-related genes are aberrant (Pesce et al., 2001). It has been reported that in mice, CBHA could improve the derivation efficiency of ESCs from SCNT embryos (Dai et al., 2010). In this process, improved expression of *pou5f1* may act. In our experiment, although still lower than the IVF embryos, the expression of *pou5f1* in CBHA-SCNT embryos was significantly increased compared to the control SCNT embryos. For the importance of embryo development, the two best studied imprinted genes, *igf2* and *igf2r*, have been discussed here. It has been reported that these two genes are related to EGA and involved in fetal growth regulation and are essential for normal development (Lee et al., 2006). Aberrant expression patterns of these imprinted genes may be responsible for the abnormal development found in fetuses and offspring originating from SCNT embryos (Latham et al., 1994; Yang et al., 2005). After CBHA treatment, we observed that the expression of *igf2* increased to the same level as the IVF embryos, yet the expression of *igf2r* was not affected. It can be concluded that CBHA can improve the transcription activity of these genes, which should be related to the more "open" chromatin configuration. CBHA "corrects" the expression of the pivotal genes. It is known that histone deacetylase (*hdacs*) and histone acetylase (*hats*) mediate histone acetylation-deacetylation (Peserico et al., 2011). Previous studies on IVF and *in vivo* porcine embryos have reported that *hdac2* exhibited a low level of expression at 4-cell stage, which steadily increased to its maximum level at the blastocyst stage, however this trend was reversed in SCNT embryos (Kumar et al., 2007; Peserico et al., 2011). Indeed, our results showed that the highest expression of *hdac2* occurred in the IVF blastocyts. However, we also observed that the expression of *hdac2* was hardly affected by the CBHA treatment. This may be because the *hdac2* was not the target protein of CBHA and it is not the only deacetylase to regulate histone deacetylation.

Nevertheless, unlike the improved development *in vitro*, full term development efficiency was hardly affected by the CBHA treatment. The cloning efficiency between CBHA-SCNT and the control SCNT groups is not significantly different. One reason may be that the sample volume was not sufficient to show the difference between the two groups. Another may be because we transferred the embryos at the 2-cell stage. While the *in vivo* and *in vitro* conditions for pre-implanted embryo development are different, the *in vivo* environment may correct the aberrant development from the 2-cell stage onwards to some extent. Another possibility is that the donor cells we transferred are in a good state, therefore the embryos are well developed *in vivo* even without the CBHA treatment. Thus in our experiment, the piglet birth rate is much higher than other reports (0–3.7%) (Zhao et al., 2010), regardless of the CBHA-SCNT (4.2%) or the control SCNT groups (4.4%).

To sum up, our present study indicates that, as a HDACi, CBHA can improve the *in vitro* development of pig nuclear transfer embryos. The presence of CBHA can increase the global acetylation level in AcH3K9, AcH3K18, and AcH4K16. It influences the modification of chromatin structure and improves the expression of important genes in development and imprinting, rendering a more similar expression profile to IVF ones. It enhances the nuclear reprogramming and *in vitro* development of pig SCNT embryos, which may provide better qualified embryos for ES cell derivation in the case of low quality donors. Further studies are needed to discuss the effect of CBHA on chromatin configuration modifications.

MATERIALS AND METHODS

Chemicals

All chemicals were purchased from Sigma Chemical Co. (St. Louis, MO), unless otherwise stated.

Preparation of somatic cells for SCNT

Porcine fetal fibroblast cells were established from a 35-day old fetus. Cells at passage 2–5 were grown to 90% confluency in Dulbecco's Modified Eagle Medium (DMEM; Gibco BRL, Life Technologies, Grand Island, USA) and supplemented with 10% fetal bovine serum (Gibco BRL, Paisley, UK). Fibroblast cells were dissociated with 0.25% trypsin-EDTA (Invitrogen, Carlsbad, CA, USA) at 37°C for about 3–5 min, then transferred to a centrifuge tube with Hepes-buffered Tissue Culture Medium 199 (TCM-199; BioWhittake, Walkersville, MD, USA) and supplemented with 2% cattle serum (CS; Danish Veterinary Institute, DTU, Frederiksberg, Denmark). The suspension was stored in ice until use.

In vitro maturation of oocytes

Ovaries were collected from prepubertal gilts at a local slaughterhouse and transported to the laboratory at 37°C. Follicles between 3 mm and 6 mm in diameter were aspirated with an 18-gauge needle attached to a 10-mL syringe. The cumulus-oocyte complexes (COCs) in the follicular fluid were allowed to settle by gravity. The COCs were rinsed in Hepes-buffered Tyrode's medium containing 0.01% PVA three times. Only the COCs with multiple layers of intact cumulus cells and uniform ooplasm were selected for *in vitro* maturation (IVM). After washing three times in IVM medium, a group of 50 COCs was placed into each well of twenty-four-well cell culture plates containing 500 µL of IVM medium. The COCs were matured for 42 h at 38.5°C and 5% CO_2 in air with 100% humidity. Matured COCs were then vortexed in 0.1% hyaluronidase in Hepes-buffered Tyrode's medium containing 0.01% PVA for 3 min to remove the cumulus cells. Only the matured oocytes having an extruded first polar body (PB) with uniform cytoplasm were used for the SCNT or IVF embryos.

Somatic cell nuclear transfer

Matured oocytes used for SCNT were placed in manipulation medium (MAN) supplemented with 7.5 µg/mL Cytochalasin B, and overlaid with warm mineral oil. Oocytes were enucleated by aspirating the polar body (PB), metaphase II chromosomes, and a small amount of surrounding cytoplasm using a beveled glass pipette with an inner diameter of 17–20 µm. A single intact donor cell was injected into the perivitelline space and placed adjacent to the recipient cytoplasm. Karyoplast-cytoplast complexes (KCCs) were placed into embryo culture medium until fusion and activation occurred. The fusion and activation of the KCCs were accomplished with two direct current pulses (1 s interval) of 1.2 kV/cm for 30 microseconds provided by Eppendorf Multiporator in fusion medium (0.3 mol/L mannitol, 1.0 mmol/L $CaCl_2$, 0.1 mmol/L $MgCl_2$, and 0.5 mmol/L Hepes [pH adjusted to 7.0–7.4]). Oocytes were then incubated for 20 min in PZM3 and evaluated for fusion under a stereomicroscope. Only the fused embryos were placed into four-well cell culture plates containing 500 µL of PZM3 at 38.5°C and 5% CO_2 with 100% humidified air.

Post-activation treatment and embryo culture

Stock solutions of CBHA were dissolved in dimethyl sulfoxide (DMSO) at 100 mmol/L and stored at −80°C. Following electrical activation, the SCNT embryos were treated with various concentrations (from 0.02 µmol/L to 400 µmol/L) of CBHA for different times (from 16 h to 24 h) in PZM3. After treatment, embryos were washed three times and then transferred into a four-well cell culture plate containing 500 µL PZM3 medium. They were cultured at 38°C and 5% CO_2 with 100% humidity either overnight or for 6 days. Cleavage and blastocyst formation were evaluated on days 2 and 7, respectively, with the day of SCNT designated day 0.

Embryo transfer

Day 2 SCNT embryos were transferred to the oviducts of surrogates on the day of, or 1 day after, the onset of estrus. Pregnancy was diagnosed on day 25, and the surrogates were checked regularly at 2-week intervals by ultrasound examination. All of the cloned piglets were delivered by either natural birth or cesarean section on day 117 of gestation and hand raised. All animals were treated according to preapproved institutional animal care and use protocols.

Blastocyst cell number counting

Expanded day 6 blastocysts derived from SCNT were selected for cell number comparison. After being fixed in 4% paraformaldehyde in PBS for 15 min at room temperature, embryos were mounted on slides in mounting medium containing 4,6-diamidino-2-phenylindole. At least 10 oocytes or embryos were processed for each separate sample, and the experiments were replicated three times. Slides were analyzed under an epifluorescent microscope (LEICA DM2500) equipped with a digital camera.

Detection of histone acetylation in SCNT embryos

Embryos treated without (control) or with the optimal CBHA concentration were collected at different stages of development: 1-cell stage (4 h after fusion); 2-cell stage (28–40 h after fusion); and blastocyst (140–150 h after fusion); and fixed in 4% paraformaldehyde in phosphate buffered saline (PBS) for 1 h at 4–8°C. Subsequently, embryos were washed in PBS containing 0.1% Tween 20 (TPBS) for 30 min and permeabilized by 1% Triton X-100 in PBS overnight at 4–8°C. After washing, unspecific antibody binding was blocked by incubation in 2% bovine serum albumin (BSA) in PBS for 1 h, followed by 1 h incubation with primary antibody at RT (anti-acH3K9, 1:500; anti-acH4K16, 1:500; anti-acH3K18, 1:500). After washing, the embryos were incubated for 1 h with Alexa Flour 594-conjugated secondary antibody (1:100, Jackson, USA). Immunofluorescent labeling of AcH3K9, AcH3K18, and AcH4K16 in TSA treated and control embryos were evaluated under Zeiss LSM 510 Laser Scanning Microscope System with the same exposure times and adjustments of the microscope. Fluorescence was measured by analyzing the embryo pictures with Image J software. Average pixel intensity value was measured in 5 random nucleoplasm regions (excluding the nucleolar regions) subtracting the average value measured in the same process within the cytoplasm. Negative control was performed by omitting the primary antibody and led in all cases to the lack of labeling.

Quantitative PCR

At day 15, seven blastocysts were used per sample. The total RNA of the embryos was isolated using the Phenol-chloroform extraction protocol. The mRNA levels were quantified using SYBR Green Jump Start Taq Ready Mix on a real-time PCR detection system at the following thermal cycling conditions: 94°C for 2 min, followed by 40 PCR cycles of 94°C for 15 s, 60°C for 30 s, and 72°C for 30 s. Transcripts were quantified in 3 replicates for each sample and calculated relative to the transcription of the housekeeping gene, β-actin, in every sample. The $2^{-\Delta\Delta CT}$ was used to quantify the relative mRNA levels. The primers used in the quantitative PCR are listed in the Table S1.

Statistical analysis

Data were analyzed using SAS (9.1) with one-way ANOVA. A probability of $P < 0.05$ was considered to be statistically significant.

ACKNOWLEDGEMENTS

This study was supported by grants from the National Natural Science Foundation of China (Grant Nos.90919060 to Q.Z. and 31201112 to T.H.), and the "Strategic Priority Research Program" of the Chinese Academy of Sciences (Grant No. XDA01020101 to Q.Z.).

We thank Yuanxuan Zhao and Zechen Li for nuclear transfer and *in vitro* fertilization assistance; Xiangnan Miao for embryo transfer; and Shiwen Li for confocal microscopic studies.

ABBREVIATIONS

CBHA, *m*-carboxycinnamic acid bishydroxamide; COCs, cumulus-oocyte complexes; EGA, embryonic gene activation; ESCs, embryonic stem cells; HDACi, histone deacetylase inhibitors; iPSC, induced pluripotent stem cells; KCCs, karyoplast-cytoplast complexes; PBS, phosphate buffered saline; SCNT, somatic cell nuclear transfer.

COMPLIANCE WITH ETHICS GUIDELINES

Yuran Song, Tang Hai, Ying Wang, Runfa Guo, Wei Li, Liu Wang, Qi Zhou declare that they have no conflict of interest.

All experiments comply with all of the relevant institutional and national animal welfare laws.

REFERENCES

Baguisi A, Behboodi E, Melican DT, Pollock JS, Destrempes MM et al (1999) Production of goats by somatic cell nuclear transfer. Nat Biotechnol 17:456–461

Bendixen E, Danielsen M, Larsen K, Bendixen C et al (2013) Advances in porcine genomics and proteomics—a toolbox for developing the pig as a model organism for molecular biomedical research. Brief Funct Genomics 9(3):208–219

Bock C, Kiskinis E, Verstappen G et al (2011) Reference maps of human ES and iPS cell variation enable high-throughput characterization of pluripotent cell lines. Cell 144:439–452

Chawengsaksophac K, Graaff W, Beck F et al (2004) Cdx2 is essential for axial elongation in mouse development. PNAS 101:7641–7645

Chesne P, Adenot PG, Viglietta C, Baratte M, Boulanger L et al (2002) Cloned rabbits produced by nuclear transfer from adult somatic cells. Nat Biotechnol 20:366–369

Dai XP, Hao J, Hou XJ, Hai T, Zhou Q et al (2010) Somatic nucleus reprogramming is significantly improved by m-carboxycinnamic acid bishydroxamide, a histone deacetylase inhibitor. J Biol Chem 285:31002–31010

Deshmukh RS, Ostrup O, Hyttel P et al (2012) Early aberrations in chromatin dynamics in embryos produced under in vitro conditions. Cell Reprogram 14:225–234

Ekser B, Ezzelarab M, Hara H, Windt D, Wijkstrom M, Bottino R, Trucco M, Cooper DKC et al (2012) Clinical xenotransplantation: the next medical revolution? Lancet 379:672–683

Galli C, Lagutina I, Crotti G, Colleoni S, Turini P et al (2003) A cloned horse born to its dam twin (vol 424, p 635, 2003). Nature 425:680

Gock H, Nottle M, Lew AM, d'Apice AJ, Cowan P et al (2011) Genetic modification of pigs for solid organ xenotransplantation. Transplant Rev 25:9–20

Han YM, Kang YK, Koo DB, Lee KK et al (2003) Nuclear reprogramming of cloned embryos produced in vitro. Theriogenology 59:33–34

Kato Y, Tani T, Sotomaru Y, Kurokawa K, Kato J et al (1998) Eight calves cloned from somatic cells of a single adult. Science 282:2095–2098

Kim YJ, Ahn KS, Kim M, Shim H et al (2011) Comparison of potency between histone deacetylase inhibitors trichostatin A and valproic acid on enhancing in vitro development f porcine somatic cell nuclear transfer embryos. In Vitro Cell Dev Biol Anim 47:283–289

Kishigami S, Mizutani E, Ohta H et al (2006) Significant improvement of mouse cloning technique by treatment with trichostain A after somatic nuclear transfer. Biochem Biophys Res Commun 340:183–189

Kumar BM, Jin HF, Kim JG, Ock SA, Hong Y, Balasubramanian S et al (2007) Differential gene expression patterns in porcine nuclear transfer embryos reconstructed with fetal fibroblasts and mesenchymal stem cells. Dev Dyn 236:435–446

Lager AE, Ragina NP, Cibelli JB et al (2008) Trichostatin A improves histone acetylation in bovine somatic cell nuclear transfer early embryos. Cloning Stem Cells 10:371–379

Latham KE, Doherty AS, Scott CD, Schultz RM (1994) Igf2r and Igf2 gene expression in androgenetic, gynogenetic, and parthenogenetic preimplantation mouse embryos: absence of regulation by genomic imprinting. Genes Dev 8:290–299

Lee JH, Hart SRL, Skalnik DG (2004) Histone deacetylase activity is required for embryonic stem cell differentiation. Genesis 38:32–38

Lee BC, Kim MK, Jang G, Oh HJ, Yuda F et al (2005) Dogs cloned from adult somatic cells. Nature 436:641

Lee E, Lee SH, Lee BC et al (2006) Analysis of nuclear reprogramming in cloned miniature pig embryos by expression of Oct-4 and Oct-4 related genes. Biochem Biophys Res Commun 348:1419–1428

Li E et al (2002) Chromatin modification and epigenetic reprogramming in mammalian development. Nature 3:662–673

Li ZY, Sun XS, Chen J, Liu XM, Wisely SM et al (2006) Cloned ferrets produced by somatic cell nuclear transfer. Dev Biol 293:439–448

Li J, Svarcova O, Villemoes K, Vajta G et al (2008) High in vitro development after somatic cell nuclear transfer and trichostatin A treatment of reconstructed porcine embryos. Theriogenology 70:800 808

Loh YH, Wu Q, Ng HH et al (2006) The Oct4 and Nanog transcription network regulated pluripotency in mouse embryonic stem cells. Nat Genet 38:431–440

Luo YL, Lin L, Bolund L, Jesen TG, Sorensen CB et al (2012) Genetically modified pigs for biomedical research. J Inherit Metab Dis 35:695–713

Maalouf W, Liu ZC, Zink D et al (2009) Trichostatin A treatment of cloned mouse embryos improves constitutive heterochromatin remodeling as well as developmental potential to term. BMC Dev Biol 9:11

Mason K, Liu ZC, Beaujean N et al (2012) Chromatin and epigenetic modifications during early mammalian development. Anim Reprod Sci 134:45–55

Mattout A, Biran A, Meshorer E et al (2011) Global epigenetic changes during somatic cell reprogramming to iPS cells. J Mol Biol 3:341–350

Menendez S, Camus S, Herreria A et al (2011) Increased dosage of tumor suppressors limits the tumorigenicity of cells without affecting their pluripotency. Aging Cell 11:41–50

Miyoshi K, Mori H, Mizobe Y, Sato M et al (2010) Benefical effects of reversine on in vitro development of miniature pig somatic cell nuclear transfer embryos. J Reprod Dev 56:291–296

Monteriro FM, Oliveira CS, Oliveira LZ, Garcia JM et al (2011) Chromatin modifying agents in the vitro production of bovine embryos. SAGE-Hindawi access to research veterinary medicine international

Niwa H, Toyooka Y, Rossant J et al (2005) Interaction between Oct3/4 and Cdx2 determines trophectodern differentiation. Cell 123:917–929

Ogura A, Inoue K, Wakayama T et al (2013) Recent advancements in cloning by somatic cell nuclear transfer. Philos Trans R Soc 368:20110329

Okita K, Ichisaka T, Yamanaka S et al (2007) Generation of germline-competent induced pluripotent stem cells. Nature 448:313–318

Onishi A, Iwamoto M, Akita T, Mikawa S, Takeda K et al (2000) Pig cloning by microinjection of fetal fibroblast nuclei. Science 289:1188–1190

Pesce M, Scholer HR et al (2001) Oct-4: gatekeeper in the beginnings of mammalian development. Stem Cells 19:271–278

Peserico A, Simone C et al (2011) Physical and functional HAT/HDAC interplay regulates protein acetylation balance. J Biomed Biotechnol 2011:371832

Polejaeva IA, Chen SH, Vaught TD, Page RL, Mullins J, Ball S, Dal Y, Boone J et al (2000) Cloned pigs produced by nuclear transfer from adult somatic cells. Nature 407:86–90

Rybouchkin A, Kato Y, Tsunoda Y et al (2006) Role of histone acetylation in reprogramming of somatic nuclei following nuclear transfer. Biol Reprod 74:1083–1089

Shin T, Kraemer D, Pryor J, Liu L, Rugila J et al (2002) A cat cloned by nuclear transplantation. Nature 415:859

Strumpf D, Mao CA, Yamanaka Y, Rossant J et al (2005) Cdx2 is required for correct cell fate specification and differentiation of trophectoderm in the mouse blastocyst. Development 132:2093–2102

Takahashi K, Yamanaka S et al (2006) Induction of pluripotent stem cells from mouse embryonic and adult fibroblast cultures by defined factors. Cell 126:663–676

Tomanck M, Kopecny V, Kanka J et al (1989) Genome reactivation in developing early embryos: an ultrastructural and autoradiaphic analysis. Anat Embryol 180:309–316

Verma N, Rettenmeier AW, Spanke SS et al (2011) Recent advances in the use of *Sus scrofa* (pig) as a model system for proteomic studies. Proteomics 11:776–793

Vignon X, Zhou Q, Renard JP et al (2002) Chromatin as a regulative architecture od the early developmental functions of mammalian embryos after fertilization or nuclear transfer. Cloning Stem Cells 4:363–377

Wakayama T, Perry ACF, Zuccotti M, Johnson KR, Yanagimachi R (1998) Full-term development of mice from enucleated oocytes injected with cumulus cell nuclei. Nature 394:369–374

Walker SC, Shin T, Zaunbrecher GM, Romano JE, Johnson GA, Bazer FW, Piedrahita JA et al (2002) A highly efficient method for porcine cloning by nuclear transfer using in vitro-matured oocytes. Cloning Stem Cells 4:105–112

Whitworth K, Prather RS et al (2010) Somatic cell nuclear transfer efficiency: how can it be improved through nuclear remodeling and reprogramming? Mol Reprod Dev 77:1001–1015

William M, Rideout III, Eggan K, Jaenisch R et al (2001) Nuclear cloning and epigenetic reprogramming of the genome. Science 293:1093–1098

Wilmut I, Schnieke AE, McWhir J, Kind AJ, Campbell KHS (1997) Viable offspring derived from fetal and adult mammalian cells (vol 385, p 810, 1997). Nature 386:200

Woods GL, White KL, Vanderwall DK, Li GP, Aston KI et al (2003) A mule cloned from fetal cells by nuclear transfer. Science 301:1063

Yang L, Palmer PC, Tian XC et al (2005) Expression of imprinted genes is aberrant in deceased newborn cloned calves and relatively normal in surviving adult clones. Mol Reprod Dev 71:431–438

Yu JY, Vodyanik MA, Smuga-Otto K et al (2007) Induced pluripotent stem cell lines derived from human somatic cells. Science 318:1916–1920

Zhao JG, Hao YH, Prather RS et al (2010) Histone deacetylase inhibitors improve in vitro and in vivo developmental competence of somatic cell nuclear transfer porcine embryos. Cell Reprogram 12:75–83

Zhou Q, Renard JP, Le Friec G, Brochard V, Beaujean N et al (2003) Generation of fertile cloned rats by regulating oocyte activation. Science 302:1179

11

Understanding the genetic and epigenetic architecture in complex network of rice flowering pathways

Changhui Sun[1], Dan Chen[1], Jun Fang[2], Pingrong Wang[1], Xiaojian Deng[1]✉, Chengcai Chu[2]✉

[1] Rice Research Institute, Sichuan Agricultural University, Chengdu 611130, China
[2] State Key Laboratory of Plant Genomics and Center for Plant Gene Research (Beijing), Institute of Genetics and Developmental Biology, Chinese Academy of Sciences, Beijing 100101, China
✉ Correspondence: xjdeng2006@aliyun.com (X. Deng), ccchu@genetics.ac.cn (C. Chu)

ABSTRACT

Although the molecular basis of flowering time control is well dissected in the long day (LD) plant Arabidopsis, it is still largely unknown in the short day (SD) plant rice. Rice flowering time (heading date) is an important agronomic trait for season adaption and grain yield, which is affected by both genetic and environmental factors. During the last decade, as the nature of florigen was identified, notable progress has been made on exploration how florigen gene expression is genetically controlled. In Arabidopsis expression of certain key flowering integrators such as *FLOWERING LOCUS C* (*FLC*) and *FLOWERING LOCUS T* (*FT*) are also epigenetically regulated by various chromatin modifications, however, very little is known in rice on this aspect until very recently. This review summarized the advances of both genetic networks and chromatin modifications in rice flowering time control, attempting to give a complete view of the genetic and epigenetic architecture in complex network of rice flowering pathways.

KEYWORDS rice, flowering time, genetic network, chromatin modifications, Arabidopsis, florigen

INTRODUCTION

Rice flowering time (heading date), which is affected by both endogenous and exogenous factors, is an important agronomic trait for regional and seasonal adaption. Heading on a proper time is the most critical step for grain production. Precocious flowering reduces the vegetative phase and leads to reduction of biological yield. On the other hand, delayed flowering could cause low seed setting percentage in cold late autumn or delay next planting season, which both results in production loss.

Florigen is produced in the leaf under inductive day length conditions and transported to the shoot apex where it triggers flowering transition (Cajlachjan, 1937; Corbesier et al., 2007; Tamaki et al., 2007). Unlike only one florigen gene *FLOWERING LOCUS T* (*FT*) in Arabidopsis, rice evolves two florigen genes, *Heading date 3a* (*Hd3a*) and *RICE FLOWERING LOCUS T 1* (*RFT1*), and at least two flowering pathways are developed to control the expression of florigens, the *Heading date 1* (*Hd1*) pathway which is conserved between rice and Arabidopsis, and the *Early heading date 1* (*Ehd1*) pathway which is unique to rice (Doi et al., 2004) (Fig. 1). Numerous studies reveal that a large number of rice genes regulate flowering time through the two flowering integrators.

In Arabidopsis, some flowering regulators such as *FLC* and *FT* are reported to be regulated by various chromatin modifications (He, 2009; Liu et al., 2010). However, little is known in rice in this field. Recently, we characterized a major histone methyltransferase (HMTase) gene *SET DOMAIN GENE 724* (*SDG724*), which is required for Histone H3 lysine 36 (H3K36) methylation, promotes rice heading, indicating that rice flowering could also be regulated by chromatin modifications (Sun et al., 2012). In the past two years, more and more molecular genetic studies gave the clues on the chromatin modification mechanism in rice flowering pathways, we summarize here most recent advances towards understanding of genetic networks and epigenetic chromatin modifications in rice flowering time control.

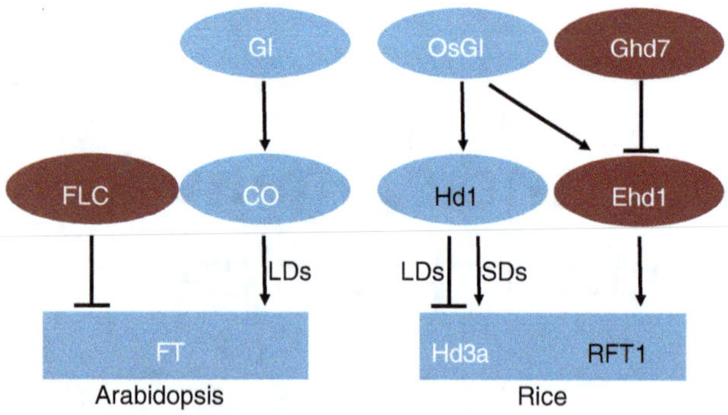

Figure 1. Comparison of core-flowering-pathways in rice and Arabidopsis. *OsGI/GI-Hd1/CO-Hd3a/FT* pathway is conserved between rice and Arabidopsis. *CO*. accelerates flowering under LD, however, *Hd1* promotes flowering under SD and represses it under LD. Besides, *Ghd7* and *Ehd1* in rice, and *FLC* in Arabidopsis are unique flowering integrators, respectively. *FLC* and *Ghd7* are major flowering suppressors, while *Ehd1* acts as a flowering promoter.

TWO FLORIGEN GENES *HD3A* AND *RFT1* IN RICE

Florigen, which has been hypothesized by many physiological studies, is believed to be produced in leaves by the inductive photoperiod, then moves to the shoot apical meristem (SAM) and triggers flowering transition. But this florigen has been eluded identification since it was first proposed for 70 years (Cajlachjan, 1937). In 2007, it was firstly revealed that *FT* encoded protein in Arabidopsis, is a leaf-derived long-distance signal directed to floral transition (Corbesier et al., 2007).

In rice, there are 13 *FT* homologs in the genome (Chardon and Damerval, 2005), *Hd3a* and *RFT1* are two of them which were confirmed to act as florigen genes (Komiya et al., 2008; Komiya et al., 2009; Tamaki et al., 2007). By fusing *Hd3a* or *RFT1* with GFP, it was demonstrated that Hd3a or RFT1 protein was expressed in vascular tissue of leaves, and could be moved to SAM where they started flowering induction. As Hd3a-GFP was only detected in the SAM of plants grown under short day conditions (SD), RFT1-GFP was merely detected under long day conditions (LD) (Komiya et al., 2009; Tamaki et al., 2007). On the other hand, *Hd3a-RNAi* (*RNA interference*) plants significantly delayed heading date under SD but not LD, *RFT1-RNAi* plants flowering was obviously delayed under LD but not SD oppositely. Furthermore, rice with knockout of both Hd3a and RFT1 caused at least 300 days late flowering under both SD and LD (Komiya et al., 2009). All these data demonstrated that, unlike Arabidopsis, rice has two florigen genes, *Hd3a* and *RFT1*, *Hd3a* is responsible for flowering under inductive SD, whereas *RFT1* is responsible for flowering transition under non-inductive LD. Although *Hd3a* and *RFT1* are located in some chromosome and separated by only 11.5 kb in the genome, the fine-tuning of long day flowering by the H3K36me2/3 level of *RFT1* but not *Hd3a* via SDG724, therefore, *RFT1* and *Hd3a* which have functionally diverged to control flowering time under LD and SD conditions are partly due to a fine-tuned epigenetic mechanism (Sun et al., 2012).

FLORIGEN REGULATED NETWORK

How flowering pathways are regulated differs in plants. In Arabidopsis, flowering is controlled by a small number of large-effect genes such as *FLC* (Salome et al., 2011), whereas in maize is controlled by many additive small-effect quantitative trait loci (QTLs) (Buckler et al., 2009). Interestingly, rice combines both regulatory manners, including a few large-effect factors, such as *Hd1*, *Ehd1*, and *Grain number, plant height and heading date 7* (*Ghd7*), in addition to some small-effect QTLs and genes (Ebana et al., 2011; Tsuji et al., 2013) (Table 1).

So far, quite a number of QTLs controlling rice heading date (Hd) were identified and characterized using different segregating populations derived from crossing a japonica cultivar (*Nipponbare*) and an indica cultivar (*Kasalath*) (Lin et al., 1998; Yano et al., 1997). These QTLs include the major loci controlling photoperiodic flowering responses, *Hd1* (Yano et al., 1997; Yano et al., 2000), *Hd2/Ghd7.1/OsPRR37* (*Oryza sativa Pseudo-Response Regulator 37*) (Koo et al., 2013; Liu et al., 2013; Shibaya et al., 2011; Yamamoto et al., 2000; Yan et al., 2013), *Hd3a* (Kojima et al., 2002), *Hd4/Ghd7* (*Ghd7* for short) (Koo et al., 2013; Xue et al., 2008), *Hd5/Days to heading 8/Grain number, Plant height, and Heading date 8/Late Heading Date 1* (*Hd5/DTH8/Ghd8/LHD1*) (Dai et al., 2012; Fujino et al., 2013; Yan et al., 2011). Furthermore, backcross progenies derived from the same original cross allowed identification of other QTLs, such as *Hd6/CK2* (*CASEIN KINASE 2*) (Ogiso et al., 2010; Takahashi et al., 2001; Yamamoto et al., 2000), *Hd14/Ehd1* (Doi et al., 2004), *Hd16/EL1* (*Early flowering 1*) (Dai and Xue, 2010; Hori et al., 2013; Shibaya et al., 2011), *Hd17/OsELF3/EF7/OsEF3/Hd3b* (*Hd17/Oryza sativa Early Flowering 3/Early Flowering 7/Oryza sativa Early Flowering 3/Hd3b*, *OsELF3* for short) (Hori et al., 2012; Matsubara et al., 2012; Saito et al., 2012; Yang et al., 2013; Zhao et al., 2012). Additionally, using rice near isogenic lines and mutants, more genes implicated in controlling flowering time have been identified and positioned into a regulatory network (Brambilla and Fornara, 2013; Itoh and Izawa, 2013; Tsuji et al., 2011, 2013) (Fig. 2).

HD1-DEPENDENT PATHWAY

There is a similar molecular system for florigen control in Arabidopsis and rice (Izawa, 2007; Tsuji et al., 2011). *Hd1* and *Hd3a* in rice are homologs of *CONSTANS* (*CO*) and *FT* in Arabidopsis, respectively. As in Arabidopsis, *Hd1* acts upstream of *Hd3a* (Kojima et al., 2002; Yano et al., 2000), and overexpression of a rice ortholog of Arabidopsis

Table 1. Flowering regulators in rice

Gene name	Pathways	Gene ID	Note	For short
RFT1		*Os06g0157500*	Similar to phosphatidylethanolamine-binding protein	
Hd3a		*Os06g0157700*	Similar to phosphatidylethanolamine-binding protein	
Hd1		*Os06g0275000*	CCT domain protein	
Hd14/Ehd1		*Os10g0463400*	B-type response regulator	*Ehd1*
RCN1		*Os11g0152500*	Similar to phosphatidylethanolamine-binding protein	
Hd6/CK2	*Hd1*	*Os03g0762000*	Casein kinase II alpha subunit	*Hd6*
ETR2	*Hd1, RCN1*	*Os04g0169100*	Ethylene receptor	
OsLF	*Hd1*	*Os05g0541400*	Atypical HLH protein	
SPL11	*Hd1, Hd3a*	*Os03g0275900*	E3 ubiquitin ligase	
Spin1	*Hd1, Hd3a*	*Os07g0227400*	K homology domain protein	
OsGI	*Hd1, Ehd1*	*Os01g0182600*	Circadian clock related GIGANTEA protein	
SE5	*Hd1, Ehd1*	*Os06g0603000*	Heme oxygenase	
OsphyB	*Hd1, Ehd1*	*Os03g0309200*	Phytochrome	
Hd17/OsELF3/EF7/OsEF3/Hd3b	*Hd1, Ehd1*	*Os06g0142600*	ELF3 homolog protein	*OsELF3*
Hd4/Ghd7	*Ehd1*	*Os07g0261200*	CCT domain protein	*Ghd7*
Ehd3	*Ehd1*	*Os08g0105000*	Plant homeo domain (PHD) finger	
Hd16/EL1	*Ehd1*	*Os03g0793500*	Casein kinase I	*EL1*
OsMADS50	*Ehd1*	*Os03g0122600*	MADS box protein	
OsMADS56	*Ehd1*	*Os10g0536100*	MADS box protein	
OsLFL1	*Ehd1*	*Os01g0713600*	B3 domain transcription factor	
OsCOL4	*Ehd1*	*Os02g0610500*	CCT domain protein	
Hd5/Ghd8/DTH8	*Ehd1*	*Os08g0174500*	HAP3 subunit	*DTH8*
Ehd2/OsID1/RID1	*Ehd1*	*Os10g0419200*	C2H2 zinc-finger protein	*RID1*
Ehd4	*Ehd1*	*Os03g0112700*	CCCH-type zinc finger protein	
OsMADS51	*Ehd1*	*Os01g0922800*	MADS box protein	
OsCO3	*Florigen*	*Os09g0240200*	CONSTANS-like protein	
DTH2	*Florigen*	*Os02g0724000*	CONSTANS-like protein	
Hd2/OsPRR37/Ghd7.1	*Florigen*	*Os03g0112700*	CCT domain protein	*OsPRR37*
OsDof12	*Florigen*	*Os03g0169600*	DNA-binding with one finger protein	

Green font, flowering promoter; red font, flowering supressor; purle font, promoter under SDs and supressor under LDs.

GIGANTEA (*GI*) which acts upstream of *CO*, namely *OsGI*, increased the expression of *Hd1* in the transgenic plants, followed by suppressing *Hd3a* expression, resulting in late flowering under both SD and LD (Hayama et al., 2003). Differently, *CO* merely promotes *FT* expression, *Hd1* plays a more enigmatic role in rice, which promotes flowering under SD, but represses flowering under LD (Hayama et al., 2003; Komiya et al., 2008; Lin et al., 2000; Tamaki et al., 2007). These results indicate that the core photoperiodic pathway composed of the three key flowering genes *OsGI/GI-Hd1/CO-Hd3a/FT* is conserved between rice and Arabidopsis, but its function has diverged during evolution to produce opposite flowering responses. While the photoperiodic pathway in Arabidopsis merely accelerates flowering under LD, in rice, it promotes flowering under SD and represses flowering under LD (Takahashi and Shimamoto, 2011).

The reversible mechanism that *Hd1* functions as either an activator or suppressor of *Hd3a* involves the action of the red-light photoreceptor phytochrome B (phyB), since mutations in *phyB* or phytochrome chromophore synthesis, such as *photoperiod sensitivity 5* (*se5*), attenuate this conversion and maintain *Hd1* as an activator under any photoperiodic conditions. On the other hand, though *Hd1*-overexpressing plants delay flowering, Hd1 protein levels in these plants are not significantly altered (Andres et al., 2009; Ishikawa et al., 2011; Izawa et al., 2002), thus it is speculated that LD light signals may modify the protein of Hd1 or Hd1 complex through phytochrome but not its expression levels, and convert it into a repressor of flowering. Therefore, uncovering

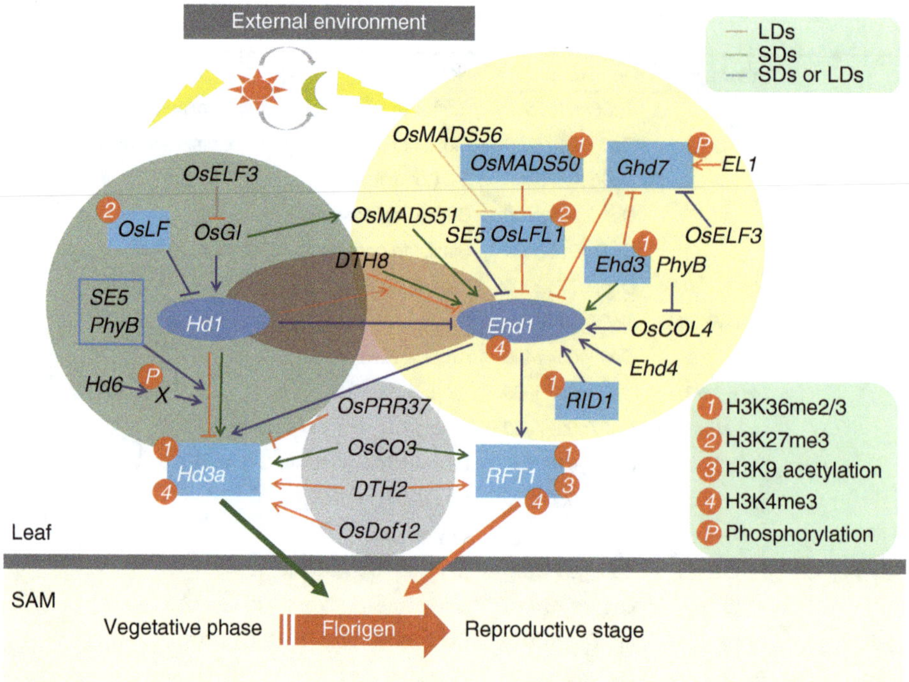

Figure 2. Complex flowering time control network in rice. Rice flowering network is formed by two florigen genes *Hd3a* and *RFT1*, and four regulation modules, including *Hd1*-dependent pathway, *Ehd1*-dependent pathway, crosstalk between *Hd1* and *Ehd1* pathway, and flowering regulators independent of *Hd1* and *Ehd1*. The first three signals come together to regulate *Hd1* and *Ehd1* and affect florigen gene expression; the last module may directly control the expression of florigen genes independent of *Hd1* and *Ehd1*. Besides, expressions of *Ehd3*, *RID1*, *OsMADS50*, *Hd3a*, and *RFT1* can be affected by H3K36me2/3; *OsLFL1* and *OsLF* transcriptions can be mediated by H3K27me3. Finally, all those florigen signals are transported from leaves to SAM and trigger flowering transition there. All the gene names for short are showed in Table 1.

of the biochemical function of Hd1 protein and the molecular nature of its dual activity will provide exciting insight into the control of photoperiodic flowering in rice.

Recently, it is deduced that Hd1 protein activity is possible affected by an additional posttranslational regulatory factor, *Hd6*, which encodes a CK2 α-subunit (Ogiso et al., 2010; Takahashi et al., 2001). The delay flowering effect of Hd6 is observed only when Hd1 is functional, however, Hd1 is not phosphorylated by Hd6 *in vitro* (Ogiso et al., 2010), suggesting that Hd6 phosphorylates unknown substrates that cooperate with Hd1 in the LD floral suppression pathway.

EHD1 DEPENDENT PATHWAY

In 2004, a novel regulatory *Ehd1*-pathway which is not presented in Arabidopsis, is discovered in rice (Doi et al., 2004). *Ehd1*, encoding a B-type response regulator, is a floral promoter, and rice variety Taichung 65 (T65) without functional *Ehd1* allele delays flowering under both LD and SD (Doi et al., 2004). As it has been shown that *Ehd1* contributes to flowering time by its expression levels (Takahashi et al., 2009), thus fine-tuning of *Ehd1* expression is crucial for rice flowering at suitable time, and several flowering regulators have been identified to participate in this regulation.

Ghd7, which is important for increasing rice productivity and adaptability, is a major regulator of *Ehd1* and could delay flowering by repressing *Ehd1* under LD (Takahashi et al., 2009; Xue et al., 2008). As *Ghd7* encodes a CCT (CONSTANS, CO-like, and TOC1) domain protein, which shows very low homology to Arabidopsis genome, the *Ghd7-Ehd1* may be a unique pathway in rice (Koo et al., 2013; Xue et al., 2008). Further study shows that *Ghd7* and *Ehd1* can respectively set a daylength threshold for *Hd3a* expression, which is usually observed in SD plants but not in LD plants (Itoh et al., 2010; Takimoto and Ikeda, 1961), and this capacity of discernment in critical day length in rice greatly enriches the daylength-dependent regulated mechanism of florigen gene expression.

Until now, at least three genes, *Early heading date 3* (*Ehd3*), *ELF3*, and *Hd16/EL1*, were identified to control *Ghd7* expression in *Ehd1*-pathway. *Ehd3* encodes a plant homeodomain (PHD) finger protein and is identified as one repressor of *Ghd7*. Generally, *Ghd7* transcript reaches its highest level after seeding for two weeks, and then the expression is gradually reduced to a basal level, but in *ehd3* mutants, *Ghd7* expression level is always higher and delays heading date for more than one year under LD. Interestingly, under SD, *Ehd3* could promote *Ehd1* expression regardless of *Ghd7*, suggesting a perplexed role of *Ehd3* (Matsubara et al., 2011).

ELF3 in Arabidopsis is responsible for generating circadian rhythm and regulating photoperiodic flowering, consistently, its homolog in rice *OsELF3* is also required to sustain the robust oscillation, and lesions in *OsELF3* delay flowering under both SD and LD (Saito et al., 2012; Yang et al., 2013; Zhao et al., 2012). Under SD, *OsELF3* promotes flowering mainly by repressing *Ghd7*, because late flowering of *oself3* mutants can be rescued if *Ghd7* but not *Hd1* is mutated. Under LD, *oself3* mutants increase *OsGI* and *Ghd7* expression, thus up-regulate *Hd1* and repress *Ehd1* expression, respectively, indicating that *OsELF3* influences photoperiodic flowering in both *Hd1* and *Ehd1* pathways (Brambilla and Fornara, 2013; Saito et al., 2012).

Hd16/EL1, encoding a casein kinase I protein, is associated with the gibberellin-mediated flowering transition (Dai and Xue, 2010). Deficient in *Hd16* weakens rice photoperiod sensitivity, but increases *Ehd1*, *Hd3a*, and *RFT1* expression under LD. Though the expression level of *Ghd7* is not significantly altered in *el1* mutants, the biochemical data indicate that *Hd16* acts as a flowering repressor by phosphorylation of *Ghd7* (Dai and Xue, 2010; Kwon et al., 2013).

OsLFL1 (*Oryza sativa LEC2 and FUSCA3 Like 1*) encodes a putative B3 transcription factor, knockdown of *OsLFL1* does not affect flowering time, while ectopic overexpression of *OsLFL1* decreases *Ehd1* expression and results in late flowering (Peng et al. 2007, 2008). *OsLFL1* is controlled by two members of MIKC-type MADS-box family, *OsMADS50* and *OsMADS56*. Both *osmads50* mutants and *OsMADS56*-overexpressing plants, which produce increased *OsLFL1* expression, show late flowering phenotype (Lee et al., 2004; Ryu et al., 2009). Interestingly, OsMADS56 can interact with OsMADS50 *in vitro*, suggesting that the two MADS-box members tend to form a heterodimer complex and function antagonistically through *OsLFL1-Ehd1* pathway under LD (Ryu et al., 2009).

As mentioned in *Hd1*-pathway, phytochrome is probably a primary cause of *Hd1*-dependent suppression of rice flowering, but underlying molecular mechanism of phytochrome in *Ehd1*-pathway is not well understood. Recent studies showed that *SE5* and *phyB* also suppress *Ehd1* expression, and the *phyB*-mediated suppression of *Ehd1* is confirmed to be repressed by a CONSTANS-like (COL) gene *OsCOL4* (*Oryza sativa CONSTANS-like 4*) (Andres et al., 2009; Komiya et al., 2009; Lee et al., 2010). *OsCOL4* expression is decreased in *osphyB* mutants, and *osphyB oscol4* double mutants flower is similar to *osphyB* single mutants, indicating that *OsCOL4* functions downstream of *OsphyB* (Lee et al., 2010).

Besides the above regulators, *Ehd1* expression is also modulated by other four flowering factors independently. *Indeterminate 1* (*ID1*) is one of them, which expresses in leaf but induces flowering in the shoot meristem. *ID1* has been once thought to be involved in the florigen synthesis in maize (Colasanti et al., 2006; Colasanti et al., 1998), and its regulated mechanism has been exhibited in rice. Lesions in rice *RID1* (*Early heading date 2/OsINDETERMINATE 1/Rice INDETERMINATE 1, Ehd2/OsID1/RID1, RID1* for short) lead to extremely late flowering phenotype, as well as decreased expression of *Ehd1* and downstream florigen genes under both SD and LD (Matsubara et al., 2008; Park et al., 2008; Wu et al., 2008).

Ehd4 (*Early heading date 4*), encoding a CCCH-type zinc finger transcriptional regulator, is expressed mostly in immature leaves and shows a similar diurnal expression pattern of *Ehd1* under both SD and LD. *Ehd4* up-regulates the expression of the florigen genes *Hd3a* and *RFT1* through *Ehd1*. Strikingly, *Ehd4* is highly conserved in both wild rice and cultivated rice, but homologs cannot be found in other species, suggesting that *Ehd4* is unique flowering regulator in *Oryza* genus differed from other grass members during evolution (Gao et al., 2013).

OsMADS51 is another MADS box gene, other than *OsMADS50* and *OsMADS56*, it acts downstream of *OsGI*, transmits a promotion signal from *OsGI* to *Ehd1* under SD. Though its null mutants showed late flowering phenotype followed by decreased expression of *Ehd1* and *Hd3a*, ectopic expression of *OsMADS51* causes early flowering, accompanying with increased expression *Ehd1* and *Hd3a* (Kim et al., 2007).

Hd5/DTH8/Ghd8/LHD1 encodes a putative HEMEACTIVATOR PROTEIN 3 (HAP3) subunit of a CCAAT-box binding protein (HAP complex) that binds to CCAAT boxes in yeast and animals. Similar to *Hd1*, *Hd5/DTH8/Ghd8/LHD1* delayed flowering in rice under LD and promotes flowering under SD, but by regulating expression of *Ehd1* (Dai et al., 2012; Lin et al., 2003; Wei et al., 2010; Yan et al., 2011).

Most interestingly, though *Hd5/DTH8/Ghd8/LHD1* suppresses rice heading though *Ehd1*, genetic analysis implies that *Hd5* requires functional *Hd1* to repress flowering under LD (Nonoue et al., 2008), rising a question what is the relationship between *Hd1* and *Ehd1*. Recent findings indicate that transcript level of *Ehd1* is down-regulated in *Hd1*-overexpression transgenic lines, suggesting that, to some degree, *Hd1* is an upstream regulator of *Ehd1* expression, but how this crosstalk works is still undefined (Ishikawa et al., 2011).

FLOWERING REGULATORS INDEPENDENT OF *HD1* AND *EHD1*

Besides *Ehd1*, T65 also bears a loss-of-function allele of *Hd1*, but it could still flower in time and serves as a commercial rice variety, so there are must some other regulators independent of *Hd1* and *Ehd1* in rice flowering network (Doi et al., 2004). *OsCO3* and *DTH2* are two of them, and promote flowering by regulating florigen genes. Though both of them are COL genes, they function under different photoperiodic conditions. Expressions of *Hd3a* and *FT-like* genes are decreased in the *OsCO3*-overexpressing plants under SD without altered expression of other florigen upstream regulators, suggesting that *OsCO3* primarily controls flowering time under SD by negatively regulating the expression of florigen genes, independent of other known SD-promotion pathways (Kim et al., 2008). For *DTH2*, both association analysis and transgenic experiments indicate that two functional nucleotide polymorphisms that correlated with early heading and increased reproductive fitness under natural LD in northern Asia. Further combined population genetics and network analyses suggest that *DTH2* probably represents a target of artificial selection for adaptation to LD during rice domestication and improvement, demonstrating an important role of minor effect quantitative trait loci in crop adaptation and breeding (Wu et al., 2013).

Although some *PRR* genes are major components of the circadian oscillator, a rice *PRR* gene *Hd2/Ghd7.1/OsPRR37* may down-regulate *Hd3a* expression independent of any known pathways to suppress flowering under LD. As lesions in *Hd2/Ghd7.1/OsPRR37* cause early flowering phenotype, the japonica varieties harboring nonfunctional alleles of both *Ghd7* and *Hd2/Ghd7.1/OsPRR37* flower extremely early under natural LD, and make these varieties adapt to the northernmost rice cultivation regions. Further study implied that natural variations in *Hd2/Ghd7.1/OsPRR37* have contributed to the expansion of rice cultivation to temperate and cooler regions (Koo et al., 2013; Liu et al., 2013; Yan et al., 2013).

Different from *Hd2/Ghd7.1/OsPRR37*, *OsDof12* is LD-specific flowering repressor and encodes a DNA-binding with one finger (Dof) transcription factor which is involved in a variety of biological processes of plants. The transcriptions of *OsDof12* can express at different development stages, but strongly inhibited by dark treatment. *OsDof12*-overexpressing plants flower earlier in consistent with the up-regulation of *Hd3a* independent of other flowering genes under LD but not SD (Li et al., 2009).

CHROMATIN MODIFICATIONS REGULATE FLOWERING IN RICE

Chromatin, which is composed by complexing DNA with histone, carries not only genetic, but also epigenetic information. In Arabidopsis, the expression of a major flowering repressor *FLC* is regulated by a number of active and repressive chromatin modifications, such as histone tails methylation, acetylation, ubiquitination etc. In addition, histone modifications can also directly regulate the expression of florigen gene *FT*, and the regulation manner of *FLC* and *FT* provides a paradigm for control of developmental regulators through chromatin modifications (He, 2009). Currently, not so many data are available about that in rice, but molecular genetic studies indicated that rice flowering control also undergoes the complex chromatin modifications (Table 2).

ACTIVE CHROMATIN MODIFICATIONS AND RICE FLOWERING

S-Adenosyl-l-methionine is a universal methyl group donor involved in numerous transmethylation reactions, including histone methylation. Knockdown of rice *S-Adenosyl-l-methionine synthetase* (*SAMS*) *1*, *2*, and *3* greatly reduced the expression of *Ehd1*, *Hd3a*, *RFT1* and led to a late flowering phenotype. Moreover, the histone H3K4me3 and symmetric DNA methylation at these genes was significantly reduced, suggesting an association between epigenetic modification and flowering in rice, but more research are required on this relationship (Li et al., 2011).

We have demonstrated that *SDG724*, a histone methyltransferase gene which belongs to SET domain family Class II (Ng et al., 2007), affected flowering time by mediating H3K36 methylation in rice. *SDG724* loss-of-function mutant *lvp1* showed a late flowering phenotype under both LD and SD, which was associated with the suppressed expression of *RFT1* and *Hd3a*. Interestingly, only the chromosomal region of *RFT1*, but not *Hd3a*, reduced the level of H3K36me2/3 modifications which associated with the transcriptionally active chromatin state, although the two florigenic genes are closely linked in the genome and separated by only 11.5 kb (Sun et al., 2012). This similar regulated way in *RFT1* is also found in a previous report that *RFT1* expression can be promoted through another active

Table 2. Chromatin modification regulators in rice

Gene name	Pathways	Gene ID	Note	Modifications	Target genes
SDG724	*Ehd1*	*Os09g0307800*	SET domain group protein	H3K36me2/3	*RFT1, OsMADS50*
SDG725	*Ehd1*	*Os02g0554000*	SET domain group protein	H3K36me2/3	*Ehd3, Ehd2, OsMADS50, Hd3a, RFT1*
LC2/OsVIL3	*Hd1*	*Os02g0152500*	Plant homeo domain (PHD) finger	H3K27me3	*OsLF*
OsVIL2	*Ehd1*	*Os12g0533500*	Plant homeo domain (PHD) finger	H3K27me3	*OsLFL1*
OsEMF2b	*Ehd1*	*Os04g0162100*	C2H2 zinc-finger protein, interact with *OsVIL2*	H3K27me3	*OsLFL1*
OsTrx1	*Ehd1*	*Os09g0134500*	SET domain group protein, interact with *Ehd3*	Unkown	Unkown

histone modification H3K9 acetylation around the transcriptional start site of its chromatin in *Hd3a*-RNAi transgenic plants (Komiya et al., 2008). In conclusion, both of the two findings suggest an epigenetic regulation mechanism through *RFT1*. In addition, *SDG724* also affects the histone modification state at *OsMADS50* chromosomal region, thus all the results suggest a LD floral promotion pathway mediated by H3K36me2/3 deposition through *OsMADS50-Ehd1-RFT1* pathways in rice (Sun et al., 2012).

Coincidentally, another member of Class II in SET domain family (Ng et al., 2007), *SDG725*, is also involved in promoting rice flowering through H3K36me2/3. In *SDG725* knockdown plants, the expression levels of *Ehd3*, *RID1*, *OsMADS50*, *OsMADS51*, *Ehd1*, *Hd3a*, and *RFT1* were all drastically reduced, but the *Ghd7* expression was increased, under either SD or LD. Different from *SDG724*, *SDG725* is required for deposition of H3K36me2/3 at more flowering gene loci, such as *Ehd3*, *RID1*, *OsMADS50*, *Hd3a*, and *RFT1*. Thus, *SDG724* and *SDG725* regulate both overlapped and specific flowering genes by mediating H3K36me2/3 deposition and promote rice flowering, which are different to the previously known function of these epigenetic marks in Arabidopsis flowering (Sui et al., 2012; Xu et al., 2008; Zhao et al., 2005).

Very recently, another homolog of *SDG724*, *OsTrx1*, which belongs to SET domain family Class III (Ng et al., 2007), might activate or maintain the active transcribed states of target genes, and was reported to delay flowering time under LD through *Ghd7* pathway but not *OsMADS50* and *Hd1* pathways (Choi et al., 2014). Though expression of *Ehd3* that functions upstream of *Ghd7* is unchanged in *ostrx1* mutants, it was proved that OsTrx1 could bind to Ehd3 *in vitro*. Further study showed that PHD motif of OsTrx1 could bind to native histone H3 and the C-terminal end of SET domain of OsTrx1 has histone H3 methyltransferase activity, thus OsTrx1 and Ehd3 tend to form a complex to methylate downstream genes, but further studies are needed to illuminate its function in detail (Choi et al., 2014).

REPRESSIVE CHROMATIN MODIFICATIONS AND RICE FLOWERING

Arabidopsis *VILs* (*VIL*, *VERNALIZATION INSENSITIVE*), *VIN3*, and *VRN5* are components of PRC2 (Polycomb Repressive Complex 2), mediating the H3K27 trimethylation at the *FLC* locus to repress its expression and hence to induce flowering. In rice, a *VIL* homolog gene *LC2/OsVIL3* is considered as a possible component of PRC2 complex, and *lc2* mutants display late flowering along with the reduced expression of *Hd1* and *Hd3a* under SD. Furthermore, consistent with the result that *OsLF* (*Oryza sativa Late Flowering*) directly repressed *Hd1* expression (Zhao et al., 2011), *LC2/OsVIL3* binds to the promoter region of *OsLF* and represses the *OsLF* expression via H3K27me3 methylation, thus eventually promotes flowering (Wang et al., 2012).

OsVIL2 may be another *VILs* member in rice PRC2 complex, and mutations in *OsVIL2* cause late flowering under both SD and LD. Different from *LC2/OsVIL3*, the late flowering phenotype is associated with increased *OsLFL1* and reduced *Ehd1*, *Hd3a*, *RFT1* expression. Furthermore, OsVIL2 can bind to native histone H3 *in vitro* and is directly associated with *OsLFL1* chromatin *in vivo*, and H3K27me3 is significantly reduced on *OsLFL1* chromatin in *osvil2* mutants compared to the wild type, indicating that *OsVIL2* epigenetically represses *OsLFL1* expression to promote flowering in rice. Besides, OsVIL2 can physically interact with OsEMF2b, which may be also a member of PRC2. Similar to *osvil2*, a null mutation of *OsEMF2b* caused late flowering by increasing *OsLFL1* and decreasing *Ehd1* expression (Wang et al., 2012; Yang et al., 2012).

In short, similar to Arabidopsis, *LC2/OsVIL3*, *OsVIL2*, and *OsEMF2b* may function together with PRC2 to induce flowering by affecting histone modification H3K27me3, but their target flowering genes are different, indicating that a diverse flowering pathway regulated by PRC2 in rice flowering.

CONCLUSION AND PERSPECTIVES

Heading date is an important agronomic trait that determining rice to grow in different regions and seasons. In last two decades, tremendous progress has been made by the study of QTLs and genes controlling rice flowering, which not only identified the nature of the mobile signal florigen, but also unveiled a complex genetic network that controls florigen in rice. Hd3a and RFT1, two florigens regulated respectively rice flowering under SD and LD, are mainly controlled through *Hd1* and *Ehd1* pathways. However, as mentioned in various T65 with lack of both, rice also develops some additional pathways that could induce rice flowering.

Histone modification is very important for defining transcriptional regulation expression, thus plays a fundamental role in plant growth and development, as well as responding to various environmental conditions. These modification marks are dynamically "written" and "erased", and then specifically recognized by the "readers" and instruct specific biological process, such as flowering. Very recently, a large number of studies have revealed that various 'active' histone modifications, H3K4 methylation, H2B monoubiquitination, H3K36me2/me3, histone deacetylation, and 'repressive' chromatin modifications, H3K4 demethylation, H3K9 methylation, H3K27 methylation, histone arginine methylation, are involved in modulating *FLC* expression in Arabidopsis. Though the regulation of *FLC* expression via chromatin modification provides a paradigm in flowering gene expression, whether there exists a major flowering regulator such as *FLC* in rice is still unknown (Fig. 1). Rice possibly has some new routes in its flowering control. In rice, a number of studies revealed the difference in chromatin modification mechanism in the past two years, 'active' H3K36me2/3, H3K4me3, H3K9 acetylation and 'repressive' H3K27me3 modifications mediate flowering time through *Hd1* and *Ehd1* dependent pathways, and our finding about *SDG724* also suggests a LD floral promotion pathway that could be mediated via an epigenetic regulation of florigen *RFT1* itself. All these data suggest that the target flowering genes of chromatin modifications are dispersed in both conserved and unique flowering pathways in rice. Taken together, all the progress in rice, along with Arabidopsis, provides a complete evolutionarily comparative view of genetic and epigenetic flowering mechanisms in plants until now.

Furthermore, in rice, some histone modification participators tend to function under SD and LD, but others like to function mainly under SD or LD, thus unveiling of histone modification mechanism in rice flowering might set a solution to verify the relationships between particular histone modifications and photoperiod environments. On the other hand, as a LD plant, Arabidopsis flowering is accelerated by LD, but SD plant rice flowers earlier under SD than under LD, further study will be helpful to distinguish the function and evolutionary process of histone modification in various photoperiodic plants. Thus, it would be of great interest to identify more chromatin modification regulators and their target genes in rice flowering in future.

ACKNOWLEDGMENTS

This study was financially supported by the National Natural Science Foundation of China (Grant Nos. 31371602 and 91335107) and Specialized Research Fund for the Doctoral Program of Higher Education (20125103120008).

ABBREVIATIONS

FLC, *FLOWERING LOCUS C*; *FT*, *FLOWERING LOCUS T*; H3K36, Histone H3 lysine 36; Hd, heading date; HMTase, histone methyltransferase; LD, long day; *RFT1*, *RICE FLOWERING LOCUS T 1*; SAM, shoot apical meristem; SD, short day.

COMPLIANCE WITH ETHICS GUIDELINES

Changhui Sun, Dan Chen, Jun Fang, Pingrong Wang, Xiaojian Deng, and Chengcai Chu declare that they have no conflict of interest.

This article does not contain any studies with human or animal subjects performed by the any of the authors.

REFERENCES

Andres F, Galbraith DW, Talon M, Domingo C (2009) Analysis of *PHOTOPERIOD SENSITIVITY5* sheds light on the role of phytochromes in photoperiodic flowering in rice. Plant Physiol 151:681–690

Brambilla V, Fornara F (2013) Molecular control of flowering in response to day length in rice. J Integr Plant Biol 55:410–418

Buckler ES, Holland JB, Bradbury PJ, Acharya CB, Brown PJ, Browne C, Ersoz E, Flint-Garcia S, Garcia A, Glaubitz JC, Goodman MM, Harjes C, Guill K, Kroon DE, Larsson S, Lepak NK, Li H, Mitchell SE, Pressoir G, Peiffer JA, Rosas MO,

phytochromes in photoperiodic flowering in rice. Plant Physiol 151:681–690

Brambilla V, Fornara F (2013) Molecular control of flowering in response to day length in rice. J Integr Plant Biol 55:410–418

Buckler ES, Holland JB, Bradbury PJ, Acharya CB, Brown PJ, Browne C, Ersoz E, Flint-Garcia S, Garcia A, Glaubitz JC, Goodman MM, Harjes C, Guill K, Kroon DE, Larsson S, Lepak NK, Li H, Mitchell SE, Pressoir G, Peiffer JA, Rosas MO,

Rocheford TR, Romay MC, Romero S, Salvo S, Sanchez Villeda H, da Silva HS, Sun Q, Tian F, Upadyayula N, Ware D, Yates H, Yu J, Zhang Z, Kresovich S, McMullen MD (2009) The genetic architecture of maize flowering time. Science 325:714–718

Cajlachjan MC (1937) Concerning the hormonal nature of plant development processes. Compt Rend Acad Sci URSS 16:227–230

Chardon F, Damerval C (2005) Phylogenomic analysis of the PEBP gene family in cereals. J Mol Evol 61:579–590

Choi SC, Lee S, Kim S-R, Lee Y-S, Liu C, Cao X, An G (2014) Trithorax group protein OsTrx1 controls flowering time in rice via interaction with Ehd3. Plant Physiol 164(3):1326–1337

Colasanti J, Yuan Z, Sundaresan V (1998) The indeterminate gene encodes a zinc finger protein and regulates a leaf-generated signal required for the transition to flowering in maize. Cell 93:593–603

Colasanti J, Tremblay R, Wong AY, Coneva V, Kozaki A, Mable BK (2006) The maize *INDETERMINATE1* flowering time regulator defines a highly conserved zinc finger protein family in higher plants. BMC Genomics 7:158

Corbesier L, Vincent C, Jang S, Fornara F, Fan Q, Searle I, Giakountis A, Farrona S, Gissot L, Turnbull C, Coupland G (2007) FT protein movement contributes to long-distance signaling in floral induction of Arabidopsis. Science 316:1030–1033

Dai C, Xue HW (2010) Rice *early flowering1*, a CKI, phosphorylates DELLA protein SLR1 to negatively regulate gibberellin signalling. EMBO J 29:1916–1927

Dai X, Ding Y, Tan L, Fu Y, Liu F, Zhu Z, Sun X, Sun X, Gu P, Cai H, Sun C (2012) *LHD1*, an allele of *DTH8*/*Ghd8*, controls late heading date in common wild rice (*Oryza rufipogon*). J Integr Plant Biol 54:790–799

Doi K, Izawa T, Fuse T, Yamanouchi U, Kubo T, Shimatani Z, Yano M, Yoshimura A (2004) *Ehd1*, a B-type response regulator in rice, confers short-day promotion of flowering and controls *FT-like* gene expression independently of *Hd1*. Genes Dev 18:926–936

Ebana K, Shibaya T, Wu J, Matsubara K, Kanamori H, Yamane H, Yamanouchi U, Mizubayashi T, Kono I, Shomura A, Ito S, Ando T, Hori K, Matsumoto T, Yano M (2011) Uncovering of major genetic factors generating naturally occurring variation in heading date among Asian rice cultivars. Theor Appl Genet 122:1199–1210

Fujino K, Yamanouchi U, Yano M (2013) Roles of the *Hd5* gene controlling heading date for adaptation to the northern limits of rice cultivation. Theor Appl Genet 126:611–618

Gao H, Zheng XM, Fei G, Chen J, Jin M, Ren Y, Wu W, Zhou K, Sheng P, Zhou F, Jiang L, Wang J, Zhang X, Guo X, Wang JL, Cheng Z, Wu C, Wang H, Wan JM (2013) *Ehd4* encodes a novel and *Oryza*-genus-specific regulator of photoperiodic flowering in rice. PLoS Genet 9:e1003281

Hayama R, Yokoi S, Tamaki S, Yano M, Shimamoto K (2003) Adaptation of photoperiodic control pathways produces short-day flowering in rice. Nature 422:719–722

He Y (2009) Control of the transition to flowering by chromatin modifications. Mol Plant 2:554–564

Hori K, Kataoka T, Miura K, Yamaguchi M, Saka N, Nakahara T, Sunohara Y, Ebana K, Yano M (2012) Variation in heading date conceals quantitative trait loci for other traits of importance in breeding selection of rice. Breed Sci 62:223–234

Hori K, Ogiso-Tanaka E, Matsubara K, Yamanouchi U, Ebana K, Yano M (2013) *Hd16*, a gene for casein kinase I, is involved in the control of rice flowering time by modulating the day-length response. Plant J 76:36–46

Ishikawa R, Aoki M, Kurotani K, Yokoi S, Shinomura T, Takano M, Shimamoto K (2011) Phytochrome B regulates *Heading date 1* (*Hd1*)-mediated expression of rice florigen *Hd3a* and critical day length in rice. Mol Genet Genomics 285:461–470

Itoh H, Izawa T (2013) The coincidence of critical day length recognition for florigen gene expression and floral transition under long-day conditions in rice. Mol Plant 6:635–649

Itoh H, Nonoue Y, Yano M, Izawa T (2010) A pair of floral regulators sets critical day length for Hd3a florigen expression in rice. Nat Genet 42:635–638

Izawa T (2007) Adaptation of flowering-time by natural and artificial selection in Arabidopsis and rice. J Exp Bot 58:3091–3097

Izawa T, Oikawa T, Sugiyama N, Tanisaka T, Yano M, Shimamoto K (2002) Phytochrome mediates the external light signal to repress *FT* orthologs in photoperiodic flowering of rice. Genes Dev 16:2006–2020

Kim SL, Lee S, Kim HJ, Nam HG, An G (2007) *OsMADS51* is a short-day flowering promoter that functions upstream of *Ehd1*, *OsMADS14*, and *Hd3a*. Plant Physiol 145:1484–1494

Kim SK, Yun CH, Lee JH, Jang YH, Park HY, Kim JK (2008) *OsCO3*, a *CONSTANS-LIKE* gene, controls flowering by negatively regulating the expression of *FT-like* genes under SD conditions in rice. Planta 228:355–365

Kojima S, Takahashi Y, Kobayashi Y, Monna L, Sasaki T, Araki T, Yano M (2002) *Hd3a*, a rice ortholog of the Arabidopsis *FT* gene, promotes transition to flowering downstream of *Hd1* under short-day conditions. Plant Cell Physiol 43:1096–1105

Komiya R, Ikegami A, Tamaki S, Yokoi S, Shimamoto K (2008) *Hd3a* and *RFT1* are essential for flowering in rice. Development 135:767–774

Komiya R, Yokoi S, Shimamoto K (2009) A gene network for long-day flowering activates *RFT1* encoding a mobile flowering signal in rice. Development 136:3443–3450

Koo BH, Yoo SC, Park JW, Kwon CT, Lee BD, An G, Zhang Z, Li J, Li Z, Paek NC (2013) Natural variation in *OsPRR37* regulates heading date and contributes to rice cultivation at a wide range of latitudes. Mol Plant 6:1877–1888

Kwon CT, Yoo SC, Koo BH, Cho SH, Park JW, Zhang Z, Li J, Li Z, Paek NC (2013) Natural variation in *Early flowering1* contributes to early flowering in japonica rice under long days. Plant Cell Environ 37:101–112

Lee S, Kim J, Han JJ, Han MJ, An G (2004) Functional analyses of the flowering time gene *OsMADS50*, the putative *SUPPRESSOR OF OVEREXPRESSION OF CO 1/AGAMOUS-LIKE 20* (*SOC1/AGL20*) ortholog in rice. Plant J 38:754–764

Lee YS, Jeong DH, Lee DY, Yi J, Ryu CH, Kim SL, Jeong HJ, Choi SC, Jin P, Yang J, Cho LH, Choi H, An G (2010) *OsCOL4* is a constitutive flowering repressor upstream of *Ehd1* and downstream of *OsphyB*. Plant J 63:18–30

Li D, Yang C, Li X, Gan Q, Zhao X, Zhu L (2009) Functional characterization of rice *OsDof12*. Planta 229:1159–1169

Li W, Han Y, Tao F, Chong K (2011) Knockdown of SAMS genes encoding S-adenosyl-l-methionine synthetases causes methylation alterations of DNAs and histones and leads to late flowering in rice. J Plant Physiol 168:1837–1843

Lin S, Sasaki T, Yano M (1998) Mapping quantitative trait loci controlling seed dormancy and heading date in rice, *Oryza sativa* L., using backcross inbred lines. Theor Appl Genet 96:997–1003

Lin HX, Yamamoto T, Sasaki T, Yano M (2000) Characterization and detection of epistatic interactions of 3 QTLs, Hd1, Hd2, and Hd3, controlling heading date in rice using nearly isogenic lines. Theor Appl Genet 101:1021–1028

Lin H, Liang Z-W, Sasaki T, Yano M (2003) Fine mapping and characterization of quantitative trait loci *Hd4* and *Hd5* controlling heading date in rice. Breed Sci 53:51–59

Liu C, Lu F, Cui X, Cao X (2010) Histone methylation in higher plants. Annu Rev Plant Biol 61:395–420

Liu T, Liu H, Zhang H, Xing Y (2013) Validation and characterization of *Ghd7.1*, a major quantitative trait locus with pleiotropic effects on spikelets per panicle, plant height, and heading date in rice (*Oryza sativa* L.). J Integr Plant Biol 55:917–927

Matsubara K, Yamanouchi U, Wang ZX, Minobe Y, Izawa T, Yano M (2008) *Ehd2*, a rice ortholog of the maize *ID1* gene, promotes flowering by upregulating *Ehd1*. Plant Physiol 148:1425–1435

Matsubara K, Yamanouchi U, Nonoue Y, Sugimoto K, Wang ZX, Minobe Y, Yano M (2011) *Ehd3*, encoding a plant homeodomain finger-containing protein, is a critical promoter of rice flowering. Plant J 66:603–612

Matsubara K, Ogiso-Tanaka E, Hori K, Ebana K, Ando T, Yano M (2012) Natural variation in *Hd17*, a homolog of Arabidopsis *ELF3* that is involved in rice photoperiodic flowering. Plant Cell Physiol 53:709–716

Ng DW, Wang T, Chandrasekharan MB, Aramayo R, Kertbundit S, Hall TC (2007) Plant SET domain-containing proteins: structure, function and regulation. Biochim Biophys Acta 1769:316–329

Nonoue Y, Fujino K, Hirayama Y, Yamanouchi U, Lin SY, Yano M (2008) Detection of quantitative trait loci controlling extremely early heading in rice. Theor Appl Genet 116:715–722

Ogiso E, Takahashi Y, Sasaki T, Yano M, Izawa T (2010) The role of casein kinase II in flowering time regulation has diversified during evolution. Plant Physiol 152:808–820

Park SJ, Kim SL, Lee S, Je BI, Piao HL, Park SH, Kim CM, Ryu CH, Park SH, Xuan YH, Colasanti J, An G, Han CD (2008) Rice *Indeterminate 1* (*OsId1*) is necessary for the expression of *Ehd1* (*Early heading date 1*) regardless of photoperiod. Plant J 56:1018–1029

Peng LT, Shi ZY, Li L, Shen GZ, Zhang JL (2007) Ectopic expression of *OsLFL1* in rice represses *Ehd1* by binding on its promoter. Biochem Biophys Res Commun 360:251–256

Peng LT, Shi ZY, Li L, Shen GZ, Zhang JL (2008) Overexpression of transcription factor *OsLFL1* delays flowering time in Oryza sativa. J Plant Physiol 165:876–885

Ryu CH, Lee S, Cho LH, Kim SL, Lee YS, Choi SC, Jeong HJ, Yi J, Park SJ, Han CD, An G (2009) *OsMADS50* and *OsMADS56* function antagonistically in regulating long day (LD)-dependent flowering in rice. Plant Cell Environ 32:1412–1427

Saito H, Ogiso-Tanaka E, Okumoto Y, Yoshitake Y, Izumi H, Yokoo T, Matsubara K, Hori K, Yano M, Inoue H, Tanisaka T (2012) *Ef7* encodes an ELF3-like protein and promotes rice flowering by negatively regulating the floral repressor gene *Ghd7* under both short- and long-day conditions. Plant Cell Physiol 53:717–728

Salome PA, Bomblies K, Laitinen RA, Yant L, Mott R, Weigel D (2011) Genetic architecture of flowering-time variation in *Arabidopsis thaliana*. Genetics 188:421–433

Shibaya T, Nonoue Y, Ono N, Yamanouchi U, Hori K, Yano M (2011) Genetic interactions involved in the inhibition of heading by heading date QTL, *Hd2* in rice under long-day conditions. Theor Appl Genet 123:1133–1143

Sui P, Shi J, Gao X, Shen WH, Dong A (2012) H3K36 methylation is involved in promoting rice flowering. Mol Plant 6:975–977

Sun C, Fang J, Zhao T, Xu B, Zhang F, Liu L, Tang J, Zhang G, Deng X, Chen F, Qian Q, Cao X, Chu C (2012) The histone methyltransferase *SDG724* mediates H3K36me2/3 deposition at *MADS50* and *RFT1* and promotes flowering in rice. Plant Cell 24:3235–3247

Takahashi Y, Shimamoto K (2011) *Heading date 1* (*Hd1*), an ortholog of Arabidopsis *CONSTANS*, is a possible target of human selection during domestication to diversify flowering times of cultivated rice. Genes Genet Syst 86:175–182

Takahashi Y, Shomura A, Sasaki T, Yano M (2001) *Hd6*, a rice quantitative trait locus involved in photoperiod sensitivity, encodes the alpha subunit of protein kinase CK2. Proc Natl Acad Sci USA 98:7922–7927

Takahashi Y, Teshima KM, Yokoi S, Innan H, Shimamoto K (2009) Variations in Hd1 proteins, *Hd3a* promoters, and *Ehd1* expression levels contribute to diversity of flowering time in cultivated rice. Proc Natl Acad Sci USA 106:4555–4560

Takimoto A, Ikeda K (1961) Effect of twilight on photoperiodic induction in some short day plants. Plant Cell Physiol 2:213–229

Tamaki S, Matsuo S, Wong HL, Yokoi S, Shimamoto K (2007) Hd3a protein is a mobile flowering signal in rice. Science 316:1033–1036

Tsuji H, Taoka K, Shimamoto K (2011) Regulation of flowering in rice: two florigen genes, a complex gene network, and natural variation. Curr Opin Plant Biol 14:45–52

Tsuji H, Taoka K, Shimamoto K (2013) Florigen in rice: complex gene network for florigen transcription, florigen activation complex, and multiple functions. Curr Opin Plant Biol 16:228–235

Wang J, Hu J, Qian Q, Xue HW (2012) LC2 and OsVIL2 promote rice flowering by photoperoid-induced epigenetic silencing of *OsLF*. Mol Plant 6:514–527

Wei X, Xu J, Guo H, Jiang L, Chen S, Yu C, Zhou Z, Hu P, Zhai H, Wan J (2010) *DTH8* suppresses flowering in rice, influencing plant height and yield potential simultaneously. Plant Physiol 153:1747–1758

Wu C, You C, Li C, Long T, Chen G, Byrne ME, Zhang Q (2008) *RID1*, encoding a Cys2/His2-type zinc finger transcription factor, acts as a master switch from vegetative to floral development in rice. Proc Natl Acad Sci USA 105:12915–12920

Wu W, Zheng XM, Lu G, Zhong Z, Gao H, Chen L, Wu C, Wang HJ, Wang Q, Zhou K, Wang JL, Wu F, Zhang X, Guo X, Cheng Z, Lei C, Lin Q, Jiang L, Wang H, Ge S, Wan J (2013) Association of functional
acts as a master switch from vegetative to floral development in rice. Proc Natl Acad Sci USA 105:12915–12920

Wu W, Zheng XM, Lu G, Zhong Z, Gao H, Chen L, Wu C, Wang HJ, Wang Q, Zhou K, Wang JL, Wu F, Zhang X, Guo X, Cheng Z, Lei C, Lin Q, Jiang L, Wang H, Ge S, Wan J (2013) Association of functional
nucleotide polymorphisms at *DTH2* with the northward expansion of rice cultivation in Asia. Proc Natl Acad Sci USA 110:2775–2780

Xu L, Zhao Z, Dong A, Soubigou-Taconnat L, Renou JP, Steinmetz A, Shen WH (2008) Di- and tri- but not monomethylation on histone H3 lysine 36 marks active transcription of genes involved in flowering time regulation and other processes in *Arabidopsis thaliana*. Mol Cell Biol 28:1348–1360

Xue W, Xing Y, Weng X, Zhao Y, Tang W, Wang L, Zhou H, Yu S, Xu C, Li X, Zhang Q (2008) Natural variation in *Ghd7* is an important regulator of heading date and yield potential in rice. Nat Genet 40:761–767

Yamamoto T, Lin H, Sasaki T, Yano M (2000) Identification of heading date quantitative trait locus *Hd6* and characterization of its epistatic interactions with *Hd2* in rice using advanced backcross progeny. Genetics 154:885–891

Yan WH, Wang P, Chen HX, Zhou HJ, Li QP, Wang CR, Ding ZH, Zhang YS, Yu SB, Xing YZ, Zhang QF (2011) A major QTL, *Ghd8*, plays pleiotropic roles in regulating grain productivity, plant height, and heading date in rice. Mol Plant 4:319–330

Yan W, Liu H, Zhou X, Li Q, Zhang J, Lu L, Liu T, Zhang C, Zhang Z, Shen G, Yao W, Chen H, Yu S, Xie W, Xing Y (2013) Natural variation in Ghd7.1 plays an important role in grain yield and adaptation in rice. Cell Res 23:969–971

Yang J, Lee S, Hang R, Kim SR, Lee YS, Cao X, Amasino R, An G (2012) OsVIL2 functions with PRC2 to induce flowering by repressing *OsLFL1* in rice. Plant J 73:566–578

Yang Y, Peng Q, Chen GX, Li XH, Wu CY (2013) *OsELF3* is involved in circadian clock regulation for promoting flowering under long-day conditions in rice. Mol Plant 6:202–215

Yano M, Harushima Y, Nagamura Y, Kurata N, Minobe Y, Sasaki T (1997) Identification of quantitative trait loci controlling heading date in rice using a high-density linkage map. Theor Appl Genet 95:1025–1032

Yano M, Katayose Y, Ashikari M, Yamanouchi U, Monna L, Fuse T, Baba T, Yamamoto K, Umehara Y, Nagamura Y, Sasaki T (2000) *Hd1*, a major photoperiod sensitivity quantitative trait locus in rice, is closely related to the Arabidopsis flowering time gene *CONSTANS*. Plant Cell 12:2473–2484

Zhao Z, Yu Y, Meyer D, Wu C, Shen WH (2005) Prevention of early flowering by expression of *FLOWERING LOCUS C* requires methylation of histone H3 K36. Nat Cell Biol 7:1256–1260

Zhao XL, Shi ZY, Peng LT, Shen GZ, Zhang JL (2011) An atypical HLH protein OsLF in rice regulates flowering time and interacts with OsPIL13 and OsPIL15. Nat Biotechnol 28:788–797

Zhao J, Huang X, Ouyang X, Chen W, Du A, Zhu L, Wang S, Deng XW, Li S (2012) *OsELF3-1*, an ortholog of *Arabidopsis early flowering 3*, regulates rice circadian rhythm and photoperiodic flowering. PLoS ONE 7:e43705

12

The key role of *CYC2* during meiosis in *Tetrahymena thermophila*

Qianlan Xu[1,2,3], Ruoyu Wang[3], A. R. Ghanam[3,6], Guanxiong Yan[4,5], Wei Miao[4], Xiaoyuan Song[1,2,3]✉

[1] Hefei National Laboratory for Physical Sciences at the Microscale, University of Science and Technology of China, Hefei 230026, China
[2] CAS Center for Excellence in Molecular Cell Science, Shanghai 200031, China
[3] School of Life Sciences, University of Science and Technology of China, Hefei 230071, China
[4] Key Laboratory of Aquatic Biodiversity and Conservation, Institute of Hydrobiology, Chinese Academy of Sciences, Wuhan 430072, China
[5] University of Chinese Academy of Sciences, Beijing 100049, China
[6] Anatomy and Embryology Department, Suez Canal University, Ismailia 41522, Egypt
✉ Correspondence: songxy5@ustc.edu.cn (X. Song)

ABSTRACT

Meiotic recombination is carried out through a specialized pathway for the formation and repair of DNA double-strand breaks (DSBs) made by the Spo11 protein. The present study shed light on the functional role of cyclin, *CYC2*, in *Tetrahymena thermophila* which has transcriptionally high expression level during meiosis process. Knocking out the *CYC2* gene results in arrest of meiotic conjugation process at 2.5–3.5 h after conjugation initiation, before the meiosis division starts, and in company with the absence of DSBs. To investigate the underlying mechanism of this phenomenon, a complete transcriptome profile was performed between wild-type strain and *CYC2* knock-out strain. Functional analysis of RNA-Seq results identifies related differentially expressed genes (DEGs) including *SPO11* and these DEGs are enriched in DNA repair/mismatch repair (MMR) terms in homologous recombination (HR), which indicates that *CYC2* could play a crucial role in meiosis by regulating *SPO11* and participating in HR.

KEYWORDS cyclin, meiosis, RNA-Seq, *Tetrahymena thermophila*, homologous recombination

Qianlan Xu and Ruoyu Wang have contributed equally to this work.

INTRODUCTION

Tetrahymena thermophila, one of typical ciliate protozoans, is a unicellular eukaryotic species characterized by its nuclear dimorphism. As each cell contains two differentially developed and functionally distributed nuclei, a diploid germ line nucleus (micronucleus) and a polyploid somatic nucleus (macronucleus) (Collins, 2012). *Tetrahymena* also has two types of life cycle, vegetative growth cycle and meiotic conjugation cycle. In meiotic conjugation cycle, only the micronucleus takes part as one of the similar biological processes which are evolutionarily conserved with multicellular organisms meiosis, except for the extra complicated procedures of several successive mitotic cycles in conjugation after meiotic segregation (Collins, 2012).

Meiosis is a highly conserved process in sexually reproducing eukaryotes. It is a special mode of mitosis, during which the parental diploid chromatins duplicate once followed by two rounds of precise halving of the genome in succession to generate haploid gametes. During gametes production in most gamogenetic species, homologous recombination (HR) occurs in the prophase of meiosis I before the chromosomes segregation starts. At the same time, the chiasmata were formed between aligned homologous chromosomes as a stable physical connection which serves for maintaining the accuracy of chromosome equal segregation (Petronczki et al., 2003).

The recombination is the most protrusive and important process during meiosis I prophase by virtue of its reshuffling function through merging the two parental alleles, generating more diversified progenies, which is of evolutionary significance (Kauppi et al., 2004). Meiotic recombination has at its heart the formation and subsequent repair of DNA double strand breaks (DSBs) (Keeney, 2001). DSB formation is catalyzed by Spo11, which appears to act via a topoisomerase-like reaction to generate a transient, covalent protein-DNA intermediate (Keeney et al., 1997a). The repair of any given meiotic DSB can result in either reciprocal exchange of the chromosome arms flanking the break (a crossover), or no exchange of flanking arms (a noncrossover or parental configuration).

During the early stage of conjugation, two different mating types of *T. thermophila* cells approach each other till the pair formed. Then a rather stable junction formed followed by a course of shape changing of micronuclei and two successions of chromosome segregations. In addition, crescent stage was found to be the analogous to the bouquet stage in multicellular organisms when extremely elongated micronuclei (crescent) were formed at approximately 3 h after conjugation initiation (Loidl and Mochizuki, 2009). This is of great importance for both homologous recombination in prophase and DNA rearrangement during the growth of new macronuclei (Mochizuki and Gorovsky, 2004). On account of the unique life cycle, nuclear dimorphism as well as convenience of inducing meiosis initiation and observable distinctive characteristics summarized at each stage in meiosis, *T. thermophila* could be used as a great and unique research model of meiosis process like *Saccharomyces cerevisiae*.

Cyclin domain is a helical domain for protein recognition, which exists mostly in cyclin superfamily proteins but also in transcription factor II B and Retinoblastoma with different copies and forms of distribution (Gibson et al., 1994). Cyclin domain equips cyclin proteins with key function of conditioning the enzyme activity of cyclin-dependent kinases, which participate in the molecular regulation of the events of cell cycle and transition into the next phases (Zhang et al., 1999). *T. thermophila* possesses 26 known cyclin homologs which are classified into different cyclin groups of functions. Each of *T. thermophila* cyclin protein exhibits unique profile of mRNA expression (TGD website: http://ciliate.org/) (Miao et al., 2009). Among them, 23 of the cyclin homologs have sharp peaks at different time points of conjugation. *CYC2*, an N-terminal cyclin domain containing protein, was grouped into canonical ally and was identified as cyclin D-like proteins using phylogenetic analysis tools (Stover and Rice, 2011). Also, named as COI5 (conjugation-induced gene 5) (Woehrer et al., 2015). *CYC2* gene has no expression at all during logarithm growth when there are only micronuclei mitosis and macronuclei amitosis, as well as in starvation condition. While its expression begins when the meiosis starts about 2 h after the initiation of conjugation and remains high till the end of meiosis II, suggesting its probable role in meiosis process (Miao et al., 2009).

Moreover, yeast B-type cyclins (CLB5 and CLB6) have been reported to have key roles in the initiation of homologous chromosome recombination and the formation of synaptonemal complex during meiosis prophase (Devault et al., 2008; Henderson et al., 2006).

In the advent of Next Generation Sequencing (NGS), Whole-genome transcriptional profile has become a great tool in transcriptome studies. Comparing with microarray methods, deep RNA sequencing (RNA-Seq) has been more widely used in different cell biology processes due to advantages of unbias, high-throughput, and sensitivity. The first transcriptome of *T. thermophila* was sequenced and released in 2012 (Wang et al., 2009; Xiong et al., 2012), which identified untranslated regions (UTR), novel transcripts and alternative splicing successfully and re-annotated *T. thermophila*'s genome more accurately. Performing RNA-Seq after knocking out genes could provide us a comprehensive understanding of the eliminated gene's function in transcriptional level both in mammals and ciliates (Gao et al., 2013). This rapid and efficient method could be used as a powerful way to study further molecular mechanisms of interested genes.

While there has been a significant progress in understanding the behaviors of the *T. thermophila* proteins during meiosis and the possible interactions among them, this study was planned to uncover the functional role of cyc2p (*CYC2* protein) in a comprehensive and precise way and its possible involvement in meiotic recombination.

RESULTS

Transcriptional expression profile of *CYC2*

CYC2 gene was selected from candidate genes which shared the similar variation tendency in mRNA expression profile with meiosis associated genes already known according to the analysis result from microarray and related RNA-Seq data. The mRNA expression profile is available on TGD website (http://ciliate.org/) where the microarray analysis results of genes at whole genome scale of *T. thermophila* are displayed (Miao et al., 2009). During logarithmic growth, when there are micronuclei mitosis and macronuclei amitosis, and starvation condition, *CYC2* gene doesn't express at all. *CYC2* gene starts to express at the second hour after conjugation initiation and keeps a high expression level till the end of meiosis II (Fig. 1A).

Deletion of *CYC2* caused the conjugation process arrest before crescent stage and initiation of meiosis I

The *CYC2* gene knock-out (KO) strains (Δ*CYC2*) of two mating types (CU428 and B2086 II) were previously constructed and well-tested (Hayashi and Mochizuki, 2015). We've also confirmed successful *CYC2* KO by whole cell extracts PCR at *CYC2* genomic locus (Fig. 1B) and the extreme low mRNA expression level of *CYC2* was also proved by RT-qPCR (Fig. 1C).

For phenotype assessment, Δ*CYC2* strains were starved for 24 h and mixed with equal amount of each mating type of

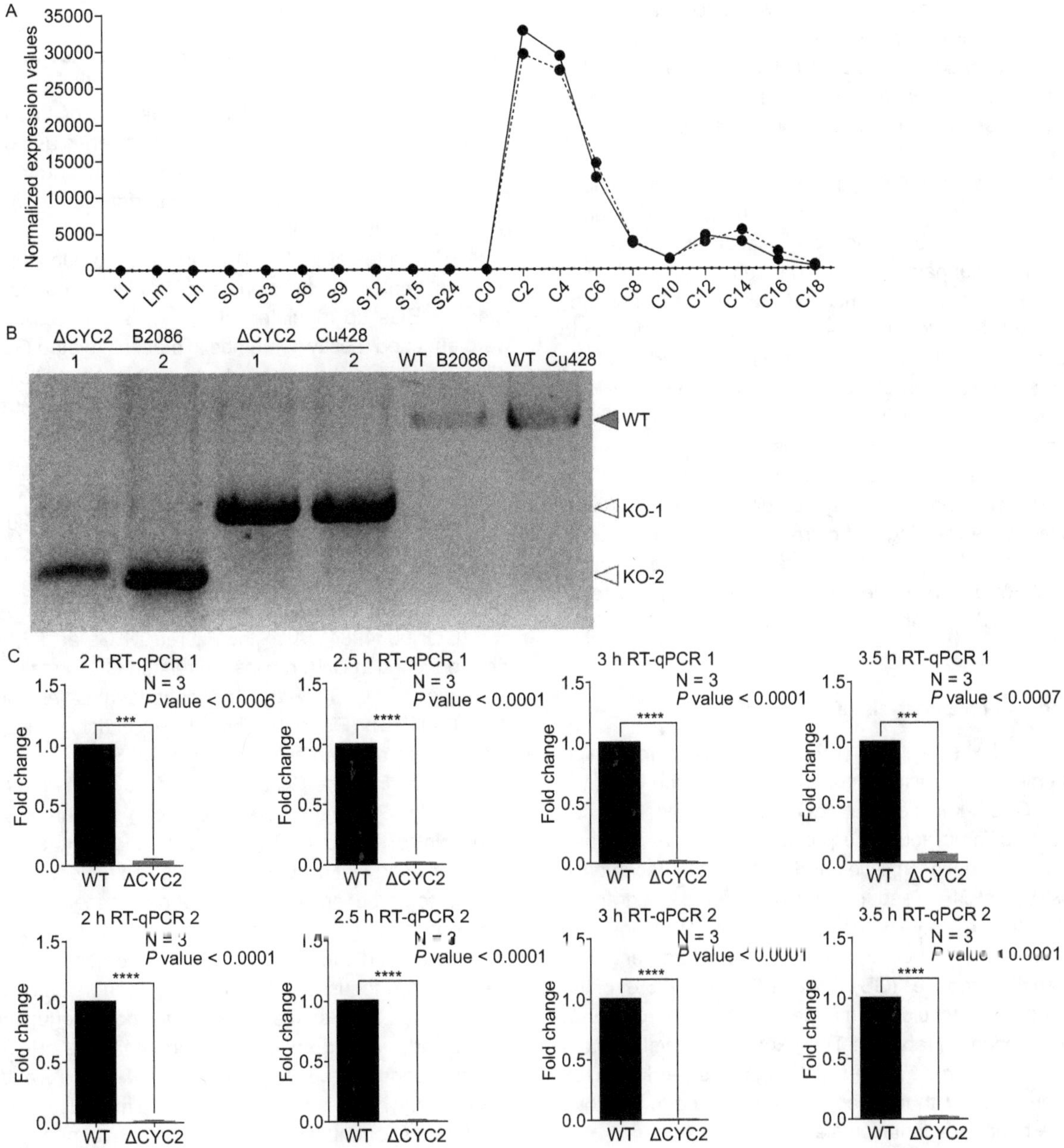

Figure 1. Transcriptional expression profile of *CYC2* (TTHERM_00079530) in *T. thermophila* and confirmation of *CYC2* knock-out strains. (A) The line chart was generated with microarray data of mRNA products of all time points (three GROWTH time points, seven STARVATION time points and ten conjugation time points) in life cycle of *T. thermophila*. The solid and dashed lines denoted the expression values normalized by two different methods. For vegetatively growing cells, L-l, L-m, and L-h respectively correspond to ~1×10^5 cells/mL, ~3.5×10^5 cells/mL, and ~1×10^6 cells/mL. Samples collected at 0, 3, 6, 9, 12, 15, and 24 h after starvation began were respectively referred to as S-0, S-3, S-6, S-9, S-12, S-15, and S-24, and at 0, 2, 4, 6, 8, 10, 12, 14, 16, and 18 h after mixing equal volumes of B2086 and CU428 cells for conjugation initiation referred to as C-0, C-2, C-4, C-6, C-8, C-10, C-12, C-14, C-16, and C-18. It is manifested that *CYC2* expressed at transcriptional level exclusively during conjugation and abundantly during 2–4 h after conjugation initiation followed by additional expression around 12–14 h. (B) The *CYC2* knock-out strains (Δ*CYC2*) of two different mating types (B2086 and Cu428) were confirmed by whole cell extract PCR. The difference of PCR products lengths between strains of two mating types (marked as KO-1 and KO-2) was reasonable for a deleted region larger than target sequence, and it was often observed during the process of co-deletion (Hayashi and Mochizuki, 2015). (C) The fold changes of mRNA expression level of paired strains of Δ*CYC2* at 4 time points after conjugation initiation were examined by RT-qPCR using two different primer sets in CDS ('qPCR mRNA check-1' primers for RT-qPCR 1 and 'qPCR mRNA check-2 exon1' primers for RT-qPCR 2) (Fig. S1). N number of biological repeats.

strains to induce meiosis. Compared with wild type (control), we observed that almost none of the KO strain pairs had generated exconjugants which are the progeny cells after conjugation completely finished. Therefore, *CYC2* would be an essential gene in regulation cell cycle during meiosis process.

The time line for progression of nuclear events in conjugation was assessed by DAPI staining (Fig. 2A). We found that at 1–2 h after the conjugation initiation, the KO strains could form normal pairs as wild type; and micronuclei could slightly migrate from macronuclei at around 2 h followed by stretching a little bit longer. The micronuclei were unable to form a crescent shape as wild type strains did in the beginning of meiosis. This process of stagnation persists till 4.5 h after conjugation initiation while the wild type pairs almost complete the meiosis process (Fig. 2B).

As the time goes on the KO strains pairs separated into single cells and the percentage of paired ones gradually decreased while the single unpaired cells increased. Eventually the majority of the KO pairs disconnected. These results showed us a meiosis disable phenotype after knocking out *CYC2* gene.

RNA-Seq and data analysis of ΔCYC2

In order to test whether the *CYC2* gene affects the regulation of some meiosis associated genes at transcription level, we collected *CYC2* knocked out (ΔCYC2) and wild type cell samples at different four time points—2 h, 2.5 h, 3 h, and 3.5 h after conjugation initiation and perform a transcriptomic analysis by mRNA deep sequencing of RNA extraction products utilizing PolyA enrichment method. We obtained about 180 million paired-end reads, with a total length of more than 45 gigabases (GB). About 173 million (96%) of the paired-end reads could be mapped to the *T. thermophila* reference genome (Table 1). The expression level of each gene was represented as FPKM (Fragments per kilobases per million reads) with mapped reads according to previous reference genome annotation file.

The identified transcriptomes were compared between Δ*CYC2* strains and wild type strains at 2 h, 2.5 h, 3 h, 3.5 h four time stages, 1918, 1663, 1489, 1305 differentially expressed genes (DEGs) were identified (FDR < 0.05) respectively. We identified 1296 up-regulated DEGs, 622 down-regulated DEGs at 2 h stage, 803 up-regulated DEGs, 860 down-regulated DEGs at 2.5 h stage, 1001 up-regulated DEGs and 488 down-regulated DEGs at 3 h stage and 539 up-regulated DEGs, 706 down-regulated DEGs at 3.5 h stage (Fig. 3A). In addition to that, there was 479 genes differentially expressed at all four stages, and 895 genes differentially expressed at both 2 h and 2.5 h stages, 967 genes differentially expressed at both 2.5 h and 3 h stages, 927 genes differentially expressed at both 3 h and 3.5 h stages (Fig. 3B).

Gene ontology enrichment analysis

The significantly up and down regulated genes at the four different time points were screened thoroughly according to the requirements that the binary logarithm of fold change should be no less than 2 for up regulated genes and no more than −2 for down regulated genes and *P* value no more than 0.05. The filtered data was used to perform the gene ontology (GO) enrichment analysis separately utilizing the ClueGO (Bindea et al., 2009) in three main groups: biological process, molecular function, and cellular component. The filtered DEGs up regulated at 2, 2.5, 3, and 3.5 h stages were enriched mostly in metabolic process related GO terms (Fig. S2), such as carbohydrate derivative metabolic process (GO: 1901135), carboxylic acid metabolic process (GO: 0019752). The filtered DEGs down-regulated at 2 h stage were enriched mainly in DNA MMR GO terms (Fig. 4A), like mismatch repair (GO: 0006298), mismatched DNA binding (GO: 0006298). At 2.5 h and 3 h stage, the filtered DEGs were enriched both in DNA MMR related GO terms and DNA replication related GO terms (Fig. 4B and 4C), such as DNA replication (GO: 0006298), cellular response to DNA damage (GO: 0034984). Also the filtered DEGs at 3.5 h stage were enriched in DNA replication related GO terms like DNA replication (GO: 0006298), DNA-dependent DNA replication (GO: 0006262) (Fig. 4D). The full GO enrichment analysis results were supplemented in Table S2–S5.

Together, from the GO analysis results of the up regulated DEGs (Fig. S2) there were little strong enrichment at any time point of the four. The results at each of the four time points showed that biological process and molecular function almost centered on various metabolic process of organic compounds and that the protein products of these genes were dispersed into different cellular components without well-detected focus. In sharp contrast to the situation in the up regulated DEGs, the significantly down-regulated part manifested a strong enrichment from the angle of biological process and molecular function in DNA MMR and DNA replication, while it is mainly enriched in nucleus in aspect of cellular component, which may indicate that knocking out of *CYC2* may result in DNA MMR and DNA replication related genes' dysfunction, further causing Δ*CYC2* strains' meiosis disability.

The affected meiosis associated genes in ΔCYC2 strain

The 60 best-conserved *Tetrahymena* meiotic proteins used as a criteria for a possible meiosis-specific function was previously manifested (Mochizuki et al., 2008). We observed that 20 of these 60 proteins (Table 2) were significantly affected upon KO of cyc2p indicating that around one third of the known meiosis-associated genes were transcriptionally associated with *CYC2*. Among these genes, *SPO11* (TTHERM_00627090), which plays a significant role in meiotic homolog recombination as it catalyzes DSB formation (Mochizuki et al.,

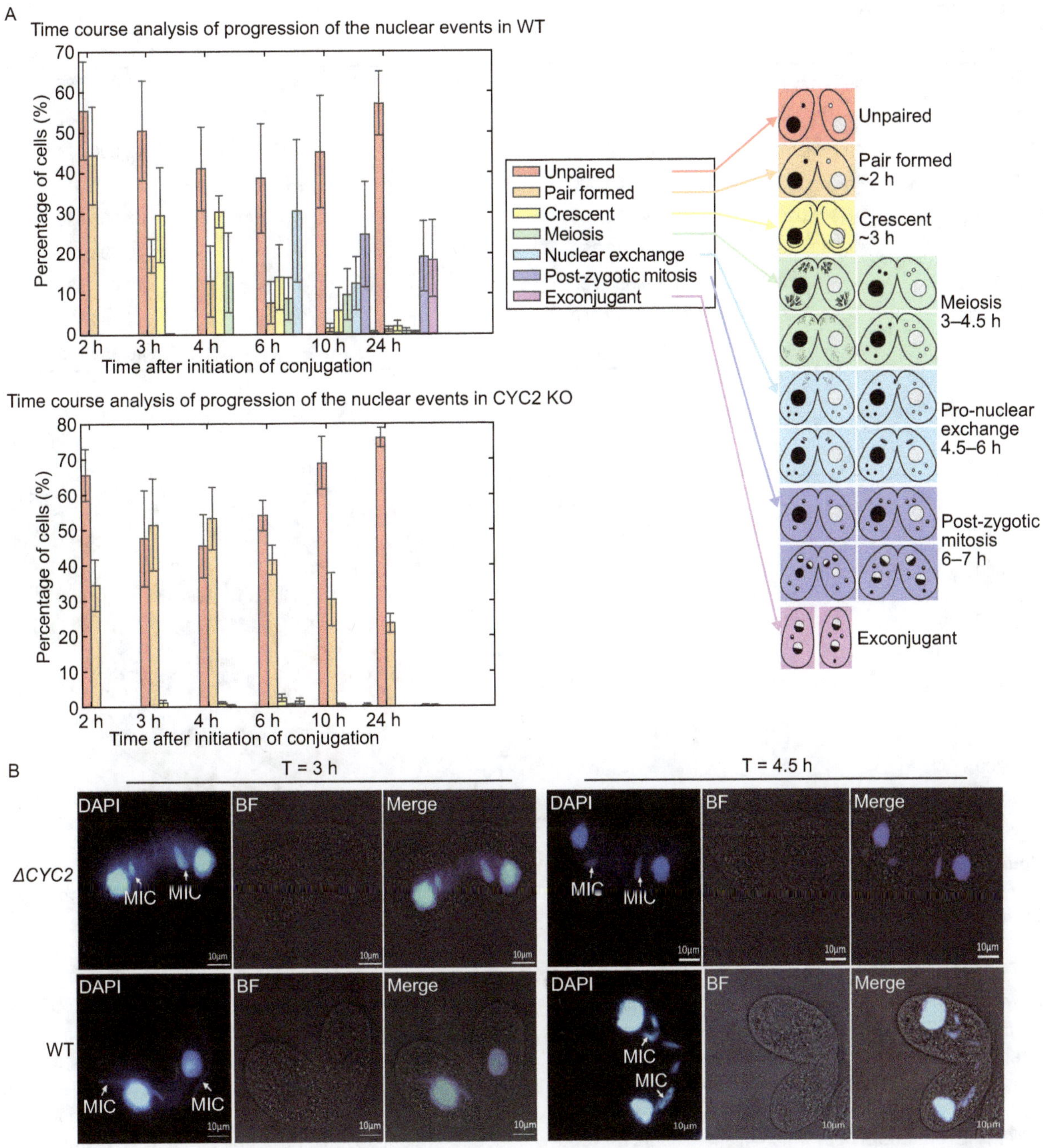

Figure 2. Phenotypic analysis of Δ*CYC2*. (A) The time course analysis of the progression of the nuclear events showed that the loss of *CYC2* resulted in an early conjugation arrest where the conjugants were not able to pass through the crescent stage successfully and lack of meiosis division. The conjugating process was artificially separated into seven parts according to the morphology characteristics (unpaired, pair formed, crescent, meiosis divisions, pro-nuclear exchange, post-zygotic mitosis, and exconjugant) and the related graphical representation were highlighted in different color at the right side. The 4 independent biological repeats of cell samples were collected respectively at 2, 3, 4, 6, 10, and 24 h after conjugation initiation. At least 100 pairs of cells were counted in each repeat. The percentage of each morphologic status was calculated and displayed in the histographs above. (B) The status of cell cycle arrest of Δ*CYC2* before crescent stage was captured during microscopic observation of cells stained with DAPI at 3 h when crescent shape formed in wild type conjugants and at 4.5 h when meiosis II was finished.

Table 1. Statistics of RNA-seq reads

Sample name	Clean reads	Clean bases	Read length	Q20 (%)	Unique mapped reads
WT-2 h	44,363,694	5,545,461,750	125	96.60	42,626,886
WT-2.5 h	44,950,340	5,618,792,500	125	96.46	43,577,402
WT-3 h	45,183,552	5,647,944,000	125	96.06	43,300,024
WT-3.5 h	45,196,512	5,649,564,000	125	95.87	43,577,316
Δ*CYC2*-2 h	45,394,020	5,674,252,500	125	96.14	43,683,066
Δ*CYC2*-2.5 h	45,283,310	5,660,413,750	125	96.24	43,598,652
Δ*CYC2*-3 h	48,036,470	6,004,558,750	125	97.46	42,755,586
Δ*CYC2*-3.5 h	44,986,790	5,623,348,750	125	95.22	43,280,830
Total	363,394,688	45,424,336,000			346,399,762

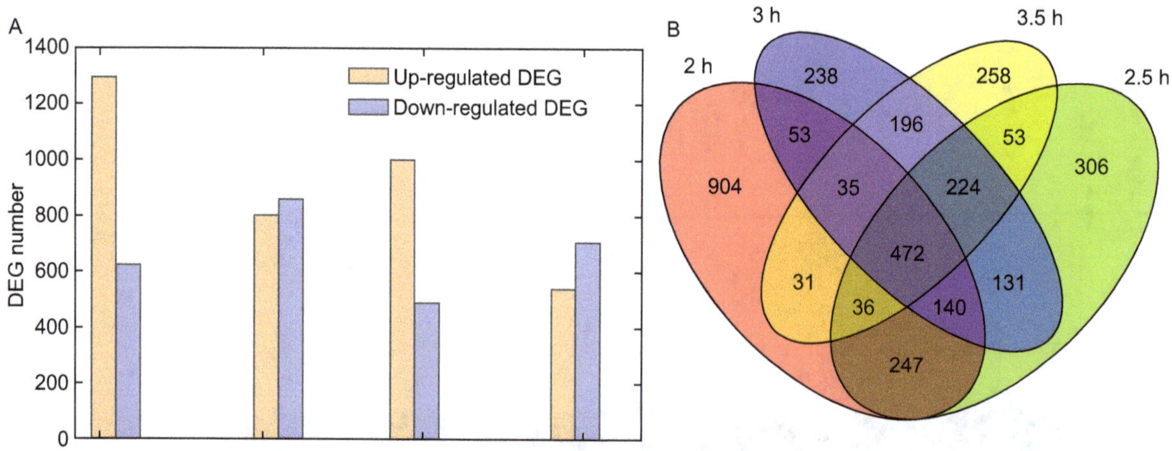

Figure 3. Overview of differentially expressed genes (DEGs). (A) DEG number of *CYC2* KO strains and wild type strains at four time stages (2, 2.5, 3, and 3.5 h after conjugation initiation). (B) Venn diagram identifying transcriptome DEG features between KO strains and wild type strains at four different time stages.

2008), was drastically down regulated at the fold change of −16. *IME4* (TTHERM_00962190), a homolog of yeast gene that is required for entry into meiosis (Shah and Clancy, 1992) was significantly up regulated at the 2.5 h stage. *CYC24* (TTHERM_00842480), *CYC6* (TTHERM_00194440), and *CYC18* (TTHERM_00827080) which might account for regulation of cell cycle progression in meiosis (Mochizuki et al., 2008) were up regulated at 3.5 h stage.

The following genes were also down regulated in Δ*CYC2*: *RAD51*, TTHERM_00142330; *DMC1*, TTHERM_00459230; *SGS1*, TTHERM_01030000; *HOP2*, TTHERM_00794620; *MND1*, TTHERM_00300660) which responsible for strand exchange during crossing over and (*RFA1*, TTHERM_00106890; *MRE11*, TTHERM_00721450; *EXO1*, TTHERM_01179960; *PMS2*, TTHERM_01109940; *TMLH1*, TTHERM_00127000; *MSH6*, TTHERM_00194810; *MSH3*, TTHERM_00426230. They were reported to be responsible for DNA damage repair especially mismatch repair (Mochizuki et al., 2008). The expression of *ESP1* (TTHERM_00297160) and a putative Aurora protein kinase (TTHERM_00684590), which may involve in spindle assembly and chromosome separation at the anaphase of meiosis (Mochizuki et al., 2008), were also down regulated. The large percentage of significantly regulated genes in the listed meiotic specific genes added weight to our view point that *CYC2* was probably involved in the meiosis associated regulation network.

Disable meiotic DSB formation in Δ*CYC2* strain

The presence of the phosphorylated histone H2A.X variant (gamma-H2A.X) is a marker for the location of DSBs appears on the meiotic chromosomes of *T. thermophila* (Song et al., 2007). We test the formation of meiotic DSBs by performing immunofluorescence of gamma-H2AX. In wild type, meiotic DSBs begin before crescent stage when micronuclei haven not elongated (Fig. 5A) and also were seen at crescent stage (Fig. 5B). In contrast, we did not detect any DSB signal in micronuclei of *CYC2* KO strains, which also could not elongate

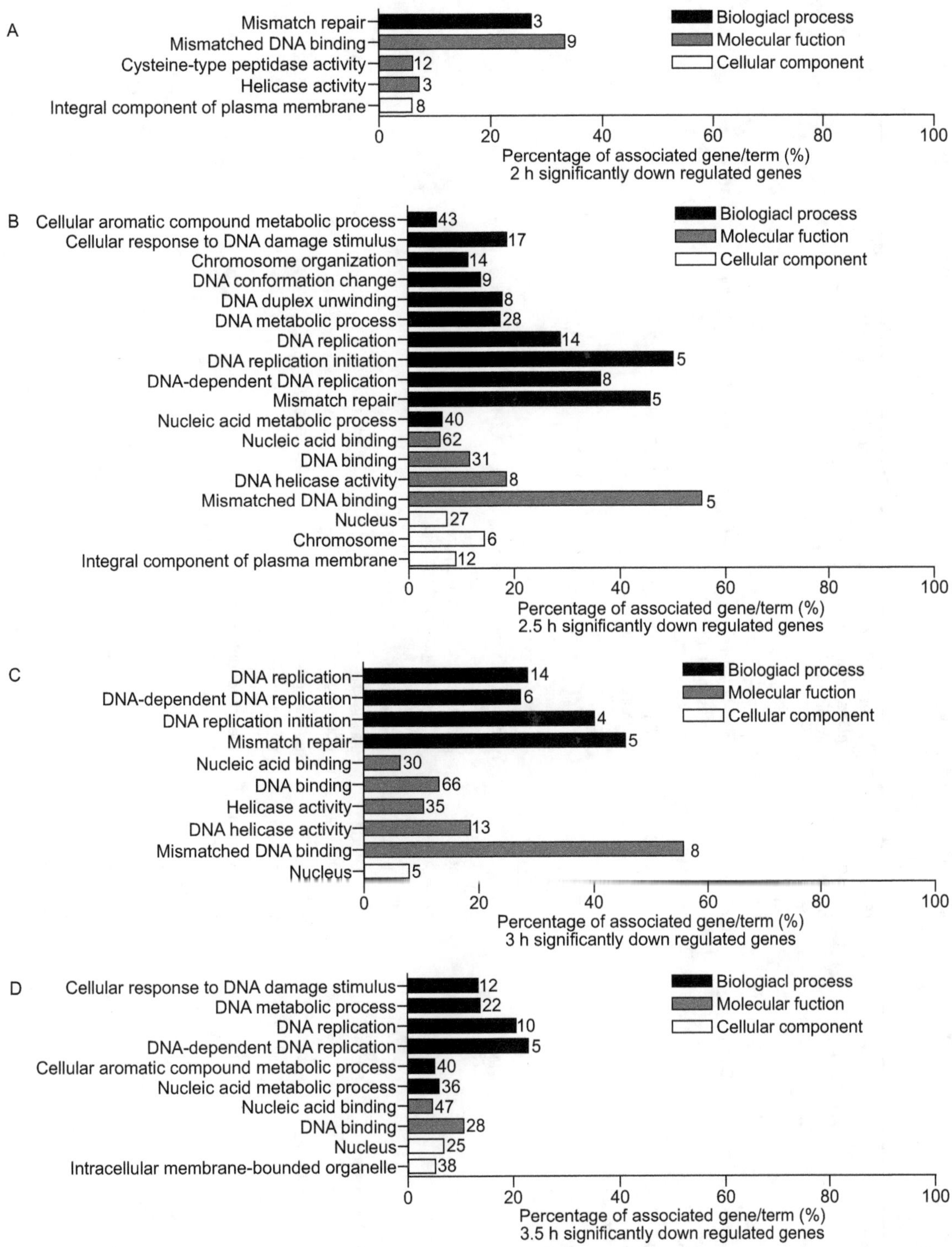

Figure 4. Gene ontology (GO) enrichment analysis of down regulated genes. The filtered significantly down regulated genes (Log2 (fold change) < −2 and *P* value < 0.05) of each time point were subjected to GO enrichment analysis separately in biological process, molecular biology, and cellular function. (A) Percentage of associated genes/term (%) at 2 h. (B) Percentage of associated genes/term (%) at 2.5 h. (C) Percentage of associated genes/term (%) at 3 h. (D) Percentage of associated genes/term (%) at 3.5 h. The length of bars from each histogram indicates the percentage of associated genes for each term. The number on top of each bar means the number of associated genes.

Table 2. Significantly regulated meiosis associated genes after *CYC2* KO

Gene ID	Meiotic functions	Family name	Standard name of *T. thermophila*	Discription	2 h	2.5 h	3 h	3.5 h
TTHERM_00962190	Meiotic induction	Ime4	IME4 (Homolog of budding yeast IME4 (Inducer of meiosis))	In yeast, required for entry into meiosis\nMT-A70 family protein	0.463063	1.95484	1.51113	0.963928
TTHERM_00842480	Regulation of cell cycle progression	Cyclin	CYC24 (CYClin)	Cyclin, N-terminal domain containing protein	0.81999	1.72414	2.11614	2.01071
TTHERM_00194440	Regulation of cell cycle progression	Cyclin	CYC6 (CYClin)	Cyclin, N-terminal domain containing protein	3.27593	2.20212	2.18551	2.44223
TTHERM_00827080	Regulation of cell cycle progression	Cyclin	CYC18 (CYClin)	Cyclin, N-terminal domain containing protein	2.70469	2.74254	2.16665	2.41128
TTHERM_00627090	DSB formation	Spo11/ Rec12	SPO11 (Ortholog of budding yeast SPO11)	Type IIB DNA topoisomerase family protein required for meiotic DNA DSBs. Required for the elongation of meiotic nuclei and full chromosome pairing in *Tetrahymena*	−4.28045	−4.81505	−4.71778	−4.31692
TTHERM_00459230	Strand exchange	Dmc1	DMC1 (DMC1 homolog)	meiosis-specific RecA homolog	−4.15141	−4.85911	−4.68173	−4.49637
TTHERM_00721450	Recombinational repair	Mre11/ Rad32	MRE11 (Homolog of budding yeast MRE11 and of fission yeast RAD32)	Ser/Thr protein phosphatase family protein. Mre11p is required for the repair of meiotic double-strand breaks and full chromosome pairing	−0.555912	0.128007	1.20542	1.88526
TTHERM_01179960	DSB repair	Exo1	EXO1 (Homolog of budding yeast, fission yeast and mouse EXO1)	XPG I-region family protein involved in DSB repair	−0.224175	−1.9432	−2.19387	−2.02446

Table 2. continued

Gene ID	Meiotic functions	Family name	Standard name of *T. thermophila*	Discription	2 h	2.5 h	3 h	3.5 h
TTHERM_00011650	DSB repair	Rad10	RAD10 (Homolog of budding yeast RAD10)	In yeast, single-stranded DNA endonuclease (with Rad1p), cleaves single-stranded DNA during nucleotide excision repair and double-strand break repair	1.56953	−0.692742	−2.38111	−1.50885
TTHERM_01109940	Mismatch repair	Pms1	PMS2 (DNA mismatch repair also called pms1)	DNA mismatch repair protein, C-terminal domain containing protein; homolog to human PMS2	−2.05342	−2.63375	−2.38925	−1.71021
TTHERM_00194810	Mismatch repair	Msh6	MSH6 (homolog to human protein MSH6)	MutS domain III family protein	−1.80791	−2.75677	−2.25144	−1.89783
TTHERM_00426230	Mismatch repair	Msh3	MSH3 (homolog to human protein MSH3)	MutS domain III family protein	−0.461977	−2.16236	−2.1011	−1.67332
TTHERM_01030000	Regulates crossing over	Sgs1/Rqh1	SGS1 (Homolog of budding yeast Sgs1 and fission yeast rqh1)	ATP-dependent DNA helicase, RecQ family protein involved in DNA joint molecule resolution dHJ dissolution and crossing over	−2.14875	−2.88759	−2.53104	−1.75497
TTHERM_00794620	Strand exchange	Hop2/ Meu13	HOP2 (Meiosis-specific homolog of budding yeast Homologous Pairing 2)	HOP2 has a role in chiasmata and meiotic bivalent formation. There exists a ubiquitously expressed paralog, TTHERM_01190440 (HOPP2), which is essential for vegetative growth	−3.07189	−3.80692	−2.88744	−2.01004

Table 2. continued

Gene ID	Meiotic functions	Family name	Standard name of *T. thermophila*	Discription	2 h	2.5 h	3 h	3.5 h
TTHERM_00300660	Strand exchange	Mnd1/ Mcp8	MND1 (Homolog of budding yeast MND1 (meiotic non disjunction) and fission yeast MCP7)	In budding yeast, the Mnd1 protein forms a complex with Hop2 to promote homologous chromosome pairing and meiotic DSB repair. Mnd1 requires Hop2 to localize to chromosomes	−0.581187	−2.04827	−1.82826	−1.75667
TTHERM_00441940	APC regulator	Fzr1/CDH2	FZY9 (Fizzy)	fizzy/CDC20/CDH1 family protein; homolog of *S. cerevisiae* CDH1, a cell cycle regulated activator of the anaphase-promoting complex/cyclosome (APC/C), which directs ubiquitination of various targets	−0.0326336	0.687104	1.94776	2.2213
TTHERM_00297160	–	Cut1	ESP1 (Extra Spindle Pole bodies)	Separase protein, required for mitotic and meiotic chromosome segregation	−1.64457	−3.70644	−3.01699	−2.45854
TTHERM_00158460	–	Mei2	RRM68 (RNA recognition motif-containing protein 68)	RNA recognition motif 2 family protein	−1.76957	−2.30628	−1.55013	−1.17294
TTHERM_00684590	–	Aurora kinases	None	Protein kinase domain containing protein. Sequence similarity to the Aurora protein kinase family	−2.31013	−3.86966	−3.97498	−3.74671
TTHERM_00991560	–	–	None	Protein phosphatase 2A regulatory B subunit (B56 family)	3.49729	5.07489	2.99617	2.78042

Numbers in last four columns (2 h, 2.5 h, 3 h, 3.5 h) mean log2 fold change of KO/WT RPKM of related genes, bolded numbers are significant statistically.

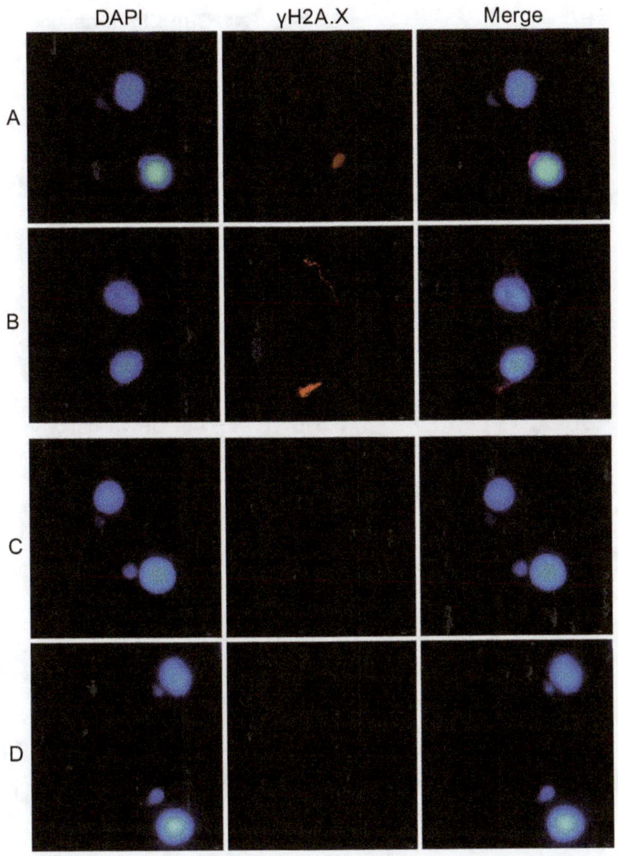

Figure 5. Immunostaining of wild type and *CYC2* KO strains. (A) Gamma-H2AX signals occupied the whole wild type micronuclei before crescent stage. (B) Gamma-H2AX signals occupied the whole wild type meiotic micronuclei during crescent stage. (C and D) There were no gamma H2AX signals in *CYC2* KO micronuclei, which could not elongate fully during conjugation.

into crescent (Fig. 5C and 5D). These results suggest that meiotic DSB formation was totally disrupted after knocking out of the *CYC2* gene.

Validation of the RNA-Seq results

From the significantly regulated genes of four time points after conjugation initiation we randomly selected 9 genes together with *SPO11* gene to check the validity and credibility of RNA-Seq results by comparing the mRNA expression profile of 3 independent biological repeats of conjugants of wild type and *CYC2* KO strains, using reverse transcription quantitive PCR (RT-qPCR) with *RPL11* as the internal control (Fig. 6). The list of these 9 genes with some annotation details and the binary logarithm of fold changes of being regulated can be viewed in supplement archive. Despite some slight differences, the general expression profiles were quite consistent between the RNA-Seq and RT-qPCR data, convincingly validating the reliability of the transcriptomic changes after knocking out *CYC2*.

DISCUSSION

T. thermophila is a great model organism for studying meiosis process, since its meiosis property make synchronization easy and high-efficient. Previous work mainly conducted traditional biochemistry and cell biology methods to study meiosis process of this ciliate (Howard-Till et al., 2011). In our research, we highlight on meiosis process by combining phenotype analysis and transcriptome-wide deep sequencing with traditional biochemistry methods. RNA-Seq analysis was performed to determine the effects of *CYC2* KO on genome-wide genes' expression levels during meiosis. Plenty of DEGs between *CYC2* KO and wild type strains were functionally enriched in DNA mismatch repair and DNA replication field, revealing that cyc2p, a cell cycle related protein, would be essential to meiosis by participating in DSBs formations, DNA mismatch repair and replication.

On this experimental design, Δ*CYC2* displayed conspicuously defect of meiotic micronuclei elongation and meiotic DSBs formation in compatible with previously published phenotype of *SPO11* knockout strains (Δ*SPO11*) during the prophase of meiosis I (Mochizuki et al., 2008; Zhang et al., unpublished data). Ample evidence indicates that DSBs are initiators of meiotic recombination. We reasoned that the absence of DSBs formation in Δ*CYC2* strain was due to the drastic reduction in *SPO11* expression, which is responsible for formation of meiotic DSBs as catalyzing the DNA cleavage via its Type II B topoisomerase-like transesterase activity with the help of a series of other related proteins (Bergerat et al., 1997; Cao et al., 1990; Keeney et al., 1997b; Lam and Keeney, 2015; Sun et al., 1989; Szostak et al., 1983).

Some previous research reported that spo11p in mice and budding yeast might function independently of DSBs induction to assist homolog alignment (Boateng et al., 2013, Loidl, 2013). Thus, spo11p has very crucial roles in inducing DNA self-generated DSBs, which is indispensable for homolog chromatin crossover formation at synaptonemal complex state, and possibly also in promoting homologs to approach each other to be aligned faithfully. That may be reasonable to explain the act in *T. thermophila* of the germline nuclear elongation since the homolog recognition is easy to achieve when chromosome become a long thin bouquet in crescent stage.

However, although *SPO11* is the catalytic center of the meiotic recombination initiation mechanism, and is also highly expressed at 2–4 h after conjugation initiation and involved in DSBs formations, *CYC2* gene's expression has not been affected by knocking out *SPO11* according to TGFD (*Tetrahymena* Functional Genomics Database) meiosis RNA-Seq data (Fig. S3) and (Zhang et al., unpublished data). This could indicate that *CYC2* may play a key role in regulating *Spo11* and further influencing the whole downstream process of meiosis.

Additionally, HR is associated very closely with DNA repair including mismatch repair (Spies and Fishel, 2015). Previous study has demonstrated that the recombinases Rad51 and Dmc1 play important roles in DNA break repair

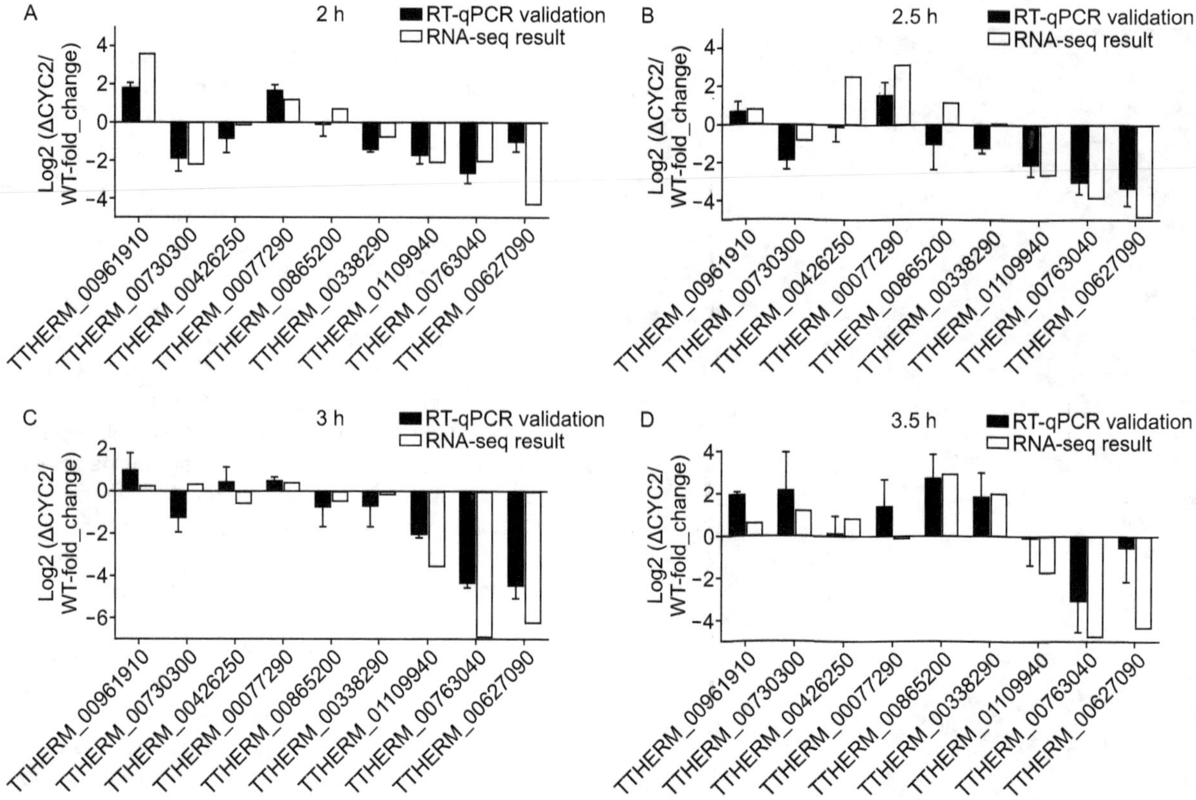

Figure 6. RT-qPCR verification of the RNA-Seq results. The 8 randomly selected genes (TTHERM_00961910; TTHERM_00730300; TTHERM_00426250; TTHERM_00077290; TTHERM_00865200; TTHERM_00338290; TTHERM_01109940; TTHERM_00763040) plus *CYC2* TTHERM_00079530) and *SPO11* (TTHERM_00627090) were performed with RT-qPCR. The primers were designed to across an intron to eliminate potential confluence from residue genomic DNA contamination. cDNA templates were from of 3 independent biological repeats of cell samples respectively at 2 h (A), 2.5 h (B), 3 h (C), and 3.5 h (D) after conjugation initiation. The tendencies of mRNA expression changes were in accordance with the RNA-Seq analysis results, which well-validated of the sequencing data.

and recombination partner choice in the meiosis of *Tetrahymena* (Howard-Till et al., 2011). Also during the streamlined meiosis of *Tetrahymena*, Msh family protein function in synaptonemal complex independent chiasma formation (Shodhan et al., 2014). A very interesting phenomenon we observed is that several key meiotic recombination proteins associated with DNA repair (*MRE 11*, *RAD51*, *DMC1*, *Msh3*, and *Msh6*), which were all down regulated in *CYC2* KO (Table 2), were not affected after ablating *SPO11* according to TGFD meiosis RNA-Seq data (Zhang et al., unpublished data). Also the recombination defect in *SPO11* mutants can be partially rescued by production of DSBs from an exogenous source such as ionizing radiation (Loidl and Mochizuki, 2009). These data may confirm the involvement of *SPO11* in DSBs formation without effects on meiotic recombination machinery proteins. Meanwhile these could indicate that DNA repair (including MMR) genes down regulated in Δ*CYC2* may be also directly regulated by *CYC2* rather than caused through *SPO11* down-regulated.

In the present study we concluded that the meiotic disabled phenotypes and the absence of DSBs in Δ*CYC2* strains due to the down regulation of *SPO11* but the exact mechanism how *CYC2* actively control *SPO11* at the right time and place remains unknown. Also, these data will encourage us to unravel the pathway how *CYC2* affect meiotic recombination process during meiosis I prophase. Another challenge is the need to fully catalog *CYC2* in *T. thermophila* and consider the following questions concerning the initiation of meiotic recombination, for example, how does *CYC2* actively regulate *SPO11* to promote DSBs formation? How does *CYC2* regulate the recombination machinery genes or proteins? Does *CYC2* act as a meiosis specific transcription factor?

MATERIALS AND METHODS

T. thermophila strains and culture conditions

WT strains CU428 and B2086 II were provided by P. J. Bruns (Cornell University, Ithaca, NY, USA, now available through the

National *Tetrahymena* Stock Center, http://tetrahymena.vet.cornell.edu/index.html). Cells were grown in super proteose peptone (1× SPP) medium (1% proteose peptone, 0.2% glucose, 0.1% yeast extract, and 0.003% EDTA ferric sodium salt) at 30°C with vigorous shaking (rotation speed usually at 180–220 rpm), unless otherwise specifically requested. For starvation, log phase cells were washed and resuspended in 10 mmol/L Tris (pH 7.4) at 30°C without shaking for 24 h unless specified otherwise. Cells were counted using a hemocytometer to control the concentration of the strains (Gorovsky et al., 1975).

Phenotypic analysis

Two mating types (CU428 VII and B2086 II) of *T. thermophila* were fully starved by washing and resuspending in 10 mmol/L Tris (pH 7.4) in flasks without shaking for 20–24 h at 30°C, followed by mixing almost equal quantities of each to artificially start the conjugation process. Several time points interested were chose and the corresponding cell samples were collected, fixed with formaldehyde and stained with DAPI. Each sample was observed by Olympus fluorescence microscope to be classified to different stages by the developmental characteristics of both the Micronucleus and Macronucleus during conjugation and counted many enough cells to conclude the statistical number proportion.

Transcriptome sequencing

Cells for each of the samples with the amount of approximately 1 × 10^7 (cells) for each samples were washed once with 1× PBS (centrifuge at 450 ×*g* for 2 min) and spun down to pellet followed by immediate freezing in liquid nitrogen. After the total RNA extraction and DNase I treatment, magnetic beads with Oligo (dT) are used to isolate mRNA. Mixed with the fragmentation buffer, the mRNA is fragmented into short fragments. Then cDNA is synthesized using the mRNA fragments as templates. Short 6/8 fragments are purified and resolved with EB buffer for end reparation and single nucleotide A (adenine) addition. After that, the short fragments are connected with adapters. The suitable fragments are selected for the PCR amplification as templates. During the QC steps, Agilent 2100 Bioanaylzer and ABI Step One Plus Real-Time PCR System are used in quantification and qualification of the sample library. At last, the libraries are sequenced using Illumina HiSeq 2000.

Data processing, identified DEGs and GO enrichment analysis

Raw sequencing reads were trimmed by removing Illumina adapter sequences and low quality bases. The STAR program with default parameters was used to map the sequenced reads to the reference genome and to find those that spanned exon–exon junctions. Transcripts were assembled using the Cufflinks software and gene expressions were based on fragments per kilobase of exon model per million mapped reads (FPKM) values. FPKM of different samples were compared by using Cuffdiff to identify DEGs. The up-regulated and down-regulated DEGs were analyzed for gene ontology (GO) through ClueGO (v 2.1.7) software. The RNA-Seq data used in this study have been deposited in NCBI Gene Expression Ommbius (GSE79286).

Indirect immunofluorescence microscopy

1.5 mL of mating cells with 2 × 10^5 cells/mL density in 10 mmol/L Tris, pH 7.5 were mixed with 5 μL of partial Schaudin's fixative (two parts saturated $HgCl_2$ to one part 100% ethanol) and incubated for 5 min at room temperature (RT). Cells were gently pelleted (130 ×*g* for 30 s) and washed once with 3 mL of RT methanol and resuspended in 1 mL of RT methanol. 50 μL of cells suspension was spread onto a coverslip and air dried for 30 min. Cells were stained with anti-γH2A.X (1:100) (Anti-phospho-H2A.X (Ser139) Mouse Ab (Millipore 05-636-I) Antibody) followed by incubation with AlexaFluor 568 goat anti-mouse immunoglobulin G (IgG; 1:400) (Invitrogen). Nuclei were stained with the DNA-specific dye 4′,6′-diamidino-2-phenylindole (DAPI; Roche) at 10 ng/mL for 10 min. Images were obtained with an Olympus IX73 fluorescence microscope.

Isolation of total cellular RNA

The total cellular RNA of all samples in this study were extracted using Eastep™ Universal RNA Extraction Kit (Promega, Cat.No. LS1030) followed by reverse transcription using GoScript™ Reverse Transcription System Kit (Promega, Cat.No.A5000) and the step by step procedures are based on the given product manuals all available online.

RT-qPCR

The cDNA products from RNA extraction and reverse transcription underwent real time quantitive PCR to quantify the mRNA expression level of selected genes using AceQ® qPCR SYBR® Green Master Mix (Vazyme, Cat.No.Q111-03) on the equipment CFX Connect Bio-Rad. The procedures in details and the PCR thermocycle protocol are based on the given product manuals all available online.

ACKNOWLEDGMENTS

We sincerely thank Dr. Kazufumi Mochizuki for providing us *CYC2* KO strain, and Dr. Xiangting Wang for the support in the process of searching for gamma-H2AX antibodies worked for immune-fluorescence staining. This work was supported by grants from the Projects of International Cooperation and Exchanges Ministry of Science and Technology of China (No. 2013DFG32390). X.S is a recipient of the Young Thousand Talents program (KJ2070000026).

ABBREVIATIONS

DEGs, differentially expressed genes; DSB, double strand break; HR, homologous recombination; KO, knock out; MMR, mismatch repair

COMPLIANCE WITH ETHICS GUIDELINES

Qianlan Xu, Ruoyu Wang, A.R. Ghanam, Guanxiong Yan, Wei Miao, and Xiaoyuan Song declare that they have no conflict of interest. This article does not contain any studies with human or animal subjects performed by the any of the authors.

REFERENCES

Bergerat A, de Massy B, Gadelle D, Varoutas PC, Nicolas A, Forterre P (1997) An atypical topoisomerase II from Archaea with implications for meiotic recombination. Nature 386:414–417

Bindea G, Mlecnik B, Hackl H, Charoentong P, Tosolini M, Kirilovsky A, Fridman WH, Pages F, Trajanoski Z, Galon J (2009) ClueGO: a cytoscape plug-into decipher functionally grouped gene ontology and pathway annotation networks. Bioinformatics 25:1091–1093

Boateng KA, Bellani MA, Gregoretti IV, Pratto F, Camerini-Otero RD (2013) Homologous pairing preceding SPO11-mediated double-strand breaks in Mice. Dev Cell 24:196–205

Cao L, Alani E, Kleckner N (1990) A pathway for generation and processing of double-strand breaks during meiotic recombination in *S. cerevisiae*. Cell 61:1089–1101

Collins K (2012) Perspectives on the ciliated protozoan *Tetrahymena thermophila*. Method Cell Biol 109:3–7

Devault A, Gueydon E, Schwob E (2008) Interplay between S-cyclin-dependent kinase and Dbf4-dependent kinase in controlling DNA replication through phosphorylation of yeast Mcm4 N-terminal domain. Mol Biol Cell 19:2267–2277

Gao S, Xiong J, Zhang CC, Berquist BR, Yang RD, Zhao M, Molascon AJ, Kwiatkowski SY, Yuan DX, Qin ZH et al (2013) Impaired replication elongation in *Tetrahymena mutants* deficient in histone H3 Lys 27 monomethylation. Gene Dev 27:1662–1679

Gibson TJ, Thompson JD, Blocker A, Kouzarides T (1994) Evidence for a protein domain superfamily shared by the cyclins, Tfiib and Rb/P107. Nucleic Acids Res 22:946–952

Gorovsky MA, Yao MC, Keevert JB, Pleger GL (1975) Isolation of micro- and macronuclei of *Tetrahymena pyriformis*. Methods Cell Biol 9:311–327

Hayashi A, Mochizuki K (2015) Targeted gene disruption by ectopic induction of DNA elimination in *Tetrahymena*. Genetics 201:55–64

Henderson KA, Kee K, Maleki S, Santini PA, Keeney S (2006) Cyclin-dependent kinase directly regulates initiation of meiotic recombination. Cell 125:1321–1332

Howard-Till RA, Lukaszewicz A, Loidl J (2011) The recombinases Rad51 and Dmc1 play distinct roles in DNA break repair and recombination partner choice in the meiosis of *Tetrahymena*. Plos Genet 7:e1001359

Kauppi L, Jeffreys AJ, Keeney S (2004) Where the crossovers are: recombination distributions in mammals. Nat Rev Genet 5:413–424

Keeney S (2001) Mechanism and control of meiotic recombination initiation. Curr Top Dev Biol 52:1–53

Keeney S, Giroux CN, Kleckner N (1997a) Meiosis-specific DNA double-strand breaks are catalyzed by Spo11, a member of a widely conserved protein family. Cell 88:375–384

Keeney S, Giroux CN, Kleckner N (1997b) Meiosis-specific DNA double-strand breaks are catalyzed by Spo11, a member of a widely conserved protein family. Cell 88:375–384

Lam I, Keeney S (2015) Mechanism and regulation of meiotic recombination initiation. Cold Spring Harb Perspect Biol 7:a016634

Loidl J (2013) The hidden talents of SPO11. Dev Cell 24:123–124

Loidl J, Mochizuki K (2009) *Tetrahymena* meiotic nuclear reorganization is induced by a checkpoint kinase-dependent response to DNA damage. Mol Biol Cell 20:2428–2437

Miao W, Xiong J, Bowen J, Wang W, Liu YF, Braguinets O, Grigull J, Pearlman RE, Orias E, Gorovsky MA (2009) microarray analyses of gene expression during the *Tetrahymena thermophila* life cycle. Plos One 4:e4429

Mochizuki K, Gorovsky MA (2004) Small RNAs in genome rearrangement in *Tetrahymena*. Curr Opin Genet Dev 14:181–187

Mochizuki K, Novatchkova M, Loidl J (2008) DNA double-strand breaks, but not crossovers, are required for the reorganization of meiotic nuclei in *Tetrahymena*. J Cell Sci 121:2148–2158

Petronczki M, Siomos MF, Nasmyth K (2003) Un menage a quatre: the molecular biology of chromosome segregation in meiosis. Cell 112:423–440

Shah JC, Clancy MJ (1992) Ime4, a gene that mediates mat and nutritional control of meiosis in *Saccharomyces-cerevisiae*. Mol Cell Biol 12:1078–1086

Shodhan A, Lukaszewicz A, Novatchkova M, Loidl J (2014) Msh4 and Msh5 function in SC-independent chiasma formation during the streamlined meiosis of *Tetrahymena*. Genetics 198:983–993

Song XY, Gjoneska E, Ren QH, Taverna SD, Allis CD, Gorovsky MA (2007) Phosphorylation of the SQ H2A.X motif is required for proper meiosis and mitosis in *Tetrahymena thermophila*. Mol Cell Biol 27:2648–2660

Spies M, Fishel R (2015) Mismatch repair during homologous and homeologous recombination. Cold Spring Harb Perspect Biol 7: a022657

Stover NA, Rice JD (2011) Distinct cyclin genes define each stage of ciliate conjugation. Cell Cycle 10:1699–1701

Sun H, Treco D, Schultes NP, Szostak JW (1989) Double-strand breaks at an initiation site for meiotic gene conversion. Nature 338:87–90

Szostak JW, Orrweaver TL, Rothstein RJ, Stahl FW (1983) The double-strand-break repair model for recombination. Cell 33:25–35

Wang Z, Gerstein M, Snyder M (2009) RNA-Seq: a revolutionary tool for transcriptomics. Nat Rev Genet 10:57–63

Woehrer SL, Aronica L, Suhren JH, Busch CJL, Noto T, Mochizuki K (2015) A *Tetrahymena* Hsp90 co-chaperone promotes siRNA loading by ATP-dependent and ATP-independent mechanisms. EMBO J 34:559–577

Xiong J, Lu XY, Zhou ZM, Chang Y, Yuan DX, Tian M, Zhou ZG, Wang L, Fu CJ, Orias E et al (2012) Transcriptome analysis of the model protozoan, *Tetrahymena thermophila*, using deep RNA sequencing. Plos One 7:e30630

Zhang H, Adl SM, Berger JD (1999) Two distinct classes of mitotic cyclin homologues, Cyc1 and Cyc2, are involved in cell cycle regulation in the ciliate *Paramecium tetraurelia*. J Eukaryot Microbiol 46:585–596

13

Structural insights into the assembly of the 30S ribosomal subunit *in vivo*: functional role of S5 and location of the 17S rRNA precursor sequence

Zhixiu Yang[1], Qiang Guo[1], Simon Goto[2], Yuling Chen[1], Ningning Li[1], Kaige Yan[1], Yixiao Zhang[1], Akira Muto[2], Haiteng Deng[1], Hyouta Himeno[2], Jianlin Lei[1✉], Ning Gao[1✉]

[1] Ministry of Education Key Laboratory of Protein Sciences, Center for Structural Biology, School of Life Sciences, Tsinghua University, Beijing 100084, China
[2] Department of Biochemistry and Molecular Biology, Faculty of Agriculture and Life Science, Hirosaki University, Hirosaki 036-8561, Japan
✉ Correspondence: jllei@tsinghua.edu.cn (J. Lei), ninggao@tsinghua.edu.cn (N. Gao)

ABSTRACT

The *in vivo* assembly of ribosomal subunits is a highly complex process, with a tight coordination between protein assembly and rRNA maturation events, such as folding and processing of rRNA precursors, as well as modifications of selected bases. In the cell, a large number of factors are required to ensure the efficiency and fidelity of subunit production. Here we characterize the immature 30S subunits accumulated in a factor-null *Escherichia coli* strain (Δ*rsgA*Δ*rbfA*). The immature 30S subunits isolated with varying salt concentrations in the buffer system show interesting differences on both protein composition and structure. Specifically, intermediates derived under the two contrasting salt conditions (high and low) likely reflect two distinctive assembly stages, the relatively early and late stages of the 3′ domain assembly, respectively. Detailed structural analysis demonstrates a mechanistic coupling between the maturation of the 5′ end of the 17S rRNA and the assembly of the 30S head domain, and attributes a unique role of S5 in coordinating these two events. Furthermore, our structural results likely reveal the location of the unprocessed terminal sequences of the 17S rRNA, and suggest that the maturation events of the 17S rRNA could be employed as quality control mechanisms on subunit production and protein translation.

Zhixiu Yang and Qiang Guo have contributed equally to this work.

KEYWORDS RsgA, RbfA, ribosome assembly, cryo-EM, quantitative mass spectrometry

INTRODUCTION

Fast growing bacteria have a tremendous demand for the ribosome so as to produce proteins required for various cellular activities in a timely manner. In *Escherichia coli*, there are around 30 known assembly factors *in vivo*, and some of them possess well-defined enzymatic functions, such as rRNA modification and processing. Genetic perturbation of different assembly factors confers similar phenotypes in general, often characterized by accumulation of rRNA precursors and immature assembly intermediates [reviewed in (Shajani et al., 2011)]. However, further functional elucidation of assembly factors in molecular detail has been staggered, mainly due to three reasons. The first is that translation is such a fundamental process that the interference of ribosome assembly by genetic manipulations of assembly genes often causes disparate secondary cellular disorders. The second is that the assembly process is highly efficient *in vivo*, and most factors play important, but non-essential roles, which makes it impossible to isolate factor-specific intermediates for functional assays. As a result, assembly intermediates from factor-null strains are in principle a heterogeneous set of presumably related intermediates. The last is that the assembly of ribosomal proteins is highly coupled with rRNA maturation, such as rRNA folding, modification and processing (Shajani et al., 2011), and disruption of a single assembly event might elicit a cascade of assembly defects.

Nevertheless, the *in vitro* reconstitution of ribosomal subunit free of assembly factors has remained to be a classical biochemical system for decades, with a focus on studying the binding interdependence among ribosomal proteins [for examples, see (Grondek and Culver, 2004; Mizushima and Nomura, 1970; Rohl and Nierhaus, 1982)] and metastable rRNA conformational transitions [for examples, see (Calidas and Culver, 2011; Holmes and Culver, 2004, 2005; Powers et al., 1993; Ramaswamy and Woodson, 2009; Stern et al., 1989)] during the assembly process. More recently, time-resolved techniques from pulse-labeling based quantitative mass-spectrometry (QMS) (Mulder et al., 2010; Talkington et al., 2005) and synchrotron X-ray footprinting (Adilakshmi et al., 2008) have further generated a large amount of real-time data for various protein binding and rRNA folding events in the process of the *in vitro* 30S subunit assembly. These kinetic data, together with previous knowledge, have stitched a general picture that the assembly of the 30S subunit *in vitro* is a highly branched process, featured with interconnected parallel assembly pathways (Mulder et al., 2010), and the rRNA folding is tightly coupled with protein binding, displaying an apparent "induced fit" strategy (Adilakshmi et al., 2008).

With a wealth of *in vitro* information at hand, in recent years two complimentary approaches were taken to tackle the *in vivo* 30S assembly process. One is to quantify ribosomal protein levels in presumably different assembly intermediates accumulated *in vivo* (from genetically or chemically perturbed strains) through QMS (Clatterbuck Soper et al., 2013; Guo et al., 2013; Jomaa et al., 2011; Leong et al., 2013; Sykes et al., 2010). The other is to use structure-probing methods to investigate rRNA conformations of these *in vivo* intermediates (Clatterbuck Soper et al., 2013; Guo et al., 2013; Jomaa et al., 2011; Leong et al., 2013; Roy-Chaudhuri et al., 2010). Through the integration of the compositional and structural data with the known protein binding interdependences described in the Nomura map (Mizushima and Nomura, 1970), as well as with the roughly determined binding orders of ribosomal proteins (Chen and Williamson, 2013), it starts to become practical to dissect these *in vivo* data to identify rate-limiting steps of the 30S assembly and to uncover molecular roles of assembly factors and ribosomal proteins.

Previously, we have constructed an *E. coli* mutant strain (Guo et al., 2013), in which genes of two assembly factors, RbfA and RsgA, are deleted. RbfA is a cold-shock protein (Jones and Inouye, 1996) involved in the maturation of the 30S subunit (Bylund et al., 1998; Dammel and Noller, 1995; Goto et al., 2011; Xia et al., 2003). RbfA binds to the neck of the 30S subunit (Datta et al., 2007) and its function was genetically linked to the maturation of the 5′ end of the 16S rRNA (Dammel and Noller, 1995). RsgA is a 30S subunit-dependent GTPase also involved in the maturation of the 30S subunit (Daigle and Brown, 2004; Himeno et al., 2004). Importantly, RsgA is one of the few examples among all assembly factors whose molecular role has been biochemically established, which is to promote the timely release of RbfA from the mature or nearly mature 30S subunit in a GTP-dependent manner (Goto et al., 2011; Guo et al., 2011). Disruption of these two genes simultaneously confers phenotypes similar to those of single gene disruptants (Bylund et al., 1998; Dammel and Noller, 1995; Himeno et al., 2004; Jomaa et al., 2011; Xia et al., 2003), such as accumulation of immature 30S subunits and the 17S rRNA precursor in the cell (Goto et al., 2011; Guo et al., 2013).

In the present work, we characterize the immature 30S subunits isolated from this Δ*rsgA*Δ*rbfA* strain. The immature intermediates were prepared using two different buffers varying in salt concentration and subjected to the compositional and structural analyses separately. Interestingly, the two sets of the immature 30S subunits are very distinct in both the protein composition and the 3D structure. The exposure to the high salt buffer does not destroy the known binding interdependences between the ribosomal proteins. But, it appears to have a destabilizing effect preferentially on the tertiary proteins in the 3′ domain. Structural data indicate that protein level of the 3′ domain is proportional to its rigidity, displaying an apparent coupling between the binding of proteins and the assembly of the rRNA secondary elements. Especially, the deficiency of S5 in structures from the high salt treated sample is correlated with a dramatic rotation of the head domain, as well as an alternative conformation of the 5′ end of the 16S rRNA, which suggests an underrecognized role of S5 in coordinating the 5′ end maturation and the 3′ domain assembly. Furthermore, our data reveal the possible location of the predicted helical stem formed by basepairing between the extra sequences in the 5′ and 3′ ends of the 17S rRNA. Overall, our results not only provide structural insights into the *in vivo* maturation of the 30S subunit, particularly at both early and late stages of the 3′ domain assembly, but also demonstrate the presence of multiple rRNA maturation checkpoints along the assembly pathway.

RESULTS

Protein composition of the immature 30S subunits isolated with high salt buffer from the Δ*rbfA*Δ*rsgA* strain

Previously, we have determined protein levels of the immature 30S subunits isolated from the Δ*rbfA*Δ*rsgA* strain, using a relatively low salt concentration (150 mmol/L NH_4Cl) in the

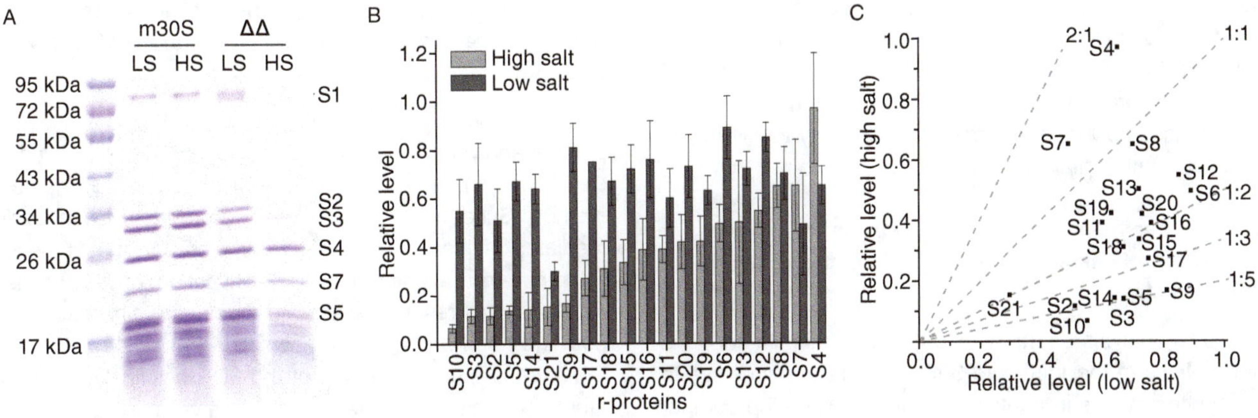

Figure 1. Protein compositions of the Δ*rbfA*Δ*rsgA* immature 30S subunits isolated under different salt conditions. (A) SDS-PAGE analysis of mature 30S (m30S) and Δ*rbfA*Δ*rsgA* (ΔΔ) immature 30S subunits isolated under low salt (LS) and high salt (HS) conditions. (B) QMS analysis of relative protein levels of the Δ*rbfA*Δ*rsgA* immature 30S subunits. (C) Comparison of relative levels of individual r-proteins under low salt and high salt conditions. The primary (1°), secondary (2°) and tertiary (3°) binding proteins are colored red, green and blue, respectively. The compositional data for the Δ*rbfA*Δ*rsgA* immature 30S subunits isolated under low salt condition are from a previous study (Guo et al., 2013).

sucrose cushion (Guo et al., 2013). Notably, a previous study showed that the salt concentration during purification could largely change the protein profile of isolated *in vivo* 50S assembly intermediates (Jiang et al., 2006). And it is known that the salt concentration is an important factor for reconstitutions of ribosomal subunits in *in vitro*. Typical *in vitro* 30S reconstitutions were carried out in a buffer containing 330 mmol/L KCl [for examples, see (Adilakshmi et al., 2008; Culver and Noller, 1999; Talkington et al., 2005)], and even higher salt concentrations were used in earlier studies (Traub and Nomura, 1968). Therefore, in order to interpret the compositional data based on isolated samples in a physiological context, it is necessary to compare the *in vivo* 30S intermediates obtained under different experimental conditions. It is conventional to use high salt buffers (for example, 300–1000 mmol/L NH_4Cl) to remove bound factors or other undesired proteins to obtain active ribosomes or ribosomal subunits through centrifugation (Blaha et al., 2000; Rodnina and Wintermeyer, 1995; Selmer et al., 2006) or chromatography (Jelenc, 1980; Maguire et al., 2008) based purification techniques. In the present work, we decided to try a relatively harsh condition (1 mol/L NH_4Cl) in the sucrose cushion to purify the immature 30S subunits from the Δ*rbfA*Δ*rsgA* strain. This would enable us to understand the effect of salt concentration on the subunit assembly and to compare the data acquired with different experimental conditions (Guo et al., 2013; Jomaa et al., 2011; Leong et al., 2013).

For comparison, mature 30S subunits, which were derived from 70S ribosomes of the parental strain in a dissociating buffer containing 2 mmol/L Mg^{2+}, were also prepared similarly with two different salt buffers (150 mmol/L vs 1 mol/L NH_4Cl). As shown in Fig. 1A, based on the protein gel analysis, the mature 30S subunit tolerates the high salt exposure well; the protein levels of the mature 30S subunits that have undergone two different treatments are virtually identical. In contrast, immature 30S subunits show apparently reduced levels for some proteins, such as S1, S2, S3 and S5, when exposed to high salt buffer (Fig. 1A). To quantitatively analyze the protein composition of the immature 30S subunits isolated under high salt condition, we employed a previously established QMS method (Guo et al., 2013). Among all proteins, S1 was not included in the QMS analysis, as it dissociates readily from the 30S subunit during centrifugation-based purification. For a similar reason, although S21 was included in the QMS analysis, it is not discussed in the following sections.

As expected, high salt treatment has an obvious effect on most of the proteins (Fig. 1B), rendering an average occupancy of 0.36 (Table S3). Nevertheless, protein levels in this high salt treated sample still show a clear pattern, with the 3′ domain proteins mostly underrepresented. Particularly, the levels of proteins S10, S3, S2, S14, S5, S21 and S9 are severely reduced, less than 20% of those in the mature 30S subunit (Fig. 1B and Table S1). In fact, all of these proteins are tertiary or secondary proteins, and most of them (such as S10, S3, S2, S14 and S9) are localized in the 3′ domain. Also, in the high salt treatment sample, several primary binding proteins such as S17, S15 and S20 also show reduced levels to some extents (less than 50%) (Fig. 1B).

Therefore, taking into consideration the known binding interdependencies of the 30S proteins (Mizushima and Nomura, 1970), it can be concluded that the *in vivo* intermediates isolated with high salt buffers are a heterogeneous set of particles, and it may include intermediates that lack primary proteins in the 5′ and central domains. A major population of this set of *in vivo* particles isolated under high salt condition might still represent snapshots of 30S assembly intermediates trapped in early stages of the 3′ domain assembly, characterized by extremely low levels of secondary and tertiary proteins in the 3′ domain.

High salt treatment preferentially weakens the binding of tertiary proteins in the head domain of the immature 30S subunits

Our QMS analysis on the high salt treated sample confirms that proteins in the immature 30S subunits are sensitive to salt concentrations in purification buffers. When the compositional data of the immature 30S subunits acquired under two salt conditions are plotted against each other, a recognizable pattern becomes apparent (Fig. 1C). The first observation is that high salt treatment decreases the occupancies of most of the proteins in the immature 30S subunits (Fig. 1C). The only exception is that two primary proteins S4 and S7 show relatively higher levels in the high salt treated sample. The second observation is that exposure to the high salt buffer affects the protein levels in a non-uniform manner. Previously, we reported that S21, S7 and S2 are the most underrepresented proteins (ranging from 0.3 to 0.5) in the low salt treated samples (Fig. 1B) (Guo et al., 2013). However, in the high salt treated sample, the most underrepresented one is S10, followed by S3, S2, S5, S14, S21 and S9 (lower than 0.20), clearly displaying a different pattern. In terms of the extent of protein level changes, the high salt treatment appears to have a more marked effect on tertiary proteins (Fig. 1C). In fact, all of the tertiary proteins except S12 and S11 have the largest changes in their relative levels, especially those in the 3′ domain (S10, S3, S2 and S14), as they are seen to have approximately a 5-fold change in occupancy. In addition, S9, a secondary binder in the 3′ domain, is also among the most affected group. The rest of secondary proteins are in the moderately affected group, with up to 2-fold change in occupancy. As to the primary proteins, they are the least affected, with slight changes of occupancy in both directions. Thus, taken together, the compositional difference in the two samples reveals that the high salt treatment preferentially facilitates dissociation of tertiary proteins in the 3′ head domain of the immature 30S subunits.

It is known that prior binding of late proteins in the 3′ domain could arrest the assembly process in certain stages, resulting in different kinetically trapped assembly intermediates (Mulder et al., 2010; Talkington et al., 2005). The immature 30S subunits isolated under low salt condition are low in S7 (Fig. 1B), and therefore they are likely a set of intermediates that are kinetically trapped due to slow binding of S7. The salt sensitivity of the 3′ domain tertiary proteins in the immature 30S subunits indicates that the premature association of these late proteins is probably non-native and relatively weak.

Cryo-EM structures of the immature 30S subunits isolated under low salt condition from the Δ*rbfA*Δ*rsgA* strain

Similar to our previous work, we applied cryo-EM single particle analysis to the low salt treated immature 30S particles from the Δ*rbfA*Δ*rsgA* strain. A multi-structure refinement method (Scheres, 2012) was used to probe structural intermediates of the immature 30S particles. On the 2D level, reference-free alignment and classification of all particles results in well resolved class averages, with fine details on both the head and body domains of the 30S subunits (Fig. S1A). With extensive tuning of the 3D classification parameters, the data were finally split into five classes, and the resulting five density maps are in a resolution of 13–18 Å. Similar to previous cryo-EM studies on the *in vivo* immature 30S subunits (Guo et al., 2013; Jomaa et al., 2011; Leong et al., 2013), no significant densities could be found beyond the 5′ and 3′ ends of the 16S rRNA, indicating that the terminal sequences of the 17S rRNA are flexible in these structures. However, these maps show a number of interesting differences (Movie S1). Firstly, S7 and S2 are seriously underrepresented in these structures (Fig. 2), which is highly consistent with the QMS-based compositional analysis (Fig. 1B). In fact, compared with the mature 30S map (Fig. 2A), none of structures has completely resolved S7. It is either totally (Fig. 2B) or partially (Fig. 2C–F) missing from the density maps. As to S2, it is relatively solid in two maps (Fig. 2C and 2F), but present in apparently sub-stoichiometric level in the rest three maps (Fig. 2B, 2D and 2E). Notably, S2 appears to have no strict binding dependence on S7, agreeing with our assumption that these structures represent intermediates that are arrested by prior binding of late proteins ahead of S7. Secondly, a very dramatic conformational difference lies at the 3′ minor domain of the 16S rRNA (h44 and h45). Structures of two classes (I-b and I-c), accounting for 40% of the immature 30S particles, show relatively rigid conformation for h44 and h45. In contrast, in the rest three maps (I-a, I-d and I-e), h44 is highly flexible, in completely undocked positions. Thirdly, the 30S platform region (h22–h24 of the central domain) appears to be in an open conformation in the I-a structure (Fig. 2B and 2G). This indicates that the late-stage assembly of the 30S subunit also involves the reorientation of the central domain of the 16S rRNA, consistent with a very recent X-ray footprinting data (Clatterbuck Soper et al., 2013). Lastly, although all the five structures have well resolved, rigid head domains, they appear in different rotational positions. This is most evident in the temperature maps of the 16S rRNA in these structures (Fig. 2G). Temperature maps were constructed by coloring the flexibly fitted atomic models of these structures according to their distance deviations from the structure of the mature 30S subunit. As shown, the head rotations in these maps are in a range of 10–18 Å, which is in sharp contrast to the immature 30S structures from the Δ*rimM* strain (Guo

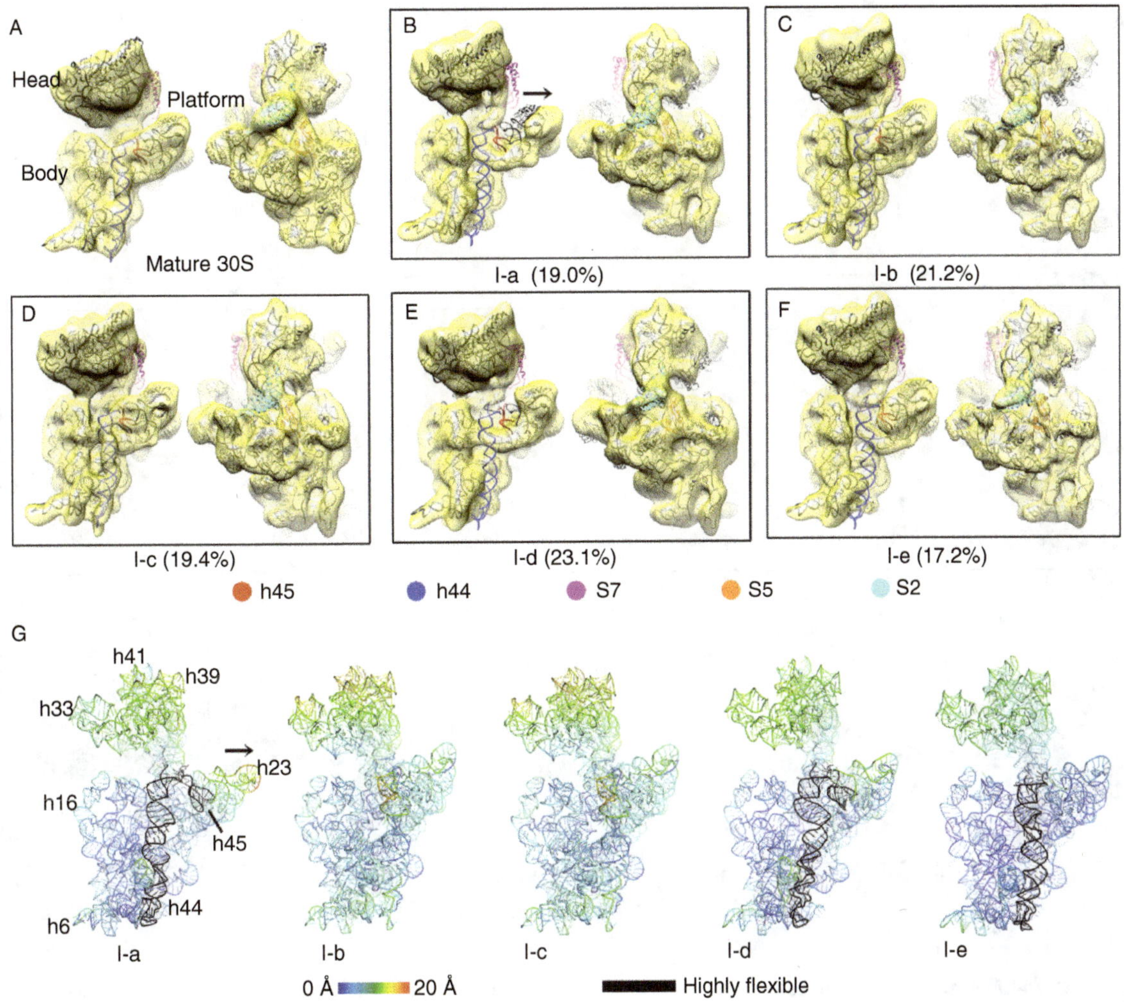

Figure 2. Cryo-EM structures of low salt treated Δ*rbfA*Δ*rsgA* immature 30S subunits. Density maps of the mature 30S (A) and low salt treated immature 30S subunits (B–F) are superimposed with the atomic model of the mature 30S subunit and displayed from both intersubunit and solvent views. h45, h44, S7, S5 and S2 are colored in red, blue, magenta, orange and cyan, respectively. Particle ratio of each structure is shown. (G) Distance deviations of the 16S rRNA in the five immature 30S structures from that of the mature 30S subunit are displayed as temperature maps, colored according to the scale bar. The highly flexible helices, such as h44 and h45 are colored black. Open conformation of the platform region in I-a is indicated by a black arrow (B and G).

et al., 2013), with rotations in a much larger scale (up to 40 Å). Besides the large domain rotation, the temperature maps clearly show that the upper portion of h44 in the two structures with relatively ordered h44 (I-b and I-c, Fig. 2G) also deviates from the mature conformation, indicating that the decoding center in these immature 30S structures is not fully assembled.

In summary, in terms of protein composition and structural features, the immature 30S subunits obtained with low salt buffer is quite similar to the late-stage Group IV particles identified in a previous time-resolved structural study (Mulder et al., 2010), confirming that they are a set of late-stage intermediates that are close to completion.

Cryo-EM structures of the immature 30S subunits isolated under high salt condition from the Δ*rbfA*Δ*rsgA* strain

Similarly, we analyzed the immature 30S particles isolated under high salt condition using the identical image classification method. On the 2D level, class averages already show significant difference from the low salt treated particles, with a large portion of particles displaying extremely mobile head domains (Fig. S1B). On the 3D level, the particles were first separated into five classes. These five classes show very large structural differences (Figs. 3 and S2). Structures of the two classes (II-a and II-b) do not reveal any reliable

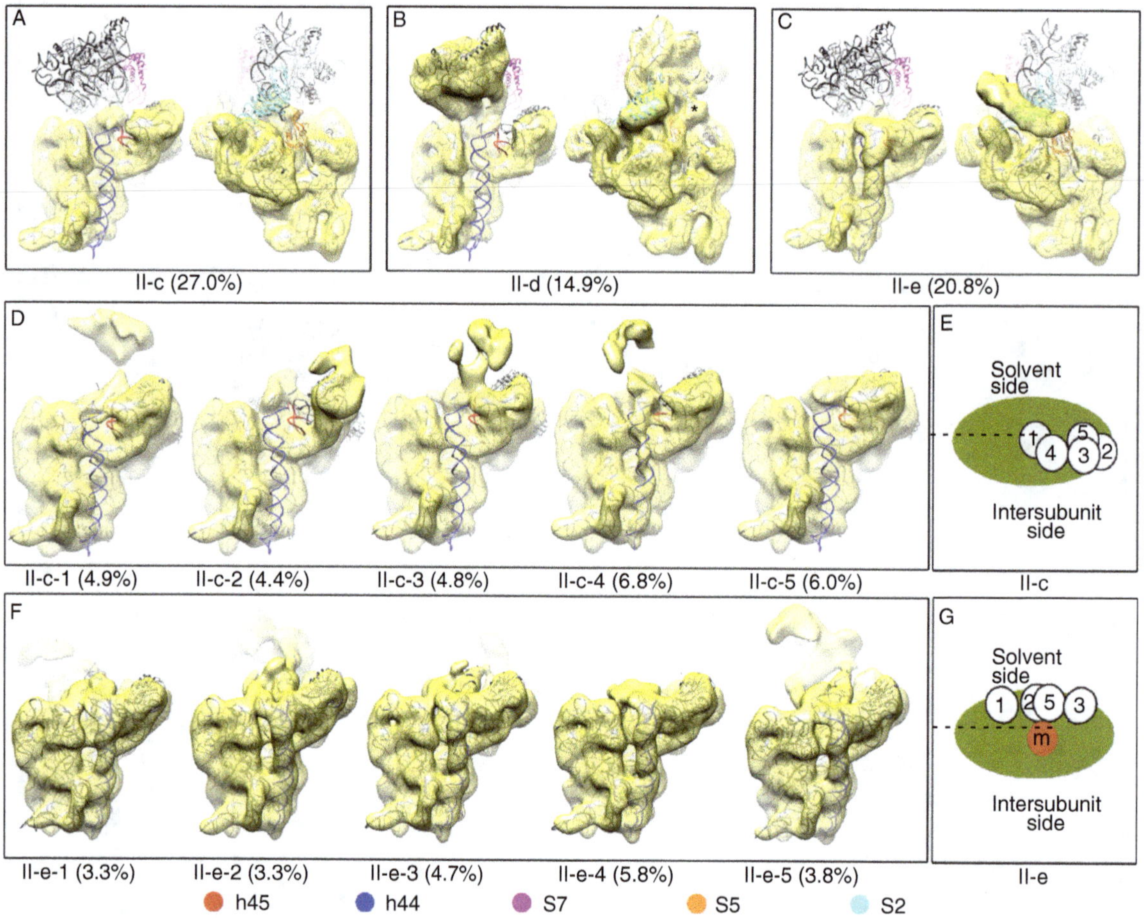

Figure 3. Cryo-EM structures of high salt treated Δ*rbfA*Δ*rsgA* immature 30S subunits. (A–C) Three representative density maps of high salt treated particles are superimposed with the atomic model of the mature 30S subunit and displayed from intersubunit and solvent views. (D) Subclass density maps of the II-c group are displayed from intersubunit view with the globally fitted atomic model of the mature 30S subunit (head domain excluded) superimposed. (F) Subclass density maps of the II-e group are displayed from intersubunit view with the flexibly fitted atomic model of the mature 30S subunit (head domain excluded) superimposed. The relative head domain orientations of the II-c and II-e subclass density maps are indicated as cartoon models in panel (E) and (G), respectively. In these models, the body domain is represented as a large green oval and the centers of the head domains as small white ovals with subclass number. The head domain position of the mature 30S subunit as a reference is shown in (G) as a red oval labeled with "m". The head domain position of the II-e-4 map is not given in (G) since it has extremely low density in the head domain region.

structural details (Fig. S2), and are virtually biased global averages. The 2D reference-free classification of the particles from these two classes reveals that they lack the general defined shape of the 30S subunit (Fig. S2), suggesting they are a wide variety of heterogeneous particles, probably including low quality particles, aggregated particles, as well as very early-stage particles that are deficient in primary proteins in the 5′ and central domains. In contrast, the rest three classes (II-c, II-d and II-e) have well-resolved structural details (Fig. 3A–C), which indicates that they are truly populated structural intermediates. Among these three classes, the II-d structure is similar to structures from the low salt treated samples, with highly flexible h44 and h45 but a relatively rigid head domain (Fig. 3B). Interestingly, compared with the low salt structures, a small density blob appears at the mRNA entrance channel, which could be residual density for the 5′ leader sequence of the 17S rRNA (Figs. 3B and S3). The other two (II-c and II-e, Fig. 3A and 3C), in sharp contrast, display extremely mobile head domains, resulting in a cluster of smeared densities at the head domain location (Movies S2 and S3). There is also an apparent difference between the II-c and II-e structures: one is with a disordered 3′ minor domain (II-c, Fig. 3A), and the other with a well resolved 3′ minor domain (II-e, Fig. 3C). Based on the general structure features of the three class structures (Fig. 3A–C), we could group the high salt treated particles into two major populations: one with an extreme flexibility at the 3′ domain (II-c and II-e) and the other being typical late-stage particles (II-d) (Fig. S4).

The smeared densities at the head domain of two of these structures (II-c and II-e) suggest additional structural heterogeneity within these groups. Therefore, we further classified each of these two groups into five subclasses. In the subclass structures from the class II-c, different conformations of h44 are revealed (Fig. 3D). One of the five subclasses (II-c-4) shows partially resolved densities for h44, while the remaining four (20.1% particles in total) show completely undocked h44. Interestingly, in one of the subclass structure (II-c-2), the platform of the 30S subunit is again in an open conformation. Other than the conformational differences at the 3′ minor domain, the head domains of these subclass structures are clearly in different positions (Fig. 3E). Although the densities of head domains are relatively low compared with the body domain, recognizable shape of a typical head domain can be seen in lower threshold (Fig. S5 and Movie S2).

The subclass structures of the class II-e also reveal a number of interesting structural observations (Fig. 3F). Firstly, all five subclass structures have well-resolved h44 and h45. However, h44 in these structures is in different conformations, with its upper portion significantly bent, at varying extents, toward the 50S subunit direction, compared with the mature 30S structure (Fig. S6), indicating that the observed docking of h44 in these structures is probably premature. Secondly, the head domains in these maps are also at different positions (Fig. 3G, Fig. S5 and Movie S3). However, in terms of direction of the head rotation, subclass structures from the II-c and II-e groups are sharply different. While the head domain in the II-c subclass structures appears to swivel around the connecting hinge (h28) between the body domain and head domain (Fig. 3E and Movie S2), the head domain of the subclasses from the II-e group is almost bent 90 degree toward the back of the 30S subunit, and rotates in a totally different plane (Fig. 3G, Fig. S5 and Movie S3).

Therefore, our structural data on the high salt treated particles are consistent with the QMS data, indicating that many of the tertiary proteins in the 3′ domain are in fact absent from these intermediates. Consequently, loss of proteins at different rRNA locations greatly changes both the overall and local conformations of the 17S rRNA, particularly the 3′ head and minor domains.

DISCUSSION

Effect of salt on the 30S subunit assembly

In the present work, we characterized the *in vivo* assembly intermediates from the Δ*rbfA*Δ*rsgA* strain isolated under two different salt conditions. Our results indicate that proteins in the *in vivo* intermediates are sensitive to salt treatment. While the exposure to 1 mol/L NH_4^+ largely changes both the protein profile and the rRNA conformation of the immature 30S particles, it has minimal effect on the mature 30S subunits. Especially, it is extremely effective in removing the 3′ domain tertiary proteins. This observation might have functional implications. Previous kinetic data show that the *in vitro* assembly of the 30S subunit often encounters kinetic traps in the 3′ domain, such as those caused by prior binding of late proteins, and predicts that assembly factors function to subvert these kinetic traps (Bunner et al., 2010; Mulder et al., 2010; Talkington et al., 2005). Our data further indicate that the premature associations of the 3′ domain proteins in these kinetically trapped intermediates are relatively weak, compared with those in the 5′ and central domains. This would mean that the requirement of non-physiologically high salt concentration in *in vitro* experiments is to minimize those relatively weak and non-productive associations of late proteins. In other words, the salt in the *in vitro* experiments partially compensates the functional role of assembly factors, by preventing the formation of kinetic traps. Indeed, a very interesting observation is that deletion of assembly factor genes confers salt resistance to *E. coli* cells (Hase et al., 2013; Hase et al., 2009), which also seems to suggest an equivalency between assembly factors and the salt ions.

The compositional differences in the immature 30S subunits obtained with different buffers also indicate that the QMS data based on purified samples might not truly reflect the situations inside the cell in every detail. One has to take into consideration the purification condition when it comes to the integration of data from different sources. It has drawn our attention that, in terms of individual protein levels, our previous QMS data of the immature 30S particles from the *rimM* null strain (Guo et al., 2013) show some discrepancies with another independently obtained QMS data from a similar strain (Leong et al., 2013). Besides the possible methodological errors introduced in mass spectrometry, the buffer systems (150 mmol/L vs 500 mmol/L NH_4Cl in sucrose cushion) might be the major source of these differences (Table S3). Nevertheless, both our and their data are in general consistent with the current model of the ordered 5′ to 3′ assembly of the 30S subunit, showing that disruption of *rimM* causes the assembly arrest at the 3′ domain of the 16S rRNA (Guo et al., 2013).

The intracellular K^+ concentration of the log phase *E. coli* cells is around 200 mmol/L (Schultz and Solomon, 1961), which is similar to our low salt condition (150 mmol/L NH_4Cl). This indicates that our compositional data of the low salt treated samples (Guo et al., 2013) might largely retain the *in vivo* information. In terms of functional role of RbfA, our structural data on the low salt derived 30S particle from the Δ*rbfA*Δ*rsgA* strain is in consistent with recent findings that immature 30S subunits accumulated in different assembly factor-null strains share common structural features, such as the flexibility of the 3′ major and minor domains (Clatterbuck Soper et al., 2013; Guo et al., 2013; Jomaa et al., 2011;

Leong et al., 2013). This indicates the presence of a common bottleneck during the late-stage assembly, and suggests that factors contribute not only to a single assembly event but to a global structural reorganization, probably in a concerted manner (Clatterbuck Soper et al., 2013; Leong et al., 2013).

As to the high salt derived particles, it appears that they might not have a direct functional relevance to RbfA and RsgA, due to the high salt treatment. However, our compositional data show that the high salt treatment does not destroy the relationship in the Nomura map, and the relative levels of the primary, secondary and tertiary proteins in these particles are exactly in a descending order (Fig. 1C). Furthermore, our structural classification successfully distinguishes the populations of the high salt washed particles in different conformational groups (Fig. 3). Consistently, both the protein composition and structural features of the two major groups correlate well with the previously reported *in vitro* assembly intermediates, corresponding to the Group II and IV particles, respectively (Mulder et al., 2010), suggesting that these structures might virtually reflect intermediates at different temporal stages. Altogether, our results indicate that the high salt treatment has largely changed the structures of the *in vivo* intermediates, and likely returned some of the late-stage particles to the entry stage of the 3′ domain assembly. Therefore, by varying the salt concentration in the sucrose cushion, we might have captured the 30S assembly intermediates at two distinctive stages.

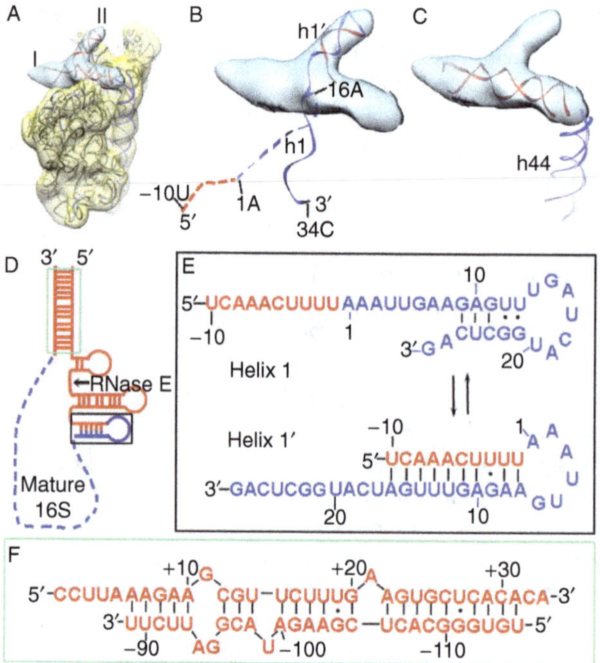

Figure 4. Visualization of the extra 5′ and 3′ ends of the 17S rRNA. (A) The density map of the II-e-3 subclass is displayed in transparent yellow, with flexibly fitted atomic model superimposed. A "T" shaped extra piece of densities (composed of two rods, I and II) is colored light blue. (B) Compatibility of the rod II with the alternative conformation of the 5′ end helix (h1′). The mature and precursor sequences are colored blue and red, respectively. Residues −10 to −1 of the leader sequences and 1–16 of the mature 16S rRNA in its native conformation are shown as dashed line. As indicated, the 5′ leader sequences together with the mature sequences (residues 1–16) could flip almost 180 degree and fold into an alternative structure (h1′). Also see Movie S4. (C) Compatibility of the rod I with the predicted helical stem formed between the precursor sequences of the 5′ and 3′ ends of the 17S rRNA. A segment of h44 in the mature 30S structure is also shown, indicating the continuity between the rod I and h44. For illustration, a standard helix of ~27 bp is fitted in the map. (D) Secondary structure prediction of the complete 17S rRNA. Sequences of the mature 16S rRNA are indicated as blue dashed line and the precursor sequences are colored red. The prediction was performed with a standalone version of RNA Structure (Version 4.6) using free energy minimization (Mathews et al., 2004). (E) Competing conformations of the 5′ end of the 16S rRNA, as two equilibrating states (h1 and h1′). The alternative h1′ is formed by basepairing between residues 7–16 and −1 to −10 of the 5′ leader sequences. (F) Predicted helical stem formed between the unprocessed 5′ and 3′ precursor sequences at the termini of the 17S rRNA.

Possible location of the precursor sequences in the 5′ and 3′ ends of the 17S rRNA

A very important observation from the structures of high salt treated samples is that in three subclass structures of the II-e group (II-e-2, II-e-3 and II-e-5), we found a piece of well resolved additional mass, located on the shoulder region of the 30S subunit (Fig. 4A). This mass is like a deformed letter "T", composed of a long rod with an orthogonal short rod (Fig. 4A). The two rods (I and II) of this T-shaped density cluster match well with the diameter of an rRNA helix, suggesting they are from a long helix (I) and a relatively vertical short helix (II). Careful examination of this additional mass reveals that it is not from the known fragments of the 16S rRNA (Movie S4). In terms of the density continuity, the longer rod clearly connects to the h44-h45 (Fig. 4C), while the shorter one is near the location of h1 in the mature 30S structure (Fig. 4B). Previously, we showed that a majority of the rRNA species in the immature 30S particles from the *ΔrbfAΔrsgA* strain is unprocessed at both ends (Guo et al., 2013), being the cleavage product of RNase III. Therefore, it is highly likely this extra mass is from the precursor sequences of the 17S rRNA.

Previous genetic and biochemical data have established a helical switching model for the maturation of the 5′ end of the 17S rRNA (Dammel and Noller, 1993; Young and Steitz, 1978). In this model, the 5′ mature sequence of the 17S rRNA would form a different secondary structure (h1′) other than that seen in the mature 16S form (h1) (Fig. 4E), with base paring between residues 7 to 16 and −1 to −10 of the leader sequence (Young and Steitz, 1978). Disturbing the transition from h1′ to the mature h1, such as mutations on protein S5 (G28D) (Roy-Chaudhuri et al., 2010) or on the 16S rRNA (C23U) (Dammel and Noller, 1993) would introduce defects in the 17S rRNA processing and the 30S subunit assembly. Additionally, overexpression of RbfA, which binds to the neck of the 30S subunit, at a helical junction of h28, h27, h18 and h1 (Datta et al., 2007), could suppress the assembly defects in the C23U rRNA mutant strain (Dammel and Noller, 1995), which is in further support of this helical switching model for the maturation of the 17S rRNA 5′ end. As to the maturation of the 3′ precursor sequences, based on the secondary structure prediction, multiple rRNA duplexes could potentially form from the extra nucleotides of the 17S rRNA (115 nucleotides at the 5′ end and 33 nucleotides at the 3′ end) (Fig. 4D). Especially, the 5′ and 3′ ends of the 17S rRNA have complementary sequences (Fig. 4F), and could form a long helix (Young and Steitz, 1978).

The shorter rod of this extra mass could exactly accommodate a helical stem of ~10 base pairs (Fig. 4B and Movie S4), matching well with the predicted alternative helix (h1′) in the 5′ end of the 17S rRNA (Dammel and Noller, 1993; Roy-Chaudhuri et al., 2010; Young and Steitz, 1978). The longer rod of this extra mass is roughly equivalent to an rRNA helix of 26 base pairs (Fig. 4C). Since this longer rod is close to the very 3′ end of the 16S rRNA, it is likely the rRNA duplex formed between the two unprocessed ends of the 17S rRNA (Movie S4). Therefore, we might directly visualize an earlier conformation of the 17S rRNA that is with base paired precursor sequences from the 5′ and 3′ ends. Although it is not possible to further characterize structural details of this extra mass, our data from the high salt treated sample reveals, at least partially, the location of the precursor sequences of the 17S rRNA, confirming the presence of complementary base pairing within the 5′ end, as well as between the 5′ and 3′ ends in the 17S rRNA precursor.

S5 couples the 5′ end maturation of the 17S rRNA with the 3′ domain assembly of the 30S subunit

S5 is a tertiary protein that binds at a helical junction (Culver et al., 1999), formed by h1, h2, h28, h36 and h44 (Fig. 5A), including rRNA segments from all four domains of the 16S rRNA. Previous work has implicated an important role of S5 in translation, showing that several *ram* mutations could be mapped to S5 (Ito and Wittmann, 1973; Kirthi et al., 2006; Piepersberg et al., 1975). Recently, in a series of studies it was shown that the miscoding phenotype of one S5 *ram* mutation (G28D), as well as a few other mutant strains (Δ*rng*, Δ*rpsO*, Δ*rimM* and Δ*ksgA*), is a consequence of cellular defects in ribosome assembly, particularly the faulty maturation of the 5′ end of the 17S rRNA (Kirthi et al., 2006; Roy-Chaudhuri et al., 2010; Roy-Chaudhuri et al., 2008), which have demonstrated a functional link between the decreased ribosome fidelity and the maturation defect of the 5′ end of the 17S rRNA. The site of this specific S5 mutation (G28D) is located in a protruding loop inserted into the interface of several rRNA helices, including h28, h36, h44 and h1 (Fig. 5A), manifesting an essential role of S5 in the conformational maturation of the 17S rRNA.

Very interestingly, we found that the dramatic head rotation observed in the high salt treated samples (II-c and II-e) is well correlated with the absence of S5 in these structures (Fig. 3A and 3C, Fig. 5, Movies S2 and S3). Specifically, we observed a correspondence between the local structural difference at the S5-h1-h28 interface and global conformational difference of the 17S rRNA (Fig. 5). In all structures from the low salt treated sample, S5 is well represented (Fig. 2B–F and Fig. S3), and the head domain in these structures is relatively rigid, although in slightly different rotational position (~10–15 Å in scale). The 5′ mature sequence of the 17S rRNA in these structures also appears to be in a mature conformation, but the leader sequence at the 5′ end is highly flexible (Fig. S3). In contrast, only one structure of the high salt treated sample displays well resolved densities for S5, and the 5′ mature sequence in the 17S rRNA and the head domain are again relatively rigid in this structure (Figs. 5D and S3). Other structures from the high salt treated sample are clearly depleted in S5 (Figs. 3 and 4), and they display very dramatic motions at the head domain and the 3′ minor domain. Detailed analysis indicates that local conformational differences on h28 and h1 might account for structural dynamics of the head domain. As shown, in the two S5-deficient structures with extremely mobile head domain, h28 is not well resolved (Fig. 5C and 5E). Other than that, in these S5-deficient structures, the very 5′ end of the 16S rRNA (single strand region, residues 1–8), as well as the first helix h1, is also highly flexible (Fig. 5C and 5E), indicating that the 5′ end of the 16S rRNA in these structures is in a conformation distinctive from its mature state. This also means that the central pseudoknot h2, formed by basepairing between one extension strand of h28 and the terminal loop of h1 (Fig. 5A), is not formed in these structures.

Altogether, these data suggest the binding of S5 facilitate the transition from h1′ to h1 and the stabilization of the native S5-h1-h28 interface. Consequently, the 3′ head domain acquires rigidity from a stabilized h28 and commits to further assembly. Thus, S5 appears to be an important player at the relatively early stage (probably the entry stage) of the 3′ domain assembly. An integrated role of S5 is to couple the maturation of the 5′ end of the 17S rRNA with the assembly of the 3′ domain, through the modulation of the S5-h1-h28 interface. Notably, this view is consistent with previous kinetic and time resolved structural data showing that S5 binds prior to most of the 3′ domain proteins, roughly at the entry stage of the 3′ domain assembly (Chen and Williamson, 2013; Mulder et al., 2010).

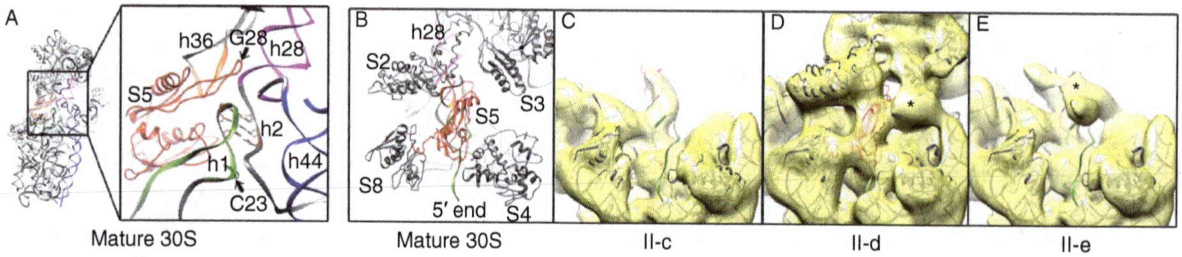

Figure 5. Role of S5 in coupling the 17S rRNA 5′ end maturation with the assembly of the 30S 3′ domain. (A) A zoom-in view of S5 and its surrounding components in the mature 30S structure. S5, h1, h2, h28, h36 and h44 are colored red, forest green, sienna, magenta, orange and blue, respectively. Residue G28 in S5 and C23 in the 16S rRNA, which are important for the 5′ end maturation of the 17S rRNA are indicated by arrows. C23 and nucleotides forming a pseudoknot h2 are highlighted with side chains. (B) Close-up solvent view of the atomic model of the mature 30S subunit, showing S5 and its neighbors, such as S2, S3, S4, S8 and h1, with neighboring ribosomal proteins colored black and h1 in forest green. (C–E) Density maps of II-c (C), II-d (D) and II-e (E), superimposed with atomic model of the mature 30S subunit, are displayed from the same view as (B). For clarification, components of the head domain of the atomic model and S5 are removed in (C) and (E) since there are only very limited densities at the head domain location in these maps. As shown in (C–E), the presence of S5 is coupled with the rigidity of the head domain, as well as with the mature h1 conformation. The asterisks in (D and E) denote the residual densities of the 5′ leader sequences in the 17S rRNA.

Quality control of the 30S assembly

It is generally believed that the maturation of the 17S rRNA would complete before the 30S subunit is incorporated into the 70S ribosomes. Recent biochemical and structural data from ours and others are consistent with this view, showing that the major rRNA species in the immature 30S particles from factor-null strains are unprocessed at both the 5′ and 3′ ends (Clatterbuck Soper et al., 2013; Davies et al., 2010; Goto et al., 2011; Guo et al., 2013; Jomaa et al., 2011; Leong et al., 2013), and especially these immature 30S particles are highly dynamic at the 3′ minor domain (h44) [(Boehringer et al., 2012; Guo et al., 2013; Jomaa et al., 2011; Leong et al., 2013) (Clatterbuck Soper et al., 2013) and current study]. The 3′ minor domain hosts a number of intersubunit contact sites that are essential for the 50S association, which explains why substantial levels of immature 30S particles accumulate in such a variety of mutant strains. This suggests that the rate-limiting docking of h44, which is also observed on the mature 16S rRNA in *in vitro* reconstitution experiments (Adilakshmi et al., 2008; Dutca and Culver, 2008), could provide an intrinsic quality control on subunit production.

However, there are also lines of evidence showing that the immature 30S subunits containing 17S rRNA precursor are also capable of being incorporated into 70S ribosomes and polysomes (Clatterbuck Soper et al., 2013; Davies et al., 2010; Li et al., 1999; Mangiarotti et al., 1974; Roy-Chaudhuri et al., 2010), and an alternative idea is that the final maturation of the 17S rRNA takes place on the 70S or polysomes (Mangiarotti et al., 1974). Although the exact processing status of the 17S rRNA contained in these 70S ribosomes is not conclusively analyzed in each of these cases, some of them appear to be completely unprocessed at both ends (Davies et al., 2010; Li et al., 1999). Nevertheless, the above two models might not be exclusive, because the majority of the 17S rRNA is still in the free 30S fractions. Substantial incorporation of immature 30S subunit into the 70S ribosome itself could be a cellular disorder due to the processing defect in the rRNA maturation. Indeed, characterization of some types of these 17S rRNA-containing ribosomes reveals a defect in translational fidelity (Davies et al., 2010; Roy-Chaudhuri et al., 2010), and more importantly, it was recently demonstrated that the immature 30S subunits in these 70S ribosomes could trigger degradation of defective 70S ribosomes by YbeY and RNase R (Jacob et al., 2013). Therefore, these results show that bacterial cells possess multiple quality control systems that make use of the rRNA maturation at different stages to ensure the integrity of the 70S ribosome.

MATERIALS AND METHODS

Isolation of the immature and mature 30S subunits

Mature 30S subunits were purified from *E. coli* A19 strain (Hfr, *rna-19*, *gdhA2*, *his-95*, *relA1*, *spoT1*, *metB1*), and immature 30S subunits from an A19 derivative strain (Δ*rsgA*Δ*rbfA*) (Guo et al., 2013). Both the mature and immature 30S subunits were prepared under two salt conditions, with a sucrose cushion containing 150 mmol/L (low salt) or 1 mol/L NH_4Cl (high salt). The low salt purification procedures were previously described (Guo et al., 2013). The high salt purification is similar with modifications. Specifically, cells (A19, A19-Δ*rsgA*Δ*rbfA*) were cultured in LB medium at 37°C to reach log phase and harvested by centrifugation in a JLA 10.500 rotor (Beckman Coulter) at 5,500 rpm for 20 min. Cell pellets were resuspended in a lysis buffer [10 mmol/L Tris-HCl (pH 7.8), 10 mmol/L $MgCl_2$, 60 mmol/L NH_4Cl, 0.5 mmol/L $CaCl_2$, 0.1 mmol/L ethylenediamine tetraacetic acid (EDTA), 1 mmol/L dithiothreitol (DTT)] and ground with 0.5 mm glass beads (Scientific Industries) for 30 min at 4°C. Cell debris was removed by centrifugation in a JA 25.50 rotor (Beckman Coulter) at 15,500 rpm for 1 h, and the supernatant was gently transferred onto the top of 5 mL sucrose cushion [30% sucrose, 10 mmol/L Tris-HCl (pH 7.8), 10 mmol/L $MgCl_2$, 0.1 mmol/L EDTA, 1 mmol/L DTT, 1 mol/L NH_4Cl] and centrifuged in a 70 Ti rotor (Beckman Coulter) for 18 h at 28,000 rpm. The pellet after centrifugation was then dissolved in a gradient buffer [10 mmol/L Tris-HCl (pH 7.8), 10 mmol/L $MgCl_2$, 60 mmol/L NH_4Cl, 0.1 mmol/L EDTA, 1 mmol/L DTT], layered on the top of 10%–40% sucrose gradient in the same buffer, and centrifuged in a SW32 rotor (Beckman Coulter) at 30,000 rpm for 7 h. The gradients were initially prepared using the Gradient Master 108 (BioComp), and analyzed with A254 absorbance using a Teledyne ISCO fractionation system. The resulting 30S and 70S peaks were collected. The 70S ribosomes from the parental strain (A19) were separated into 30S and 50S subunits by changing the buffer into the dissociation buffer (same as gradient buffer, except 2 mmol/L $MgCl_2$). Mature 30S subunits were then obtained through centrifugation with a 10%–40% gradient (2 mmol/L $MgCl_2$).

Quantitative mass spectrometry

In order to quantitatively determine abundance of ribosomal proteins in the high salt treated immature 30S subunits, a TMT-based (tandem mass tags) (Thompson et al., 2003) quantitative mass spectrometry (QMS) was applied, as previously described (Guo et al., 2013). Briefly, equal amounts of samples (exactly the same A260 units) were separated on a 15% SDS-PAGE gel. The protein bands except S1 were excised from the gel, reduced with 10 mmol/L of DTT and alkylated with 55 mmol/L iodoacetamide. After the overnight in gel digestion by trypsin (Promega), the peptides were recovered by 1% TFA (trifluoroacetic acid) solution. Peptides were then labeled with TMT reagents (Thermo, Pierce Biotechnology) according to the manufacturer's instruction (TMT 130 and 131 for the immature and mature 30S subunits, respectively). For LC-MS/MS analysis, the TMT-labeled peptides were separated by a 60-min gradient elution at a flow rate of 0.250 μL/min with an EASY-nLCII™ integrated nano-HPLC system (Proxeon), which is directly interfaced with a Q Exactive mass spectrometer (Thermo Scientific). The analytical column was a fused silica capillary column (75 μm ID, 150 mm length; packed with C-18 resin). Mobile phase A consisted of 0.1% formic acid and mobile phase B consisted of 100% acetonitrile and 0.1% formic acid. The Q Exactive mass spectrometer was operated in the data-dependent acquisition mode using the Xcalibur 2.1.3 software and there was a single full-scan mass spectrum in the Orbitrap (400–1800 m/z, 30,000 resolution) followed by 10 MS/MS scans in the quadrupole collision cell using the higher energy collision dissociation. The MS/MS spectra from each LC-MS/MS run were searched against the selected database using Proteome Discovery searching engine. The searching parameters are listed below: peptides ms tolerance of 20 ppm; ms/ms tolerance of 20 mmu; carbamidomethylation of Cys, TMT of lysine and peptide N terminal as the fixed modification, oxidation on Met as the variable modification. Peptides with high confidence were used for protein identification and MS/MS spectra for all matched peptides were manually interpreted and confirmed. Ratios of 130:131 for each of the ribosomal proteins were calculated with outliers removed by Grubbs' test (http://graphpad.com/quickcalcs/Grubbs1.cfm). Only ratios with two or more tryptic peptides from the same protein were used to calculate the means and the standard deviations. The statistics of the TMT-based QMS measurements were summarized in Table S1.

Cryo-sample preparation and data collection

Quantifoil 2/4 grids (Quantifoil Micro Tools GmbH) were coated with a thin layer of continuous carbon (K950X, EMITech) and glow-discharged in a Harrick Plasma Cleaner (PDC-32G) for 30 s. Grids were prepared with an FEI Vitrobot Mark IV at 4°C as previously described (Guo et al., 2011). The final concentration for blotting was ~60 nmol/L. Cryo-grids were examined in an FEI Titan Krios, and the data were collected under low-dose conditions (~20 $e^-/Å^2$) on an FEI Eagle 4 k × 4 k CCD using AutoEMation (Lei and Frank, 2005). The images of the low salt and high salt treated particles were collected at a nominal magnification of 59,000×, but with different acceleration voltages, at 300 KV and 120 KV, respectively.

Image processing

4365 and 5590 micrographs were obtained for low salt and high salt treated immature 30S samples, respectively. After screening for ice contamination and astigmatism, 3238 and 4210 micrographs, for the low salt and high salt treated samples, respectively, were kept for further processing. All the micrographs were decimated by a factor of two, resulting in 3 Å and 2.62 Å in effective pixel size, for low salt and high salt intermediates, respectively. Contrast transfer function (CTF) parameters estimation and particle picking were done using SPIDER software package (Shaikh et al., 2008a). Raw particles were windowed using a local cross-correlation function based method (Rath and Frank, 2004). Automatically picked particles were manually verified using a correspondence analysis based method (Shaikh et al., 2008b). Finally, 219,169 and 160,264 particles, for low salt and high salt intermediates, respectively, were kept and subjected to single particle analysis. To explore the structural heterogeneity, classification was applied both at two dimensional (2D) and three dimensional (3D) levels, using RELION, an empirical Bayesian approach (Scheres, 2012). Both samples were first classified into 200 classes at 2D level in 25 iterations with an angle step of 5 degree, and bad particles were excluded from further analysis. At the 3D level, different parameters and class numbers have been tested. The initial model in 3D classification was generated by low-pass filtering (40 Å) of a previous cryo-EM map of the mature 30S subunit (Guo et al., 2011). The particles were finally split into five groups in fifty iterations until the classification was stabilized. To further determine head orientation of the class II-c and II-e particles, these particles were subjected to additional rounds of 3D classification using RELION, resulting in five subclasses for each group. Each subclass was further refined using SPIDER according to standard reference projection matching procedures (Shaikh et al., 2008a). The final resolutions of the refined structures were estimated using FSC (Fourier shell correlation) 0.5 cutoff criterion (Fig. S7). The statistics of image processing and the general features of the density maps were summarized in Table S2.

Flexible fitting and map analysis

A crystal structure of the 30S subunit (PDB ID: 3OFA) (Dunkle et al., 2010) was used as the initial model. For structures derived from low salt treated sample, the crystal structure of the 30S subunit was docked in each map and subjected to a flexible fitting method based on molecular dynamics simulation (MDFF) (Trabuco et al., 2008), using NAMD (Phillips et al., 2005). The simulations were performed with molecules placed in vacuum, monitored by the global correlation coefficients between the density map and the simulated model. For class II-e and its subclasses (II-e-1, II-e-2, II-e-3, II-e-4 and II-e-5), due to the low densities at the 3′ head domain of the 30S subunit, only coordinates of the body domain (including S4, S6, S8, S11, S12, S15, S16, S17, S18, S20 and S21, nucleotides 2–940 and 1375–1501 of the 16S rRNA) were subjected to the MDFF. After fitting, the coordinates of the rRNA in all simulated atomic models were extracted and aligned to that of the mature 30S subunit, using the nucleotides 2–900 as the reference for alignment. Temperature maps were generated in PyMOL (Schrodinger, 2010) through calculating the deviation of the rRNA coordinates in each fitted model from that of the mature 16S rRNA. The scripts for temperature map building are accessible at http://pldserver1.biochem.queensu.ca/rlc/work/pymol/. PyMOL and Chimera (Pettersen et al., 2004) were used for structural analysis and figure preparation.

ACCESSION CODES

The cryo-EM maps have been deposited in the EMDataBank (EMDB codes 5900, 5904, 5905, 5906, 5907, 5908, 5909 and 5910, for the I-a, I-b, I-c, I-d, I-e, II-c-3, II-d and II-e-3 structures, respectively).

ACKNOWLEDGEMENTS

We thank the Tsinghua National Laboratory for Information Science and Technology for providing computation resource. This work was supported by the National Basic Research Program (973 Program) (Nos. 2010CB912401, 2010CB912402 and 2013CB910404); the National Natural Science Foundation of China (Grant No. 31170677); Beijing Higher Education Young Elite Teacher Project (YETP0131); Tsinghua University (20131089278) and the Japan Society for the promotion of Science (23380054).

COMPLIANCE WITH ETHICS GUIDELINES

Zhixiu Yang, Qiang Guo, Simon Goto, Yuling Chen, Ningning Li, Kaige Yan, Yixiao Zhang, Akira Muto, Haiteng Deng, Hyouta Himeno, Jianlin Lei and Ning Gao declare that they have no conflict of interest.

This article does not contain any studies with human or animal subjects performed by the any of the authors.

REFERENCES

Adilakshmi T, Bellur DL, Woodson SA (2008) Concurrent nucleation of 16S folding and induced fit in 30S ribosome assembly. Nature 455:1268–1272

Blaha G, Stelzl U, Spahn CM, Agrawal RK, Frank J, Nierhaus KH (2000) Preparation of functional ribosomal complexes and effect of buffer conditions on tRNA positions observed by cryoelectron microscopy. Methods Enzymol 317:292–309

Boehringer D, O'Farrell HC, Rife JP, Ban N (2012) Structural insights into methyltransferase KsgA function in 30S ribosomal subunit biogenesis. J Biol Chem 287:10453–10459

Bunner AE, Beck AH, Williamson JR (2010) Kinetic cooperativity in *Escherichia coli* 30S ribosomal subunit reconstitution reveals additional complexity in the assembly landscape. Proc Natl Acad Sci USA 107:5417–5422

Bylund GO, Wipemo LC, Lundberg LA, Wikstrom PM (1998) RimM and RbfA are essential for efficient processing of 16S rRNA in *Escherichia coli*. J Bacteriol 180:73–82

Calidas D, Culver GM (2011) Interdependencies govern multidomain architecture in ribosomal small subunit assembly. RNA 17:263–277

Chen SS, Williamson JR (2013) Characterization of the ribosome biogenesis landscape in *E. coli* using quantitative mass spectrometry. J Mol Biol 425:767–779

Clatterbuck Soper SF, Dator RP, Limbach PA, Woodson SA (2013) In vivo X-ray footprinting of pre-30S ribosomes reveals chaperone-dependent remodeling of late assembly intermediates. Mol Cell 52:506–516

Culver GM, Noller HF (1999) Efficient reconstitution of functional *Escherichia coli* 30S ribosomal subunits from a complete set of recombinant small subunit ribosomal proteins. RNA 5:832–843

Culver GM, Heilek GM, Noller HF (1999) Probing the rRNA environment of ribosomal protein S5 across the subunit interface and inside the 30 S subunit using tethered Fe(II). J Mol Biol 286:355–364

Daigle DM, Brown ED (2004) Studies of the interaction of *Escherichia coli* YjeQ with the ribosome in vitro. J Bacteriol 186:1381–1387

Dammel CS, Noller HF (1993) A cold-sensitive mutation in 16S rRNA provides evidence for helical switching in ribosome assembly. Genes Dev 7:660–670

Dammel CS, Noller HF (1995) Suppression of a cold-sensitive mutation in 16S rRNA by overexpression of a novel ribosome-binding factor, RbfA. Genes Dev 9:626–637

Datta PP, Wilson DN, Kawazoe M, Swami NK, Kaminishi T, Sharma MR, Booth TM, Takemoto C, Fucini P, Yokoyama S et al (2007) Structural aspects of RbfA action during small ribosomal subunit assembly. Mol Cell 28:434–445

Davies BW, Kohrer C, Jacob AI, Simmons LA, Zhu J, Aleman LM, Rajbhandary UL, Walker GC (2010) Role of *Escherichia coli* YbeY, a highly conserved protein, in rRNA processing. Mol Microbiol 78:506–518

Dunkle JA, Xiong L, Mankin AS, Cate JH (2010) Structures of the *Escherichia coli* ribosome with antibiotics bound near the peptidyl transferase center explain spectra of drug action. Proc Natl Acad Sci USA 107:17152–17157

Dutca LM, Culver GM (2008) Assembly of the 5′ and 3′ minor domains of 16S ribosomal RNA as monitored by tethered probing from ribosomal protein S20. J Mol Biol 376:92–108

Goto S, Kato S, Kimura T, Muto A, Himeno H (2011) RsgA releases RbfA from 30S ribosome during a late stage of ribosome biosynthesis. EMBO J 30:104–114

Grondek JF, Culver GM (2004) Assembly of the 30S ribosomal subunit: positioning ribosomal protein S13 in the S7 assembly branch. RNA 10:1861–1866

Guo Q, Yuan Y, Xu Y, Feng B, Liu L, Chen K, Sun M, Yang Z, Lei J, Gao N (2011) Structural basis for the function of a small GTPase RsgA on the 30S ribosomal subunit maturation revealed by cryoelectron microscopy. Proc Natl Acad Sci USA 108:13100–13105

Guo Q, Goto S, Chen Y, Feng B, Xu Y, Muto A, Himeno H, Deng H, Lei J, Gao N (2013) Dissecting the in vivo assembly of the 30S ribosomal subunit reveals the role of RimM and general features of the assembly process. Nucleic Acids Res 41:2609–2620

Hase Y, Yokoyama S, Muto A, Himeno H (2009) Removal of a ribosome small subunit-dependent GTPase confers salt resistance on *Escherichia coli* cells. RNA 15:1766–1774

Hase Y, Tarusawa T, Muto A, Himeno H (2013) Impairment of ribosome maturation or function confers salt resistance on *Escherichia coli* cells. PloS One 8:e65747

Himeno H, Hanawa-Suetsugu K, Kimura T, Takagi K, Sugiyama W, Shirata S, Mikami T, Odagiri F, Osanai Y, Watanabe D et al (2004) A novel GTPase activated by the small subunit of ribosome. Nucleic Acids Res 32:5303–5309

Holmes KL, Culver GM (2004) Mapping structural differences between 30S ribosomal subunit assembly intermediates. Nat Struct Mol Biol 11:179–186

Holmes KL, Culver GM (2005) Analysis of conformational changes in 16 S rRNA during the course of 30 S subunit assembly. J Mol Biol 354:340–357

Ito T, Wittmann HG (1973) Amino acid replacements in proteins S5 and S12 of two *Escherichia coli* revertants from streptomycin dependence to independence. Mol Gen Genet 127:19–32

Jacob AI, Kohrer C, Davies BW, RajBhandary UL, Walker GC (2013) Conserved bacterial RNase YbeY plays key roles in 70S ribosome quality control and 16S rRNA maturation. Mol Cell 49:427–438

Jelenc PC (1980) Rapid purification of highly active ribosomes from *Escherichia coli*. Anal Biochem 105:369–374

Jiang M, Datta K, Walker A, Strahler J, Bagamasbad P, Andrews PC, Maddock JR (2006) The *Escherichia coli* GTPase CgtAE is involved in late steps of large ribosome assembly. J Bacteriol 188:6757–6770

Jomaa A, Stewart G, Martin-Benito J, Zielke R, Campbell TL, Maddock JR, Brown ED, Ortega J (2011) Understanding ribosome assembly: the structure of in vivo assembled immature 30S subunits revealed by cryo-electron microscopy. RNA 17:697–709

Jones PG, Inouye M (1996) RbfA, a 30S ribosomal binding factor, is a cold-shock protein whose absence triggers the cold-shock response. Mol Microbiol 21:1207–1218

Kirthi N, Roy-Chaudhuri B, Kelley T, Culver GM (2006) A novel single amino acid change in small subunit ribosomal protein S5 has profound effects on translational fidelity. RNA 12:2080–2091

Lei J, Frank J (2005) Automated acquisition of cryo-electron micrographs for single particle reconstruction on an FEI Tecnai electron microscope. J Struct Biol 150:69–80

Leong V, Kent M, Jomaa A, Ortega J (2013) Escherichia coli rimM and yjeQ null strains accumulate immature 30S subunits of similar structure and protein complement. RNA 19:789–802

Li Z, Pandit S, Deutscher MP (1999) RNase G (CafA protein) and RNase E are both required for the 5′ maturation of 16S ribosomal RNA. EMBO J 18:2878–2885

Maguire BA, Wondrack LM, Contillo LG, Xu Z (2008) A novel chromatography system to isolate active ribosomes from pathogenic bacteria. RNA 14:188–195

Mangiarotti G, Turco E, Ponzetto A, Altruda F (1974) Precursor 16S RNA in active 30S ribosomes. Nature 247:147–148

Mathews DH, Disney MD, Childs JL, Schroeder SJ, Zuker M, Turner DH (2004) Incorporating chemical modification constraints into a dynamic programming algorithm for prediction of RNA secondary structure. Proc Natl Acad Sci USA 101:7287–7292

Mizushima S, Nomura M (1970) Assembly mapping of 30S ribosomal proteins from *E. coli*. Nature 226:1214

Mulder AM, Yoshioka C, Beck AH, Bunner AE, Milligan RA, Potter CS, Carragher B, Williamson JR (2010) Visualizing ribosome biogenesis: parallel assembly pathways for the 30S subunit. Science 330:673–677

Pettersen EF, Goddard TD, Huang CC, Couch GS, Greenblatt DM, Meng EC, Ferrin TE (2004) UCSF Chimera—a visualization system for exploratory research and analysis. J Comput Chem 25:1605–1612

Phillips JC, Braun R, Wang W, Gumbart J, Tajkhorshid E, Villa E, Chipot C, Skeel RD, Kale L, Schulten K (2005) Scalable molecular dynamics with NAMD. J Comput Chem 26:1781–1802

Piepersberg W, Bock A, Wittmann HG (1975) Effect of different mutations in ribosomal protein S5 of Escherichia coli on translational fidelity. Mol Gen Genet 140:91–100

Powers T, Daubresse G, Noller HF (1993) Dynamics of in vitro assembly of 16 S rRNA into 30 S ribosomal subunits. J Mol Biol 232:362–374

Ramaswamy P, Woodson SA (2009) S16 throws a conformational switch during assembly of 30S 5′ domain. Nat Struct Mol Biol 16:438–445

Rath BK, Frank J (2004) Fast automatic particle picking from cryo-electron micrographs using a locally normalized cross-correlation function: a case study. J Struct Biol 145:84–90

Rodnina MV, Wintermeyer W (1995) GTP consumption of elongation factor Tu during translation of heteropolymeric mRNAs. Proc Natl Acad Sci USA 92:1945–1949

Rohl R, Nierhaus KH (1982) Assembly map of the large subunit (50S) of *Escherichia coli* ribosomes. Proc Natl Acad Sci USA 79:729–733

Roy-Chaudhuri B, Kirthi N, Kelley T, Culver GM (2008) Suppression of a cold-sensitive mutation in ribosomal protein S5 reveals a role for RimJ in ribosome biogenesis. Mol Microbiol 68:1547–1559

Roy-Chaudhuri B, Kirthi N, Culver GM (2010) Appropriate maturation and folding of 16S rRNA during 30S subunit biogenesis are critical for translational fidelity. Proc Natl Acad Sci USA 107:4567–4572

Scheres SH (2012) A Bayesian view on cryo-EM structure determination. J Mol Biol 415:406–418

Schrodinger LLC (2010) The PyMOL molecular graphics system, version 1.3r1

Schultz SG, Solomon AK (1961) Cation transport in *Escherichia coli*. I. Intracellular Na and K concentrations and net cation movement. J Gen Physiol 45:355–369

Selmer M, Dunham CM, Murphy FVT, Weixlbaumer A, Petry S, Kelley AC, Weir JR, Ramakrishnan V (2006) Structure of the 70S ribosome complexed with mRNA and tRNA. Science 313:1935–1942

Shaikh TR, Gao H, Baxter WT, Asturias FJ, Boisset N, Leith A, Frank J (2008a) SPIDER image processing for single-particle reconstruction of biological macromolecules from electron micrographs. Nat Protoc 3:1941–1974

Shaikh TR, Trujillo R, LeBarron JS, Baxter WT, Frank J (2008b) Particle-verification for single-particle, reference-based reconstruction using multivariate data analysis and classification. J Struct Biol 164:41–48

Shajani Z, Sykes MT, Williamson JR (2011) Assembly of bacterial ribosomes. Annu Rev Biochem 80:501–526

Stern S, Powers T, Changchien LM, Noller HF (1989) RNA-protein interactions in 30S ribosomal subunits: folding and function of 16S rRNA. Science 244:783–790

Sykes MT, Shajani Z, Sperling E, Beck AH, Williamson JR (2010) Quantitative proteomic analysis of ribosome assembly and turnover in vivo. J Mol Biol 403:331–345

Talkington MW, Siuzdak G, Williamson JR (2005) An assembly landscape for the 30S ribosomal subunit. Nature 438:628–632

Thompson A, Schafer J, Kuhn K, Kienle S, Schwarz J, Schmidt G, Neumann T, Johnstone R, Mohammed AK, Hamon C (2003) Tandem mass tags: a novel quantification strategy for comparative analysis of complex protein mixtures by MS/MS. Anal Chem 75:1895–1904

Trabuco LG, Villa E, Mitra K, Frank J, Schulten K (2008) Flexible fitting of atomic structures into electron microscopy maps using molecular dynamics. Structure 16:673–683

Traub P, Nomura M (1968) Structure and function of *E. coli* ribosomes. V. Reconstitution of functionally active 30S ribosomal particles from RNA and proteins. Proc Natl Acad Sci USA 59:777–784

Xia B, Ke H, Shinde U, Inouye M (2003) The role of RbfA in 16S rRNA processing and cell growth at low temperature in *Escherichia coli.* J Mol Biol 332:575–584

Young RA, Steitz JA (1978) Complementary sequences 1700 nucleotides apart form a ribonuclease III cleavage site in *Escherichia coli* ribosomal precursor RNA. Proc Natl Acad Sci USA 75:3593–3597

14

Transcriptome analyses of insect cells to facilitate baculovirus-insect expression

Kai Yu[1,2], Yang Yu[1,2], Xiaoyan Tang[1,2], Huimin Chen[1,2], Junyu Xiao[2,3]✉, Xiao-Dong Su[1,2]✉

[1] Biodynamic Optical Imaging Center, School of Life Science, Peking University, Beijing 100871, China
[2] State Key Laboratory of Protein and Plant Gene Research, Peking University, Beijing 100871, China
[3] Peking-Tsinghua Center for Life Sciences, Peking University, Beijing 100871, China
✉ Correspondence: junyuxiao@pku.edu.cn (J. Xiao), xdsu@pku.edu.cn (X.-D. Su)

ABSTRACT

The High Five cell line (BTI-TN-5B1-4) isolated from the cabbage looper, *Trichoplusia ni* is an insect cell line widely used for baculovirus-mediated recombinant protein expression. Despite its widespread application in industry and academic laboratories, the genomic background of this cell line remains unclear. Here we sequenced the transcriptome of High Five cells and assembled 25,234 transcripts. Codon usage analysis showed that High Five cells have a robust codon usage capacity and therefore suit for expressing proteins of both eukaryotic- and prokaryotic-origin. Genes involved in glycosylation were profiled in our study, providing guidance for engineering glycosylated proteins in the insect cells. We also predicted signal peptides for transcripts with high expression abundance in both High Five and Sf21 cell lines, and these results have important implications for optimizing the expression level of some secretory and membrane proteins.

KEYWORDS High Five cell line, baculovirus-insect cell system, codon usage, glycosylation, signal peptide

INTRODUCTION

The baculovirus-insect cell expression system is one of the most popular platforms for recombinant protein expression. It is widely used for protein structure and function studies in academic laboratories, and facilitates massive protein production in industry (Kost et al. 2005). The two common cell lines in this binary system are Sf21 (IPLB-Sf21AE) from *Spodoptera frugiperda* (Vaughn et al. 1977), and High Five (BTI-TN-5B1-4) from ovarian tissues of *Trichoplusia ni* (cabbage looper) (Wickham et al.; Davis et al. 1992).

Protein expression in insect cells has several advantages such as high expression level and easy manipulation. In addition, difficult proteins especially eukaryotic proteins that need posttranslational processing usually fold better in insect cells than in the *E. coli* expression system (Brondyk 2009). However, compared to the mammalian cells, post-translational modifications are still limited in insect cells, with glycosylation as the most significant example (Jarvis 2003; Kost et al. 2005). Due to the defect in glycosylation, functions of some recombinant glycoproteins are impaired (Xu and Ng 2015). For example, the insect cells cannot produce sialylated N-linked glycans. In the past two decades, various efforts were made to import the mammalian glycosylation pathway related genes into the insect cells to engineer the required glycosylation modification (Castilho 2015). For example, Hollister first reported in 1998 that an engineered Sf9 cell line expressing the B4GALT1 gene could produce foreign glycoproteins with terminally galactosylated N-glycans (Hollister et al. 1998). It was also reported in 2001 that both Sf9 and High Five cells were engineered to produce sialylated proteins by adding the ST6GAL1 gene (Hollister and Jarvis 2001; Breitbach and Jarvis 2001). Hollister et al. (Hollister et al. 2002) later transformed a set of other genes to generate the SfSWT-1 cell line which produce biantennary, terminally α-2,6- and α-2,3- sialylated N-glycans. More work has been recently done to obtain more powerful insect cell lines.

Next generation sequencing technology is widely used recently in biological studies. Genomes and transcriptomes of different species are sequenced, which generate high-input information for genomic studies and molecular modifications. We believe similar information can also be explored to provide guidance to engineer new strains of insect cell line for expressing proteins with mammalian type of posttranslational modifications. Genome and transcriptome of Sf21 cell line have already been reported in 2014 and 2015, respectively (Kakumani et al. 2014; Kakumani et al. 2015); however, the global genomic information of the High Five cell line is still unknown.

We constructed and sequenced an mRNA library of the High Five cell line, assembled a reference transcriptome for function and expression studies. We analyzed some protein-expression-related problems by comparing our High Five transcriptome with the reported Sf21 transcriptome (Kakumani et al. 2015). In addition, we extracted codon usage information from their coding sequences and compared it with other expression systems and model species. We also annotated transcripts that may have glycosylation-related functions, and evaluated their expression abundance to generate the global view of glycogenes in High Five and Sf21 cell lines. High expression transcripts, which have predicted signal peptide sequences, were analyzed for predicting highly efficient signal peptide sequences for secretory protein expression.

RESULTS AND DISCUSSION

Reference transcriptome assembly

Considering the genome size of several reference-ready insects, a total 49.5 million 101 bp paired-end reads were sequenced and yield 4.95 Gb bases raw data, 48 million clean reads were kept after low quality reads were filtered. After reads trimming with Trimmomatic, we used Trinity pipeline to do the *de novo* transcriptome assembly and obtained 31,068 transcripts with an N50 value of 2,276 bp (Haas et al. 2013; Bolger et al. 2014).

In total, 39.4 Mb bases are assembled and the average transcript length is 1,269 bp. To reduce the redundancy of the assembly, cd-hit-est was used and transcript number was reduced to 25,234 under 90% sequence identity threshold. These transcripts data sets are the so-called 'unigene'. 13,732 coding peptide sequences are predicted with Trans-Decoder. Detailed statistics numbers are shown in Table 1.

All clean reads have been submitted to NCBI SRA database under accession number SRP068276. Assembly version in this paper has been submitted to NCBI TSA database (GEEM01000000).

Assembly assessment

To evaluate the assembly quality, we employed several strategies for quality assessment. First, we used bwa (v0.7.10) (Li and Durbin 2010) to align all clean reads back to the assembly. 97.4% of the reads could be aligned and 95.3% are properly paired, indicating that the completeness of our assembly is high and very reliable. All transcripts were then aligned to SwissProt to check the proportion of transcripts that may be full-length or near full-length. As shown in Fig. 1A, about 6000 transcripts, with more than 10% of its contig length, could be aligned to a homolog in SwissProt. Among them, about 1/3 of the transcripts are fully aligned and more than 5000 have at least 30% sequence overlapping with known homolog. The result could be underestimated as the homologs of different species' genes have different proportion in the SwissProt database, but we believe our data reached our expectation.

Taking expression values into consideration, we recalculated the N50 value after low expression contigs were eliminated (Fig. 1B) and plotted expression value distribution pattern in Fig. 1C. Ex in Fig. 1B means a subset of top x% highly expressed transcript, the ExN50 reached the max length at E95, showing that 8,147 transcripts are in the top 95% expression subset with the minimum TPM of 6.1. From this, we can conclude from these data that most of the extremely high expression transcripts in High Five cells are in the range from 600 to 1000 bp. Longer transcripts have more regular expression level. As shown in Fig. 1C, the transcript number was reduced to 10,047 after transcripts of low TPM values (below 5) were eliminated.

Function annotation

To annotate functions of transcripts and coding peptides, we searched homologous genes in SwissProt, TrEMBL90 and NCBI nr databases with blast. Among the 25,234 transcripts, 13,492 got blast hits in nr database, 9,639 and 13,767 have similar sequences in SwissProt and TrEMBL90 database, respectively. With 13,732 coding peptides, 9,427 and 12,198 got alignments in SwissProt and TrEMBL90 database, respectively.

While executing GO annotation and EggNOG annotation, we also used protein sequence predicted from published

Table 1. Assembly statistics information

	Raw assembly	Duplicate removed assembly
Trinity 'genes'	27,389	24,000
Trinity transcripts	31,068	25,234
GC content (%)	40.84	40.71
Median contig length (bp)	722	622
Average contig length (bp)	1269.6	1160.9
Total assembled bases (bp)	39,444,068	29,294,166

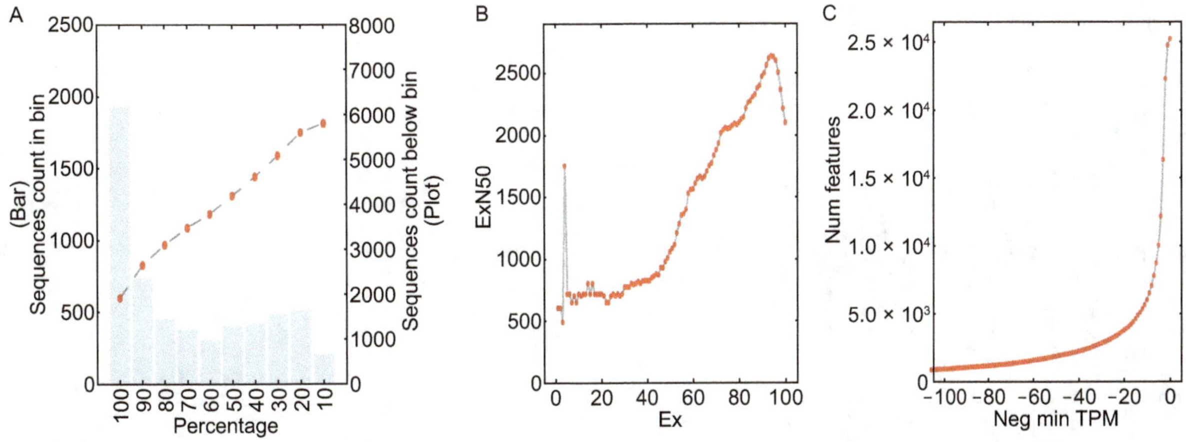

Figure 1. Assembly quality assessment. (A) Full-length transcript assessment. Bin on x-axis represent the percentage of the hit's length included in the alignment to the Trinity transcript. Left y-axis with bar plot is the transcript count in each bin and right y-axis with point plot is the accumulate count below that bin. (B) N50 of subset of transcript by decreasing the expression level. Ex is the top most expressed transcripts that represent x% of the data. ExN50 is the length of a transcript while the total length of transcripts shorter that it reached 50% of total length of all transcripts in this dataset. (C) Transcript count with a threshold of negative minimum TPM value.

Sf21 transcriptome data with the same analysis procedural. In High Five transcriptome, 13,447 transcripts have been annotated for GO terms and 13,459 have been annotated in EggNOG database. More specific function classification data are shown in Fig. 2 and Fig. 3, and detailed annotations for each transcript are described in supplemental file S1. From the global pattern of these figures, we can tell that the differences in GO terms and EggNOG categories between High Five and Sf21 cells are quite similar. Transcript numbers are higher for some function classes, including those related to intracellular trafficking, secretion, vesicular transport, posttranslational modification, protein turnover, transcription, translation, etc. Possessing plentiful genes with these functions make these insect cell lines an ideal host for protein production.

Codon usage among recombinant protein expression systems

Among different organisms, synonymous codons are usually utilized with different frequencies; a phenomenon generally referred to as the 'codon usage bias'. Codon usage bias is a major factor that affects the level of recombinant protein expression (Holm 1986). We collected CDS sequences from the Sf21 and High Five cells and compared their codon usage preference to other commonly used expression hosts, including the prokaryotic system *E. coli* BL21, the eukaryotic expression system *S. cerevisiae* and mammalian system CHO. We also compared the codon preference of insect cells with five other species including human, mouse, drosophila, zebrafish and arabidopsis. RSCU (relative synonymous codon usage) values are used to compare the use of synonymous codons. RSCU values of all amino acids in 10 species' CDS sequence were calculated separately for downstream studies as shown in Fig. 4.

With the characteristic of RSCU value, all x synonymous codon RSCU values of an amino acid always get a sum up value equals to x. For amino acids only encoded by one or two codons, there is no extreme distribution in it, but others are quite different. We calculated the RSCU range value of a set of codon related to one amino acid. By comparing this value, we can tell in which species this set of codon have greater bias. For example, the range value of arginine, isoleucine, leucine and proline in BL21 reached 2.29, 1.32, 2.81 and 1.00, while the minimum codon's RSCU is only 0.13, 0.21, 0.21 and 0.49. This situation exactly indicates that optimization of codon usage is of great significance. For instance, if BL21 is used to produce recombinant protein, failure to avoid these minimum codons may dramatically reduce the expression level.

In comparison with the range and standard deviation values in all species, no matter which amino acid you are using, both High Five and Sf21's range value are at a relatively low level. The homogeneity of codon usage in baculovirus-insect system could be an advantage for protein expression. Coding sequence cloned from most species could be normally expressed in insect cells without codon optimization. This robust property of codon usage in baculovirus-insect system made it a good platform for both eukaryotic and prokaryotic recombinant protein expression. But some previous publications claim that codon with lower RSCU value is intended to slow down the translation speed in order to produce well-folded proteins (Chaney and Clark 2015). More experimental evidence is required to show whether the usage of codon with relative higher RSCU values could become a disadvantage for expression of proteins of complex folding.

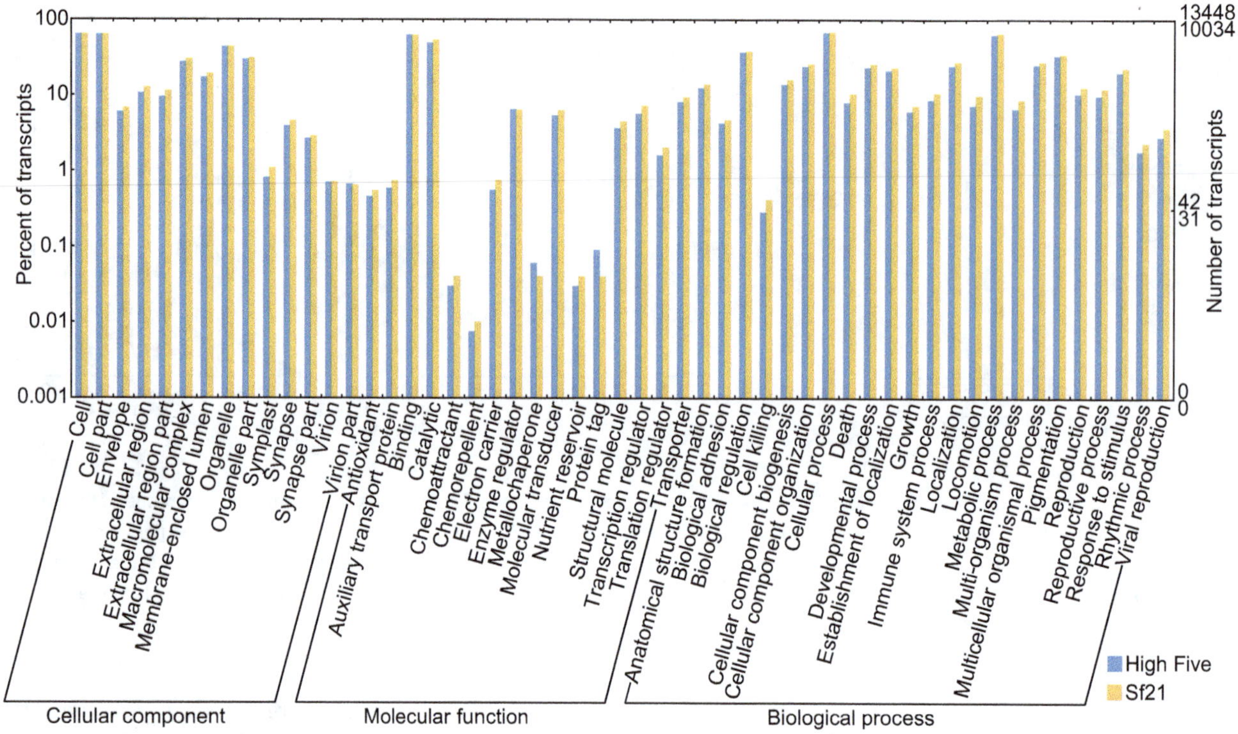

Figure 2. Gene Ontology of High Five and Sf21 transcriptome. Summarized in three main GO categories: Cellular component, Molecular function and Biological process. Right y-axis is the transcript count in that function item, left y-axis is the corresponding percentage of transcripts number.

Glycogene profile in protein expression insect cells

Post-translational modification is one of the most important characteristics of baculovirus-insect expression system. But the truncated N-glycosylation pathways in insect cells limit its application on some glycoprotein expression (Jarvis 2003). Several glyco-engineering modifications have been reported in the past two decades. Some modifications require importation of glycogenes into baculovirus-insect system. Glycosylation is mediated with complicated pathways and a number of genes are involved. Without a global gene map of the insect cells, we cannot thoroughly understand glycosylation related problems. Since GGDB and CAZy databases included genes associated with glycan synthesis procedural, we used them as references to find homologs of glycogenes in our High Five transcriptome and previous Sf21 transcriptome.

Here we identified 69 glycogenes in the High Five transcriptome and 72 in Sf21, with an overlap of 66 genes. Those genes are marked in blue with their expression value in Fig. 5A, and detailed informations are described in supplemental file S2. Glycogenes can be classified into several types according to their functions. Total gene counts of each type in High Five and Sf21 are shown in Fig. 5B. Insect cells have more or less homologs among most types. But for sialyltransferases and N-acetylgalactosaminyltransferases, no similar transcript was found in these two cell lines. That is the main reason why baculovirus-insect cannot produce complete mammalian N-glycosylation proteins. O-glycan modification is more complicated and not well studied. From the gene matrix, we found that some O-glycosylation required enzymes are detectable, such as OGT, POFUT1/2, XYLT1, POMT2, etc. Previous study suggests that O-mannosylation in insect species may occur more frequently than what is currently believed (Vandenborre et al. 2011). Understanding the profile of glycogenes in insect cells would be helpful for more detailed research on glycosylation.

We also compared the glycogenes' functioning mechanism and structure status in High Five and Sf21 (Fig. 5C). More than 50% of the glycosyltransferases function as inverting mechanism by catalyzing group transfer with inversion at anomeric reaction center of substrate, and less than 30% are for retentions. About 30% glycosyltransferase consist of two closely abutting β/α/β Rossmann domains, 20% consist of two β/α/β Rossmann domains that face each other and is flexibly linked. Remaining part have not yet been experimentally determined or studied (Lairson et al. 2008).

Here we complemented the glycogene database with data from baculovirus-insect expression system related cell lines. Our data would be valuable for introducing supplemental mammalian glycogesnes into insect cell lines and more efficiently modifying their glycosylation properties.

Highly expressed signal peptide containing genes

Baculovirus-insect system is a good platform for secretory protein expression. When there is a signal peptide fused to

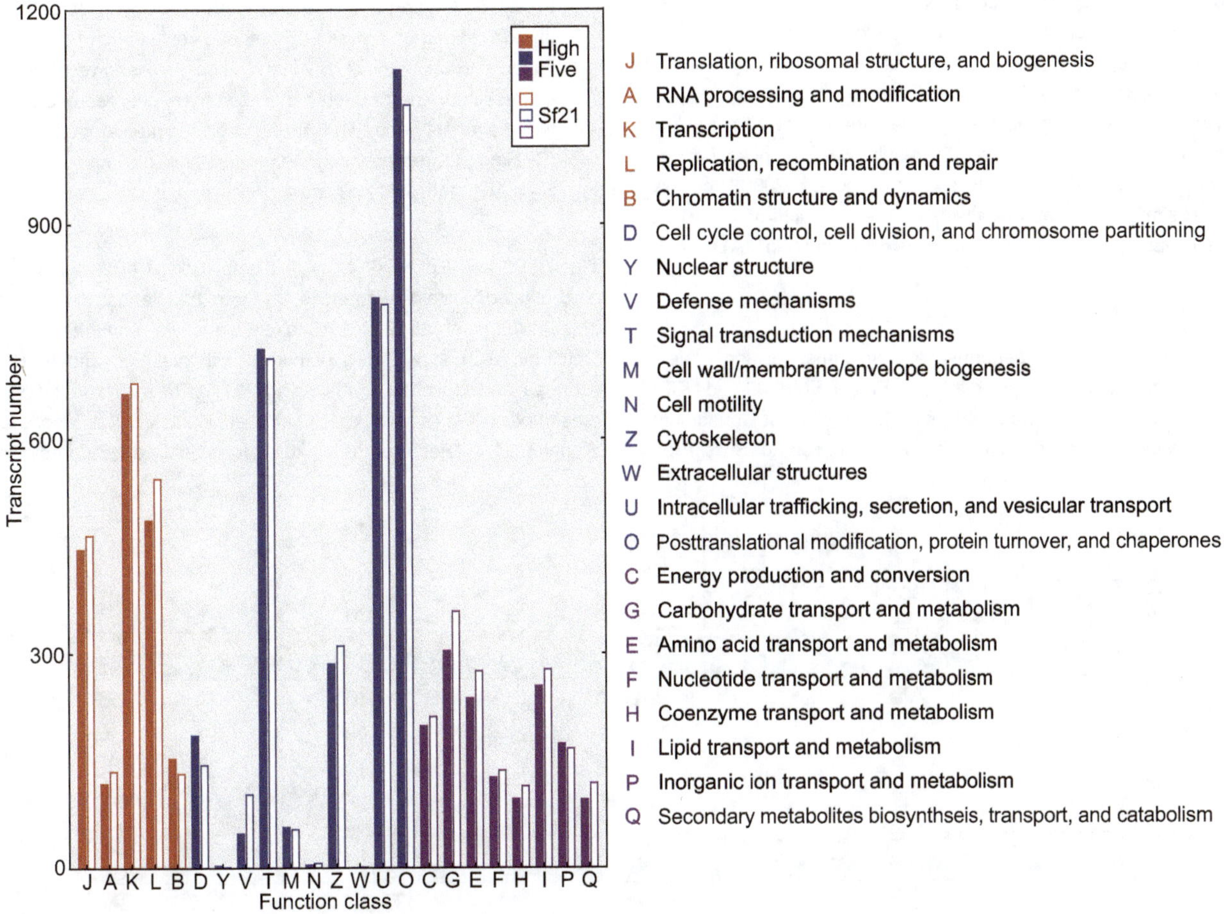

Figure 3. Transcript number in each EggNOG function classes. Divided into 3 parts by colors. (Red) Information storage and processing; (Blue) Cellular processes and signaling; (Purple) Metabolism.

the recombinant protein, product usually secreted to the outside environment through the secretory pathway. Optimization of signal peptide would be helpful to get a better yield (Olczak and Olczak 2006). Here we used SignalP software (Petersen et al. 2011) to predict all possible signal peptide in all protein sequences, and then sorted them with their transcript expression value. Fig. 6 showed all predicted transcripts that may have signal peptide sequence in High Five and Sf21 transcriptome. We identified signal peptide sequences from top 100 expressed transcripts and believe they are good candidates for higher protein production. Related peptide sequence, CDS sequence and functional annotation are described in supplementary file S3. Because higher expression value usually means higher protein amount, secretion efficiency could be closely linked with the amount of the signal peptide containing protein. Moreover, signal peptide sequence from insect itself is of the best choice because endogenous secretory signal peptide is more efficient than exogenous signal peptides (von Heijne and Abrahmsén 1989; Soejima et al. 2013). Combined with expression value and corresponding function of these proteins, we believe this approach would be useful and simplify the complexity of finding good signal peptides for protein expression.

MATERIALS AND METHODS

RNA extraction and library construction

High Five cell line in this study was purchased from Life technologies, USA. Cells were cultured in serum free medium containing 0.5% Penicillin-Streptomycin (Gibco 15140-122) for 24 h at 27°C in suspension at a shaking speed of 110 rpm. When cell density reached 2.4×10^6/mL, 2×10^7 cells were collected by centrifugation. Total RNA was extracted with QIAGEN RNeasy Mini Kit (QIAGEN, 74104) immediately after cell collection. RNA quality was examined using Agilent 2100 RNA chip (standard setting). Poly-A tailed RNA was enriched and used to construct sequencing library with Illumina TruSeq RNA Sample Prep Kit (Illumina, RS-122-2001), following standard instruction. RNA-seq library was sequenced on Illumina HiSeq 2000 platform.

Assembly and statistics

After we got the raw reads, quality control was performed to remove poor-quality reads with in-house QC script. Then we used Trimmomatic (v0.32) (Bolger et al. 2014) software to trim low quality ends of all reads. Reference transcriptome was assembled with trimmed reads using Trinity (v2.0.6) (Grabherr et al. 2011; Haas et al. 2013).

To reduce the assembly redundancy, cd-hit-est (v4.6.1) (Fu et al. 2012) was used to cluster reads with 90% identity. At this time, the remaining contigs are the so called 'unigenes'. Coding peptide sequences were predicted with TransDecoder.

We used the SwissProt database to check the integrity of transcripts to evaluate the quality of our assembly result. All transcript sequences were aligned to SwissProt using blastx (Camacho et al. 2009), and only the most similar target was kept with the e-value cutoff of 1×10^{-20}. Length coverage of aligned transcript was examined (Fig. 1A).

Function annotation

Function annotation was performed at both transcript and protein level. All transcripts were aligned to SwissProt, TrEMBL90 and NCBI nr database with blastx (Camacho et al. 2009; UniProt Consortium 2015). Predicted protein sequences were aligned to SwissProt, TrEMBL90 and EggNOG 4.1 with blastp (Powell et al. 2014). We also aligned protein sequences to Pfam28 database with hmm-search (Finn et al. 2014). Subsequently, we annotated Gene Ontology (GO) and related pathways with blast2go and KOBAS (Conesa and Götz 2008; Xie et al. 2011). The GO annotation is presented in figure with WEGO (Ye et al. 2006). We also used SignalP (v4.1) (Petersen et al. 2011) for signal peptide prediction.

Figure 5. Glycogene profile of High Five and Sf21 cell line. (A) Heatmap represents the gene constituent of each species. Blue mark of Sf21 and High Five showed the expression value of each gene. Red mark only represent they have this gene. (B) Bar plot of glycogene categories. (C) Ring plot representing the properties of glycogenes in High Five and Sf21 cell lines. ►

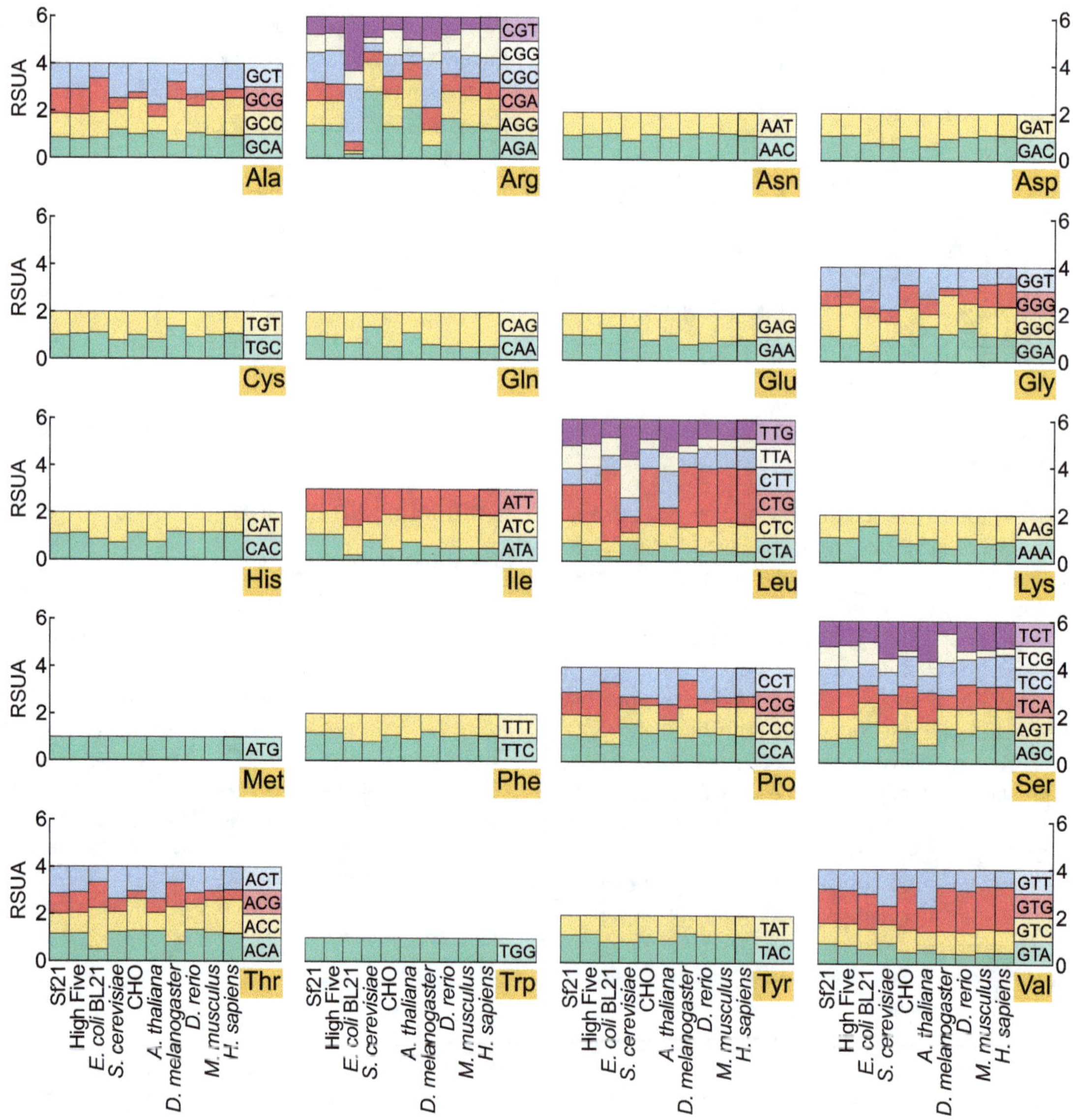

Figure 4. Codon usage of 20 amino acids across 10 different species. Each subplot is an amino acid, x-axis is different species and y-axis is the RSCU value of codons coding that amino acid.

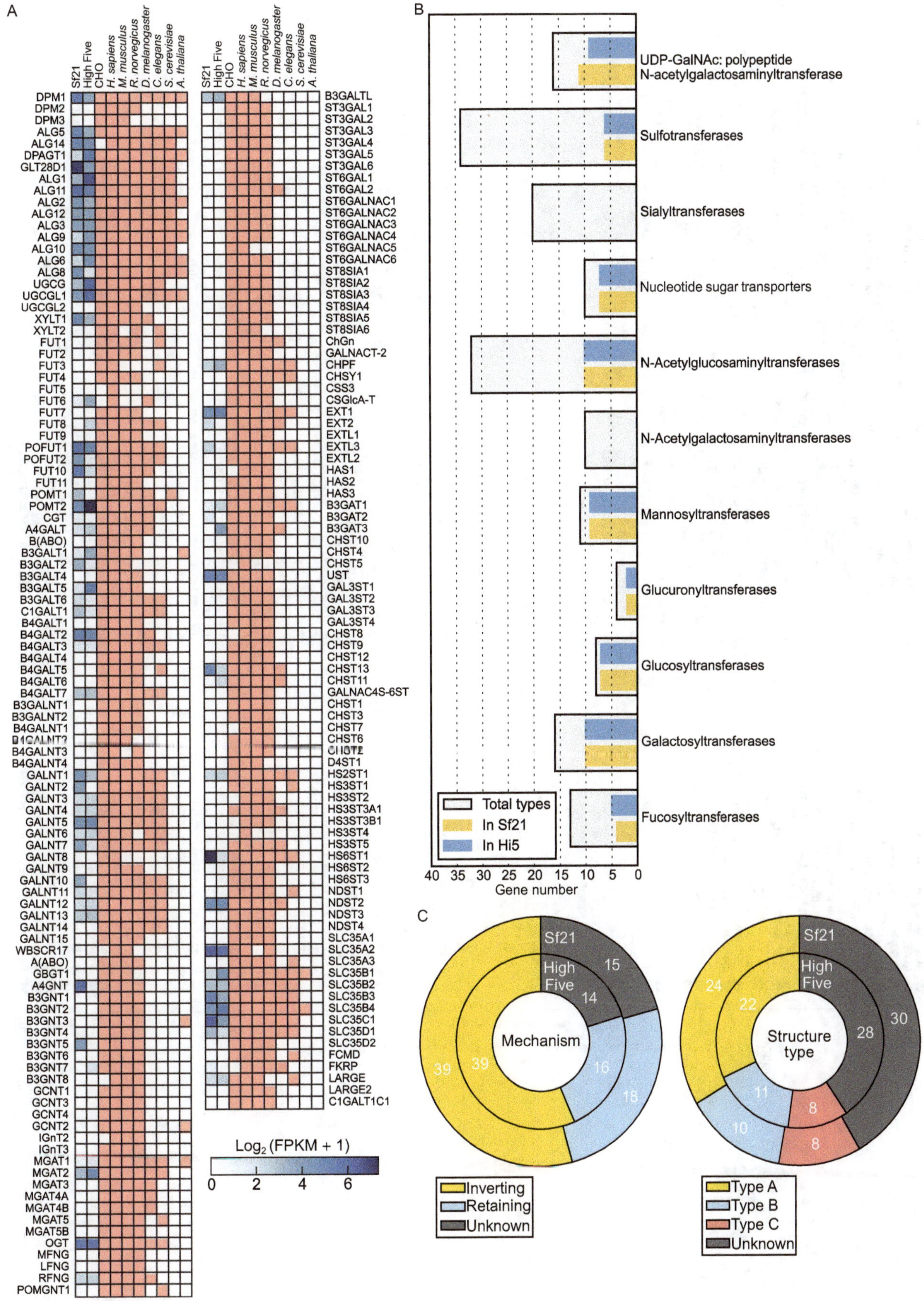
A
Sf21
High Five
CHO
H. sapiens
M. musculus
R. norvegicus
D. melanogaster
C. elegans
S. cerevisiae
A. thaliana
Log2 (FPKM + 1)
0 2 4 6
B
UDP-GalNAc: polypeptide N-acetylgalactosaminyltransferase
Sulfotransferases
Sialyltransferases
Nucleotide sugar transporters
N-Acetylglucosaminyltransferases
N-Acetylgalactosaminyltransferases
Mannosyltransferases
Glucuronyltransferases
Glucosyltransferases
Galactosyltransferases
Fucosyltransferases
Total types
In Sf21
In Hi5
40 35 30 25 20 15 10 5 0
Gene number
C
Mechanism
Sf21
High Five
15
14
16
18
39
39
Inverting
Retaining
Unknown
Structure type
24
22
30
28
11
10
8
8
Type A
Type B
Type C
Unknown

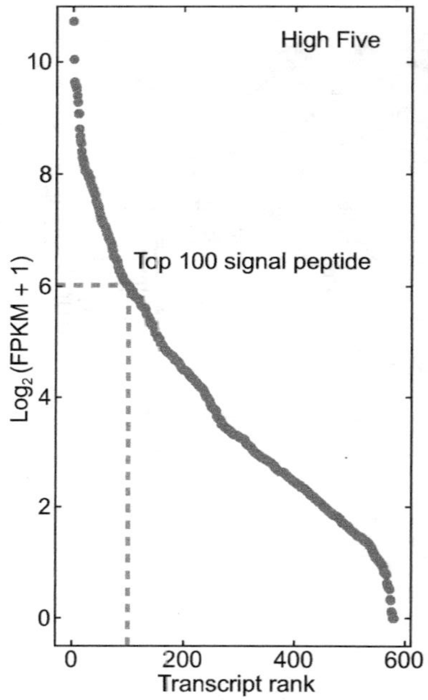

01. (0.53) MIATLLYLLIHLRRA
02. (0.84) MRLFIFALLTVCAHA
03. (0.83) MLKMRWSILALACVVVAVYA
04. (0.59) MIFLLYNLHLALN
05. (0.86) MKSVAVFLCLVAAASA
06. (0.81) MWSKFVVMLAVSCVANA
07. (0.88) MFEFRILLCSLLIISCTIA
08. (0.81) MKSLVLAVVSLLAISSINC
09. (0.89) MKVILFTAIALLGAVLA
10. (0.69) MKGVIFVCLLSLSVSFSLP
11. (0.88) MSRLRFCLLLAVFCSCTQYGLG
12. (0.86) MMTSVSLLVALLVSLQCVFQTSA
13. (0.82) MKITLILGIIASFSLALA
14. (0.77) MGAFAAALGFILLVAAARS
15. (0.67) MYLVILFAATALLGSSRS
16. (0.60) MKVWGAHGEFCARHQWEVIVATLALLACAAS
17. (0.84) MVMKYVICLSALAAVAFA
18. (0.85) MYLRALLLLSLSTLCLS
19. (0.68) MEGFKALILLQLVYLCGC
20. (0.58) MKQIIFCVIVALSSVIA
21. (0.89) MAPLFLIFAVFISATLG
22. (0.67) MKLIYLTLLVLPAALLCFQSANT
23. (0.79) MCRTECVLALVALSFGVAHS
24. (0.86) MKRGLLVLLAMCSLACA
25. (0.89) MELFRISSLFALFLILDVTSA
26. (0.92) MKMLILAVSCLLAITGSLA
27. (0.77) MSGPARARFLALHLVPATLLLLLRNPTPTSA
28. (0.87) MILNIKMFAVLSVILLPLFLSQQTEA
29. (0.85) MGRSLAVLGCLCALAALAHS
30. (0.74) MLLAFGRSIGLIFFVCLLSCVVS
31. (0.90) MKYFILLAVGVLLLSGWSHA
32. (0.87) MAVTMALFWVSLVLSVIGANA
33. (0.87) MQKLRVFLLALAALCISCEA
34. (0.71) MLNGKFLLCVVLVAAWNEVQP
35. (0.68) MEVHYLVFVLCAIWQGTNA
36. (0.77) MAGAGGFSRRFLHVFLDCLLLLQLAEYVPTASA
37. (0.81) MKSWGLLLVAALAIGSSFG
38. (0.81) MRLHMIAVLTLLGVVNYAAG
39. (0.87) MFRMRLPFVLVIILMIHVSVS
40. (0.83) MKSSWRTFSFAILFAASLLNTAHS
41. (0.55) MAPVPSLFITLILSIVLSVNA
42. (0.82) MIRSRGDTMSRTLARLLLALSIVAQCRA
43. (0.84) MSVQRVRTFRVLLVVLAFVNTSRA
44. (0.92) MGLLNIVLFLVLVAACCA
45. (0.50) MARVHTVPQFLHFFLYFSFVLCCFTSFVSA
46. (0.83) MSSIKLNLVLCSVILSILVCVNT
47. (0.84) MIYSTWISQFYSVLLLICSVITYVQS
48. (0.76) MFAYLNALIVLCFCFKNVAS
49. (0.87) MSAKLLSLILIIATAVTA
50. (0.59) MALAHGGQLVLVALALALLTLG
51. (0.90) MSLYVVCALLLLTPIRA
52. (0.81) MDSQVSVVIIAFLCILGTNA
53. (0.84) MELKTVNLFLLIALSICFVHNAIA
54. (0.92) MYKLNVFHLLFLATIVQVALA
55. (0.89) MNSILLVFAGILAVCLPASA
56. (0.76) MLRSLFVALAVSAVFA
57. (0.68) MKQLCIVVALVSVFGLSFQ
58. (0.75) MESLSRIVLGILILQIYVHA
59. (0.78) MCYYIIFVVFSIFVTDCTG
60. (0.79) MKFKLLLFLTIILTYYEGEA
61. (0.83) MAKIVFLLVLLCISASFA
62. (0.83) MTLRCVFAVLALAGATFA
63. (0.78) MFNSYFLGILICVAGSATA
64. (0.68) MSLKMSSVALLAYFTLNLIYCNA
65. (0.59) MMSRIYCSIFLLGLLLPLTTA
66. (0.83) MHIYSQTLVFVLLIVSCVLC
67. (0.77) MGARSRELVLLLAVCALALSEPVSKA
68. (0.76) MKQLLSIFFSVSTFSYVLS
69. (0.69) MMMSFSTNFCISLSFLTIVITCIVLSNDNYAQA
70. (0.61) MARVPGSVYVLLLCSFVTETVLS
71. (0.63) MFKWAVFAVLALGSVLGPMREVEA
72. (0.80) MRSLILIFLVLYLGYEVHG
73. (0.93) MILRMMSFCVFLFFLFSFNQVLC
74. (0.62) MLRLLVLTAITAA
75. (0.86) MFSPRRTILLGVILACTIIVVLG
76. (0.85) MKTILLVVTQILILASVES
77. (0.59) MAVLADINRIFSVLLSIIVITRCNA
78. (0.52) MHVAVVLVLMALVCVEG
79. (0.79) MEVKAWVILTLTALSIHGSCG
80. (0.78) MLFPKTALFYGFLIALLLEPISG
81. (0.53) MVDFVASVVLVALVLELAG
82. (0.70) MRGSWVICLCVVKLALC
83. (0.87) MLLKTLVFSLLAVAVLA
84. (0.84) MMLRILFILIVTIVKVTHS
85. (0.86) MYKIIVIALFITLVHG
86. (0.80) MSILIKMFFRVICLYLLFTTQISC
87. (0.56) MGKLLIVFLAFAATCVA
88. (0.79) MYLNNVISSVLLLIATSFSITSC
89. (0.60) MFRKVVLGSVGVAALIPAIGAA
90. (0.84) MFFSPFIITILLSLVNGILS
91. (0.87) MAGMLWALLILQLFFNAEG
92. (0.51) MNSLLRMFCISCLRSRCLS
93. (0.75) MGQARYIACALLMLCPLALT
94. (0.71) MEHALSFLLSLLLFSLLLQLLCT
95. (0.75) MKYHLLSIILVITLSISCES
96. (0.60) MNKYIYCFVLSAVLASVRA
97. (0.85) MYCFIVLCLVLILLQMNINSSSC
98. (0.59) MAFCRVVFPLLVLSIVSCSA
99. (0.84) MRTMASVWFSVLVLVSVQQVFG
100. (0.66) MLEVINGFLLVYFVVLCSLSLLVPQLVKPISA

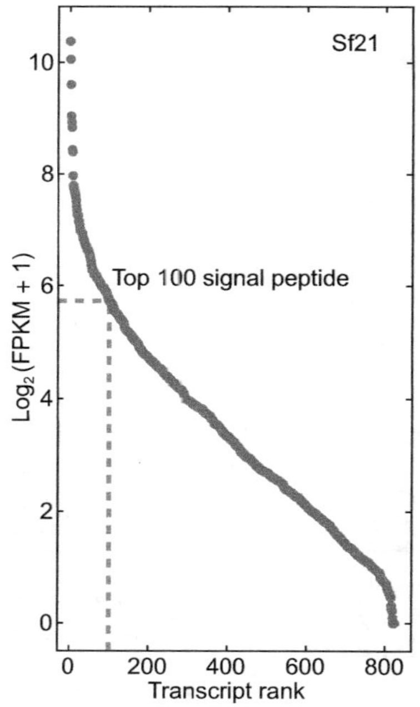

01. (0.50) MRYSLTYCSALLKISFLFLMA
02. (0.89) MKGVAVLLCLVAGACA
03. (0.48) MMIVSKRAAWLLLLTTRGYNRCCA
04. (0.48) MLLLLSQLLGQVVKG
05. (0.78) MNMFEPRAGRVVLLFIITLQLFEFCAS
06. (0.82) MSTVLLAVIAVILCVLFRILNVNS
07. (0.84) MSGPARARFIALQLVPATLLLLSWTPTPSSA
08. (0.92) MKVIILAVSCLLAVSGALA
09. (0.79) MFSAAIFMCFATFFILSSNA
10. (0.65) MSKSVLCFVIAVLCFASVNT
11. (0.81) MKSLVLAVVSLLAISSINC
12. (0.82) MIKMRWSILALACVVVAVYA
13. (0.73) MRSVSVVLILVVAAMASS
14. (0.55) MHVIIMDGWSHLLLVCNVGTQLAYI
15. (0.71) MGERSRELILLLAVCALALS
16. (0.82) MVTKYAICLFAIATVAFA
17. (0.88) MKAILFALAGLLAVCLPASA
18. (0.73) MSDWYWRVIALKATFALTFIFCNVRA
19. (0.80) MDFRILLLWMFAVVTRNYA
20. (0.89) MLGSLKIVLLLGVIYLCRA
21. (0.88) MDLKFIFCLFLAFISSASS
22. (0.69) MLISKQSASVSRALNFGKMKCSALLLVLMLCVLSDA
23. (0.90) MAWIYLILSVFISTAVA
24. (0.83) MAGAGGILRRFLLVFLDCLLLLQLAQY
25. (0.90) MRGRAAQIAALVILCSLLGTTHA
26. (0.51) MKHWTAWMVTAFIFVVLVGVEELTA
27. (0.86) MELFKVSSLFALFFFLDVTSA
28. (0.89) MCKLSVFHLLFLATIVQVTIA
29. (0.85) MTLPCVFAVLALVGATFA
30. (0.73) MEVHYLVFVLSVLWHGTDA
31. (0.49) MFQTILYTVVSLVIVQA
32. (0.78) MVFLSLSTIIFILSLRAVTC
33. (0.88) MSRLRFCFLLAVLCSCLQNGYG
34. (0.47) MSKLCKVIFLYIRLAFDLMVDGIFS
35. (0.84) MMLVSVVVTALLVLRSIEG
36. (0.76) MWRVLLALGAACATARA
37. (0.77) MKILICITLLLSIFANYDEA
38. (0.67) MYKNMWCVFVAILGLAAG
39. (0.90) MLRIAVFVFVALLCTVMC
40. (0.90) MTNTLVFCLIALFCCTDLSFA
41. (0.88) MYLFVIHLSMLLSVLA
42. (0.85) MSLTMKSVYLLVLVSVCQA
43. (0.48) MQFARSSCCRLYYNIVIFQLVLHICIERVLP
44. (0.78) MRIAALFALFAIGLA
45. (0.83) MGAFAAALGFLLLIAAARS
46. (0.61) MRLDIIALLILYLLNININIC
47. (0.84) MHFQSIAAILFLTLINYAVC
48. (0.87) MRVILFTAIALLGAALA
49. (0.73) MYSLTVVGVLLAVCGGAFS
50. (0.83) MALFSRKIYLLGAFLVLVNAISVVHS
51. (0.57) MVGLCTSFFLYVSVVYS
52. (0.78) MVASRATFVALVCALALASA
53. (0.84) MAKLVSLVLCVVLVYMISA
54. (0.85) MSVRRLRSFRVLLVVLVFANTSRA
55. (0.95) MGLVMYRLWMWCVALALAALWFAPAQA
56. (0.77) MLRSLFVALAVSAVFA
57. (0.83) MKPAFISICKIALFVLVFMSVISHVVIG
58. (0.84) MNLGHAVVLFNAIFIALVNS
59. (0.74) MKHFLILFTIVQSTLS
60. (0.70) MHLLIVLSAIVYASVDIVQG
61. (0.83) MRPAFVVILVVLLNQCDA
62. (0.63) MKLLYLTLLVLPAALLCFQSANTILARA
63. (0.66) MPIRKEVFWIVIICIISQCYG
64. (0.79) MKVLVVLAGLLAISAA
65. (0.78) MTAYNFRSLLLAVIVLSNCITWTAS
66. (0.78) MAVARAVLVLLASTAQA
67. (0.81) MTIAYIHISVILLFASAAHA
68. (0.89) MKLLSSLLFVLMILHCSYA
69. (0.50) MISCWMSSISSSASS
70. (0.79) MHGSARALGLALAVLALAYARA
71. (0.74) MFKLLVLILCLGFNYCLIES
72. (0.76) MDLTASLLTFTMFWKLSWS
73. (0.80) MELRRFTLCMALFVLAFAHCSES
74. (0.72) MKYILVSIILVLTFSNSCKS
75. (0.46) MLTMVPIFVVSWAGRLMA
76. (0.57) MFRMAQQAFFCTLALS
77. (0.59) MHKIFSFIGSIFFSFSSFSG
78. (0.76) MMSGLVRSMSVSTVTLVFCCLALMCTG
79. (0.47) MEIPKLMILLCCSYYGFA
80. (0.80) MVLNIKMFGVLSVFLLPLLLSQETKA
81. (0.56) MESRSSSLHRLSLDFFLSLFLSNSLLA
82. (0.78) MAFCRVVFPLLVLSVISCYA
83. (0.78) MYLRISRNFYFVLFFAFALVSVTT
84. (0.90) MRPIIFFKFLLLFSLFALTFS
85. (0.82) MWKIILFCTFLSVSQA
86. (0.49) MKMLISLNKKSLITLYFSFLRPCLVLS
87. (0.85) MCKTQCVLALLALSVGVAHS
88. (0.77) MHVVIPLLLVWALCGVRG
89. (0.92) MKITLLLVAVCLAVSLVSA
90. (0.46) MATMLNGAAVMDAALLLIAGNES
91. (0.67) MQRYLLLYTFVLIVPFAYQ
92. (0.64) MCQLMFYIVVSIFSVLKTATG
93. (0.81) MVPNLNLLHVICFTLLFGLCVRTHA
94. (0.60) MDAILAALVALMALAAAMA
95. (0.82) MLKKLLKVCAVFMIFFKCAAS
96. (0.78) MFRIIFSSLLITTVLC
97. (0.54) MAYESSIVYTLIICFFITDTALS
98. (0.92) MQRLLKILVLAIVCTAVRA
99. (0.74) MFTTIFMLLYFLNFESVSS
100. (0.85) MYFYALFLLLLSTCTISHG

◀ **Figure 6. Highly expressed predicted signal peptides.** Plot on the left are the expression values of those transcripts which have a predicted signal peptide. Sequences on the right are the top 100 signal peptide sequence with the signalP score in the brackets.

Since some protein sequences are 5′ truncated or predicted starting from an upstream point, we manually checked the prediction of top expression of transcripts containing signal peptides.

To compare the constituent of glycogenes, we collected glycogene sequences from GGDB as reference (Narimatsu 2004). Published Sf21 transcriptome was downloaded from TSA database with accession No. GCTM00000000 (Kakumani et al. 2015). Protein sequences were aligned to this reference with blastp. Blast results of glycogenes were manually checked by aligning sequences to the TrEMBL90 and nr databases, to eliminate false positive hits. Classification by function, mechanism and structure were based on information from GGDB and CAZy database (Narimatsu 2004; Lombard et al. 2014).

Codon usage evaluation

We used our predicted coding sequences in codon usage evaluation. CDS sequences from other species were downloaded from Ensembl and NCBI genome databases. Only those sequences longer than 100 amino acids (300 bp CDS sequence) were used for calculation. All codons were counted with our script by shifting a simulated reading frame from 5′ end to 3′ end. With the count data of all codons, we calculated RSCU value according to the formula as described (Cannarozzi and Schneider 2012).

Expression analysis

To evaluate the transcript abundance, Bowtie (v1.0.0) and RSEM (v1.2.15) were used for sequence alignment and calculation of TPM (Transcripts Per Million) and FPKM (Fragments Per Kilobase of transcript per Million mapped reads) (Langmead et al. 2009; Li and Dewey 2011). This value could measure the transcripts abundance with transcript length, thus we can compare it among different genes or samples.

ACKNOWLEDGMENTS

We thank Dr. Yun Zhang and the High-Throughput Sequencing Center of Peking University for assistance with Illumina sequencing. This work was supported by the National Basic Research Program (973 Program) (No. 2014CB910102 to X.S.).

ABBREVIATIONS

GO, Gene Ontology; TPM, Transcripts Per Million; FPKM, Fragments Per Kilobase of transcript per Million mapped reads; RSCU, relative synonymous codon usage.

COMPLIANCE WITH ETHICS GUIDELINES

Kai Yu, Yang Yu, Xiaoyan Tang, Huimin Chen, Junyu Xiao and Xiao-Dong Su declare that they have no conflict of interest. This article does not contain any studies with human or animal subjects performed by the any of the authors.

REFERENCES

Bolger AM, Lohse M, Usadel B (2014) Trimmomatic: a flexible trimmer for Illumina sequence data. Bioinformatics 30:2114–2120. doi:10.1093/bioinformatics/btu170

Breitbach K, Jarvis DL (2001) Improved glycosylation of a foreign protein by Tn-5B1-4 cells engineered to express mammalian glycosyltransferases. Biotechnol Bioeng 74:230–239. doi:10.1002/bit.1112

Brondyk WH (2009) Selecting an appropriate method for expressing a recombinant protein. Methods Enzymol 463:131–147. doi:10.1016/S0076-6879(09)63011-1

Camacho C, Coulouris G, Avagyan V et al (2009) BLAST+: architecture and applications. BMC Bioinformatics 10:421. doi:10.1186/1471-2105-10-421

Cannarozzi GM, Schneider A (eds) (2012) Codon Evolution. Oxford University Press

Castilho A (ed) (2015) Glyco-Engineering. Springer New York, New York, NY

Chaney JL, Clark PL (2015) Roles for Synonymous Codon Usage in Protein Biogenesis. Annu Rev Biophys 44:143–166. doi:10.1146/annurev-biophys-060414-034333

Conesa A, Götz S (2008) Blast2GO: A comprehensive suite for functional analysis in plant genomics. Int J Plant Genomics 2008:619832. doi:10.1155/2008/619832

Davis TR, Trotter KM, Granados RR, Wood HA (1992) Baculovirus Expression of Alkaline Phosphatase as a Reporter Gene for Evaluation of Production, Glycosylation and Secretion. Bio/Technology 10:1148–1150. doi:10.1038/nbt1092-1148

Finn RD, Bateman A, Clements J et al (2014) Pfam: the protein families database. Nucleic Acids Res 42:D222–D230. doi:10.1093/nar/gkt1223

Fu L, Niu B, Zhu Z et al (2012) CD-HIT: accelerated for clustering the next-generation sequencing data. Bioinformatics 28:3150–3152. doi:10.1093/bioinformatics/bts565

Grabherr MG, Haas BJ, Yassour M et al (2011) Full-length transcriptome assembly from RNA-Seq data without a reference genome. Nat Biotechnol 29:644–652. doi:10.1038/nbt.1883

Haas BJ, Papanicolaou A, Yassour M et al (2013) De novo transcript sequence reconstruction from RNA-seq using the Trinity platform for reference generation and analysis. Nat Protoc 8:1494–1512. doi:10.1038/nprot.2013.084

Hollister JR, Jarvis DL (2001) Engineering lepidopteran insect cells for sialoglycoprotein production by genetic transformation with mammalian 1,4-galactosyltransferase and 2,6-sialyltransferase genes. Glycobiology 11:1–9. doi:10.1093/glycob/11.1.1

Hollister J, Grabenhorst E, Nimtz M et al (2002) Engineering the Protein N-Glycosylation Pathway in Insect Cells for Production of Biantennary, Complex N-Glycans †. Biochemistry 41:15093–15104. doi:10.1021/bi026455d

Hollister JR, Shaper JH, Jarvis DL (1998) Stable expression of mammalian beta 1,4-galactosyltransferase extends the N-glycosylation pathway in insect cells. Glycobiology 8:473–480

Holm L (1986) Codon usage and gene expression. Nucleic Acids Res 14:3075–3087

Jarvis DL (2003) Developing baculovirus-insect cell expression systems for humanized recombinant glycoprotein production. Virology 310:1–7. doi:10.1016/S0042-6822(03)00120-X

Kakumani PK, Malhotra P, Mukherjee SK, Bhatnagar RK (2014) A draft genome assembly of the army worm, Spodoptera frugiperda. Genomics 104:134–143. doi:10.1016/j.ygeno.2014.06.005

Kakumani PK, Shukla R, Todur VN et al (2015) De novo transcriptome assembly and analysis of Sf21 cells using illumina paired end sequencing. Biol Direct 10:44. doi:10.1186/s13062-015-0072-7

Kost TA, Condreay JP, Jarvis DL (2005) Baculovirus as versatile vectors for protein expression in insect and mammalian cells. Nat Biotechnol 23:567–575. doi:10.1038/nbt1095

Lairson LL, Henrissat B, Davies GJ, Withers SG (2008) Glycosyltransferases: structures, functions, and mechanisms. Annu Rev Biochem 77:521–555. doi:10.1146/annurev.biochem.76.061005.092322

Langmead B, Trapnell C, Pop M, Salzberg SL (2009) Ultrafast and memory-efficient alignment of short DNA sequences to the human genome. Genome Biol 10:R25. doi:10.1186/gb-2009-10-3-r25

Li B, Dewey CN (2011) RSEM: accurate transcript quantification from RNA-Seq data with or without a reference genome. BMC Bioinformatics 12:323. doi:10.1186/1471-2105-12-323

Li H, Durbin R (2010) Fast and accurate long-read alignment with Burrows-Wheeler transform. Bioinformatics 26:589–595. doi:10.1093/bioinformatics/btp698

Lombard V, Golaconda Ramulu H, Drula E et al (2014) The carbohydrate-active enzymes database (CAZy) in 2013. Nucleic Acids Res 42:D490–D495. doi:10.1093/nar/gkt1178

Narimatsu H (2004) Construction of a human glycogene library and comprehensive functional analysis. Glycoconj J 21:17–24. doi:10.1023/B:GLYC.0000043742.99482.01

Olczak M, Olczak T (2006) Comparison of different signal peptides for protein secretion in nonlytic insect cell system. Anal Biochem 359:45–53. doi:10.1016/j.ab.2006.09.003

Petersen TN, Brunak S, von Heijne G, Nielsen H (2011) SignalP 4.0: discriminating signal peptides from transmembrane regions. Nat Methods 8:785–786. doi:10.1038/nmeth.1701

Powell S, Forslund K, Szklarczyk D et al (2014) eggNOG v4.0: nested orthology inference across 3686 organisms. Nucleic Acids Res 42:D231–D239. doi:10.1093/nar/gkt1253

Soejima Y, Lee J, Nagata Y et al (2013) Comparison of signal peptides for efficient protein secretion in the baculovirus-silkworm system. Open Life Sci 8:1–7. doi:10.2478/s11535-012-0112-6

UniProt Consortium (2015) UniProt: a hub for protein information. Nucleic Acids Res 43:D204–D212. doi:10.1093/nar/gku989

Vandenborre G, Smagghe G, Ghesquière B et al (2011) Diversity in Protein Glycosylation among Insect Species. PLoS One 6: e16682. doi:10.1371/journal.pone.0016682

Vaughn JL, Goodwin RH, Tompkins GJ, McCawley P (1977) The Establishment of Two Cell Lines from the Insect Spodoptera frugiperda (Lepidoptera; Noctuidae). In Vitro 13:213–217

von Heijne G, Abrahmsén L (1989) Species-specific variation in signal peptide design. Implications for protein secretion in foreign hosts. FEBS Lett 244:439–446

Olczak M, Olczak T (2006) Comparison of different signal peptides for protein secretion in nonlytic insect cell system. Anal Biochem 359:45–53. doi:10.1016/j.ab.2006.09.003

Petersen TN, Brunak S, von Heijne G, Nielsen H (2011) SignalP 4.0: discriminating signal peptides from transmembrane regions. Nat Methods 8:785–786. doi:10.1038/nmeth.1701

Powell S, Forslund K, Szklarczyk D et al (2014) eggNOG v4.0: nested orthology inference across 3686 organisms. Nucleic Acids Res 42:D231–D239. doi:10.1093/nar/gkt1253

Soejima Y, Lee J, Nagata Y et al (2013) Comparison of signal peptides for efficient protein secretion in the baculovirus-silkworm system. Open Life Sci 8:1–7. doi:10.2478/s11535-012-0112-6

UniProt Consortium (2015) UniProt: a hub for protein information. Nucleic Acids Res 43:D204–D212. doi:10.1093/nar/gku989

Vandenborre G, Smagghe G, Ghesquière B et al (2011) Diversity in Protein Glycosylation among Insect Species. PLoS One 6: e16682. doi:10.1371/journal.pone.0016682

Vaughn JL, Goodwin RH, Tompkins GJ, McCawley P (1977) The Establishment of Two Cell Lines from the Insect Spodoptera frugiperda (Lepidoptera; Noctuidae). In Vitro 13:213–217

von Heijne G, Abrahmsén L (1989) Species-specific variation in signal peptide design. Implications for protein secretion in foreign hosts. FEBS Lett 244:439–446

15

A new glimpse of FadR-DNA crosstalk revealed by deep dissection of the *E. coli* FadR regulatory protein

Yongchang Zhang[1,2,3], Rongsui Gao[2,3], Huiyan Ye[2,3,4], Qingjing Wang[2,3], Youjun Feng[2,3]✉

[1] College of Life Sciences, Nanjing Normal University, Nanjing 210023, China
[2] State Key Laboratory for Diagnosis and Treatment of Infectious Diseases, Collaborative Innovation Center for Diagnosis and Treatment of Infectious Diseases, First Affiliated Hospital, Zhejiang University School of Medicine, Hangzhou 310003, China
[3] Department of Medical Microbiology and Parasitology, Center for Infection and Immunity, Zhejiang University School of Medicine, Hangzhou 310058, China
[4] College of Life Science and Technology, Guangxi University, Nanning 530004, China
✉ Correspondence: fengyj@zju.edu.cn (Y. Feng)

ABSTRACT

Escherichia coli (_E. coli_) FadR regulator plays dual roles in fatty acid metabolism, which not only represses the fatty acid degradation (_fad_) system, but also activates the unsaturated fatty acid synthesis pathway. Earlier structural and biochemical studies of FadR protein have provided insights into interplay between FadR protein with its DNA target and/or ligand, while the missing knowledge gap (_esp._ residues with indirect roles in DNA binding) remains unclear. Here we report this case through deep mapping of old _E. coli fadR_ mutants accumulated. Molecular dissection of _E. coli_ K113 strain, a _fadR_ mutant that can grow on decanoic acid (C10) as sole carbon sources unexpectedly revealed a single point mutation of T178G in _fadR_ locus (W60G in $FadR_{K113}$). We also observed that a single genetically-recessive mutation of W60G in FadR regulatory protein can lead to loss of its DNA-binding activity, and thereby impair all the regulatory roles in fatty acid metabolisms. Structural analyses of FadR protein indicated that the hydrophobic interaction amongst the three amino acids (W60, F74 and W75) is critical for its DNA-binding ability by maintaining the configuration of its neighboring two β-sheets. Further site-directed mutagenesis analyses demonstrated that the FadR mutants (F74G and/or W75G) do not exhibit the detected DNA-binding activity, validating above structural reasoning.

KEYWORDS FadR, fatty acid metabolism, crosstalk

INTRODUCTION

FadR, a member of the GntR super-family of transcription factors, is recognized as a global regulator (DiRusso et al., 1992, Henry & Cronan, 1992, Clark & Cronan, 2005). Not only does it exerts extensive effects on fatty acid metabolism (Clark, 1981, Nunn et al., 1983, Hughes et al., 1988), but also regulates acetate metabolism (*iclR*) (Gui et al., 1996, Maloy & Nunn, 1981) and stress response (*uspA*) (Farewell et al., 1998). The dual functions of FadR regulatory protein in lipid metabolism can be described as follows: 1) repression of expression of genes essential for fatty acid degradation system, such as *fadBA* (DiRusso et al., 1993, Iram & Cronan, 2006), *fadH* (He et al., 1997, Feng & Cronan, 2010), *fadM* (Feng & Cronan, 2009b), etc.; 2) activation of unsaturated fatty acid biosynthesis pathway via directly up-regulating the transcription of both *fabA* (Nunn et al., 1983, Henry & Cronan, 1992, Campbell & Cronan, 2001) and *fabB* (Campbell & Cronan, 2001, Feng & Cronan, 2009a). The physiological ligands of FadR regulator are long chain fatty acyl-CoA thioesters, which can neutralize and release FadR protein from its binding operators (Henry & Cronan, 1992, Cronan, 1997). Thereby FadR protein can function as a sensor of the availability of long chain fatty acid in bacterial growth environment (Cronan & Subrahmanyam, 1998,

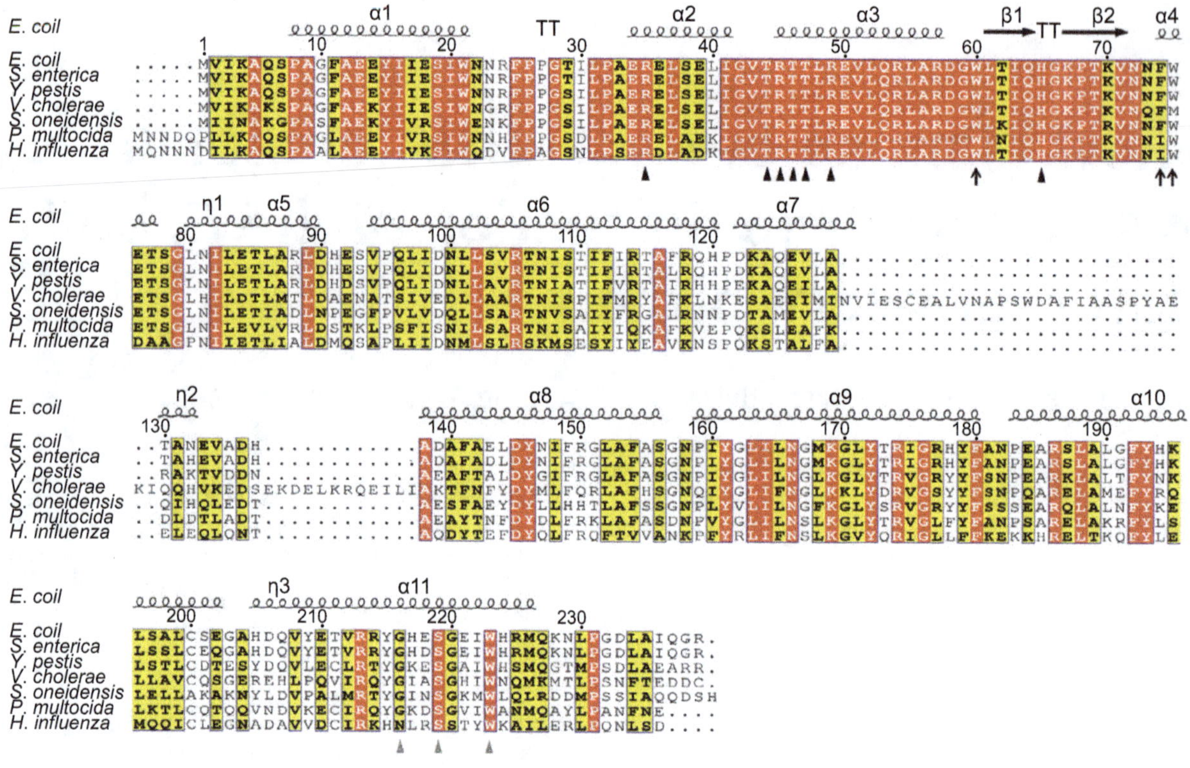

Figure 1. Sequence alignments of *E. coli* FadR regulatory protein with other homologues from six different species of γ-proteobacteria. The multiple alignments of amino acid sequences were conducted using ClustalW2 (http://www.ebi.ac.uk/Tools/clustalw2/index.html), and the resultant output was processed by program ESPript 2.2 (http://espript.ibcp.fr/ESPript/cgi-bin/ESPript.cgi), generating the final BLAST photography (Feng & Cronan, 2011b). Identical residues are in white letters with red background, similar residues are in black letters in yellow background, and the varied residues are in grey letters. As we earlier described (Feng et al., 2008), the protein secondary structure was shown in cartoon (on top), in terms of the structural architecture of *E. coli* FadR protein (PDB:1E2X) (van Aalten et al., 2000). α: alpha-helix; β: beta-sheet; T: Turn; η: coil. The DNA binding sites are indicated with black triangles (R35, T44, R45, T46, T47, R49 and 65H) (Xu et al., 2001), the ligand binding sites are shown with grey triangles (216G, 219S and 223W) (van Aalten et al., 2001), and the newly-proposed amino acids essential for DNA binding activity of FadR protein are highlighted with dark arrows (W60, F74 and W75). The FadR sequences are separately sampled from *E. coli* K12 (Accession no.: CAA30881), *S. enterica* (*Salmonella enterica*) (Accession no.: ACF63827), and *Y. pestis* (*Yersinia pestis*) (Accession no.: ZP_06207826), *V. cholerae* (*Vibrio cholerae*) (Accession no.: AAO37924), *S. oneidensis* (*Shewanella oneidensis*) (Accession no.: NP_718457), *P. multocida* (*Pasterurella multocida*) (Accession no.: AAK02132), and *H. influenza* (*Haemophilus influenza*) (Accession no.: AAC22085).

DiRusso et al., 1998, Campbell & Cronan, 2001). The phenotype of *E. coli* that wild-type strains cannot grow on decanoic acid (C10) as sole carbon source (only the *fadR* mutant can) is due to the inability of short chain fatty acyl-CoA to turn off FadR-mediated repression of *fad* regulon (Iram & Cronan, 2006).

It is well known that the fatty acid-responsive FadR regulatory protein is exclusively present in γ-proteobacteria (Iram & Cronan, 2005). The paradigm member of FadR regulator is the *E. coli fadR* protein product, a 239-residue polypeptide with solution structure of dimer (van Aalten et al., 2001, Xu et al., 2001, Feng & Cronan, 2010), and also belongs to the helix-turn-helix type transcription factor, comprising N-terminal DNA-binding domain and C-terminal ligand-binding domain (van Aalten et al., 2000, van Aalten et al., 2001, Xu et al., 2001). The accumulated crystal structures of FadR complexed with DNA target/acyl-CoA or alone have established structural basis for FadR-mediated transcription regulation of fatty acid metabolism (van Aalten et al., 2000, van Aalten et al., 2001, Xu et al., 2001). In addition to the crucial residues for ligand binding revealed by biochemical analyses (DiRusso et al., 1998), several amino acids essential for DNA binding have also been implied from FadR structural dissection (Xu et al., 2001). Additionally, the functional diversity has also been suggested by systematically biochemical documentation of a set of bacterial FadR homologues (Iram & Cronan, 2005).

Although the FadR picture seems relatively complete right now, the residues other than direct DNA-binding sites remain unclear that contributes significantly to the binding activity of

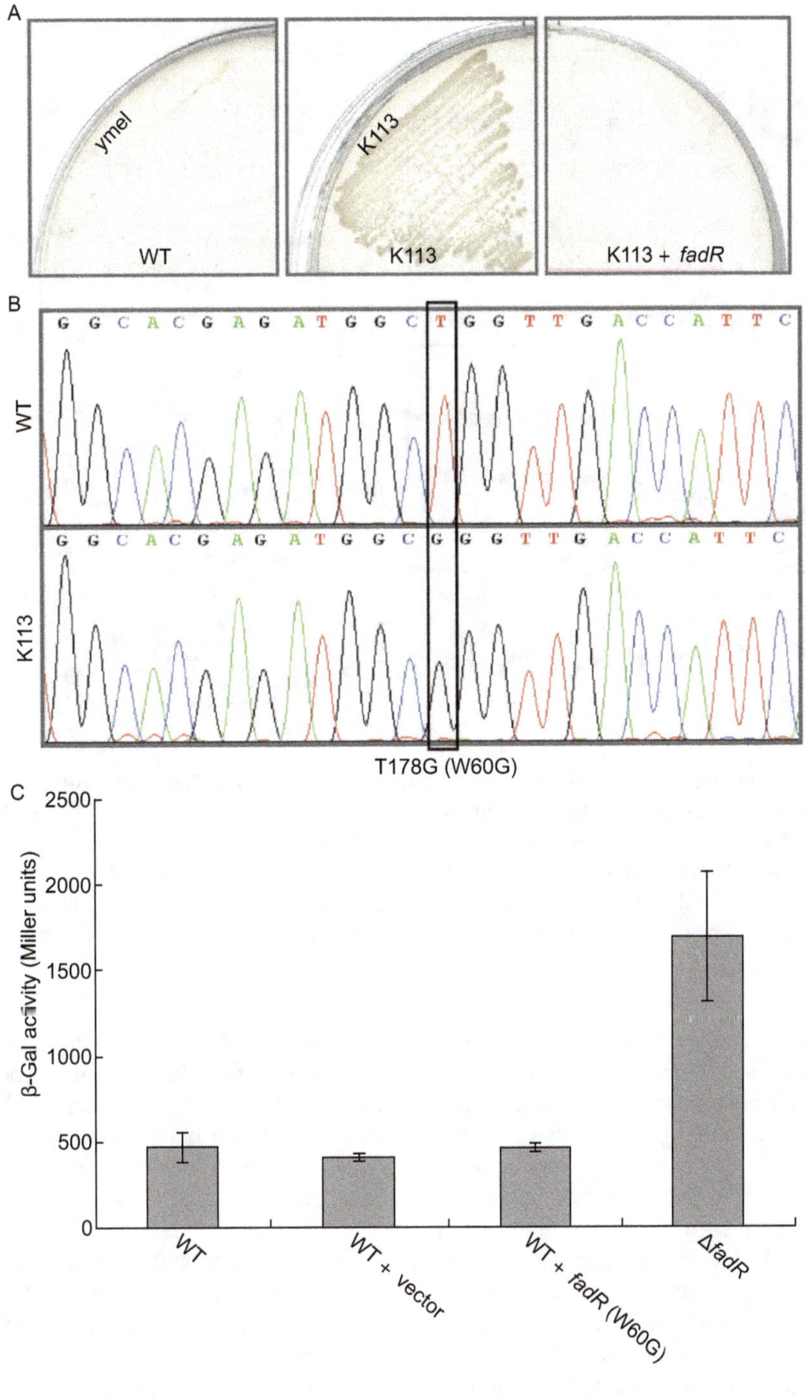

Figure 2. Determination of a single recessive mutation (W60G) present in *fadR* mutant strain K113. (A) Growth phenotype of *E. coli fadR* mutant strain K113 on minimal media with C10 as sole carbon source in comparison with those of the wild-type strain and the *fadR* complemented strain. The three strains used here included WT (ymel strain) (Feng & Cronan, 2009a), K113 (*fadR* mutant) (Feng & Cronan, 2009a, Clark et al., 1983) and the complemented strain FYJ7 (K113, *zcf-117*::Tn*10*, *fadR*$^{+}$, Tet^{R}) (Feng & Cronan, 2009a). (B) A single mutation (T178G) present in the K113 *fadR* revealed by direct DNA sequencing. The mutation of T178G at DNA level denotes the mutation of at protein level. (C) The mutation of W60G in FadR is genetically recessive. β-Gal activities were recorded in three independent experiments. All the strains assayed here contain a chromosomal *fadBA-lacZ* transcriptional fusion (Table 1). They are SI203 (WT), SI207 (Δ*fadR*), FYJ225 (WT + vector) and FYJ226 (WT + *fadR* (W60G)), respectively. Vector: pBAD24.

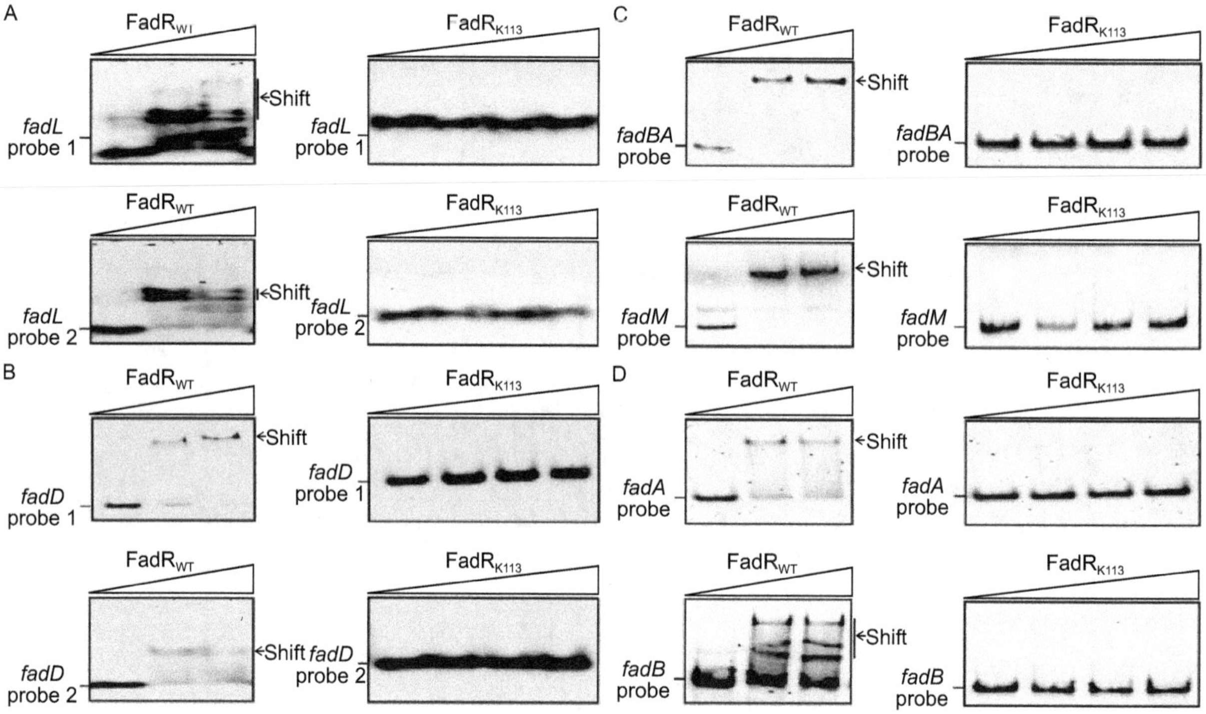

Figure 3. Mutant FadR protein ($FadR_{K113}$) losses its DNA binding activity *in vitro*. The two FadR-binding sites from both *fadL* (A) or *fadD* (B) can bind to the wild-type FadR protein ($FadR_{WT}$), whereas not the mutant version of FadR, $FadR_{k113}$ (W60G). (C) The wild-type FadR protein ($FadR_{WT}$) can bind to promoter regions covering the FadR-binding sites of *fadBA* and *fadM*, whereas $FadR_{k113}$ does not. (D) $FadR_{K113}$ fails to bind to promoter regions of *fabA* and *fabB*, two UFA biosynthetic genes whereas $FadR_{WT}$ does. All the EMSA experiments were carried out using 7% native PAGE, and a representative result is shown here. In gel shift assays, $FadR_{WT}$ is added as follows: 0, 0.5 and 1 pmol. Similarly, $FadR_{K113}$ is supplemented in 0, 0.5, 1 and 2 pmol, respectively. All the DIG-labeled probes are added to 0.1 pmol.

FadR to target promoters. Systematic analyses of *E. coli* *fadR* mutants collected in our laboratory in recent years allowed us to unexpectedly discover from K113 strain that a single point mutation (W60G) can impairs dramatically regulatory roles of FadR in lipid metabolism. Structure-guided mutagenesis analyses of FadR protein further suggested that the hydrophobic interaction amongst the three amino acids (W60, F74 and W75) might be essential for its DNA-binding ability by maintaining the configuration of its neighboring two β-sheets. Two more *fadR* super-repressor mutants were also discussed. The missing findings we reported in this paper might add new inputs into the picture of FadR-mediated regulation of bacterial lipid metabolism.

RESULTS AND DISCUSSION

Phenotypic and genetic dissection of the *fadR* mutant strain K113

FadR is a well-studied transcription factor that plays multiple roles in regulation of bacterial lipid metabolism. This type of regulator is consisted of a highly-conservative DNA-binding domain and a divergent acyl-CoA ligand-binding motif (Fig. 1). Multiple sequence alignments of a set of FadR homologues revealed its pretty conservative architecture with an exception of *V. cholerae* FadR homologue in that it seemed likely to obtain an extra-inserting sequence of 40 residues (Fig. 1). Earlier biochemical analyses revealed three amino acids at C-terminus of FadR (G216, S219 and 223W) are critical for its binding activity to acyl-CoA ligand (DiRusso et al., 1998), whereas the reported crystal structures visualized six N-terminal residues (R35, 44T, 45R, 46T, 47T, 49R and 65H) in direct contact with its cognate DNA (van Aalten et al., 2000, Xu et al., 2001).

We had occasion to sequence the original *fadR* mutation that of strain K113 of Overath and coworkers, originally called *dec-16* and now called *fadR16*. We found that the *fadR16* lesion results in substitution of glycine for tryptophan 60, a residue that plays no direct role in DNA binding. We showed that *fadR16* ($fadR\_{K113}$) was recessive to the wild-type allele. As we knew that the wild-type *E. coli* strains can grow on media with long-chain fatty acids as sole carbon source, but not on short-chain fatty acids like decanoic acid (C10) (Fig. 2A). In contrast, the *fadR* mutants usually exhibited a phenotype with capability of growing on C10 media. Like most of the other *fadR* mutants, the K113 strain

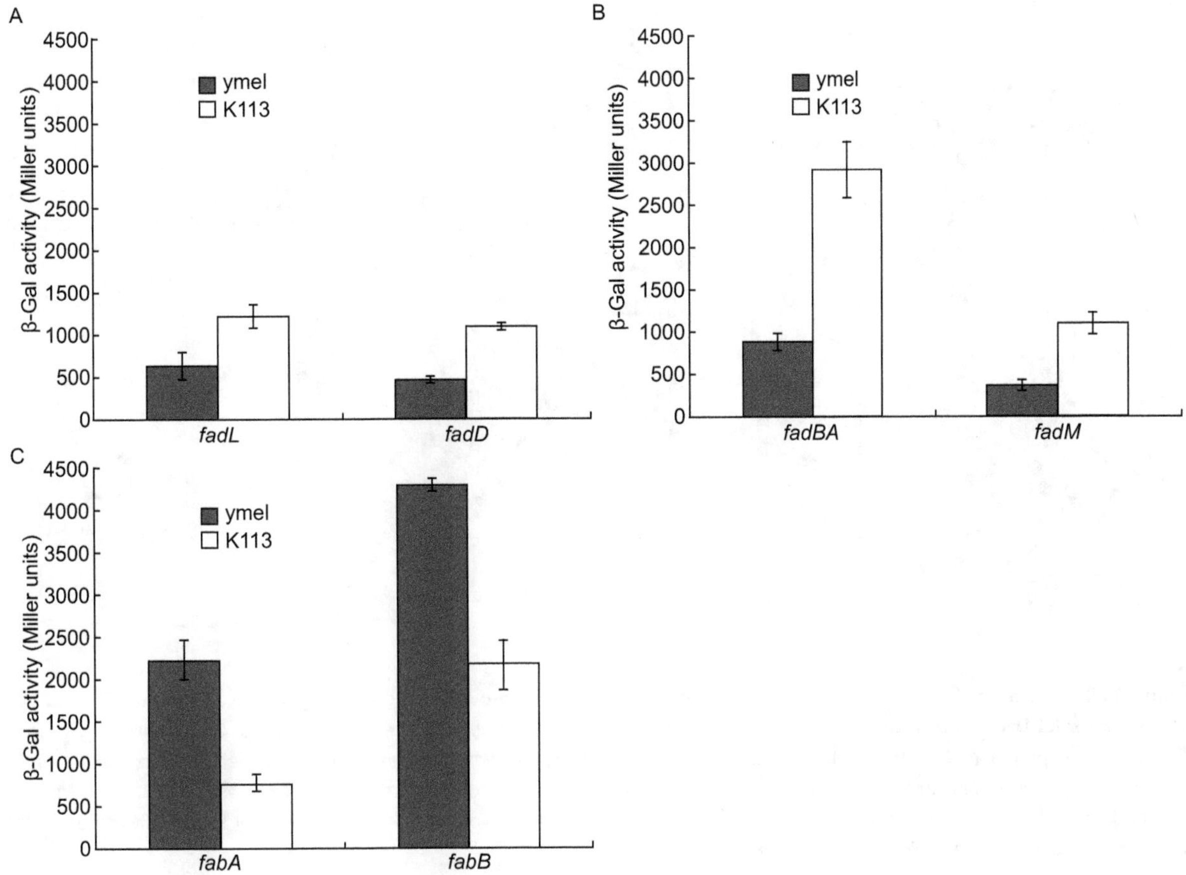

Figure 4. Regulatory dysfunction of fatty acid metabolism in K113 *fadR* strain. (A) Assays for transcriptional activities *fadL* and *fadD*, two genes of fatty acid transport system in the *fadR* strain K113. (B) Comparative analyses of expression levels of *fadBA*, a major member of beta-oxidation system and *fadM*, an auxiliary player of *fad* system in the *fadR* strain K113 relative to those of the wild-type ymel. (C) Effects on transcriptional levels of *fabA* and *fabB*, two key genes required for UFA synthesis due to *fadR* mutation in K113 strain. *E. coli* strains were grown in RB liquid media. β-Gal activities from three independent experiments are expressed in average ± standard deviations. In panel A, the strains used were FYJ185 (ymel, *fadL-lacZ* fusion), FYJ186 (K113, *fadL-lacZ* fusion), FYJ183 (ymel, *fadD-lacZ* fusion) and FYJ184 (K113, *fadD-lacZ* fusion). In panel B, the four strains included FYJ34 (ymel, *fadBA-lacZ* fusion), FYJ38 (K113, *fadBA-lacZ* fusion), FYJ35 (ymel, *fadM-lacZ* fusion) and FYJ39 (K113, *fadM-lacZ* fusion). In panel C, the strains tested referred to FYJ36 (ymel, *fabA-lacZ* fusion), FYJ 40 (K113, *fabA-lacZ* fusion), FYJ37 (ymel, *fabB-lacZ* fusion), FYJ41 (K113, *fabB-lacZ* fusion), respectively.

can grow on C10 plates (Fig. 2A), and this *fadR* phenotype can be restored to that of the parental strain ymel upon functional complementation of a copy of wild-type *fadR* gene (Fig. 2A). Comparative analyses of direct DNA sequencing results dissected firstly that this *fadR* phenotype is only attributed to the presence of a point mutation of T178G (W60G) in K113 strain (Fig. 2B).

Monitoring β-gal activity of a transcriptional *fadBA-lacZ* fusion revealed that removal of *fadR* led to >3-fold increment of expression level, whereas the introduction of a plasmid pBAD-borne mutant *fadR* ($fadR_{K113}$) into the wild-type strain could not result in any significant change of *fadBA* transcription (Fig. 2C). It was anticipated that $FadR_{K113}$ with mutation of W60G cannot compete with the wild-type version of FadR in the context of physiological interplay between FadR and *fadBA* promoter, indicating that the W60G mutation is genetically-recessive.

Functional impairment of FadR regulatory roles by a genetically recessive single mutation of W60G

To fully address the effect of the genetically-recessive single mutation of W60G on regulatory function of FadR protein, we employed the *in vitro* and *in vivo* approaches including EMSA and transcriptional fusion. The DNA binding activity of the mutant FadR protein ($FadR_{K113}$) was examined *in vitro* using gel shift assays. Eight sets of DIG-labeled DNA probes were prepared. Among them, two probes were designed for

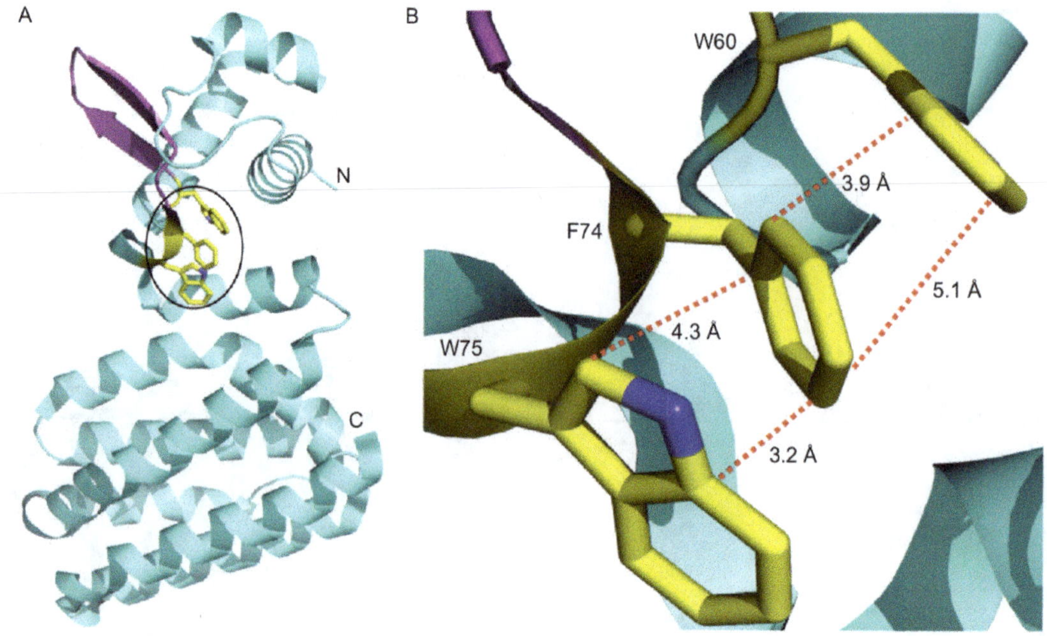

Figure 5. Structural analyses of FadR protein suggest that a hydrophobic interaction amongst W60, F74 and W75 might be important for its DNA binding activity. (A) Ribbon structure of *E. coli* FadR in monomer. N: N-terminus; C: C-terminus. α-Helix is blue, β-sheet is purple, and the three amino acids (W60, F74 and W75) that were proposed in this study is highlighted with a circle. (B) The enlarged view of the three critical residues (W60, F74 and W75). The hydrophobic bonds are expressed with dotted red lines, and the instance between two relevant atoms is labeled (angstrom). The photography was generated by Pmol software using the crystal structure's PDB file of FadR protein (PDB: 1E2X).

either *fadL* or *fadD* in that they have two FadR-binding palindromes (*fadL* probe1 plus *fadL* probe 2 (Fig. 3A) and *fadD* probe1 plus *fadD* probe 2 (Fig. 3B)) (Feng & Cronan, 2012). As we expected, the wild-type FadR protein ($FadR_{WT}$) exhibited obvious binding to *fadL* (*fadD*) promoter (Fig. 3A and 3B), however the $FadR_{K113}$ protein was not visualized to interact with them at the same protein level used in our trials (no more than 10 pmol). The promoters of *fadBA* (Feng & Cronan, 2012) and *fadM* (Feng & Cronan, 2009b), two members of β-oxidation system, also cannot be bound by $FadR_{K113}$ protein (Fig. 3C). The similar scenario was also observed with *fabA* and *fabB* (Feng & Cronan, 2011a), two important genes required for UFA biosynthetic pathway (Fig. 3D).

To further prove the dysfunction of $FadR_{K113}$ *in vivo*, we evaluated altered expression profile of a series of FadR-regulated genes in K113 strain vs. ymel strain (parental strain). As a result, we noted that 1) about 2-fold improved expression of both *fadL* and *fadD*, two genes encoding LCFA transport system in K113 strain over ymel strain (Fig. 4A), 2) 3.5-fold increment of *fadBA* expression and 2.5-fold elevated transcription of *fadM* in K113 strain relative to ymel strain (Fig. 4B), and 3) 2 to 2.5-fold decreased expression of *fabA* and *fabB* (Fig. 4C). The combined evidence demonstrated that functional impairment of $fadR_{K113}$ regulatory roles in lipid metabolism is due to a genetically-recessive single mutation of W60G in FadR protein.

A tripartite hydrophobic interaction required for DNA binding activity by FadR

Earlier structural investigations illustrated the overall architecture of *E. coli* FadR complex and its target DNA (Fig. S1) (Xu et al., 2001). The integrated evidence from structural modeling combined with genetic/biochemical data showed that the residue W60 is not in direct contact with DNA (Fig. 5A), but rather forms an aromatic cluster with two residues phenylalanine 74 and tryptophan 75 (F74 and W75). The measured distances amongst the three aromatic side chains varied from 3.2 Å to 5.1 Å (Fig. 5B) well positioned for hydrophobic bonding interactions in that it fits generally the criteria for defining a hydrophobic bond (interaction). We ambitiously hypothesize that the above hydrophobic bonds are necessary to maintain the flexibility of the two neighboring β-sheets, which in turn stabilizes configuration of its DNA binding domain (Fig. 5A). If so, mutation of any one of the three residues might well result in the loss of DNA binding activity.

To test this possibility, we prepared three more mutant FadR proteins (F74G, W75G and F74G plus W75G) using an approach of site-directed mutagenesis. In addition to the two versions of FadR protein ($FadR_{WT}$ and $FadR_{K113}$ (W60G)), we also purified the other three types of mutant FadR proteins to homogeneity (Fig. 6A). Somewhat different from that of the recombinant $FadR_{WT}$ in *E. coli* expression

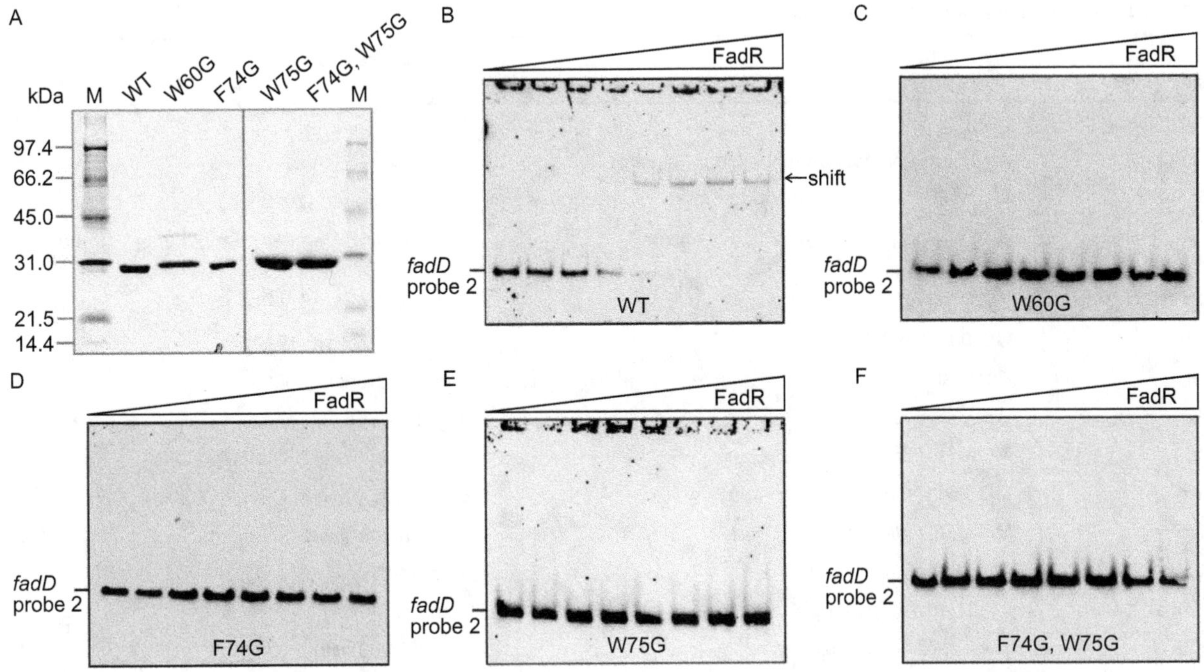

Figure 6. ***In vitro*** **preparation and functional analyses of four mutant FadR proteins plus its wild-type version.** (A) Expression and purification of four mutant FadR proteins and its wild-type. M: protein standard marker (Biorad). EMSA-based assays of binding activities of four mutant FadR proteins, $FadR_{W60G}$ (C), $FadR_{F74G}$ (D), $FadR_{W75G}$ (E) and $FadR_{F74G, W75G}$ (F) in comparison with that of the wild-type version (B). The gel shift tests were all conducted using 7.5% native PAGE, and a representative result is shown here. In these assays, FadR protein is added as follows: 0, 0.05, 0.1, 0.5, 1, 2, 5 and 10 pmol. The DIG-labeled *fadD* probe 2 is added to 0.2 pmol.

system (>70% solubility), the solubility of all the other four mutant proteins is estimated to vary from 5% to 15% (not shown). The fact that all of the mutants were appreciably less soluble than the wild-type protein is consistent with exposure of hydrophobic groups. For convenient comparison, the DIG-labeled *fadD* probe 2 with relatively high affinity to FadR protein was used in gel shift assays. As we mentioned before, $FadR_{WT}$ protein can bind to *fadD* probe 2 in a dose-dependent manner (Fig. 6B), whereas no significant binding of the mutant protein $FadR_{K113}$ (W60G) was found (Fig. 6C). Similar to what observed with $FadR_{K113}$ mutant protein, other two mutant FadR proteins with single mutation of either F74G (Fig. 6D) or W75G (Fig. 6E) did not exhibit DNA binding activity, as well as the FadR mutant with the double point-mutations of F74G plus W75G (Fig. 6F). Given the above structure-guided site-directed mutational analyses, we concluded that the aromatic cluster plays a key, but indirect, role in DNA binding. The fact that all three residues are aromatic and our prior finding that naturally occurring FadR proteins having isoleucine in place of F74 bind DNA poorly suggested that π-interactions may play a role in cluster function (Xu et al., 2001). However, we found that the F74I version of *E. coli* FadR had full DNA binding activity which ruled out π-interactions. Hence, it seemed likely that the cluster stabilizes the configuration of its DNA binding domain (Fig. 5).

CONCLUSIONS

Although the K113 *fadR* strain has been known for years, we are first to describe systemically regulatory dysfunction of lipid metabolism. Somewhat it is unexpected to note that the molecular determinant is a genetically-recessive point-mutation of W60G in *fadR* locus (Figs. 2 and 5). Relative to the known residues with direct contacting DNA operates, this hydrophobic amino acid of tryptophan at the position 60 of FadR regulatory protein represent a new functional residue in that it is without close contact to its cognate DNA, but required for capability of DNA binding.

In addition, we addressed two more *fadR* super-repressor mutant strains (TH182 and TH183, featuring that the repression of *fad* regulon expression cannot be relieved by oleate). In general consistency with the report by Hughes et al. (Hughes et al., 1988), we also found the molecular determinants of two *fadR* super-repressor strains TH182 and TH183, each of which is a single point mutation localized in C-terminal ligand binding domain (G638A, R213H in TH182 and G647A, G216E in TH183) (not shown). Ambitiously, we firstly proposed a model in which the hydrophobic interaction of W60 with other two aromatic amino acids (F74 and W75) is critical for the maintenance of configuration of the DNA binding domain of FadR protein at N-terminus (Fig. 5). It seemed likely that our findings might supplement three more new functional residues of FadR regulator.

Table 1. Strains and plasmids used in this study

Bacteria or plasmids	Relevant characteristics	References/sources
E. coli strains		
BL21 (Tuner)	An expression host for recombinant plasmids	Lab stock
MC4100	F^-, *araD*139, Δ(*argF-lac*)169	Feng & Cronan (2009b)
DH5α (λ-*pir*)	Δ*lac* host for pAH125 and its derivatives	Haldimann & Wanner (2001); Feng & Cronan (2009b); Feng & Cronan (2009a)
MC1061	Wild-type of *E. coli* K-12, Δ*lac*	Feng & Cronan (2009b)
MFH8	UB1005, *fadR*::Tn*10*	Henry & Cronan (1992)
ymel	Parental strain of K113	Feng & Cronan (2009a)
K113	ymel, *fadR16*	Clark (1981)
JT160	MC1061, *fabA-lacZ* fusion, Kan^R	Lab stock, Feng & Cronan (2011a)
JT161	MC1061, *fabB-lacZ* fusion, Kan^R	Lab stock, Feng & Cronan (2011a)
FYJ17	MC1061, *fadM-lacZ* fusion, Kan^R	Feng & Cronan (2009b)
FYJ30	JT180, *fadR*::Tn*10*, Kan^R, Tet^R	Feng & Cronan (2011a)
FYJ32	JT181, *fadR*::Tn*10*, Kan^R, Tet^R	Feng & Cronan (2011a)
FYJ34	ymel, *fadBA-lacZ* fusion, Kan^R	*P1vir*(SI203)xymel[c]
FYJ35	ymel, *fadM-lacZ* fusion, Kan^R	*P1vir*(FYJ17)xymel[c]
FYJ36	ymel, *fabA-lacZ* fusion, Kan^R	*P1vir*(JT160)xymel[c]
FYJ37	ymel, *fabB-lacZ* fusion, Kan^R	*P1vir*(JT161)xymel[c]
FYJ38	K113, *fadBA-lacZ* fusion, Kan^R	*P1vir*(SI203)xK113[c]
FYJ39	K113, *fadM-lacZ* fusion, Kan^R	*P1vir*(FYJ17)xK113[c]
FYJ40	K113, *fabA-lacZ* fusion, Kan^R	*P1vir*(JT160)xK113[c]
FYJ41	K113, *fabB-lacZ* fusion, Kan^R	*P1vir*(JT161)xK113[c]
SI203	*fadBA-lacZ* transcription fusion	Iram & Cronan (2005); Feng & Cronan (2010)
SI207	*fadBA-lacZ* transcription fusion, Δ*fadR*::Tn*10*	Iram & Cronan (2005); Feng & Cronan (2010)
FYJ104	FYJ103, *fadL-lacZ* transcription fusion	This work
FYJ158	DH5α (λ-*pir*) carrying pAH-*PfadD*	Feng & Cronan (2011a)
FYJ159	MC4100, *fadD-lacZ* transcription fusion integrated into the chromosomal λ-attB site	This work
FYJ183	ymel, *fadD-lacZ* transcriptional fusion	*P1vir*(FYJ159)×ymel[c]
FYJ184	K113, *fadD-lacZ* transcriptional fusion	*P1vir*(FYJ159)× K113[c]
FYJ185	ymel, *fadL-lacZ* transcriptional fusion	*P1vir*(FYJ104)×ymel[c]
FYJ186	K113, *fadL-lacZ* transcriptional fusion	*P1vir*(FYJ104)× K113[c]
FYJ187	MC4100 carrying pINT-ts	Feng & Cronan (2011a)
FYJ191	BL21(DE3) carrying pET28-*fadR*ec(W75G)	This work
FYJ192	BL21(DE3) carrying pET28-*fadR*_K113	This work
FYJ206	BL21(DE3) carrying pET28-*fadR*ec(F74G)	This work
FYJ207	BL21(DE3) carrying pET28-*fadR*ec(F74G/W75G)	This work
FYJ221	Topo 10 carrying pBAD-*fadR*_K113	This work
FYJ225	SI203 carrying pBAD24	This work
FYJ226	SI203 carrying pBAD-*fadR*_K113	This work
Plasmids		
pET28(a)	Commercial T7-driven expression vector	Novagen

Table 1 continued

Bacteria or plasmids	Relevant characteristics	References/sources
pBAD24	Arabinose-inducible promoter-driven expression vector, Amp^R	Guzman et al. (1995)
pBAD-*fadR*_K113	pBAD24 carrying K113 *fadR* gene, Kan^R	This work
pET28-*fadR*ec	pET28 carrying *E. coli fadR*, Kan^R	Feng & Cronan (2009b); Feng & Cronan (2009a)
pET28-*fadR*_K113	pET28 carrying K113 *fadR* gene, Kan^R	This work
pET28-*fadR*ec (F74G)	pET28 carrying *E. coli fadR* gene with a single F74G mutation, Kan^R	This work
pET28-*fadR*ec (W75G)	pET28 carrying *E. coli fadR* gene with a single W75G mutation, Kan^R	This work
pET28-*fadR*ec (F74G, W75G)	pET28 carrying *E. coli fadR* gene with the double mutations (F74G, W75G), Kan^R	This work

[a] CGSC denotes Coli Genetic Stock Center, Yale University; [b] Selection for tetracycline resistance; [c] Selection for kanamycin resistance.

Together, we believed that the deep dissection of the *E. coli* FadR regulatory protein provided a new glimpse of FadR-DNA interplay/crosstalk in the context of lipid metabolism.

MATERIALS AND METHODS

Bacterial strains and growth conditions

All the strains used here are *E. coli* K-12 derivatives (Table 1). Three types of media used for *E. coli* included LB medium (Luria-Bertani medium containing 10 g of tryptone, 5 g of yeast extract and 10 g of NaCl per liter), Rich broth (RB medium containing 10 g of tryptone, 1 g of yeast extract and 5 g of NaCl per liter) and minimal medium M9 (Iram & Cronan, 2006) supplemented with 0.4% glucose, 0.1% vitamin-free Casamino Acids, 1 mmol/L $MgSO_4$ and 0.1 mmol/L $CaCl_2$, and 0.001% thiamine were used for bacterial growth and analyses of β-galactosidase activity. If necessary, antibiotics were used as follows (in mg/liter): 100 for sodium ampicillin, 50 for kanamycin sulfate, 20 for chloramphenicol and 15 for tetracycline HCl.

Plasmids and DNA manipulations

The expression vector pET28a (Novagen) was used for protein preparation (Table 2). The K113 *fadR* was amplified using primers *fadR*ec-F2 plus *fadR*ec-R2 that carried *Eco*RI and *Sal*I restriction sites, respectively (Table 2). The PCR product was digested with these enzymes and inserted into vector pBAD24 (Guzman et al., 1995) cut with the same enzymes giving the plasmid pBAD-*fadR*_K113 (Table 1). All plasmids acquired were validated by PCR testing plus direct DNA sequencing.

$P1_{vir}$ phage transduction

$P1_{vir}$ transductions were carried out as described by Miller (Miller, 1992) with minor modifications. Strains FYJ34, FYJ35, FYJ36, FYJ37, FYJ183 and FYJ185 (Table 1) were obtained by $P1_{vir}$ transduction of strain ymel, using lysates grown on strains SI203 (*fadBA-lacZ*), FYJ17 (*fadM-lacZ*), JT160 (*fabA-lacZ*), JT161 (*fabB-lacZ*), FYJ159 (*fadD-lacZ*) or FYJ104 (*fadL-lacZ*) with selection for kanamycin resistance. Strains FYJ38, FYJ39, FYJ40 and FYJ41 were similarly obtained by transduction of strain K113 (Table 1).

Measurement of β-galactosidase

Mid-log phase cultures in LB, RB or minimal media (with or without supplementation with various carbon sources), were collected by centrifugation, washed twice with Z Buffer (Miller 1972) and assayed for β-galactosidase activity after lysis with sodium dodecyl sulfate-chloroform (Miller 1972). The data were recorded in triplicate with no less than three independent experiments.

Site-directed mutagenesis

Site-directed mutagenesis was done as we recently described with little changes (Feng et al., 2014). The PCR reaction system (25 μL) consisted of the following components: 2.5 μL of 10× pfx buffer (Invitrogen), 0.5 μL of 40 mmol/L dNTP mix (10 mmol/L each), 1.0 μL of forward/reverse primers (10 pmol/μL), 1.0 μL of pET28-*fadR*ec as template (5 ng/μL), 0.5 μL of Platinum pfx (2.5 U/μL, Invitrogen), and 17.5 μL of distilled sterilized H_2O. The reaction was performed using the program consisting of a denaturing cycle at 95°C for 5 min; 20 cycles comprised of 95°C for 50 s, 60°C for 50 s, and 68°C for 6 min and a final step of 8 min at 68°C. To remove the residual template plasmid pET28-*fadR*ec, the gel purified PCR products were digested for 1 h with DpnI (20 U/μL, NEB) at 37°C. Subsequently, they were transformed into chemically-competent cells of DH5α and the inserts of purified plasmids were verified by direct DNA sequencing. The plasmids were transformed into BL21 (Tuner) to produce the FadR mutant proteins.

Protein expression and purification

The hexahistidine-tagged *E. coli* FadR proteins were produced in *E. coli* BL21 (DE3) carrying the appropriate expression plasmids (e.g., pET28-*fadR*ec, Table 1) by induction of bacterial cultures at

Table 2. Primers used in this study

Primers	Primer sequences (5′-3′)
lacZ-R	GACCATGATTACGGATTCACTG
*fadR*_K113-F (*Bam*HI)	CG*GGATCC*ATGGTCATTAAGGCGCAAAGCC
*fadR*_K113-R (*Xho*I)	*CCGCTCGAG*TTATCGCCCCTGAATGGCTAAATC
fadL-FadR1-F	GCAACATTCC**AGCTGGTCCGACCTATA**CTCTCGCC
fadL-FadR1-R	GGCGAGAG**TATAGGTCGGACCAGCT**GGAATGTTGC
fadL-FadR2-F	CTCTCGCC**ACTGGTCTGATTTCTAA**GATGTACCTC
fadL-FadR2-R	GAGGTACATC**TTAGAAATCAGACCAGT**GGCGAGAG
fadD-FadR1-F	GAAACAGC**GGCTGGTCCGCTGTTTC**TGCATTCT
fadD-FadR1-R	AGAATGCA**GAAACAGCGGACCAGCC**GCTGTTTC
fadD-FadR2-F	GTAATTATCA**AGCTGGTATGATGAGTT**AATATTATG
fadD-FadR2-R	CATAATATT**AACTCATCATACCAGCT**TGATAATTAC
fadBA-F	ACTTCGACTC**ATCTGGTACGACCAGAT**CACCTTGCGG
fadBA-R	CCGCAAGGTG**ATCTGGTCGTACCAGAT**GAGTCGAAGT
fabA-F	TTTATTCCG**AACTGATCGGACTTGTT**CAGCGTACACGTGTTAGCTATCCTGCGTGC
fabA-R	GCACGCAGGATAGCTAACACGTGTACGCTG**AACAAGTCCGATCAGTT**CGGAATAAA
fabB-F	TCTATTAAAT**GGCTGATCGGACTTGTT**CGGCGTACAAGTGTACGCTATTGTGCATTC
fabB-R	GAATGCACAATAGCGTACACTTGTACGCCG**AACAAGTCCGATCAGCC**ATTTAATAGA
F74G-F	CGACGAAGGTGAATAATGGCTGGGAAACTTCCGGTT
F74G-R	AACCGGAAGTTTCCCAGCCATTATTCACCTTCGTCG
W75G-F	CGAAGGTGAATAATTTCGGGGAAACTTCCGGTTTAAA
W75G-R	TTTAAACCGGAAGTTTCCCCGAAATTATTCACCTTCG
F74G/W75G-F	CGACGAAGGTGAATAATGGCGGGGAAACTTCCGGTTTAAA
F74G/W75G-R	TTTAAACCGGAAGTTTCCCCGCCATTATTCACCTTCGTCG
pBAD-chk5	CTGTTTCTCCATACCCGTT
pBAD-chk3	GGCTGAAAATCTTCTCT
*fadR*ec-F2 (*Eco*RI)	AACC*GAATTC*ATGGTCATTAAGGCGCAAAGCC
*fadR*ec-R2 (*Sal*I)	CCG*GTCGAC*TTATCGCCCCTGAATGGCTAAATC
T7-F	TAATACGACTCACTATAGGG
T7-R	GCTAGTTATTGCTCAGCGG

The sequences underlined are restriction sites, and the bold letters are putative FadR binding sites.

an $OD_{600\ nm}$ of 0.8–1.0 with 0.3 mmol/L IPTG at 30°C for 3 h (Feng & Cronan, 2011b, Feng & Cronan, 2010). The cells were pelleted, washed twice with ice-cold PBS buffer (101.4 mmol/L Na_2HPO_4, 1.8 mmol/L KH_2PO_4, 137 mmol/L NaCl, 2.7 mmol/L KCl, 8% glycerol, pH7.4), dissolved in the same buffer and lysed using a French Press. The extracts were centrifuged to remove bacterial debris and the supernatants loaded onto a nickel chelate column (Qiagen). Following washing with ten column volumes of PBS buffer containing 50 mmol/L imidazole, the FadR proteins were eluted with 150 mmol/L imidazole. Appropriate eluted protein fractions were pooled and dialyzed against PBS buffer, then concentrated by ultrafiltration (30 kDa cut-off, Amicon Ultra) (Feng & Cronan, 2010). The protein purity was judged by 12% SDS-PAGE, followed by staining with Coomassie brilliant blue R250 (Sigma, St. Louis, MO).

Electrophoretic mobility shift assays

Electrophoretic mobility shift assays were performed to address interaction between FadR and series of taget DNA probes (e.g., *fadD* and *fadL* (Feng & Cronan, 2012)) essentially as previously reported (Feng & Cronan, 2011b, Feng & Cronan, 2011a, Feng & Cronan, 2009b). With an exception of *fadM* probe (PCR products) (Feng & Cronan, 2009b), all of the FadR-recognizable probes were prepared by annealing two complementary primers (Table 2) by incubation in TEN buffer (10 mmol/L Tris-HCl, 1 mmol/L EDTA, 100 mmol/L NaCl, pH 8.0) at 95°C for 5 min followed by slow cooling to 25°C and then digoxigenin (DIG) labeling by terminal transferase with DIG-ddUTP (Roche). The DIG-labeled DNA probes (either 0.1 pmol or 0.2 pmol) were incubated with either DNA binding protein in binding buffer (Roche) for 15 min at room temperature and then analyzed by native PAGE (7–7.5% PAGE for all probes).

Bioinformatic analyses

The known FadR sequences were all from the NCBI database and the multiple alignments were done using ClustalW2 (http://www.ebi.ac.uk/Tools/clustalw2/index.html), and the resultant output was processed by program ESPript 2.2 (http://espript.ibcp.fr/ESPript/cgi-bin/ESPript.cgi), generating the final BLAST version. The FadR structure was visualized using PyMol software.

ACKNOWLEDGEMENTS

This work was supported by a start-up package from Zhejiang University (YF). YF is a recipient of the "Young 1000 Talents" award. We would like to thank Dr. Zhi Li (Department of Biochemistry, UIUC) for helpful discussion on *E. coli* FadR structure.

COMPLIANCE WITH ETHICS GUIDELINES

Yongchang Zhang, Rongsui Gao, Huiyan Ye, Qingjing Wang and Youjun Feng declare that they have no conflict of interest.

This article does not contain any studies with human or animal subjects performed by the any of the authors.

REFERENCES

Campbell JW, Cronan JE Jr (2001) *Escherichia coli* FadR positively regulates transcription of the *fabB* fatty acid biosynthetic gene. J Bacteriol 183:5982–5990

Clark D (1981) Regulation of fatty acid degradation in *Escherichia coli*: analysis by operon fusion. J Bacteriol 148:521–526

Clark D, Cronan J (2005) Two-carbon compounds and fatty acids as carbon sources. In: Bock A, Curtiss III R, Karper J, Karp P, Neidhardt F, Nystrom T, Slauch J, Squiress C, Ussery D (eds) EcoSal–*Escherichia coli* and *Salmonella*: cellular and molecular biology. http://www.ecosal.org. ASM Press, Washington, DC

Clark DP, DeMendoza D, Polacco ML, Cronan JE Jr (1983) Beta-hydroxydecanoyl thio ester dehydrase does not catalyze a rate-limiting step in *Escherichia coli* unsaturated fatty acid synthesis. Biochemistry 22:5897–5902

Cronan JE Jr (1997) In vivo evidence that acyl coenzyme A regulates DNA binding by the *Escherichia coli* FadR global transcription factor. J Bacteriol 179:1819–1823

Cronan JE Jr, Subrahmanyam S (1998) FadR, transcriptional co-ordination of metabolic expediency. Mol Microbiol 29:937–943

DiRusso CC, Heimert TL, Metzger AK (1992) Characterization of FadR, a global transcriptional regulator of fatty acid metabolism in *Escherichia coli*. Interaction with the *fadB* promoter is prevented by long chain fatty acyl coenzyme A. J Biol Chem 267:8685–8691

DiRusso CC, Metzger AK, Heimert TL (1993) Regulation of transcription of genes required for fatty acid transport and unsaturated fatty acid biosynthesis in *Escherichia coli* by FadR. Mol Microbiol 7:311–322

DiRusso CC, Tsvetnitsky V, Hojrup P, Knudsen J (1998) Fatty acyl-CoA binding domain of the transcription factor FadR. Characterization by deletion, affinity labeling, and isothermal titration calorimetry. J Biol Chem 273:33652–33659

Farewell A, Kvint K, Nystrom T (1998) Negative regulation by RpoS: a case of sigma factor competition. Mol Microbiol 29:1039–1051

Feng Y, Cronan JE (2009a) *Escherichia coli* unsaturated fatty acid synthesis: complex transcription of the *fabA* gene and *in vivo* identification of the essential reaction catalyzed by FabB. J Biol Chem 284:29526–29535

Feng Y, Cronan JE (2009b) A new member of the *Escherichia coli fad* regulon: transcriptional regulation of *fadM* (*ybaW*). J Bacteriol 191:6320–6328

Feng Y, Cronan JE (2010) Overlapping repressor binding sites result in additive regulation of *Escherichia coli* FadH by FadR and ArcA. J Bacteriol 192:4289–4299

Feng Y, Cronan JE (2011a) Complex binding of the FabR repressor of bacterial unsaturated fatty acid biosynthesis to its cognate promoters. Mol Microbiol 80:195–218

Feng Y, Cronan JE (2011b) The *Vibrio cholerae* fatty acid regulatory protein, FadR, represses transcription of *plsB*, the gene encoding the first enzyme of membrane phospholipid biosynthesis. Mol Microbiol 81:1020–1033

Feng Y, Cronan JE (2012) Crosstalk of *Escherichia coli* FadR with global regulators in expression of fatty acid transport genes. PloS one 7:e46275

Feng Y, Li M, Zhang H, Zheng B, Han H, Wang C, Yan J, Tang J, Gao GF (2008) Functional definition and global regulation of Zur, a zinc uptake regulator in a *Streptococcus suis* serotype 2 strain causing streptococcal toxic shock syndrome. J Bacteriol 190:7567–7578

Feng Y, Napier BA, Manandhar M, Henke SK, Weiss DS, Cronan JE (2014) A *Francisella* virulence factor catalyses an essential reaction of biotin synthesis. Mol Microbiol 91:300–314

Gui L, Sunnarborg A, LaPorte DC (1996) Regulated expression of a repressor protein: FadR activates *iclR*. J Bacteriol 178:4704–4709

Guzman LM, Belin D, Carson MJ, Beckwith J (1995) Tight regulation, modulation, and high-level expression by vectors containing the arabinose PBAD promoter. J Bacteriol 177:4121–4130

Haldimann A, Wanner BL (2001) Conditional-replication, integration, excision, and retrieval plasmid-host systems for gene structure-function studies of bacteria. J Bacteriol 183:6384–6393

He XY, Yang SY, Schulz H (1997) Cloning and expression of the *fadH* gene and characterization of the gene product 2,4-dienoyl coenzyme A reductase from *Escherichia coli*. Eur J Biochem 248:516–520

Henry MF, Cronan JE Jr (1992) A new mechanism of transcriptional regulation: release of an activator triggered by small molecule binding. Cell 70:671–679

Hughes KT, Simons RW, Nunn WD (1988) Regulation of fatty acid degradation in *Escherichia coli*: *fadR* superrepressor mutants are unable to utilize fatty acids as the sole carbon source. J Bacteriol 170:1666–1671

Iram SH, Cronan JE (2005) Unexpected functional diversity among FadR fatty acid transcriptional regulatory proteins. J Biol Chem 280:32148–32156

Iram SH, Cronan JE (2006) The beta-oxidation systems of *Escherichia coli* and *Salmonella enterica* are not functionally equivalent. J Bacteriol 188:599–608

Maloy SR, Nunn WD (1981) Role of gene *fadR* in *Escherichia coli* acetate metabolism. J Bacteriol 148:83–90

Miller JH (1972) Experiments in molecular genetics. Cold Spring Harbor Laboratory Press, Cold Spring Harbor

Miller JH (1992) A short course in bacterial genetics. Cold Spring Harbor Laboratory, Cold Spring Harbor

Nunn WD, Giffin K, Clark D, Cronan JE Jr (1983) Role for *fadR* in unsaturated fatty acid biosynthesis in *Escherichia coli*. J Bacteriol 154:554–560

van Aalten DM, DiRusso CC, Knudsen J, Wierenga RK (2000) Crystal structure of FadR, a fatty acid-responsive transcription factor with a novel acyl coenzyme A-binding fold. Embo J 19:5167–5177

van Aalten DM, DiRusso CC, Knudsen J (2001) The structural basis of acyl coenzyme A-dependent regulation of the transcription factor FadR. Embo J 20:2041–2050

Xu Y, Heath RJ, Li Z, Rock CO, White SW (2001) The FadR.DNA complex. transcriptional control of fatty acid metabolism in *Escherichia coli*. J Biol Chem 276:17373–17379

16

Binding of *Shewanella* FadR to the *fabA* fatty acid biosynthetic gene: implications for contraction of the *fad* regulon

Huimin Zhang[1], Beiwen Zheng[1,2], Rongsui Gao[1], Youjun Feng[1,2]✉

[1] Department of Medical Microbiology & Parasitology, Zhejiang University School of Medicine, Hangzhou 310058, China
[2] State Key Laboratory for Diagnosis and Treatment of Infectious Diseases, Zhejiang University School of Medicine, Hangzhou 310058, China
✉ Correspondence: fengyj@zju.edu.cn (Y. Feng)

ABSTRACT

The *Escherichia coli fadR* protein product, a paradigm/prototypical FadR regulator, positively regulates *fabA* and *fabB*, the two critical genes for unsaturated fatty acid (UFA) biosynthesis. However the scenario in the other γ-proteobacteria, such as *Shewanella* with the marine origin, is unusual in that Rodionov and coworkers predicted that only *fabA* (not *fabB*) has a binding site for FadR protein. It raised the possibility of *fad* regulon contraction. Here we report that this is the case. Sequence alignment of the FadR homologs revealed that the N-terminal DNA-binding domain exhibited remarkable similarity, whereas the ligand-accepting motif at C-terminus is relatively-less conserved. The FadR homologue of *S. oneidensis* (referred to FadR_she) was over-expressed and purified to homogeneity. Integrative evidence obtained by FPLC (fast protein liquid chromatography) and chemical cross-linking analyses elucidated that FadR_she protein can dimerize in solution, whose identity was determined by MALDI-TOF-MS. *In vitro* data from electrophoretic mobility shift assays suggested that FadR_she is almost functionally-exchangeable/equivalent to *E. coli* FadR (FadR_ec) in the ability of binding the *E. coli fabA* (and *fabB*) promoters. In an agreement with that of *E. coli fabA*, *S. oneidensis fabA* promoter bound both FadR_she and FadR_ec, and was disassociated specifically with the FadR regulatory protein upon the addition of long-chain acyl-CoA thioesters. To monitor *in vivo* effect exerted by FadR on *Shewanella fabA* expression, the native promoter of *S. oneidensis fabA* was fused to a LacZ reporter gene to engineer a chromosome *fabA-lacZ* transcriptional fusion in *E. coli*. As anticipated, the removal of *fadR* gene gave about 2-fold decrement of *Shewanella fabA* expression by β-gal activity, which is almost identical to the inhibitory level by the addition of oleate. Therefore, we concluded that *fabA* is contracted to be the only one member of *fad* regulon in the context of fatty acid synthesis in the marine bacteria *Shewanella* genus.

Huimin Zhang and Beiwen Zheng have contributed equally to this work.

KEYWORDS FadR, *fad* regulon, *fabA*, *fabB*, contraction, *Shewanella*

INTRODUCTION

Current knowledge on the regulation of fatty acid metabolism is mostly from studies with *Escherichia coli* (*E. coli*). The *E. coli* FadR regulatory protein that is classified into the GntR family of transcription factors, acts as a global regulator controlling bacterial lipid metabolism (Henry & Cronan, 1992, Iram & Cronan, 2005). The two opposite roles played by this regulator include repression of fatty acid degradation (*fad*) system (Feng & Cronan, 2009b, Henry & Cronan, 1991, Iram & Cronan, 2005), and activation of *fabA* and *fabB*, the two genes for unsaturated fatty acid synthesis (Feng & Cronan, 2009a, Henry & Cronan, 1992, Nunn et al., 1983). In fact, the *E. coli* FadR also indirectly regulates transcription of the glyoxylate bypass operon (*aceBAK*), through direct activating the

IclR repressor (Gui et al., 1996). Very recently, My et al. (My et al., 2013) reported that *E. coli fabH* is the third gene of fatty acid synthesis pathway that can be positively-regulated by the FadR regulator.

As the paradigm FadR regulator, the *E. coli fadR* protein product behaves as a dimer (van Aalten et al., 2000), and consists of the N-terminal DNA-binding domain (Xu et al., 2001) and the ligand-interacting motif at C-terminus (van Aalten et al., 2001). The accumulated crystal structures of FadR alone and its complex with DNA/acyl-CoA defined clearly the structural basis for FadR-mediated regulatory mechanism in the context of lipid metabolism (van Aalten et al., 2001, van Aalten et al., 2000, Xu et al., 2001). In addition to the residues directly contacting target DNA (Xu et al., 2001), we also mapped three more key residues with indirect role in FadR-DNA interplay (Zhang et al., 2014). *In vitro* and *in vivo* evidence proved that long-chain fatty acid (LCFA) acyl-CoA thioesters are small molecule effectors for the FadR regulatory system (Henry & Cronan, 1992, van Aalten et al., 2001, Cronan, 1997). The mechanism by which LCFA induces *fad* expression lies in the fact that the binding of LCFA acyl-CoA to FadR protein results in the alteration of protein configuration, which in turn triggers the loss of its DNA binding ability. However, it still remains unclear why the unexpected functional diversity exists amongst the FadR regulatory proteins (Iram & Cronan, 2005). Of particular note, it is mystery that in relative to the prototypical FadR with an origin of *E. coli*, the *Vibrio cholerae* (*V. cholerae*) FadR is strikingly superior to in the regulatory amplitude, and bound its ligands appreciably stronger (Iram & Cronan, 2005). Further sequence analyses revealed that an extra 40-aa longer region present in *V. cholerae* FadR might explain the excellent performance of its regulation role in the context of lipid metabolism (Zhang et al., 2014). Unlike the scenario seen with its closely-relative *V. cholerae*, the FadR homologue from the other marine bacterium *Shewanella*, is quite similar to that of the paradigm organism *E. coli*.

The genus of *Shewanella* is a family of Gram-negative bacteria inhabiting in marine environment/ecosystem, including no less than 50 diversified species such as *S. oneidensis* and *S. algae* (Janda & Abbott, 2014). *S. oneidensis* is referred to an alternative model anaerobic microorganism with the known genome sequence (~4.9 Mb) that encodes over 4700 genes (Kolker et al., 2005, Heidelberg et al., 2002). Not only do the species of *Shewanella* bacteria (e.g., *S. putrefaciens*) act as normal components of the surface flora of fish and are involved in the spoilage of aquatic products (Parlapani et al., 2013, Li et al., 2012), but also some species like *S. algae* is recognized to be zoonotic pathogens in that they can cause opportunistic infections via occupational exposure of workers with skin and soft tissue cuts to marine products (Janda & Abbott, 2014). Given the excellent performance of *S. oneidensis* in reduction of poisonous heavy metals like iron (Cheng et al., 2013), uranium (Sheng & Fein, 2014), and even ionic mercury (Wiatrowski et al., 2006), it was believed to have the robust/potential applications into environmental bioremediation targeting toxic elements and heavy metals and development of microbial fuel cells (Fredrickson et al., 2008, Hau & Gralnick, 2007). The advantage of *Shewanella* in biotechnology is mainly attributed to the diversified metabolic capabilities that included versatile electron-transfer systems (Hau & Gralnick, 2007, Fredrickson et al., 2008). The deep-sea environment with low temperature where the *Shewanella* bacteria naturally reside/inhabit determined that some special mechanism might be evolved for their survival. Wang et al. (Wang et al., 2009) found that *Shewanella* has appreciable ability to produce various types of low-melting-point fatty acids with monounsaturated fatty acids (MUFA) included. The similar scenario was also noted in the other marine bacterium *V. cholerae*, in which relatively-high percentage of unsaturated fatty acids (UFA) is present in relative to *E. coli* (Massengo-Tiasse & Cronan, 2008, Feng & Cronan, 2011a). The physiological explanation proposed lies in that the high percentage of UFA in the bacterial membrane incorporated with phospholipid confers the better membrane fluidity, which in turn enhances its capability of cold adaptation. Although the type II fatty acid synthesis (FAS) pathway in *Shewanella* was constructed using the approach of comparative genomics (Wang et al., 2009), it seemed likely that some unusual/unclear aspects are present in the regulation of this specialized Type II FAS (Rodionov et al., 2011). Of particular note, Rodionov and coworkers (Rodionov et al., 2011) predicted that only *fabA* (not *fabB*) of *Shewanella* has a binding site for FadR protein, posing the possibility of *fad* regulon contraction.

In this paper, we integrated *in vitro* and *in vivo* approaches to address this uncommon question and reported that this is the case. As expected, *Shewanella* FadR regulates expression of *fabA* (not *fabB*) through the direct protein-DNA physical interplay. Therefore, it is reasonable that *the fabA* fatty acid synthesis gene is contracted as the only one member of *fad* regulon in the context of fatty acid synthesis in the marine bacteria *Shewanella* genus.

RESULTS AND DISCUSSION

Contraction of *fad* regulon in *Shewanella*

Different from the paradigm enteric bacterium *E. coli* that has only one chromosome of 4.64 Mb with average GC contents of, 50.8% and encodes 4498 putative genes (Blattner et al., 1997), *S. oneidensis* MR-1, the representative cousin with marine origin, not only has a chromosome (4.96 Mb, 46% GC percentage) encoding 4403 genes, but also contains a megaplasmid (0.16 Mb, 43.7% GC percentage) encoding 149 putative proteins (Heidelberg et al., 2002, Kolker et al., 2005). The similar scenario was also seen with *V. cholerae* N16961, its closely-relative with the same marine origin, in that it carries two genomes (one of which is 2.98 Mb with 47.7% GC contents and encodes 2690 genes, and other one is 1.07 Mb (46.9% GC percentage) corresponding to 1003 genes (Heidelberg et al., 2000).

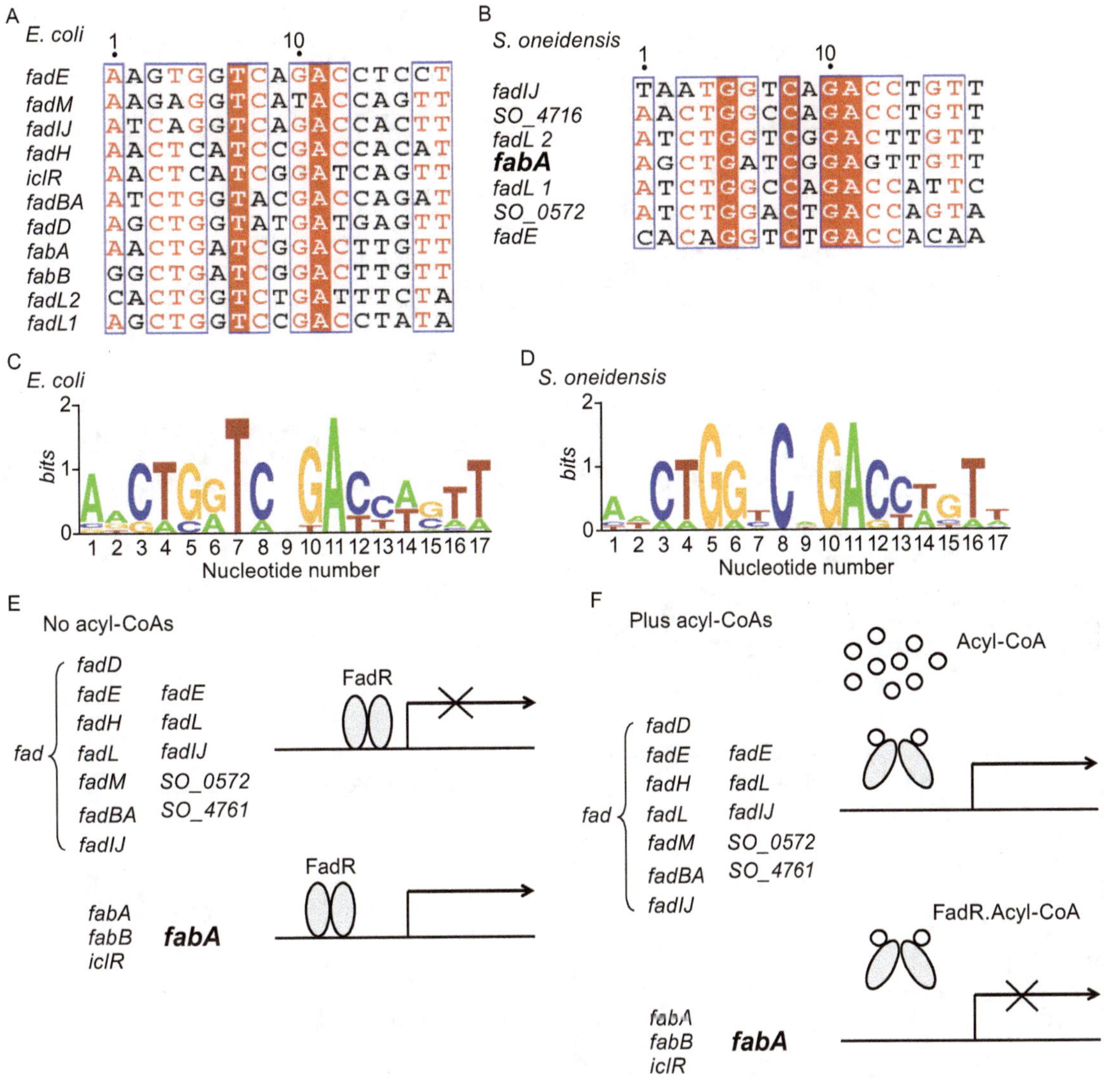

Figure 1. The working model proposed for *fad* regulon and its regulation in *Shewanella* genus. Multiple sequence alignments (A) and sequence logo (C) of the known palindromes recognized by *E. coli* FadR. Sequence analyses (B) and sequence logo (D) of the predicted FadR-binding sites of *Shewanella*. The alignment of DNA sequences was carried out using ClustalW2 (http://www.ebi.ac.uk/Tools/clustalw2/index.html), and the output was given via processing with the program ESPript 2.2. (http://espript.ibcp.fr/ESPript/cgi-bin/ESPript.cgi) (Feng & Cronan, 2011b). Identical residues are in white letters with red background, similar residues are in the form with mixture of red/black letters, and the varied residues are in black letters. Sequence logo of the FadR-binding sites was generated using the program of WebLogo (http://weblogo.berkeley.edu/logo.cgi). The sequences of the known *E. coli* FadR sites were sampled from *E. coli* K-12 MG1655 (http://regprecise.lbl.gov/RegPrecise/regulon.jsp?regulon_id=10286), and the putative *Shewanella* FadR sites were collected from *S. oneidensis* MR-1(http://regprecise.lbl.gov/RegPrecise/regulon.jsp?regulon_id=5431). (E) In the absence of a long chain acyl-CoA, *E. coli* FadR of *E. coli* and *S. oneidensis* represses the *fad* regulon genes, whereas it activates transcription of *fabA* (and/or *fabB*) with critical roles in the unsaturated fatty acid synthetic pathway. (F) Binding of long chain acyl-CoA species leads to the release of FadR protein from its operator sites. The *fad* members of *Shewanella* are in red except that *fabA* is highlighted in green. Such kind of FadR-DNA dissociation increases *fad* regulon expression whereas reduces the expression of *fabA* (and/or *fabB*). The oval denotes FadR regulatory protein whereas the small open circle represents the acyl-CoA pool.

In relative to *E. coli* that has no less than 12 *fad* members, those genes controlled by the fatty acid-responsive FadR regulator (Fig. 1A, 1C and 1E), the *fad* members seemed to be contracted in the cousin *S. oneidensis* in that only 4 well-known *fad* genes/operons (*fadE*, *fadL*, *fadIJ* and *fabA*) has the putative FadR-binding sites (Fig. 1B, 1D and 1F). Also,

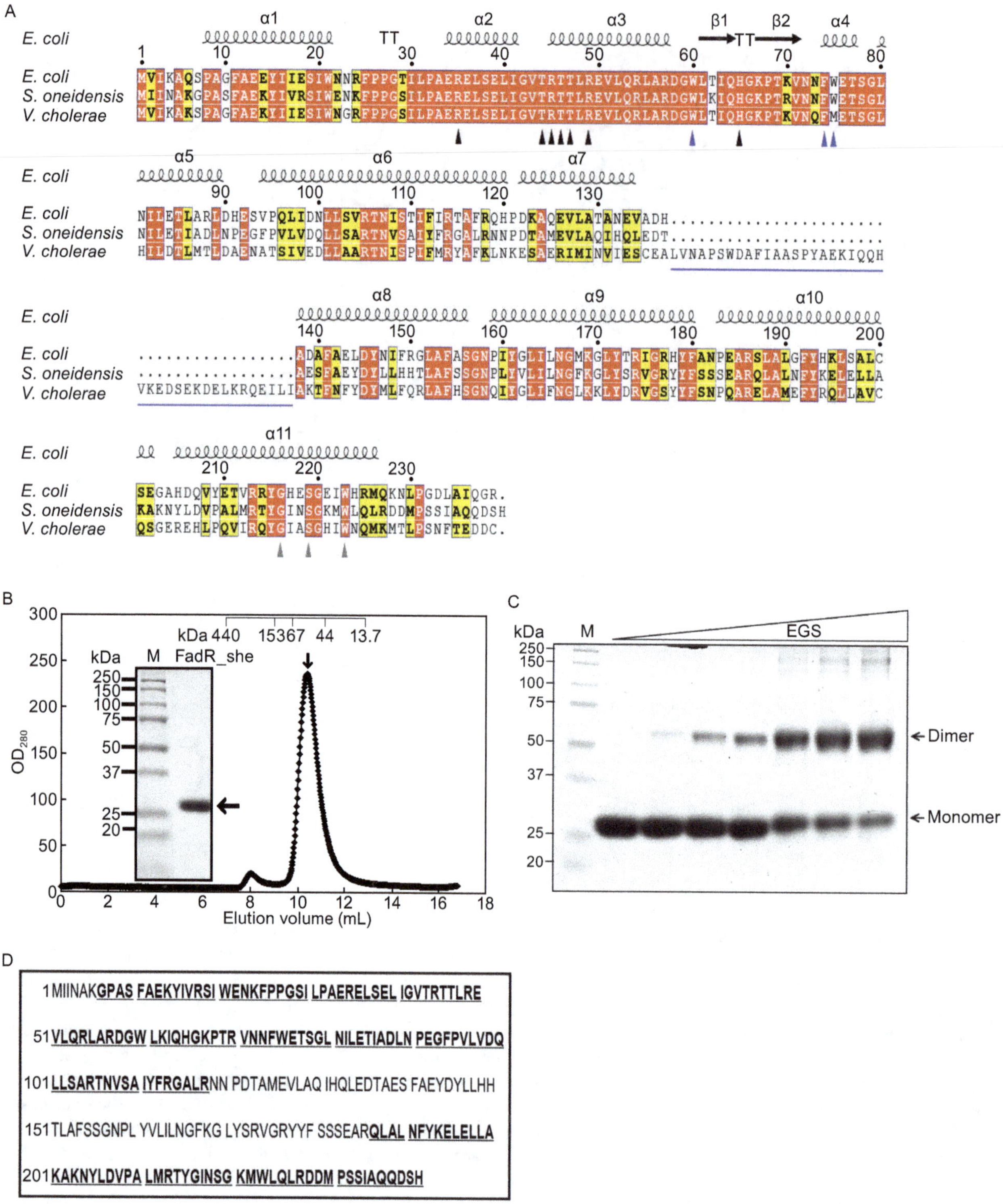
A
E. coli
S. oneidensis
V. cholerae
B
kDa M FadR_she
kDa 440 153 67 44 13.7
OD280
Elution volume (mL)
C
kDa M
EGS
Dimer
Monomer
D
1 MIINAKGPAS FAEKYIVRSI WENKFPPGSI LPAERELSEL IGVTRTTLRE
51 VLQRLARDGW LKIQHGKPTR VNNFWETSGL NILETIADLN PEGFPVLVDQ
101 LLSARTNVSA IYFRGALRNN PDTAMEVLAQ IHQLEDTAES FAEYDYLLHH
151 TLAFSSGNPL YVLILNGFKG LYSRVGRYYF SSSEARQLAL NFYKELELLA
201 KAKNYLDVPA LMRTYGINSG KMWLQLRDDM PSSIAQQDSH

◀ **Figure 2. Characterization of *S. oneidensis* FadR protein.** (A) Sequence analyses of three different FadR homologues. The multiple alignments of FadR protein sequences were performed using ClustalW2 (http://www.ebi.ac.uk/Tools/clustalw2/index.html), and the resultant output was processed by program ESPript 2.2 (http://espript.ibcp.fr/ESPript/cgi-bin/ESPript.cgi), generating the final BLAST photography (Feng & Cronan, 2011b). Identical residues are in white letters with red background, similar residues are in black letters in yellow background, the varied residues are in grey letters, and gaps are denoted with dots. In light of the structural architecture of *E. coli FadR* protein (PDB:1E2X) (van Aalten et al., 2000), the protein secondary structure was illustrated in cartoon (on top) (Zhang et al., 2014), α: alpha-helix; β: beta-sheet; T: Turn; η: coil. The seven known DNA-binding sites (R35, T44, R45, T46, T47, R49 and 65H) are highlighted with black triangles (Xu et al., 2001), the three known ligand-binding sites are shown with grey triangles (216G, 219S and 223W) (van Aalten et al., 2001), and the newly-proposed amino acids with indirect role for FadR-DNA interaction are highlighted with blue arrows (W60, F74 and W75) (Zhang et al., 2014). The extra 40-aa (138–177) longer region of *V. cholerae* FadR was underlined in blue. The FadR sequences are separately sampled from *E. coli* K12 (Accession no.: CAA30881), *V. cholerae* (*Vibrio cholerae*) (Accession no.: AAO37924) and *S. oneidensis* (Accession no.: NP_718457). (B) Gel exclusion chromatographic profile of the recombinant *S. oneidensis* FadR protein run on a Superdex 75 column (GE Healthcare). The expected peak of the target FadR was eluted at the position of 10.5 mL (highlighted with an arrow). The inset gel is the 15% SDS-PAGE photography of the collected *S. oneidensis* FadR protein sample. The mass of the monomeric *S. oneidensis* FadR is estimated to be ~27 kDa. Abbreviations: M, protein marker; OD_{280}, optical density at 280 nm; mAu, milli-absorbance units. The ruler on the top was given to describe the elution pattern of the standard proteins (Pharmacia). The standards used here included Ferritin (~440 kDa), Aldolase (153 kDa), Bovine serum albumin (~67 kDa), Ovalbumin (~44 kDa) and ribonuclease (~13.7 kDa), respectively. (C) Chemical cross-linking analyses for the purified *S. oneidensis* FadR protein. The level of EGS chemical cross-linker was illustrated with a triangle varies from 0, 0.1, 0.2, 0.5, 1.0, 1.5, to 2.0 μmol/L. (D) MS determination of the recombinant *S. oneidensis* FadR protein. The matched amino acid residues that exhibited 69% coverage to the native *S. oneidensis* FadR are given bold and underlined type.

S. oneidensis possesses two more new FadR-regulated genes (SO_4761 encoding the GNAT family of Acetyltransferease and SO_0572 representing a possible Enoyl-CoA hydratase (EC 4.2.1.17)) (Fig. 1B and 1F), somewhat suggesting the expansion of limited *fad* members in this marine bacterium. However, this observation argues the possibility of gene horizontal transfer in that the GC contents of two genes SO_4761 (45.23%) and SO_0572 (46.84%) is similar to that of the whole genome (46%). Unlike *E. coli* that encodes only one FadL fatty acid transporter (Blattner et al., 1997), *S. oneidensis* has three FadL-like homologues (FadL-1 (SO_3099, 440 aa), FadL-2 (SO_3276, 311 aa) and FadL-3 (SO_4232, 437 aa)) (Heidelberg et al., 2002), only FadL-1 of which is directly regulated by FadR regulatory protein (Fig. 1B and 1F). This situation seems unusual, but not without any precedent. The similar scenario was observed in *V. cholerae*, the other marine bacterium since three FadL orthologues are distributed in its two chromosomes (Heidelberg et al., 2000), and only FadL-2 is regulated by FadR repressor (http://regprecise.lbl.gov/RegPrecise/regulon.jsp?regulon_id=16367). It is reasonable that three FadL transporters coupled with one FadD inner-membrane protein (3FadL-1FadD) constitute a more efficient system of fatty acid uptake than the prototypical version of 1FadL-1FadD in *E. coli*. Unlike the *V. cholerae* FadL-2 that has only one FadR-recognizable site not shown), *S. oneidensis* FadL-1 exhibits two tandem FadR-binding sites (Fig. 1B), similar to the scenario seen in *E. coli* FadL (Fig. 1A). Different from the *E. coli fadD* that also carries two tandem FadR-specific palindromes, the *fadD* gene with origins of both *S. oneidensis* and *V. cholerae* seems not to be regulated by the FadR regulator in that the typical site is cryptic (Fig. 1B). In views of genomic evolutions, we anticipated that *S. oneidensis* has the relics of both *E. coli* and *Vibrio* in the context of fatty acid transporter system.

Given the significant difference of their inhabiting environments/ecological niches (*E. coli* is enteric bacterium living in fatty acids-rich gastro-intestine, whereas *Vibrio* and *Shewanella* both inhabit in the environment of fresh/salt water with poor fatty acids (Giles et al., 2011).), we believed that this kind of fatty acid uptake system might represent an evolutional/physiological advantage for these marine bacteria to scavenge the limited availability of exogenous fatty acids. Intriguingly, comparative analyses of the GC percentage, an indicator of gene horizontal transfer, showed that FadL-2 (48.61%) is appreciably lower than the average value of the whole genome (46%), implying it might be obtained by gene horizontal transfer, whereas FadL-1 (45.43%) and FadL-3 (46.73%) not (not shown).

In generally consistence with the earlier observations with *V. cholerae* (Feng & Cronan, 2011a, Rodionov et al., 2011), we noted that only the *fabA* gene from the FabA-FabB UFA synthesis pathway in *Shewanella* has a known FadR-binding site (Figs. 1B and S1A), whereas the *fabB* gene does not (Fig. S1B). It suggested the presence of an asymmetric/unparalleled FadR regulation in *Shewanella* (Fig. 1E and 1F). these findings might argue the conclusions by Shi and coauthors (Shi et al., 2015) in the case of the regulated-expression of the *fabB* gene in its closely-relative, *V. cholerae*, in that only indirect role of FadR can be assigned due to the absence of the FadR-binding site. Our observations plus the predictions by Rodionov et al. (Rodionov et al., 2011), supported the proposal that the *fad* regulon contraction is present in *Shewanella*. Given the important physiological role of the *fabA* gene in bacterial UFA synthesis, we attempted to experimentally verify this unusual hypothesis.

Characterization of *Shewanella* FadR

An earlier study (Iram & Cronan, 2005) has found that the FadR lipid metabolism regulator of *V. cholerae* has an unusual insert of 40 residues. Our results (submitted) plus Shi's observations (Shi et al., 2015) revealed an unexpected contribution of this unique inserting sequence in constituting an extra-ligand binding motif for FadR regulatory protein. The second ligand-binding site confers its excellent ability in fatty acid sensing. Given the fact that both *V. cholerae* and *S. oneidensis* are closely-related marine bacteria that shared a similar ecological niche with poor availability of fatty acids, we initially anticipated that this insert might be an indicator or relic for such kind of unparalleled regulation by FadR (Fig. 1). In fact, it is not this case. Multiple sequence alignments of three FadR proteins (FadR_ec for *E. coli*, FadR_vc for *V. cholerae*, and FadR_she for *S. oneidensis*) showed that: 1) the N-terminal DNA-binding motifs are very conserved featuring a full set of all the known residues critical for DNA binding; 2) the C-terminal ligand-interacting domains are appreciably diversified; and 3) the so-called insert of 40 residues (138–177 aa) is only present in FadR_vc (Fig. 2A). Considered the fact combined with atypical features seemed in fatty acid transport system, we favored the anticipation that *Shewanella* somewhat retains the evolutional relic that are partially observed with *E. coli* and *Vibrio*, respectively.

To further functional analyses of the above bacterial FadR proteins, we over-expressed the three types of recombinant FadR proteins (FadR_ec FadR_vc & FadR_she) and purified them to homogeneity. As expected, SDS-PAGE profile clearly showed the purified FadR_she protein migrates at the position of ~27 kDa. The FPLC profile showed that the expected peak of purified *S. oneidensis* FadR was eluted at the position of 10.5 mL (indicated with an arrow, Fig. 2B), suggesting its apparent molecular mass is more than 44 kDa, but less than 67 kDa. Given the fact that the ideal molecular weight of recombinant *S. oneidensis* FadR in momomer is ~27 kDa, we believed that the form of FadR_she in solution might be a dimer (-54 kDa). It was generally consistent with the scenario seen with the *E. coli* FadR as a dimer. Subsequently, we used chemical cross-linking assays to further prove this speculation. As we expected, appearance/formation of the dimerization for the FadR_she protein is appreciably increased upon addition of chemical cross-linker EGS (Fig. 2C). Also, it behaves in an EGS dose-dependent manner (Fig. 2C). In particular, the dimer band was excised from the SDS-PAGE and subjected to liquid chromatography mass spectrometry. As a result, the MS results confirmed this identity in that the digested peptides matched the *S. oneidensis* FadR protein with the coverage of 69% (Fig. 2D).

The FadR proteins of *E. coli* and *S. oneidensis* are functionally-equivalent

Gel shift assay was performed to detect the binding ability of FadR_ec, FadR_vc and FadR_she to the cognate DNA binding sites. As expected, EMSA-based experiments showed that the *E. coli* FadR protein (as the positive control) binds well to its own promoters of both *fabA* (Fig. 3A) and *fabB* (Fig. 3B) promoters. The fact that the FadR_vc protein gives consistently the super-shift bands for both *fabA* and *fabB* probes in the gel shift assays, is mostly attributed to the essence of its easy-forming the protein multimer (Fig. 3). Of note, FadR_she exhibited an excellent ability of interacting with the *fabA* (Fig. 3A) and *fabB* (Fig. 3B) with the origin of *E. coli*. It seemed likely that the FadR proteins of *E. coli* and *S. oneidensis* are functionally-equivalent (and/or exchangeable). It is not surprise since the FadR/FabR orthologue from other marine bacterium, *Vibrio*, also followed this rule (Feng

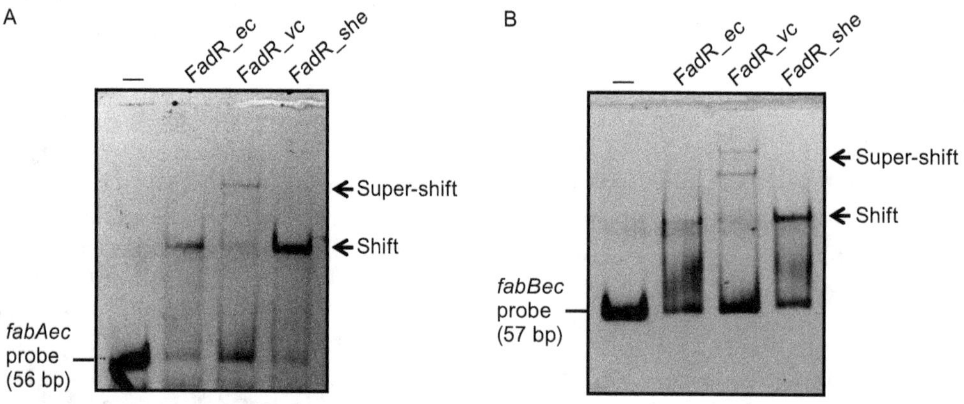

Figure 3. ***Shewanella* FadR protein is functionally-exchangeable to the paradigm *E. coli* version.** (A) EMSA-based evidence for binding of *E. coli fabA* promoter to FadR protein of three origins. (B) EMSA analyses for crosstalk of *E. coli fabA* promoter with three kinds of bacterial FadR proteins. A representative photography was given here, which were from no less than three independent EMSA experiments (7% native PAGE). Three versions of FadR protein here are FadR_ec, FadR_vc and FadR_she, respectively. In gel shift assays, the FadR protein (5 pmol) is incubated with DIG-labeled *fabA*ec (or *fabB*ec) probe (0.2 pmol). Note: An unexpected but interesting scenario "super-shift" is consistently observed in our trials with *V. cholerae* FadR.

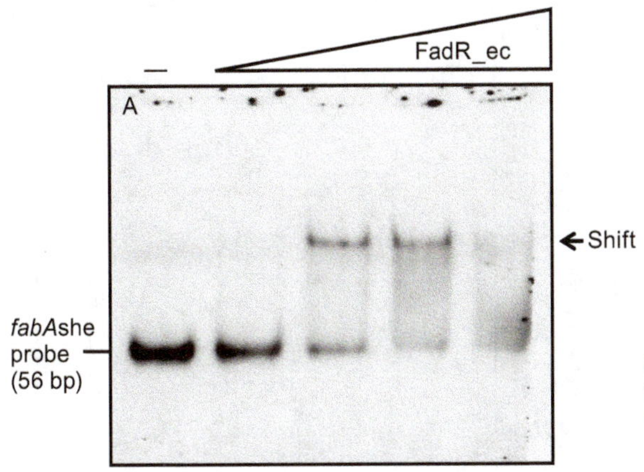

Figure 4. Evidence that *Shewanella fabA* promoter has a functional FadR-recognizable palindrome. (A) Binding of *Shewanella fabA* promoter to *E. coli* FadR protein. (B) Interplay between *Shewanella fabA* promoter and *Shewanella* FadR protein. The gel shift tests were conducted using 7% native PAGE, and a representative result is shown here. In these assays, levels of FadR protein (FadR_ec and FadR_she) added are denoted with a triangle on right hand (0.1, 0.5, 2, and 5 pmol), whereas the DIG-labeled *fabB*she probe is added to 0.2 pmol. Minus sign denotes no addition of FadR protein.

& Cronan, 2011b, Feng & Cronan, 2011a). To our knowledge, the cases of similar functional exchange of transcriptional regulators can be extended to BioR, the other GntR-type regulators implicated into the metabolism of biotin, a sulfur-containing fatty acid (Feng et al., 2013a, Feng et al., 2013b). Thereby, it makes sense that the atypical regulation by FadR in UFA synthesis of *Shewanella* is due to the cryptic site in front of *fabB* (Fig. 1), not FadR_she (Fig. 3).

S. oneidensis fabA has a functional FadR binding site, and this binding is specifically reversed by long-chain acyl-CoA

Through sequence comparison of the *fabA* and *fabB* promoter regions of *E. coli*, *V. cholerae* and *S. oneidensis*, we found that the cognate FadR-specific binding site in front of the *fabA* promoter regions of these three bacterial species are much more conservative (Fig. S1A), but that of *fabB* promoter region is not (Fig. S1B). This observation is generally consistent with the prediction by Rodionov and coworkers (Rodionov et al., 2011) that only *fab*A (not *fabB*) of *Shewanella* has a binding site for the FadR regulator. To further prove the function of this predicted site, termed *fabA* probe, we synthesized it using the approach of annealing the two complementary DNA strand. This DNA probe is digoxigenin-labeled DNA fragment of 56 bp that overlaps the candidate FadR_she binding site (Fig. S1A and Table 2). Gel shift assays confirmed that FadR_ec (Fig. 4A) and FadR_she (Fig. 4B) both can efficiently bind the *S. oneidensis fabA* promoter. In much similarity to the scenario seen with FadR_ec here (Fig. 4A), plus our former observations with FadR regulators of *E. coli* (Feng & Cronan, 2009b, Feng

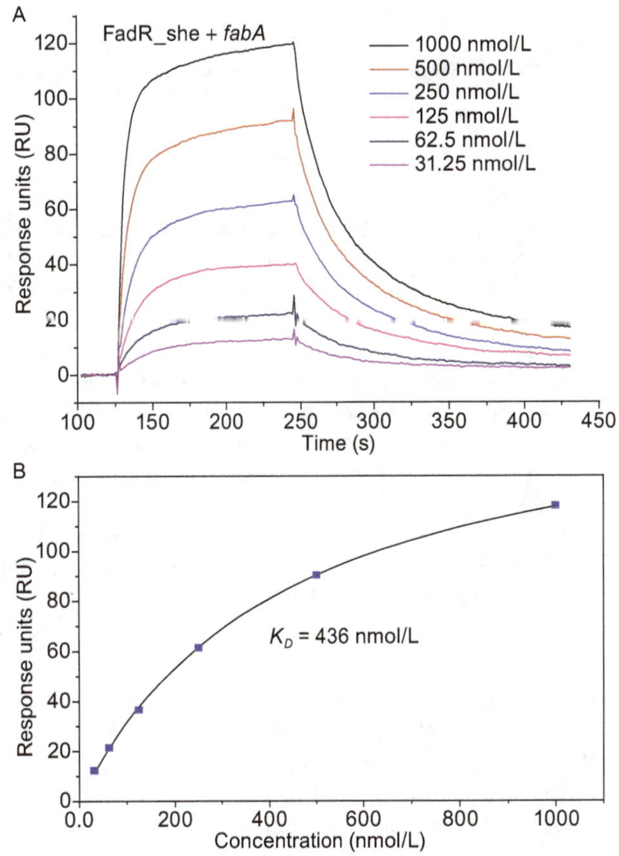

Figure 5. SPR-based dynamic analyses for binding of *fabA* to *Shewanella* FadR. (A) SPR assay for interaction between *fabA* and FadR_she. (B) Measurement of the KD value for *fabA*-FadR_she.

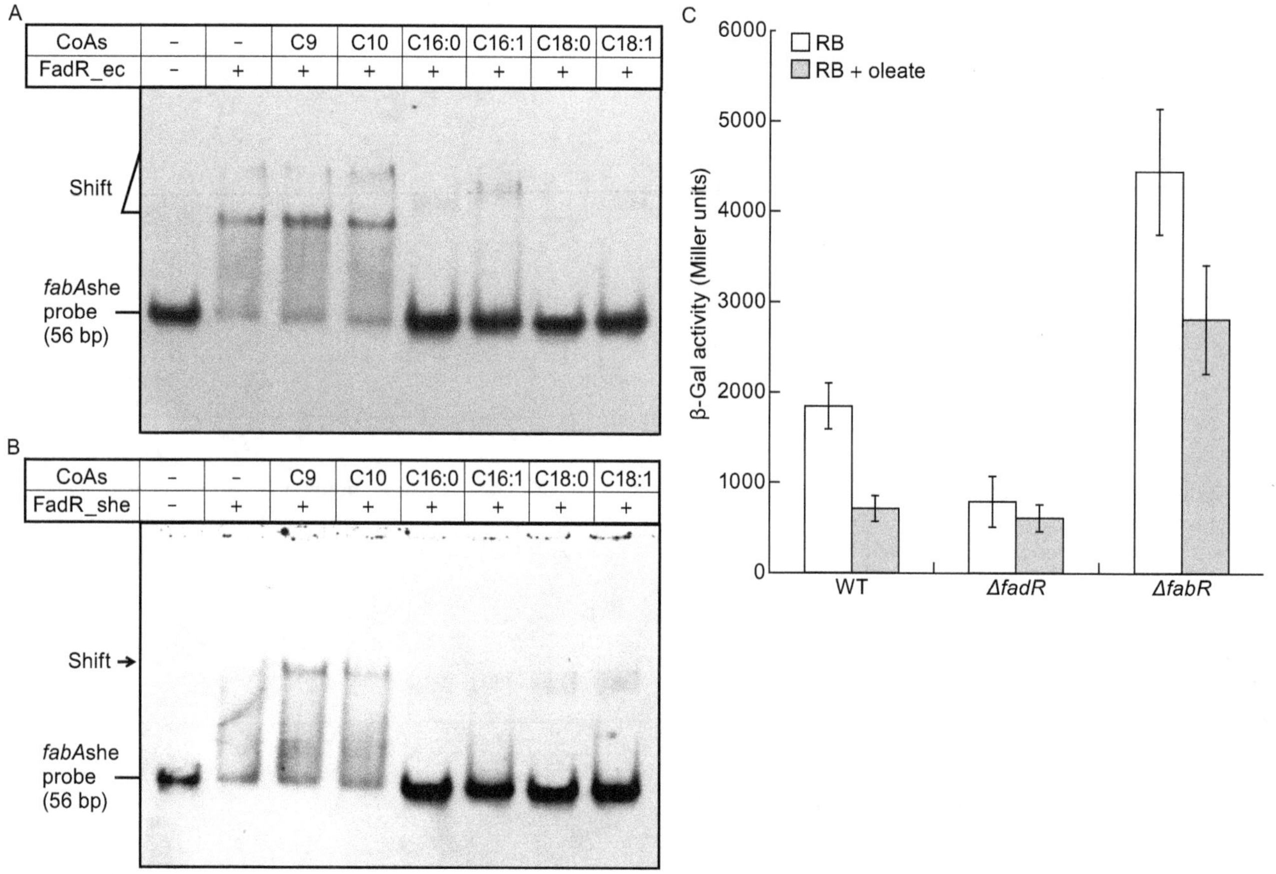

Figure 6. Role of LC fatty acyl-CoA in *fabA* expression *in vitro* and *in vivo*. (A) EMSA-based visualization for effects of medium and long chain acyl-CoA species on binding of FadR_ec to the *fabA*she probe. **(**B) Effects of different long chain acyl-CoA species on binding of FadR_she to the *fabA*she probe. In the binding reaction mixtures (10 μL in total), the FadR (~5 pmol) was incubated with 0.2 pmol of DIG-labeled *fabA*she probe. When required, acyl-CoA (~50 pmol) was added as we recently described (Feng & Cronan, 2011b). The gel shift assays were conducted for more than three times using 7% native PAGE, and the representative result is given. The shifted *fabA*she probe band is indicated with a triangle (A) or an arrow (B). Designations C9:0, nonanoyl-CoA; C10:0, decanoyl-CoA; C16:0, palmitoyl-CoA; C18:0, stearoyl-CoA; C18:1, oleoyl-CoA. Abbreviations: ec and she denote *E. coli* and *Shewanella*, respectively. Plus sign denotes addition of either FadR proteins or acyl-CoA species, whereas minus sign denotes no addition of either FadR protein or acyl-CoA species. (C) Transcription of *fabA*she is activated by FadR in *E. coli* and repressed upon oleic acid supplementation. All the *E. coli* strains used here carried a single copy of *fabA*she-*lacZ* transcriptional fusion which is integrated on chromosomes. Bacterial cultures in mid-log phase were collected for assaying the LacZ (β-gal) activity. The three strains used here included FYJ241 (WT), FYJ246 (Δ*fadR*), and FYJ247 (Δ*fabR*), respectively. As anticipated, transcription of *fabA* she is also negatively regulated by FabR in *E. coli*.

& Cronan, 2010) and *V. cholerae* (Feng & Cronan, 2011b), we also found that the *fabA*_she probe binds FadR_she protein in a dose-dependent manner (Fig. 4B). To preliminarily elucidate the kinetics of FadR_she/*fabA*_she interaction, we conducted surface plasom resonance (SPR)-based measurements. SPR results revealed that the binding affinity (K_D) of fabA_she to FadR_she is roughly 436 nmol/L (Fig. 5A and 5B), and the binding mode is 2:1 (a dimeric FadR protein is bound to a target DNA fragment (not shown).

Given the fact that long-chain (but not short chain) fatty acyl-CoA species can antagonize the DNA-binding activity of FadR with origins of *E. coli* (Henry & Cronan, 1992, Cronan, 1997) and *Vibrio* (Iram & Cronan, 2005, Feng & Cronan, 2011b), it is of much interest to test the behaviors of theses ligands in the case of *S. oneidensis* FadR. Therefore, we tested six acyl-CoA species of different acyl chain lengths. The EMSA-based competition assays showed that medium-chain acyl-CoA (C9:0; C10:0) don't interfere with the *fab*A_she binding to either FadR_ec (Fig. 6A) or FadR_she (Fig. 6B). In contrast, the long-chain acyl-CoA species (C16:0, C16:1, C18: 0 and C18:1) strongly impaired DNA binding (Fig. 6A and 6B). We believed that long-chain but not

Table 1. Strains and plasmids used in this study

Bacteria or plasmids	Relevant characteristics	Sources
Bacterial strains		
Shewanella oneidensis MR-1	A Gram-negative anaerobic bacteria which is predominantly found in deep sea anaerobic habitats	Heidelberg et al. (2002), CGSC[a]
BL21(DE3)	An expression host for recombinant plasmids	Lab stock
MFH8	UB1005, Δ*fadR::*Tn*10*	Henry & Cronan (1992)
SI91	UB1005, Δ*fabR::*Cm	Feng & Cronan (2009b)
FYJ187	MC4100/pINTts	Feng & Cronan (2011b)
FYJ189	BL21 carrying pET28a-*fabR*she	This work
FYJ214	BL21 carrying pET28a-*fadR*she	This work
FYJ193	DH5α (*λ-pir*)	Lab stock, (Feng & Cronan, 2009a, Feng & Cronan, 2011b)
FYJ236	DH5α (*λ-pir*) carrying pAH-P*fabA*she	This work
FYJ241	MC4100 with a single copy of *fabA*she-*lacZ* fusion integrated at the λ-site	This work
FYJ246	FYJ241, Δ*fadR* : Tn*10*	FYJ241/$P1_{vir}$(MFH8)
FYJ247	FYJ241, Δ*fabR* : Cm	FYJ241/$P1_{vir}$(SI91)
Plasmids		
pET28(a)	T7 promoter-driven expression vector, Kan^R	Novagen
pAH125	Promoter-less *lacZ* reporter plasmid in *E. coli*, Kan^R	Feng & Cronan (2009a), Haldimann & Wanner (2001)
pET28-*fadR*ec	Recombinant plasmid carrying *E. coli fadR*, Kan^R	Cherepanov & Wackernagel (1995), Feng & Cronan (2009b), Feng & Cronan (2009a)
pET16-*fadR*vc	Recombinant plasmid carrying *V. cholerae fadR*, Kan^R	Feng & Cronan (2011b), Iram & Cronan (2005)
pET28-*fadR*she	Recombinant plasmid carrying *Shewalle fadR*, Kan^R	This work
pAH-P*fabA*she	Recombinant plasmid carrying *Shewalle fabA* promoter region, Kan^R	This work

[a] CGSC denotes Coli Genetic Stock Center, Yale University.

medium-chain acyl-CoA can specifically interact with FadR_she ligand-binding site and release FadR_she from the *S. oneidensis fabA* promoter. In summary, the *in vitro* data accumulated suggest that long-chain acyl-CoAs regulate the *Shewanella fabA* transcription via their interaction with the FadR protein.

Expression of *Shewanella fab*A is activated by FadR, but repressed by oleate in *E. coli*

It is well known that: 1) FadR acts as an activator for expression of *fabA* and *fabB*, the two genes required for *E. coli* UFA synthesis, 2) whereas FabR is a repressor for expression of *fabA* and *fabB*. We construct three *E. coli* strains including FYJ241 (WT), FYJ246 (▵*fadR*) and FYJ247 (▵*fabR*). The three strains all carried a single copy of *fabA*she-*lacZ* transcriptional fusion on chromosomes which allows us to detect whether the *E. coli* counterpart of FadR regulatory proteins has the *in vivo* role in modulating the *Shewanella fabA* expression and to monitor the physiological effect on *fabA*_she transcription exerted by exogenous fatty acids. Measurement of the β-Gal levels of these strains showed that deletion of *fadR* eliminate its activation to *fabA*_she expression (Fig. 6C). In comparison with the wild-type strain, appreciable lower β-gal activity was seen in the ▵*fadR* mutant (Fig. 6C). In contrast, the removal of the opposite regulator, FabR repressor, gave significant increment of *fabA*_she expression (Fig. 6C). As expected, the activity of *fabA*_she promoter is inhibited by the addition of oleate and this down-regulation is dependent on the presence of FadR regulator (Fig. 6C). Thus, our results represent *in vivo* evidence that expression of *Shewanella fab*A is activated by FadR, but repressed by oleate.

CONCLUSIONS

The data reported here defined that the *Shewanella* FadR homologue is a functional regulator with the involvement of fatty acid metabolism. Also, we experimentally proved the proposal by Rodionov et al. (Rodionov et al., 2011) that *fad* regulon is contracted in *Shewanella*. Retrospectively, Wang

Table 2. Primers used in this study

Primers	Primer sequences
*fadR*_she-F	5′-CG GGATCC ATG ATT ATC AAT GCC AAA GGA CC-3′
*fadR*_she-R	5′-CCG CTCGAG CTA ATG GGA GTC CTG CTG TG-3′
*fabR*_she-F	5′-CG GGATCC ATG GGT ATT CGT GCA CAG CA-3′
*fabR*_she-R	5′-CCG CTCGAG CTA CCG ATG TTC AAC TTT ATG T-3′
*fabA*_she-BD-F	5′-GAC ATT AAT T**AG CTG ATC GGA GTT GTT** T***AG CTT ACA CGT GTT CGC T***AA TCT TGG CG-3′
*fabA*_she-BD-R	5′-CGC CAA GAT T***AG CGA ACA CGT GTA AGC T*****AA ACA ACT CCG ATC AGC T**AA TTA ATG TC-3′
*fabA*_ec-BD-F	5′-TTT ATT CCG **AAC TGA TCG GAC TTG TT**C ***AGC GTA CAC GTG TTA GCT*** ATC CTG CGT GC-3′
*fabA*_ec-BD-R	5′- GCA CGC AGG AT***A GCT AAC ACG TGT ACG CT***G **AAC AAG TCC GAT CAG TT**C GGA ATA AA-3′
*fabB*_ec-BD-F	5′-TCT ATT AAA T**GG CTG ATC GGA CTT GTT** C***GG CGT ACA AGT GTA CGC T***AT TGT GCA TTC-3′
*fabB*_ec-BD-F	5′-GAA TGC ACA AT***A GCG TAC ACT TGT ACG CC***G **AAC AAG TCC GAT CAG CC**A TTT AAT AGA -3′
*PfabA*she-F	5′-CCG GTCGAC GAG GGT TAA CGG GTA AAC AAG-3′
*PfabA*she-R	5′-AACC GAATTC GTC GAT CAT CAG CAT GTT GTC-3′
LacZ-R	5′-GAC CAT GAT TAC GGA TTC ACT G-3′

The sequences underlined are restriction sites. Putative FadR-binding sites are in bold letters, and the predicted FabR palindromes are indicated in bold and italic.

et al. (Wang et al., 2009) reported the pilot exploration of fatty acid metabolism in *Shewanella piezotolerans* with concentration on its relevance to different temperatures and pressures.

Very recently, Gao's research group also provided genetic evidence in aiming to pose its role of *fadR* into the UFA pathway in *S. oneidensis* MR-1 (Luo et al., 2014). Right now, the picture of *Shewanella* FadR seemed to be much more complete in the context of lipid metabolism. Unlike the paradigm *E. coli* in which both *fabA* and *fabB* have the FadR-binding sites, *Shewanella* only rendered the *fabA* fatty acid synthesis gene under the control by the FadR activator (Figs. 1 and 6C) and the FabR repressor (Fig. 6C). Moreover, it is in rational to functionally assign this unparalleled regulation to its unique evolutional selection and even adaptation to its growing (/neighboring) environmental/ecological niches with the availability of limited fatty acids. Taken together, we provided integrative evidence that *Shewanella* FadR binds to the *fabA* fatty acid biosynthetic gene, which is implicated into contraction of the *fad* regulon.

MATERIALS AND METHODS

Bacterial strains and growth conditions

With an exception of *Shewanella oneidensis* MR-1 (*S. oneidensis*), all of the bacterial strains used here were *E. coli* K-12 derivatives (Table 1). The media are separately LB medium (10 g of tryptone, 5 g of yeast extract and 10 g of NaCl per liter), and rich broth (RBO medium;10 g of tryptone, 1 g of yeast extract and 5 g of NaCl per liter). To measure the activity of β-galactosidase in induction assays, oleate was solubilized with Tergitol NP-40 and used at a 5 mmol/L final concentration. Antibiotics were supplemented as follows (in mg/L): sodium ampicillin, 100; kanamycin sulfate, 50; tetracycline HCl, 15; and chloramphenicol, 20.

Plasmids and DNA manipulations

The *fabA* promoter region of *S. oneidensis* was PCR amplified and directly cloned into pAH125, giving the recombinant plasmid pAH125-P*fabA*she (Table 1). Similarly, the *fadR* gene amplified from *S. oneidensis* was inserted into expression vector pET28(a), generating the chimeric plasmid pET28-*fadR*she (Table 2). To prepare three different versions of FadR proteins, the corresponding expression plasmids (pET28-*fadR*ec, pET28-*fadR*she and pET16-*fadR*vc, in Table 2) were transformed into the strain BL21 (DE3) (Feng & Cronan, 2009b). The acquired plasmids were verified by direct DNA sequencing.

Given that the replication of pAH-P*fabA*she plasmid requires the presence of *pir* protein, it thus was maintained in strain DH5α *λ-pir* (Table 1). To impart antibiotic resistance in *E. coli* MC4100 (a *lacZ* strain lacking *pir*), this plasmid must specifically integrate into the *attλ* site of bacterial chromosome in a reaction catalyzed by the pINT-ts helper plasmid, giving strain FYJ241 carrying *fabA*she-*lacZ* transcriptional fusion (Table 1). PCR assay was applied to confirm the *fabA*she-*lacZ* junction.

*P1*vir phage transductions

*P1*vir transductions were conducted as described by Miller (Miller, 1992) with little changes. The strain FYJ246 was constructed by P1vir transduction of strain FYJ241 using a lysate grown on strain MFH8 (Δ*fadR*::Tn*10*) with selection for tetracycline resistance. Similarly, strain FYJ241 was transduced by *P1*vir lysate obtained from strain SI91 (Δ*fabR::*Cm) with selection for kanamycin, giving strain FYJ247 (Table 1). The relevant genotypes of the acquired strains were proved by PCR analyses.

β-Galactosidase assays

Mid-log phase cultures grown in either LB or RB were collected by spinning, washed with Z-buffer and suspended in Z-buffer for further measurement of β-galactosidase activity (Feng & Cronan, 2009b, Feng & Cronan, 2009a, Miller, 1992). The data were recorded in triplicate in no less than three independent assays.

Expression and purification of three different FadR proteins

In addition to the FadR protein with origins of both *E. coli* and *V. cholerae*, the *S. oneidensis* FadR protein was produced in solubility via the induced expression with 0.2 mmol/L isopropyl β-D-1-thiogalactopyranoside (IPTG) at 30°C for 3.5 h. The bacterial lysis by two rounds of sonication treatment was clarified by centrifugation, and the resultant supernatant was loaded onto a nickel-ion affinity column (Qiagen). The contaminant proteins were removed with wash buffer containing 50 mmol/L imidazole, and subsequently the 6× His-tagged FadR proteins in three versions (FadR_she, FadR_ec and FadR_vc) were eluted in elution buffer containing 100 mmol/L imidazole. The protein was concentrated by ultra-filtration (30 kDa cutoff) and exchanged into 1× PBS buffer (pH 7.4) containing 10% glycerol. The purified proteins were visualized by 15% SDS-PAGE followed by staining with Coomassie Brilliant Blue R250 (Sigma, St. Louis, MO).

Size exclusion chromatography

Given the fact that both FadR_ec and FadR_vc can form dimer, we aimed to check the solution structure of FadR_she. Therefore, we used a Superdex 75 column (Pharmacia) run on an Äkta fast protein liquid chromatography system (GE Healthcare) (Feng & Cronan, 2010, Feng & Cronan, 2011a) to perform gel filtration analyses for the purified FadR_she protein. In our trials, the column effluent was monitored at a flow rate of 0.35 mL/min in running buffer (20 mmol/L Tris-HCl, 50 mmol/L NaCl, pH 7.9). The peaks of interest were collected and confirmed with 15% SDS-PAGE.

Liquid chromatography quadrupole time-of-flight mass spectrometry

The identity of the recombinant FadR_she protein we produced was confirmed using A Waters Q-Tof API-US Quad-ToF mass spectrometer connected to a Waters nano Acquity UPLC) (Feng & Cronan, 2011a). In brief, the protein band of interest was cut from 15% SDS-PAGE gel, de-stained and digested with Sequencing Grade Trypsin (G-Biosciences St. Louis, MO, 12.5 ng/μL in 25 mmol/L ammonium bicarbonate); Second, the resulting peptides were loaded on a Waters Atlantis C-18 column (0.03 mm particle, 0.075 mm × 150 mm), following the further cleaning treatment. The data dependent acquisition combined with ms/ms analysis was routinely performed (Feng & Cronan, 2011a).

Chemical cross-linking assays

To further test the solution structure of *S. oneidensis* FadR, we carried out the experiments of chemical cross-linking with ethylene glycol bis-succinimidylsuccinate (Pierce) as we described before (Feng & Cronan, 2010). In each chemical cross-linking reaction (15 μL in total), the purified FadR protein (~10 mg/mL) was separately mixed with cross-linker at different concentrations (0, 0.1, 0.2, 0.5, 1.0, 1.5 and 2.0 mmol/L), and kept 30 min at room temperature before analysis. Note: the only protein without EGS addition serves as the negative control. All the reaction products were detected using 15% SDS-PAGE.

Electrophoretic mobility shift assays

To document the function of FadR-binding site located in the *S. oneidensis fabA* promoter, gel shift assays were performed as we earlier described (Feng & Cronan, 2009b, Feng & Cronan, 2010, Feng & Cronan, 2011a) with minor modifications. Totally, three sets of DNA probes (Table 2) corresponded to *fabA*ec, *fabB*ec and *fabA*she, respectively. They were all generated by annealing two complementary oligonucleotides (e.g., *fabA*_she-BD-F plus *fabA*_she-BD-R, in Table 2) through the incubation at 95°C in TEN buffer (10 mmol/L Tris-HCl, 1 mmol/L EDTA, 100 mmol/L NaCl, pH 8.0) for 5 min followed by slow cooling to 25°C. The digoxigenin-labeled DNA probes (~0.2 pmol) were mixed with purified FadR (in appropriate concentrations) in the binding buffer (Roche) and incubated 20 min at room temperature. The DNA/protein mixtures were then analyzed by the native 7% PAGE, and directly transferred onto nylon membrane by contact blotting-aided gel transfer. Following appropriate treatments, the chemical-luminescence signals were captured by an exposure of the membrane to ECL films (Amersham).

Surface plasmon resonance

Biacore3000 instrument (GE Healthcare) was utilized to carry out the surface plasmon resonance (SPR)-based measurement. The biotinylated *fabA*_she DNA probe was immobilized by streptavidin on the chip surface. The SPR assay was run in the running buffer (20 mmol/L Tris-HCl, pH 7.5, 200 mmol/L NaCl and 0.005% Tween 20) at the flow rate of 30 μL/min. FadR_she protein in a series of dilution was injected and passed over the chip surface for 2 min. Kinetic parameters were analyzed using a global data analysis program (BIA evaluation software), and final graph was given with the Origin software.

Bioinformatic analyses

The amino acid sequences of FadR regulator are derived from *E. coli*, *V. cholerae* and *S. oneidensis* MR-1. The FadR-binding sites were all sampled from RegPrecise database (http://regprecise.lbl.gov/RegPrecise/regulon.jsp?). The multiple alignments were conducted using the program of ClustalW2 (http://www.ebi.ac.uk/Tools/clustalw2/index.html), and the resultant output was further processed by the program ESPript 2.2 (http://espript.ibcp.fr/ESPript/cgi-bin/ESPript.cgi), giving the final version of BLAST photography.

ACKNOWLEDGMENTS

This work was supported by Zhejiang Provincial Natural Science Foundation for Distinguished Young Scholars (Grant No.: LR15H190001), and the start-up package from Zhejiang University (Y.F.). Dr. Feng is a recipient of the "Young 1000 Talents" Award. We would like thank Mr. Yuan Lin for his help in SPR analyses.

ABBREVIATIONS

FAS, fatty acid synthesis; FPLC, fast protein liquid chromatography; IPTG, isopropyl β-D-1-thiogalactopyranoside; LCFA, long-chain fatty acid; UFA, unsaturated fatty acid.

COMPLIANCE WITH ETHICS GUIDELINES

Huimin Zhang, Beiwen Zheng, Rongsui Gao and Youjun Feng declare that they have no conflict of interest.

This article does not contain any studies with human or animal subjects performed by the any of the authors.

REFERENCES

Blattner FR, Plunkett G 3rd, Bloch CA, Perna NT, Burland V, Riley M, Collado-Vides J, Glasner JD, Rode CK, Mayhew GF, Gregor J, Davis NW, Kirkpatrick HA, Goeden MA, Rose DJ, Mau B, Shao Y (1997) The complete genome sequence of *Escherichia coli* K-12. Science 277:1453–1462

Cheng YY, Li BB, Li DB, Chen JJ, Li WW, Tong ZH, Wu C, Yu HQ (2013) Promotion of iron oxide reduction and extracellular electron transfer in *Shewanella oneidensis* by DMSO. PloS one 8:e78466

Cherepanov PP, Wackernagel W (1995) Gene disruption in *Escherichia coli*: TcR and KmR cassettes with the option of Flp-catalyzed excision of the antibiotic-resistance determinant. Gene 158:9–14

Cronan JE Jr (1997) In vivo evidence that acyl coenzyme A regulates DNA binding by the *Escherichia coli* FadR global transcription factor. J Bacteriol 179:1819–1823

Feng Y, Cronan JE (2009a) *Escherichia coli* unsaturated fatty acid synthesis: complex transcription of the *fabA* gene and in vivo identification of the essential reaction catalyzed by FabB. J Biol Chem 284:29526–29535

Feng Y, Cronan JE (2009b) A new member of the *Escherichia coli fad* regulon: transcriptional regulation of *fadM* (*ybaW*). J Bacteriol 191:6320–6328

Feng Y, Cronan JE (2010) Overlapping repressor binding sites result in additive regulation of *Escherichia coli* FadH by FadR and ArcA. J Bacteriol 192:4289–4299

Feng Y, Cronan JE (2011a) Complex binding of the FabR repressor of bacterial unsaturated fatty acid biosynthesis to its cognate promoters. Mol Microbiol 80:195–218

Feng Y, Cronan JE (2011b) The *Vibrio cholerae* fatty acid regulatory protein, FadR, represses transcription of *plsB*, the gene encoding the first enzyme of membrane phospholipid biosynthesis. Mol Microbiol 81:1020–1033

Feng Y, Xu J, Zhang H, Chen Z, Srinivas S (2013a) *Brucella* BioR regulator defines a complex regulatory mechanism for bacterial biotin metabolism. J bacteriol 195:3451–3467

Feng Y, Zhang H, Cronan JE (2013b) Profligate biotin synthesis in alpha-proteobacteria—a developing or degenerating regulatory system? Mol Microbiol 88:77–92

Fredrickson JK, Romine MF, Beliaev AS, Auchtung JM, Driscoll ME, Gardner TS, Nealson KH, Osterman AL, Pinchuk G, Reed JL, Rodionov DA, Rodrigues JL, Saffarini DA, Serres MH, Spormann AM, Zhulin IB, Tiedje JM (2008) Towards environmental systems biology of *Shewanella*. Nature reviews. Microbiology 6:592–603

Giles DK, Hankins JV, Guan Z, Trent MS (2011) Remodelling of the *Vibrio cholerae* membrane by incorporation of exogenous fatty acids from host and aquatic environments. Mol Microbiol 79:716–728

Gui L, Sunnarborg A, LaPorte DC (1996) Regulated expression of a repressor protein: FadR activates *iclR*. J Bacteriol 178:4704–4709

Haldimann A, Wanner BL (2001) Conditional-replication, integration, excision, and retrieval plasmid-host systems for gene structure-function studies of bacteria. J Bacteriol 183:6384–6393

Hau HH, Gralnick JA (2007) Ecology and biotechnology of the genus *Shewanella*. Annu Rev Microbiol 61:237–258

Heidelberg JF, Eisen JA, Nelson WC, Clayton RA, Gwinn ML, Dodson RJ, Haft DH, Hickey EK, Peterson JD, Umayam L, Gill SR, Nelson KE, Read TD, Tettelin H, Richardson D, Ermolaeva MD, Vamathevan J, Bass S, Qin H, Dragoi I, Sellers P, McDonald L, Utterback T, Fleishmann RD, Nierman WC, White O, Salzberg SL, Smith HO, Colwell RR, Mekalanos JJ, Venter JC, Fraser CM (2000) DNA sequence of both chromosomes of the cholera pathogen *Vibrio cholerae*. Nature 406:477–483

Heidelberg JF, Paulsen IT, Nelson KE, Gaidos EJ, Nelson WC, Read TD, Eisen JA, Seshadri R, Ward N, Methe B, Clayton RA, Meyer T, Tsapin A, Scott J, Beanan M, Brinkac L, Daugherty S, DeBoy RT, Dodson RJ, Durkin AS, Haft DH, Kolonay JF, Madupu R, Peterson JD, Umayam LA, White O, Wolf AM, Vamathevan J, Weidman J, Impraim M, Lee K, Berry K, Lee C, Mueller J, Khouri H, Gill J, Utterback TR, McDonald LA, Feldblyum TV, Smith HO, Venter JC, Nealson KH, Fraser CM (2002) Genome sequence of the dissimilatory metal ion-reducing bacterium *Shewanella oneidensis*. Nat Biotechnol 20:1118–1123

Henry MF, Cronan JE Jr (1991) *Escherichia coli* transcription factor that both activates fatty acid synthesis and represses fatty acid degradation. J Mol Biol 222:843–849

Henry MF, Cronan JE Jr (1992) A new mechanism of transcriptional regulation: release of an activator triggered by small molecule binding. Cell 70:671–679

Iram SH, Cronan JE (2005) Unexpected functional diversity among FadR fatty acid transcriptional regulatory proteins. J Biol Chem 280:32148–32156

Janda JM, Abbott SL (2014) The genus *Shewanella*: from the briny depths below to human pathogen. Crit Rev Microbiol 40:293–312

Kolker E, Picone AF, Galperin MY, Romine MF, Higdon R, Makarova KS, Kolker N, Anderson GA, Qiu X, Auberry KJ, Babnigg G, Beliaev AS, Edlefsen P, Elias DA, Gorby YA, Holzman T, Klappenbach JA, Konstantinidis KT, Land ML, Lipton MS, McCue LA, Monroe M, Pasa-Tolic L, Pinchuk G, Purvine S, Serres MH, Tsapin S, Zakrajsek BA, Zhu W, Zhou J, Larimer FW, Lawrence CE, Riley M, Collart FR, Yates JR 3rd, Smith RD, Giometti CS, Nealson KH, Fredrickson JK, Tiedje JM (2005) Global profiling of *Shewanella oneidensis* MR-1: expression of hypothetical genes and improved functional annotations. Proc Natl Acad Sci U S A 102:2099–2104

Li C, Ying Q, Su X, Li T (2012) Development and application of reverse transcription loop-mediated isothermal amplification for detecting live *Shewanella putrefaciens* in preserved fish sample. J Food Sci 77:M226–M230

Luo Q, Shi M, Ren Y, Gao H (2014) Transcription factors FabR and FadR regulate both aerobic and anaerobic pathways for unsaturated fatty acid biosynthesis in *Shewanella oneidensis*. Front Microbiol 5:736

Massengo-Tiasse RP, Cronan JE (2008) *Vibrio cholerae* FabV defines a new class of enoyl-acyl carrier protein reductase. J Biol Chem 283:1308–1316

Miller JH (1992) A short course in bacterial genetics. Cold Spring Harbor Laboratory, Cold Spring Harbor

My L, Rekoske B, Lemke JJ, Viala JP, Gourse RL, Bouveret E (2013) Transcription of the *Escherichia coli* fatty acid synthesis operon *fabHDG* is directly activated by FadR and inhibited by ppGpp. J Bacteriol 195:3784–3795

Nunn WD, Giffin K, Clark D, Cronan JE Jr (1983) Role for *fadR* in unsaturated fatty acid biosynthesis in *Escherichia coli*. J Bacteriol 154:554–560

Parlapani FF, Meziti A, Kormas KA, Boziaris IS (2013) Indigenous and spoilage microbiota of farmed sea bream stored in ice identified by phenotypic and 16S rRNA gene analysis. Food Microbiol 33:85–89

Rodionov DA, Novichkov PS, Stavrovskaya ED, Rodionova IA, Li X, Kazanov MD, Ravcheev DA, Gerasimova AV, Kazakov AE, Kovaleva GY, Permina EA, Laikova ON, Overbeek R, Romine MF, Fredrickson JK, Arkin AP, Dubchak I, Osterman AL, Gelfand MS (2011) Comparative genomic reconstruction of transcriptional networks controlling central metabolism in the *Shewanella* genus. BMC Genom 12(Suppl 1):S3

Sheng L, Fein JB (2014) Uranium reduction by *Shewanella oneidensis* MR-1 as a function of NaHCO3 concentration: surface complexation control of reduction kinetics. Environ Sci Technol 48:3768–3775

Shi W, Kovacikova G, Lin W, Taylor RK, Skorupski K, Kull FJ (2015) The 40-residue insertion in *Vibrio cholerae* FadR facilitates binding of an additional fatty acyl-CoA ligand. Nat Commun 6:6032

van Aalten DM, DiRusso CC, Knudsen J, Wierenga RK (2000) Crystal structure of FadR, a fatty acid-responsive transcription factor with a novel acyl coenzyme A-binding fold. EMBO J 19:5167–5177

van Aalten DM, DiRusso CC, Knudsen J (2001) The structural basis of acyl coenzyme A-dependent regulation of the transcription factor FadR. EMBO J 20:2041–2050

Wang F, Xiao X, Ou HY, Gai Y, Wang F (2009) Role and regulation of fatty acid biosynthesis in the response of *Shewanella piezotolerans* WP3 to different temperatures and pressures. J Bacteriol 191:2574–2584

Wiatrowski HA, Ward PM, Barkay T (2006) Novel reduction of mercury (II) by mercury-sensitive dissimilatory metal reducing bacteria. Environ Sci Technol 40:6690–6696

Xu Y, Heath RJ, Li Z, Rock CO, White SW (2001) The FadR.DNA complex. Transcriptional control of fatty acid metabolism in *Escherichia coli*. J Biol Chem 276:17373–17379

Zhang Y, Gao R, Ye H, Wang Q, Feng Y (2014) A new glimpse of FadR-DNA crosstalk revealed by deep dissection of the *E. coli* FadR regulatory protein. Protein Cell 5:928–939

17

Molecular basis for the inhibition of β-hydroxyacyl-ACP dehydratase HadAB complex from *Mycobacterium tuberculosis* by flavonoid inhibitors

Yu Dong[1,2], Xiaodi Qiu[1,2], Neil Shaw[1,3], Yueyang Xu[4], Yuna Sun[1], Xuemei Li[1], Jun Li[1]✉, Zihe Rao[1,3,4]✉

[1] National Laboratory of Biomacromolecules, Institute of Biophysics, Chinese Academy of Sciences, Beijing 100101, China
[2] University of Chinese Academy of Sciences, Beijing 100049, China
[3] Structure Biology Laboratory, Tsinghua University, Beijing 100084, China
[4] Tianjin Key Laboratory of Protein Science, College of Life Sciences, Nankai University, Tianjin 300071, China
✉ Correspondence: shyaolx@163.com (J. Li), raozh@mail.tsinghua.edu.cn (Z. Rao)

ABSTRACT

Dehydration is one of the key steps in the biosynthesis of mycolic acids and is vital to the growth of *Mycobacterium tuberculosis* (*Mtb*). Consequently, stalling dehydration cures tuberculosis (TB). Clinically used anti-TB drugs like thiacetazone (TAC) and isoxyl (ISO) as well as flavonoids inhibit the enzyme activity of the β-hydroxyacyl-ACP dehydratase HadAB complex. How this inhibition is exerted, has remained an enigma for years. Here, we describe the first crystal structures of the *Mtb*HadAB complex bound with flavonoid inhibitor butein, 2',4,4'-trihydroxychalcone or fisetin. Despite sharing no sequence identity from Blast, HadA and HadB adopt a very similar hotdog fold. HadA forms a tight dimer with HadB in which the proteins are sitting side-by-side, but are oriented anti-parallel. While HadB contributes the catalytically critical His-Asp dyad, HadA binds the fatty acid substrate in a long channel. The atypical double hotdog fold with a single active site formed by *Mtb*HadAB gives rise to a long, narrow cavity that vertically traverses the fatty acid binding channel. At the base of this cavity lies Cys61, which upon mutation to Ser confers drug-resistance in TB patients. We show that inhibitors bind in this cavity and protrude into the substrate binding channel. Thus, inhibitors of *Mtb*HadAB exert their effect by occluding substrate from the active site. The unveiling of this mechanism of inhibition paves the way for accelerating development of next generation of anti-TB drugs.

KEYWORDS *Mycobacterium tuberculosis*, hotdog fold, mycolic acid, dehydratase, flavonoid, thiacetazone, isoxyl

INTRODUCTION

Tuberculosis (TB), caused by *Mycobacterium tuberculosis* (*Mtb*), is the second most common cause of death due to a single infectious pathogen with an enormous global medical burden. In the year of 2013, approximately 30% of the world's population were latently infected with *Mtb*, and there were nearly 9 million new TB cases and 1.5 million TB deaths (WHO 2014; Jiang et al., 2014; Zumla et al., 2013). Emergence and propagation of multidrug-resistant (MDR) and extensively drug-resistant (XDR) *Mtb* strains, and the co-infection with HIV present a huge difficulty to existing TB therapy (Comas and Gagneux, 2009; Wang et al., 2013). Amongst known pathogens, the cell wall of mycobacteria is unique primarily because of its lipid content (Goldberg et al., 2012). In particular, 2-alkyl, 3-hydroxy long-chain fatty acids known as mycolic acids (MAs) and their derivates form a dense outer sheath for shielding the inner cytoskeleton of the cell wall (Carel et al., 2014). Mycobacteria's ability to survive and cause disease is dependent on the integrity of this lipid-rich layer referred to as myco-membrane (Bhatt et al., 2007; Takayama et al., 2005). Therefore, one approach to overpower mycobacteria is to scuttle the biosynthesis of mycolic acids (Jackson et al., 2013).

MAs issue from the Claisen condensation of an alkyl chain of medium length (C_{24}–C_{26}) with a long mero-mycolic chain (up to C_{60}) bearing specific biochemical modifications (Asselineau et al., 2002). The biosynthetic pathway of MAs involves two types of fatty acid-synthase system (FAS): FAS-I and FAS-II. While FAS-I is a single, large polypeptide chain folded into multiple domains that harbor the catalytic sites required for the biosynthesis of fatty acids, the FAS-II is composed of a series of discrete soluble enzymes that act successively and repetitively to elongate the fatty acid chains produced by FAS-I (Sylvain Cantaloube et al., 2011). The four enzymes operating in tandem during each cycle of elongation are: 1) β-ketoacyl-ACP synthetases (KasA and KasB), 2) β-ketoacyl-ACP reductase (MabA), 3) β-hydroxyacyl-ACP dehydratases (HadAB and HadBC complexes), and 4) *trans*-2-enoyl-ACP reductase (InhA). The importance of each of these enzymatic activities in the biosynthesis of mycolic acids is well documented. Deficiency or inactivation of any enzymes above will cease the biosynthesis of MAs. Actually, these are ideal and actual targets for drug discovery (Campbell and Cronan, 2001; Marrakchi et al., 2014): the first-line anti-TB agent isoniazide (INH) and second-line agent ethionamide (ETH) act on InhA (Banerjee et al., 1994); an antibiotic thiolactomycin (TLM) inhibits the activity of KasA and KasB (Kremer et al., 2000).

The β-hydroxyacyl-ACP dehydratase (Had) catalyzes the third step in the fatty acid elongation cycle by dehydrating β-hydroxyacyl-ACP to trans-2-enoyl-ACP (Fig. S1), which is the last piece to be identified in the mycobacterial FAS-II and exsits only in *Corynebacterineae* (Sacco et al., 2007). HadAB would take part, like KasA, in the early FA elongation cycles, leading to the formation of the intermediate-size (C_{32}–C_{42}) meromycolic chains, while HadBC, like KasB, would elongate further the intermediate-size meromycolic chains to full-size molecules (C_{52}–C_{64}) during the late elongation cycles (Gao et al., 2003; Sacco et al., 2007). Previously, flavonoid inhibitors targeting HadB (Rv0636) were shown to disrupt the biosynthesis of fatty acids, resulting in the depletion of the mycolic acid content of the Mycobacteria. Consequently, these flavonoids were shown to effectively inhibit the growth of *Mycobacterium bovis* BCG (Brown et al., 2007a). Besides flavonoids, two pro-drugs, isoxyl (ISO) and thiacetazone (TAC) (Fig. 1) used in the clinical treatment of tuberculosis, are also known to exert their anti-mycobacterial effect by stalling the dehydration step of the FAS-II elongation cycle (Belardinelli and Morbidoni, 2012; Coxon et al., 2013; Grzegorzewicz et al., 2012). Both these pro-drugs undergo activation by monooxygenase EthA, for unleashing their anti-mycobacterial potential (Dover et al., 2007; Kordulakova et al., 2007; Nishida and Ortiz de Montellano, 2011). How these drugs disrupt the dehydratase activity of the FAS-II system has remained an enigma for years. A lack of understanding of the molecular basis of this inhibition has been a major bottleneck in the development of next generation of drugs essential for targeting the mycolic acid component of mycobacteria.

Key to overcoming this impediment is the elucidation of the crystal structure of the *Mtb*HadAB complex that catalyzes the dehydration step. This is important also because the primary sequence of *Mtb*HadAB is very different from that of homologous enzymes like FabA and FabZ that catalyze similar reactions in other micro-organisms (Bloch, 1977; Brown et al., 2007a; Sacco et al., 2007). Currently, crystal structures of FabZ from four different sources, *Pseudomonas aeruginosa* (Protein Data Bank (PDB) code 1U1Z) (Kimber et al., 2004), *P. falciparum* (PDB code 1Z6B and 1ZHG) (Kostrewa et al., 2005; Swarnamukhi et al., 2006), *Helicobacter pylori* (PDB code 2GLL) (Zhang et al., 2008b) and *Campylobacter jejuni* (PDB code 3D6X) (Kirkpatrick et al., 2009) are available. All of them have a similar hexameric structure, displaying a classic "trimer of homodimers" organization. Because of the particular long substrate along with *Mtb*HadAB and failure to identify a specific mycobacterial homologue by BLAST searches using *E. coli* FabZ and FabA (Brown et al., 2007a), the structural insights obtained from these homologous enzymes cannot be extrapolated in its entirety to *Mtb*HadAB. Indeed, the first crystal structures of *Mtb*HadAB described here suggest the same and provide tantalizing glimpses into unique features of the active site of *Mtb*HadAB. Most importantly, the structures unveil an unexpected pocket near the active site that is targeted by flavonoids and other clinically used anti-mycobacterial drugs like TAC and ISO for inhibition of the dehydratase. Our structural study of *Mtb*HadAB provided the first three dimensional atomic models of the *Mtb*HadAB-inhibitor complexes and laid the foundation for new anti-TB drug development.

RESULTS

Overall structure of *Mtb*HadAB complex

The crystal structure of *Mtb*HadAB complex was solved by single wavelength anomalous dispersion (SAD) method using X-ray diffraction data collected from crystals of selenomethionine labeled protein. The crystal belonged to space group $P4_12_12$, with unit-cell parameters $a = b = 82.0$ Å, $c = 139.8$ Å, $\alpha = \beta = \gamma = 90.0°$. A Matthews coefficient of 3.56 $Å^3$ Da^{-1} (Matthews, 1968; Potterton et al., 2003), corresponding to a solvent content of 65.49%, coupled with the previous biophysical identification indicated the presence of both one molecule of HadA and HadB per asymmetric unit. The final model encompassing residues 3–146 of HadA and residues 1–142 of HadB was refined to 1.75 Å resolution with an R_{work} (R_{free}) value of 15.7% (18.5%) (Fig. 3A, Table S1).

HadA does not share any significant primary sequence identity with HadB from Blast (Fig. 2, showed by ESPript server (Robert and Gouet, 2014)). In spite of this, the overall structure of the *Mtb*HadAB complex reveals that both proteins adopt a similar hotdog fold (Fig. 3B). A central sheet

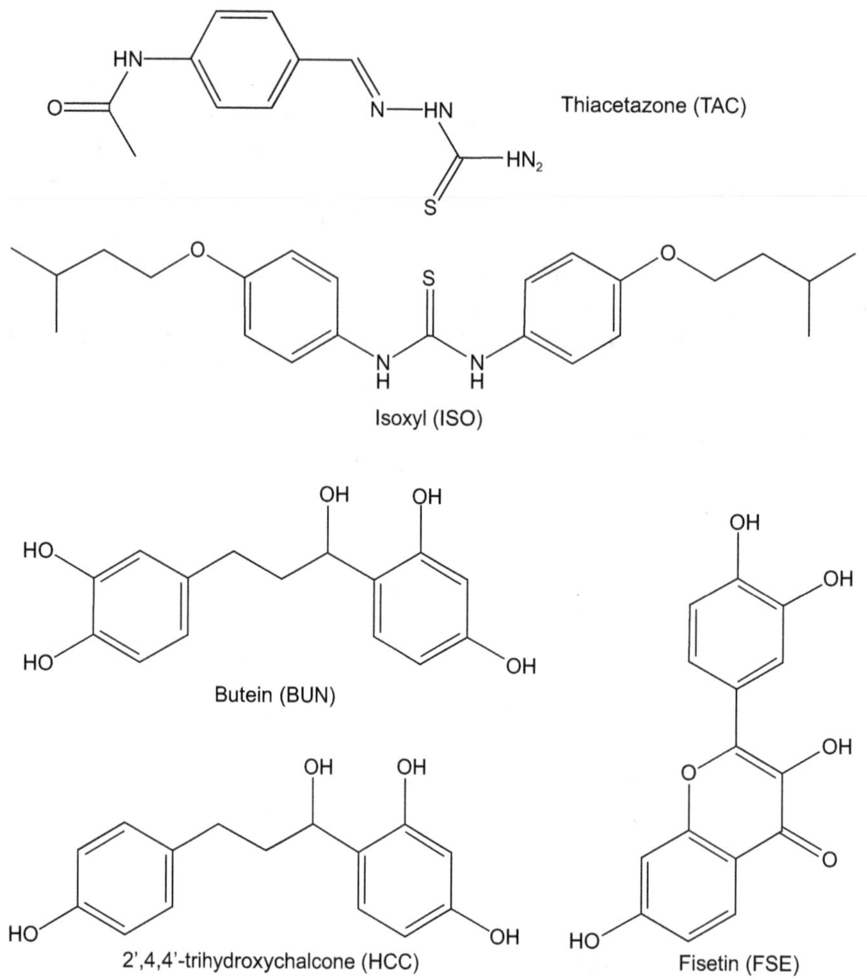

Figure 1. Chemical structures of compounds of the current study related to inhibition of *Mtb*HadAB. Thiacetazone (TAC) and isoxyl (ISO) have a sulfur containing group that could form a di-sulfide bond with Cys61 of HadA. *Mycobacterium tuberculosis* strains with C61S mutation in HadA are resistant to TAC and ISO.

composed of five strands (β1–β5) is twisted to form a concave cavity at the center. A hotdog helix, α3, (referred to as αHD hereafter) is embedded laterally in this cavity, giving the appearance of a typical hotdog fold. Helices α1 and α2 are inserted between strand β1 and αHD. Two such folds, one each contributed by HadA and HadB, sitting side-by-side, but oriented anti-parallel, result in a double hotdog fold. Thus, the overall structure of *Mtb*HadAB complex consists of a double hotdog fold (Fig. 3A).

Although the overall structures of HadA and HadB are similar, there exist several differences between the two proteins that impart distinct roles on them during catalysis (Fig. 3C and 3D). For instance, the positioning of the αHD within the concave cavity formed by the central β-sheet is different in both proteins. In HadA, the helical axis is almost perpendicular to the vertical axis of the β-strands forming the central sheet. In contrast, the αHD of HadB is placed upright, with the helical axis tilted in the direction of the vertical axis of the β-strands (Fig. 3C). This difference in the positioning of the αHD has implications for substrate binding. The orientation adopted by αHD in HadA pulls the top portion of strand β2 of HadA away from the adjacent strand β2 of HadB, creating a narrow gap between the two proteins for binding the fatty acid. Such a cavity for binding ligands is lacking in HadB. Intriguingly, electron density for a long aliphatic chain-like ligand bound by the protein during expression in *E. coli* is visible in another deep, narrow pocket that is almost perpendicular to the fatty acid binding channel (Fig. S2). Other significant differences between the structures of the two proteins include the lengths of the strands making up the central β-sheet. Notably, the length of all five β strands constituting the central sheet of HadA is longer than that of HadB. Further, the loop connecting αHD with β2 of the central sheet in HadA (residue 76–84) is longer than that in HadB (residue 75–80) (Fig. 3C). This probably facilitates formation of the substrate binding channel (Fig. 3D). Lastly, HadA is 16 residues longer at the C-terminus than HadB. These additional residues of HadA interact with the loop

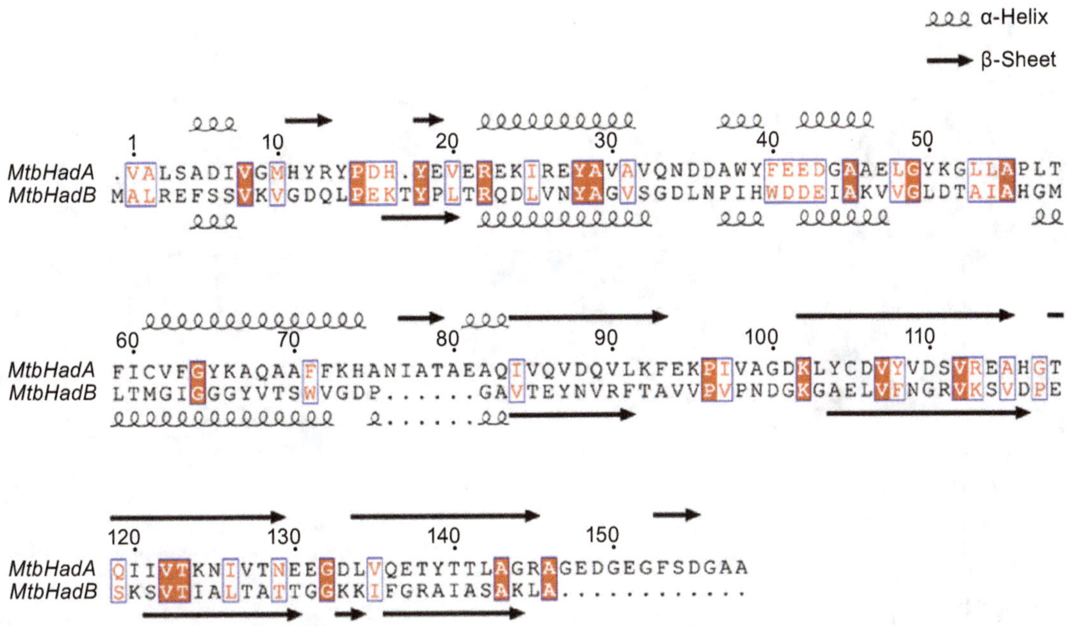

Figure 2. Sequence alignment of HadA and HadB with secondary structural elements marked near the sequence. HadA does not share any significant primary sequence identity with HadB from Blast (or only 13.57% identity calculated by ClustalW2), but their secondary structural elements distribution is similar.

connecting αHD with β2 and form a pair of short anti-parallel β-strands though it was undefined in the native structure because of poor density.

Interactions of *Mtb*HadAB hetero-tetramer and hetero-dimer

Hetero-dimerization of HadA with HadB is a prerequisite for unleashing the dehydratase activity of the complex. In agreement with this, we could not produce a stable preparation of either of the two proteins when expressed individually. A stable, hetero-oligomeric preparation of the proteins could be produced when HadA and HadB were co-expressed. Gel filtration chromatography and analytical ultracentrifugation analysis suggested that HadA and HadB formed a hetero-tetramer in solution at 16°C (Fig. S3). This result was verified in the subsequent crystal structure of the *Mtb*HadAB complex where the two proteins formed a tight hetero-dimer in an asymmetric unit. Further analysis of symmetry mates revealed that HadA and HadB formed a face-to-face dimer of hetero-dimers (Fig. S4) similar to the dimer of double hotdog fold structure of MFE-2 hydratase 2 (Koski et al., 2005), which was consistent with our analytical ultracentrifugation and gel filtration results. Two hetero-dimer of *Mtb*HadAB was symmetry related by a two-fold axis along inside the interface which buried an area of 745 Å^2 analyzed by PISA (Krissinel and Henrick, 2007). In this interface, helix α1 of HadB interacts with α1 and surrounding loops from the symmetry related HadB, while the N-terminal half of α1 of HadA was interacting with α1, β1b, and αHD from the symmetry related HadA.

A close examination of the structure of the *Mtb*HadAB heterodimer reveals that hetero-dimerization is mediated by two types of inter-molecular interactions (Fig. 4). The first type of interactions involves the main chain backbone atoms of the β2 strands from both proteins. HadA and HadB interact with each other via their β2 strands such that the sheets from the two proteins are joined together to form one single contiguous sheet (Fig. 4A and 4B). The second type of inter-molecular interactions is mainly mediated by side chains (Fig. 4C–E; Tables S2 and S3) and is observed between: 1) amino acids of αHD of HadA and the helices α1 and α2, as well as the loop connecting these helices of HadB, 2) helix α1 of HadA and HadB, 3) β2 of HadA with αHD of HadB. PISA analysis revealed that hetero-dimerization of HadA with HadB resulted in burial of 1,453 Å^2 surface area of each monomer.

Active site of *Mtb*HadAB dehydratase

Previous bioinformatics analysis (Labonte and Townsend, 2013; Sacco et al., 2007) indicated that only HadB possesses hydratase 2 motif ([YF]-x(1,2)-[LIVG]-[STGC]-G-**D**-x-N-P-[LIV]-**H**-x(5)-[AS]) among HadA, HadB, and HadC, and the main catalytic residues Asp36 and His41 of HadB are highlighted in bold. It means that in the hetero-dimer interface of *Mtb*HadAB, only one active site was confirmed in the structure while the other site was evolutionally abolished,

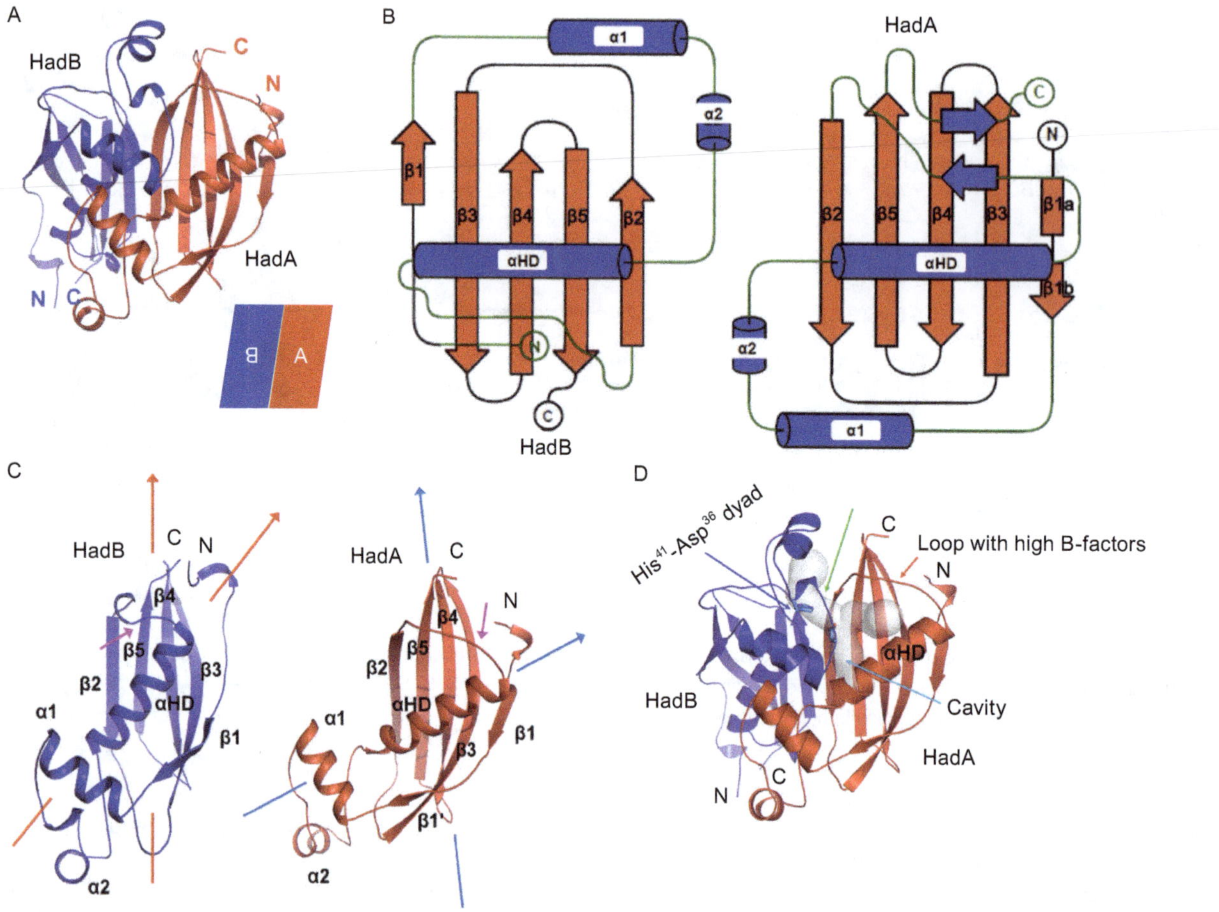

Figure 3. Overall structure of *Mtb*HadAB complex. (A) Cartoon representation of the structure of *Mtb*HadAB hetero-dimer is shown. HadA (red color) sits antiparallel to HadB (blue color). (B) Topology diagram of the structure of HadA and HadB. A short pair of strands in HadA was undefined in the native structure because of poor density. (C) Differences in positioning of αHD of HadA and HadB are shown. Direction of the axis of the β-sheet w.r.t. αHD is shown with arrows. Difference in conformation of loop connecting β2 with αHD is pointed using a magenta colored arrow. (D) Location of pockets in the active site of HadB is shown. The fatty acid binding channel and a vertical cavity traversing the fatty acid binding channel is shown in grey colored transparent surface representation. Location of catalytic dyad, His41-Asp36, is shown. Cavities were identified using CAVER software.

compared to other homodimer hotdog fold enzymes which possesses two active sites (Kimber et al., 2004; Leesong et al., 1996). The substrate of *Mtb*HadAB consists of two components—mycobacterial acyl carrier protein (AcpM) and the fatty acid (in the form of β-hydroxyacyl) (Fig. S5). The *Mtb*HadAB complex binds AcpM as well as the fatty acid during catalysis. Like other dehydratases, the fatty acid substrate binds in a narrow channel formed at the junction where HadA meets HadB (Fig. 4F). More specifically, this channel is located between αHD and the central β-sheet of HadA, a region in vicinity of helices α1 and α2 of HadB. Amino acids like Tyr65, Gln68, Ala69, Phe72, Thr79, Glu81-Ala-Gln-Ile-Val-Gln86, and Leu142 from HadA are observed lining the long, narrow channel to hold the aliphatic chain of the substrate. Asp36, Asn38, Ile40, His41, His58, Gly59, and Met60 from HadB are part of the entrance to the channel. While HadA harbors most of the residues lining this fatty acid binding channel, HadB lacks a similar channel for binding the substrate. Instead, amino acids like His41 and Asp36 of HadB are in proximity of the channel, suggesting they represent the critical catalytic dyad known to drive catalysis in hydratases (Fig. 4F). Thus, the structure of *Mtb*HadAB complex explains and provides the structural basis for the roles played by each protein. HadA functions primarily to entrench and bind the fatty acid, while HadB contributes the catalytic amino acids required for the enzymatic reaction.

Extremely long chain fatty acids (chain length > C_{16}) are hydrophobic in nature. Therefore, an acyl carrier protein, AcpM, (instead of the small molecule Coenzyme A) is covalently linked to the fatty acid chain to solubilize the substrate (Bhatt et al., 2007; Mdluli et al., 1998). Furthermore, AcpM is postulated to recognize enzymes belonging

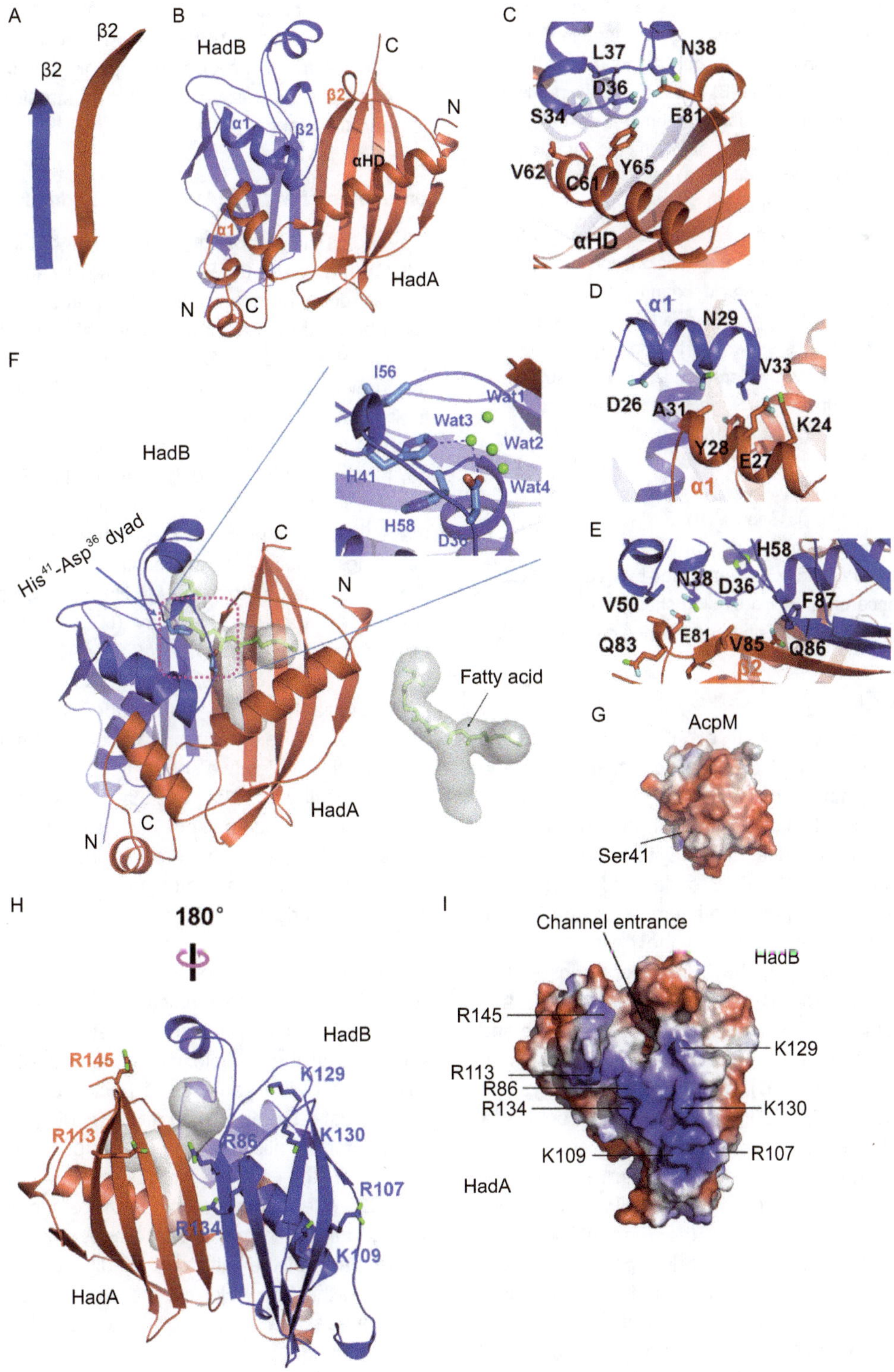
A
β2
β2
B
HadB
C
N
HadA
N
C
C
L37
N38
D36
S34
E81
V62
C61
Y65
αHD
D
α1
N29
V33
D26
A31
K24
Y28
E27
α1
F
HadB
His41-Asp36 dyad
C
N
I56
Wat1
Wat3
Wat2
H41
Wat4
H58
Fatty acid
HadA
N
C
E
H58
N38
D36
V50
F87
Q83
E81
V85
Q86
β2
G
AcpM
Ser41
H
180°
HadB
R145
K129
R113
R86
K130
R107
R134
K109
HadA
I
Channel entrance
HadB
R145
K129
R113
R86
R134
K130
K109
R107
HadA

Figure 4. Dimer interface and active site of *Mtb*HadAB complex. (A) Position of β2 strands of HadA and HadB with respect to each other is shown. Main chain atoms of both these β-strands interact with each other. (B) Location of β2 strands of HadA and HadB and other structural elements involved in dimerization is shown. (C–E) Nature of intermolecular interactions between different regions of HadA and HadB is shown. Interacting residues are shown as sticks. Distances of all the interactions between HadA and HadB are listed in Table S2. (F) His-Asp catalytic dyad (shown as sticks) contributed by HadB is in vicinity of the fatty acid channel (grey surface representation). Aliphatic carbon chain of a fatty acid molecule modeled into the channel is shown on the right hand side in stick representation. Catalytic waters bonded to His-Asp dyad are shown as spheres (inset). (G) Surface electrostatic potential representation of AcpM (PDB code 1KLP) without showing the C-terminal tail. The surface is predominantly negatively charged. (H) Location of positively charged residues on the rear side of *Mtb*HadAB is shown. (I) Surface electrostatic potential representation of the view shown in panel H. Location of positively charged amino acids contributed by HadA or HadB as well as entrance of fatty acid binding channel are marked. Blue and red colors represent positive and negative potential, respectively.

to the FAS II system (Trivedi et al., 2004), including *Mtb*HadAB, and deliver the fatty acid chain to the dehydratases. Recognition probably occurs via charge and shape complementarity. AcpM is a small globular protein consisting of a four helix bundle (residues 1–83) with a flexible C-terminal loop (residues 84–115) (PDB code 1KLP). While the globular head forges protein-protein interactions, the flexible tail binds fatty acids. The surface of the helical region is highly negatively charged (Fig. 4G). Coincidently, inspection of the structure of *Mtb*HadAB around the catalytic His-Asp dyad reveals a region that could complement the shape and charge of AcpM. This region comprising of arginines at positions 113(A), 145(A), 86(B), 107(B), 134(B) and lysines at positions 109(B), 129(B), 130(B) (A and B represent HadA and HadB, respectively) is located around the entrance of the fatty acid binding channel and could form salt bridges with glutamates and aspartates of AcpM (Fig. 4H and 4I) (Wong et al., 2002). In addition, the β-sheet near the entrance of the putative fatty acid binding channel is bent, forming a bow shaped pocket, which could hold the AcpM via shape complementarity. Such a mode of binding would be similar to that inferred for FabA from *Escherichia coli* (Nguyen et al., 2014) with one notable difference. While FabA forms a homodimer that has two active site pockets, *Mtb*HadAB has only one fatty acid binding channel. Therefore, only one molecule of AcpM binds *Mtb*HadAB (Fig. S6). Consistent with this, only one region of *Mtb*HadAB in vicinity of the lone fatty acid binding channel is compatible for binding AcpM. It is noteworthy that there is a loop of HadA with high B-factors above the substrate binding channel, and it is difficult for long fatty acid chain of substrate to insert from one end and pass through the channel. So we deduced that the chain of substrate was recognized and bound by lying into the channel through the opening of the unique loop, because this loop is very flexible in structure with a higher temperature factor.

Structures of *Mtb*HadAB-flavonoid complex

Flavonoids like butein (BUN), 2',4,4'-trihydroxychalcone (HCC) and fisetin (FSE), are known to stall biosynthesis of mycolic acids by mycobacterial FAS-II system (Fig. 1). They have been shown to particularly inhibit the activity of HadB (Brown et al. 2007b). Consequently, these flavonoids inhibit the growth of *Mycobacterium bovis* BCG, demonstrating their utility in development of anti-TB drugs. How these flavonoids inhibit HadB is currently not understood. Results of our biophysical studies on HadA and HadB show that HadB forms a tight hetero-dimeric complex with HadA. Therefore, these flavonoids probably target the *Mtb*HadAB complex for inhibiting mycolic acid synthesis. We tested the binding of these flavonoids to the *Mtb*HadAB complex using an isothermal titration calorimetry (ITC) based assay. All the three flavonoids bound the *Mtb*HadAB complex with a K_D value in the 10–15 μmol/L range (Fig. 5A).

To elucidate the binding mode of these flavonoids to *Mtb*HadAB for explaining the mechanism of inhibition of HadB by flavonoids, we solved three crystal structures of the *Mtb*HadAB-flavonoid complexes by MR method using three high resolution data sets all beyond 2.2 Å. (Fig. 5B, Table S1). All the structures including native *Mtb*HadAB shared an *r.m.s.d.* not more than 0.242 Å between each other comparing corresponding Cα atom positions. Electron density for all the ligands was clear and permitted unambiguous interpretation of the mode of binding of the flavonoids (Fig. 5C). One molecule of the flavonoid binds one molecule of the *Mtb*HadAB hetero-dimer. Remarkably, structures of the complexes reveal that all the three flavonoids bind in the same deep pocket of *Mtb*HadAB that is located just beneath the substrate binding channel (Fig. 5B). Density for an aliphatic ligand was observed bound in this pocket in the native structure of *Mtb*HadAB complex (Fig. S2). Interestingly, all the atoms of the three ligands are arranged in the same vertical plane, which is perpendicular to the plane of the fatty acid binding channel. This pocket harboring the flavonoids merges with the fatty acid binding channel near the surface of the protein, in proximity of the catalytic dyad.

Specifically, the inhibitor binding pocket is located between the central β-sheet and the αHD of HadA. Considering the fact that the chemical structures of butein and 2',4,4'-trihydroxychalcone (HCC) are almost identical; with the exception of an additional hydroxyl group on the benzene ring of butein located near the surface of the protein, it is not surprising that these two flavonoids bind *Mtb*HadAB in an almost identical manner. This observation also independently confirms the location of binding site for inhibitors. In

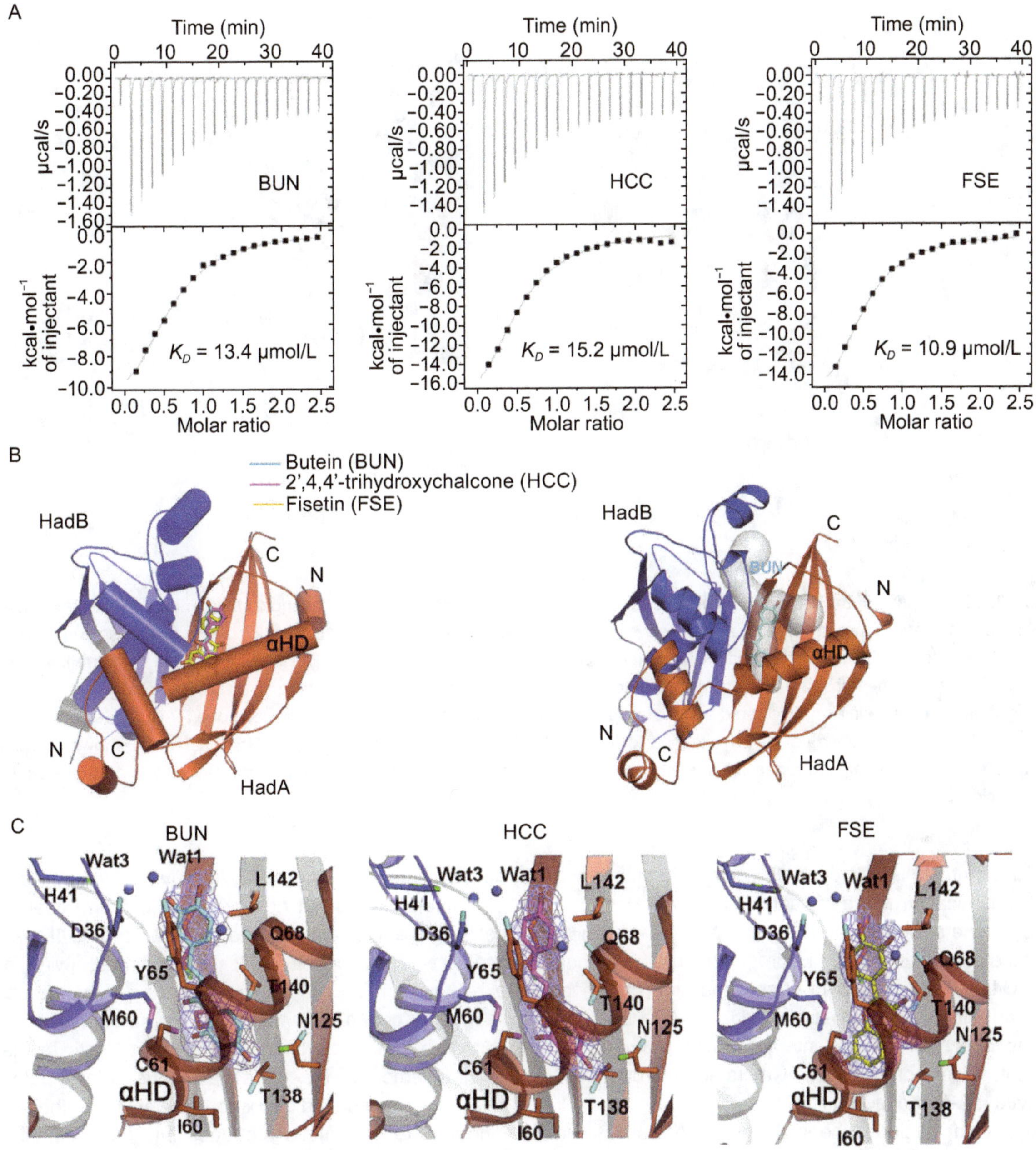

Figure 5. Binding of flavonoids to *Mtb*HadAB complex. (A) K_D values determined for the binding of flavonoids to *Mtb*HadAB complex by ITC. (B) Different flavonoids tested in the current study (shown as sticks) bind at the same location on *Mtb*HadAB. Upon superimposition of the structures, the ligands overlap. They bind in the cavity (right panel, gray colored surface representation) located just beneath the fatty acid binding channel. An example of butein (BUN; cyan sticks, right panel) bound in the cavity is shown. (C) Nature of amino acid interacting with flavonoids is shown. Flavonoids and amino acids are shown as sticks, while water is depicted as spheres. *2Fo-Fc* electron density contoured at 1σ for the three ligands is shown.

the structures of both these binary complexes, side chains of Gln86, Gln89 of β2 and Thr138, Thr140, Leu142 of β5 stack on the rear side of the ligand along its entire length, while Ile60, Cys61, Gly64, Tyr65 of αHD and Asp36 (HadB) of the catalytic loop stack on the front side (Fig. 5C). Asn125 of β4 and Gln68 of αHD together with Met60 (HadB) from another αHD further restrain the movement of ligands. The only hydrogen bond formed by these two ligands with the protein is between the hydroxyl group of the ligand buried in the protein and side chains of either Thr138 or Asn125 (Fig. S7). A noteworthy observation is the hydrogen bonding of the additional hydroxyl group of butein with the catalytic water

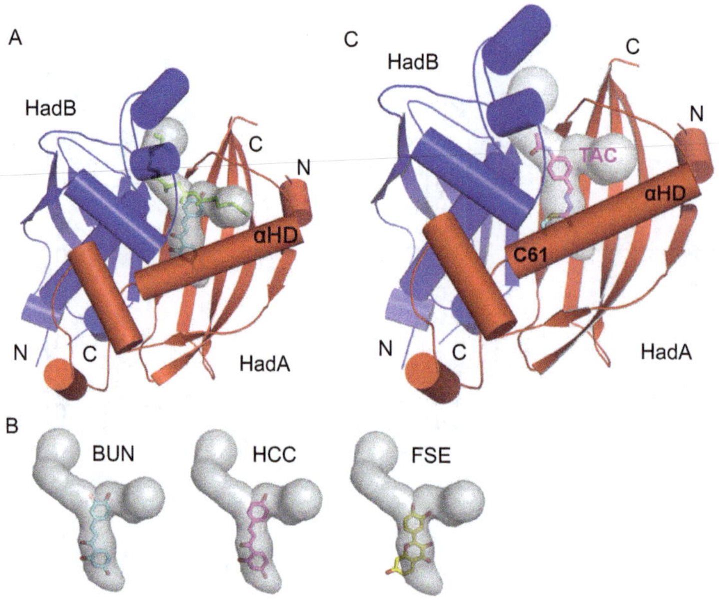

Figure 6. Mechanism of inhibition of enzymatic activity of *Mtb*HadAB complex. (A) Flavonoids bind in the cavity (gray colored surface representation) that traverses the fatty acid binding channel. An example of how binding of butein (cyan sticks) physically obstructs the placement of fatty acid (modeled as green sticks) in the substrate binding channel is shown. (B) All the three flavonoids (sticks) protrude into the substrate binding channel and perturb binding of the substrate in the active site. The cavity and substrate binding channel are shown as gray colored surface representation. (C) The inhibition model of thiacetazone (TAC, modeled as sticks) on *Mtb*HadAB. TAC is shown covalently attached to Cys61 (shown as cyan colored sticks). C61S mutation is known to confer resistance to TAC. The drug protrudes into the fatty acid binding channel (gray colored surface representation) and occludes binding of fatty acid.

Wat3 bound by the His-Asp dyad. In contrast to butein and HCC, the chemical structure of fisetin is different. However, the overall mode of binding of fisetin in the pocket is similar to that of the other two ligands. Hydrogen bonds are formed between O4 and O3' atoms of ligand and side chains of Asn125 and Gln68, respectively. In addition, O7, O3', and O4' atoms of the ligand are interacting with surrounding solvent molecules. Some of these waters are in proximity of the catalytic His-Asp dyad (Fig. S7).

The molecular basis for the inhibition of *Mtb*HadAB can be envisioned based on the structures of *Mtb*HadAB bound with flavonoids. Binding of the flavonoid to *Mtb*HadAB can physically occlude binding of the fatty acid into the substrate binding channel (Fig. 6A). All the three flavonoids protrude into the fatty acid binding channel (Fig. 6B). Notably, they are in proximity of the catalytic amino acids. Therefore, the flavonoids can scuttle binding of the fatty acid or prevent it from binding in a catalytically competent orientation. Amongst the flavonoids tested for inhibition of growth of *Mycobacterium bovis* BCG strain, butein was the most effective with an MIC value of 157 μmol/L (43 μg/mL) (Brown et al., 2007b). This can be explained by examining the structures of the binary complexes of the flavonoids with *Mtb*HadAB. The O1 hydroxyl group of butein forms a hydrogen bond with the putative catalytic water Wat3 that is hydrogen bonded to both the residues of the His-Asp dyad (Figs. 5C and S7). This interaction is not observed in HCC and fisetin. Interestingly, the IC_{50} value for the inhibition of FAS II system by fisetin was 54 μg/mL, which was almost twice of that exhibited by HCC and butein. Inspection of the structures of flavonoids bound with *Mtb*HadAB reveals that the obstruction of the fatty acid binding channel by fisetin is the least efficient amongst the three flavonoids. While butein and HCC can obstruct and almost completely block the fatty acid binding channel, fisetin protrudes into the channel; but, it only partially covers the fatty acid binding channel (Fig. 6B). Thus, crystal structures of *Mtb*HadAB bound with flavonoids illustrate that butein, fisetin, and HCC exert their inhibitory effect by obstructing the placement of the fatty acid into the substrate binding site of *Mtb*HadAB.

DISCUSSION

Inhibitors of hotdog structures have already been studied for years. A number of flavonoids and derivatives are discovered as inhibitors of β-hydroxyacyl-ACP dehydratases (Tasdemir et al., 2006). Structural studies revealed either the substrate binding site or its entrance on the dimer interface of FabZ and FabA was occupied by the inhibitors to block enzyme activity (He et al., 2009; Moynie et al., 2013; Zhang et al., 2008a, b). However, flavonoids such as butein, 2',4,4'-trihydroxychalcone, and fisetin were not found at the substrate binding channel or the entrance on the dimer interface of *Mtb*HadAB. Instead, these small molecules were inserted into a deep pocket beneath the channel at the dimer interface. Such pocket was never found in previously solved hotdog fold structures. Part of the molecules protruded out and blocked substrate binding. These findings suggest that *Mtb*HadAB is inhibited by some flavonoids in a novel way which is different from those found in other hotdog typed dehydratases.

Combining the pivotal structural insights garnered from the crystal structures of *Mtb*HadAB in complex with flavonoids with previously known information, the molecular mechanism for inhibition of mycolic acid synthesis by TAC can be envisioned. Mycobacterial strain with C61S mutation in HadA, exhibited high level of resistance to TAC and ISO (Belardinelli and Morbidoni, 2012; Gannoun-Zaki et al., 2013; Grzegorzewicz et al., 2012). Mapping of the location of this mutation on the structure of *Mtb*HadAB reveals that the mutation is located at the N-terminal end of αHD and lies at the bottom of the flavonoid binding pocket (Fig. 6C). Conceivably, the activated forms of pro-drugs TAC and ISO are highly possible to bind at this location. Resistance of C61S mutants to TAC indicates that the only sulfur containing moiety, carbothiamide of TAC, harboring the sulfur group forms a di-sulfide bond with the sulfhydryl group of Cys61. This covalent interaction is crucial for maintaining the inhibition because modeling of the drugs into the flavonoid-binding pocket reveals that there are no other hydrogen bonds formed by the drugs with the enzyme that could help hold the drug firmly into the pocket. Once bound, inhibition of the activity of *Mtb*HadAB by TAC is then accomplished by a mechanism similar to that observed for flavonoids. The distal carbonyl group of TAC protrudes into the fatty acid binding channel of *Mtb*HadAB, occluding the binding of the substrate (Fig. 6C). This abrogates mycolic acid synthesis, which adversely affects the growth of the bacterium. Consistent with this, the covalently bound activated TAC inhibits *Mycobacterium tuberculosis* with a low MIC value of 0.25 μg/mL (Coxon et al., 2013). Addition of linear alkyl groups at this end would progressively fill up the proximal end of the fatty acid binding channel, further enhancing the inhibition exerted by TAC. In agreement with this structural observation, substitution of the R_3 position of the aromatic ring with longer aliphatic groups has been shown to increase the inhibitory effect with a concomitant decrease in the MIC value from previous study. For instance, substitution by CH_2CH_2CH3, OCH_2CH_3, $OCH_2CH_2CH_3$, $O(CH_2)_3CH_3$, and OPh groups decreased the MIC 5–10 folds further. In contrast, substitution of the same position with smaller groups such as H increased the MIC from 0.25 μg/mL to 5 μg/mL (Coxon et al., 2013). Thus, inhibition by TAC can be optimized further by taking into account the nature of the binding pocket. Similar to TAC, ISO also has a sulfur containing group (Fig. 1). C61S mutation confers resistance to ISO as well (Grzegorzewicz et al., 2012). Therefore, the sulfur atom of activated ISO is also able to form a di-sulfide bond with Cys61 like TAC and bind in the flavonoid-binding pocket. In fact, both TAC and ISO could be activated first by *Mtb*EthA (Rv3854, Monooxygenase) to produce the reduced intermediates with thiol groups (Kordulakova et al., 2007; Qian and Ortiz de Montellano, 2006). And two thiol groups are able to form a di-sulfide bond. Such a mode of binding would place the bifurcated tri-carbons at the distal end of ISO into the fatty acid binding channel, obstructing fatty acid binding. Thus, we propose that through metabolism, ISO inhibits the dehydratase activity of the *Mtb*HadAB complex using a mechanism similar to TAC and flavonoids.

The hotdog fold was first identified in the structure of *Escherichia coli* FabA (PDB code 1MKA, 1MKB) (Leesong et al., 1996) and subsequently in 4-hydroxybenzoyl-CoA thioesterase from *Pseudomonas* sp. strain CSB (PDB code 1BVQ) (Benning et al., 1998). Later on, it was recognized to be widespread in eukaryotes, bacteria, and archaea, and to be involved in a range of cellular processes, from thioester hydrolysis, to phenylacetic acid degradation and transcriptional regulation of fatty acid biosynthesis (Dillon and Bateman, 2004). Therefore, the hotdog fold is both an ancient and ubiquitous motif, with members found in the three branches of life. Both HadA and HadB were predicted to be single hotdog fold structures (Castell et al., 2005; Sacco et al., 2007). Interestingly, our structural study showed a complex double hotdog fold structure of *Mtb*hadAB heterodimer which is different from previously solved hotdog fold structures. The primary sequence of HadA is very different from that of HadB. In spite of this, both proteins exhibit an almost identical hotdog fold. In the crystal structure of the *Mtb*HadAB hetero-dimer, the hotdog folds of HadA and HadB sit adjacent to each other, forming a non-canonical double hotdog fold. While HadA binds the substrate, HadB contributes the catalytically critical amino acids. This distinguishes the *Mtb*HadAB double hotdog fold from that of FabA (Fig. S8). In the crystal structures of FabA from *E. coli*, two monomers of the protein are observed forming a homo-dimer that has two catalytic sites positioned in vicinity of the dimer interface. In contrast, the 2-enoyl CoA hydratase 2 (MFE2) domain of *Candida tropicalis* (PDB code 1PN4) exhibits a double hotdog fold with only one functional catalytic site (Koski et al., 2004). With regards to this, the enzyme is similar to the *Mtb*HadAB complex (Fig. S8). A Dali search (Holm and Rosenstrom, 2010) showed that the overall structure of *Mtb*HadAB is most similar to MFE-2 hydratase 2 (Koski et al., 2005). The *r.m.s.d.* between corresponding Cα positions of the two structures is 2.0 Å. However, unlike the hotdog folds of the *Mtb*HadAB complex, the two hotdog folds of MFE2 are covalently linked via a long loop traversing the two β-sheets. Moreover, the αHD helix of the MFE2 dehydratase domain that is embedded laterally in the central β-sheet is highly distorted. There is also a fungal FAS dehydratase domain that depicts the type of the triple hotdog fold (Jenni et al., 2007; Leibundgut et al., 2007). Therefore, the non-canonical double hotdog fold exhibited by the *Mtb*HadAB complex is different from known assemblies of hotdog folds. More importantly, the unique configuration of the double hotdog fold adopted by *Mtb*HadAB results in formation of a deep pocket that vertically traverses the fatty acid binding channel. The significance of this finding for the development of new anti-TB drugs can be surmised from the structural observation that flavonoids and possibly TAC and ISO bind in this pocket, which disrupts binding of the substrate. This scuttles biosynthesis of mycolic acids and forms the basis for the anti-mycobacterial effect exerted by these compounds. The insights gained from the structures of the *Mtb*HadAB complex thus open up new avenues for accelerating development of next generation of clinically useful drugs for the treatment of all types of TB.

MATERIALS AND METHODS

Cloning

Rv0635 gene (encoding HadA) was amplified from *M. tuberculosis* H37Rv genomic DNA by polymerase chain reaction (PCR) using forward primer 5′-TTCCATATGGTGGCGTTGAGCGCAGACAT-3′ and reverse primer 5′-CCGCTCGAGTCACGCAGCGCCATCAGAAA-3′. *Nde*I and *Xho*I sites are underlined. The amplified gene was sub-cloned into the pRSFDuet-1 vector (Novagen). *Rv0636* gene (encoding HadB) from the same source was PCR-amplified using primers 5′-CGCGGATCCATGGCGCTGCGTGAGTTCAG-3′ and 5′-CCGCTCGAGCTACGCTAACTTCGCCGAGG-3′. *Bam*HI and *Xho*I sites are underlined. After restriction digestion, the amplified product was ligated into the pGEX-6p-1 vector (GE Healthcare). Recombinant plasmids pRSFDuet-1-HadA and pGEX-6p-1-HadB were co-transformed into *E. coli* BL21 (DE3) for expression of the *Mtb*HadAB complex. Clones were verified by DNA sequencing prior to expression studies.

Production of *Mtb*HadAB complex

For expression of GST-tagged *Mtb*HadAB complex, cells were grown in Terrific Broth medium supplemented with 100 μg/mL ampicilin and 50 μg/mL kanamycin at 37°C. When OD_{600} reached 0.8, the culture was induced with 0.3 mmol/L isopropyl-β-D-thiogalactopyranoside (IPTG) and the culture was further incubated at 16°C for 20 h. Cells were harvested by centrifugation at 4,200 ×*g* for 30 min and re-suspended in PBS, pH 7.4. After High voltage breaker (Avestin) treatment, the lysate was centrifuged at 30,700 ×*g* for 30 min at 4°C to remove cell debris. The supernatant was loaded onto a column containing Glutathione Sepharose 4B resin (GE Healthcare) pre-equilibrated with PBS, pH 7.4. The column was washed with buffer A (50 mmol/L Tris-HCl, pH 8.0, 300 mmol/L NaCl, and 10% (*v/v*) glycerol) to remove non-specifically bound proteins. To remove the GST-tag linked to the N-terminus of HadB, PreScission protease (GE Healthcare) was added and the mixture was incubated overnight at 4°C. The flow through containing tag-less *Mtb*HadAB complex was collected, exchanged in buffer containing 50 mmol/L Tris-HCl, pH 8.0, 50 mmol/L NaCl, and 10% (*v/v*) glycerol, and applied to a Resource Q anion exchange column (GE Healthcare) equilibrated with the same buffer. *Mtb*HadAB complex bound to the column was eluted with a 0.05–1 mol/L linear gradient of NaCl. Fractions containing *Mtb*HadAB were pooled, concentrated and further purified on a Superdex 75 16/200 gel filtration column (GE Healthcare) equilibrated with buffer containing 50 mmol/L Tris-HCl, pH 8.0, 150 mmol/L NaCl, and 10% (*v/v*) glycerol. Pure *Mtb*HadAB eluted in a single peak. The protein was concentrated to 6 mg/mL for crystallization. SeMet labeled *Mtb*HadAB was expressed in methionine-free M9 media supplemented with 60 mg/L L-SeMet. Labeled *Mtb*HadAB was purified using the same protocol as native protein. The final yield of native and SeMet labeled protein complex was 1.5 mg and 1 mg per litre culture, respectively.

Crystallization and structure determination

Crystallization screening experiments were first carried out at 16°C using vapor diffusion method. Drops were dispensed in 96-well plates by a Mosquito liquid handling system (TTP Labtech). Each crystallization drop contained 200 nL of *Mtb*HadAB protein (6 mg/mL) mixed with 200 nL of reservoir solution. Commercially available sparse matrix screens were used for screening crystallization conditions. Crystals grew in a solution containing 100 mmol/L HEPES, pH 7.5, 26% (*w/v*) PEG4000. The flavonoid inhibitors were co-crystallized with *Mtb*HadAB in the same condition. The SeMet- *Mtb*HadAB derivative crystals appeared in a different reservoir solution containing 100 mmol/L Bis-Tris, pH 6.5, 25% (*w/v*) PEG3350. All the crystals were optimized by varying the concentration of precipitant. Diffraction data were collected at 100 K on beamlines BL1A, BL5A, and BL17A at KEK (Tsukuba, Japan) after crystals were soaked in the reservoir solutions supplemented with 20% (*v/v*) glycerol for 10 s and flash frozen in liquid nitrogen. HKL2000 software (Otwinowski and Minor, 1997) was used for processing data. The structure of *Mtb*HadAB was determined by the single-wavelength anomalous diffraction (SAD) method, using the peak data from selenium derivative. All the four expected selenium sites, one in HadA and three in HadB, were located using SHELXD program (Schneider and Sheldrick, 2002; Sheldrick, 2010), and the initial phases were calculated using PHENIX (Echols et al., 2012). The model of *Mtb*HadAB complex was auto-built and then refined using native data. Manual adjustments to the model were made using COOT (Emsley and Cowtan, 2004) and refined using PHENIX. Structures of *Mtb*HadAB bound with flavonoids were solved by molecular replacement (MR) method with Phaser program (McCoy et al., 2007) in CCP4 (Potterton et al., 2003), using the native *Mtb*HadAB structure as a search template. Data collection and refinement statistics are listed in Table S1. Structural figures were prepared using PyMol (Schrodinger, 2010).

Analytical ultracentrifugation

Sedimentation velocity experiments were performed on a Beckman XL-I analytical ultracentrifuge at 16°C. Purified protein sample was diluted with buffer to 400 μL with an A_{280} absorption of about 0.7. Sample was loaded into a conventional double-sector quartz cell and mounted in a Beckman four-hole An-60 Ti rotor. Absorption data were collected at 60,000 rpm at a wavelength of 280 nm. Continuous mass distribution was calculated from the sedimentation velocity data using the SEDFIT software program.

Isothermal titration calorimetric assay

The binding affinities of butein, 2',4,4'-trihydroxychalcone and fisetin against *Mtb*HadAB were assayed using an ITC-200 microcalorimeter (MicroCal) device at 25°C. *Mtb*HadAB samples were placed in the reaction cell at a concentration of 0.025 mmol/L in buffer containing 20 mmol/L Tris-HCl, pH 8.0 and 150 mmol/L NaCl. For each binding assay, flavonoids at a concentration of 0.6 mmol/L were titrated into the *Mtb*HadAB samples. The titration consisted of an initial injection of 0.4 μL followed by 20 injections of 2 μL every 2 min. The titration data and binding plot were analyzed with program MICROCAL ORIGIN. Control plots (blank) were subtracted in every titration data.

Accession codes

The coordinates and structure factor files for native and flavonoid-bound *Mtb*HadAB have been deposited in the PDB, under accession codes 4RLJ, 4RLT, 4RLU and 4RLW, respectively.

ACKNOWLEDGEMENTS

We would like to thank staff members of the Core Facility for Protein Research of the Institute of Biophysics, CAS; especially Yi Han, Ya Wang, and Xiaoxia Yu for their technical support. This work was supported by grants from the Strategic Priority Research Program of the Chinese Academy of Sciences (grant no. XDB08020200), the National Basic Research Program (973 Program) (Nos. 2014CB542800, 2014CBA02003, 2011CB915501, and 2011CB91 0304), and the National Natural Science Foundation of China (Grant No. 813300237).

ABBREVIATIONS

BUN, butein; ETH, ethionamide; FSE, fisetin; HCC, 2',4,4'-trihydroxychalcone; INH, isoniazide; ISO, isoxyl; ITC, isothermal titration calorimetry; MAs, mycolic acids; MDR, multidrug-resistant; *Mtb*, *Mycobacterium tuberculosis*; TAC, thiacetazone; TB, tuberculosis; TLM, thiolactomycin; XDR, extensively drug-resistant.

COMPLIANCE WITH ETHICS GUIDELINES

Yu Dong, Xiaodi Qiu, Neil Shaw, Yueyang Xu, Yuna Sun, Xuemei Li, Jun Li and Zihe Rao declare that they have no conflict of interest.

This article does not contain any studies with human or animal subjects performed by the any of the authors.

REFERENCES

Asselineau C, Asselineau J, Lanéelle G, Lanéelle MA (2002) The biosynthesis of mycolic acids by Mycobacteria: current and alternative hypotheses. Prog Lipid Res 41:501–523

Banerjee A, Dubnau E, Quemard A, Balasubramanian V, Um KS, Wilson T, Collins D, de Lisle G, Jacobs WR Jr (1994) inhA, a gene encoding a target for isoniazid and ethionamide in *Mycobacterium tuberculosis*. Science 263:227–230

Belardinelli JM, Morbidoni HR (2012) Mutations in the essential FAS II beta-hydroxyacyl ACP dehydratase complex confer resistance to thiacetazone in *Mycobacterium tuberculosis* and *Mycobacterium kansasii*. Mol Microbiol 86:568–579

Benning MM, Wesenberg G, Liu R, Taylor KL, Dunaway-Mariano D, Holden HM (1998) The three-dimensional structure of 4-hydroxybenzoyl-CoA thioesterase from *Pseudomonas* sp. Strain CBS-3. J Biol Chem 273:33572–33579

Bhatt A, Molle V, Besra GS, Jacobs WR Jr, Kremer L (2007) The Mycobacterium tuberculosis FAS-II condensing enzymes: their role in mycolic acid biosynthesis, acid-fastness, pathogenesis and in future drug development. Mol Microbiol 64:1442–1454

Bloch K (1977) Control mechanisms for fatty acid synthesis in *Mycobacterium smegmatis*. Adv Enzymol Relat Areas Mol Biol 45:1–84

Brown AK, Bhatt A, Singh A, Saparia E, Evans AF, Besra GS (2007a) Identification of the dehydratase component of the mycobacterial mycolic acid-synthesizing fatty acid synthase-II complex. Microbiology 153:4166–4173

Brown AK, Papaemmanouil A, Bhowruth V, Bhatt A, Dover LG, Besra GS (2007b) Flavonoid inhibitors as novel antimycobacterial agents targeting Rv0636, a putative dehydratase enzyme involved in *Mycobacterium tuberculosis* fatty acid synthase II. Microbiology 153:3314–3322

Campbell JW, Cronan JE Jr (2001) Bacterial fatty acid biosynthesis: targets for antibacterial drug discovery. Annu Rev Microbiol 55:305–332

Cantaloube S, Veyron-Churlet R, Haddache N, Daffé M, Zerbib D (2011) The *Mycobacterium tuberculosis* FAS-II dehydratases and methyltransferases define the specificity of the mycolic acid elongation complexes. PLoS One 6:e29564

Carel C, Nukdee K, Cantaloube S, Bonne M, Diagne CT, Laval F, Zerbib D (2014) *Mycobacterium tuberculosis* proteins involved in mycolic acid synthesis and transport localize dynamically to the old growing pole and septum. PLoS One 9:e97148

Castell A, Johansson P, Unge T, Jones TA, Backbro K (2005) Rv0216, a conserved hypothetical protein from *Mycobacterium tuberculosis* that is essential for bacterial survival during infection, has a double hotdog fold. Protein Sci 14:1850–1862

Comas I, Gagneux S (2009) The past and future of tuberculosis research. PLoS Pathog 5:e1000600

Coxon GD, Craig D, Corrales RM, Vialla E, Gannoun-Zaki L, Kremer L (2013) Synthesis, antitubercular activity and mechanism of resistance of highly effective thiacetazone analogues. PLoS One 8:e53162

Dillon SC, Bateman A (2004) The Hotdog fold: wrapping up a superfamily of thioesterases and dehydratases. BMC Bioinform 5:109

Dover LG, Alahari A, Gratraud P, Gomes JM, Bhowruth V, Reynolds RC, Besra GS, Kremer L (2007) EthA, a common activator of thiocarbamide-containing drugs acting on different mycobacterial targets. Antimicrob Agents Chemother 51:1055–1063

Echols N, Grosse-Kunstleve RW, Afonine PV, Bunkoczi G, Chen VB, Headd JJ, McCoy AJ, Moriarty NW, Read RJ, Richardson DC et al (2012) Graphical tools for macromolecular crystallography in PHENIX. J Appl Crystallogr 45:581–586

Emsley P, Cowtan K (2004) Coot: model-building tools for molecular graphics. Acta Crystallogr D Biol Crystallogr 60:2126–2132

Gannoun-Zaki L, Alibaud L, Kremer L (2013) Point mutations within the fatty acid synthase type II dehydratase components HadA or HadC contribute to isoxyl resistance in *Mycobacterium tuberculosis*. Antimicrob Agents Chemother 57:629–632

Gao LY, Laval F, Lawson EH, Groger RK, Woodruff A, Morisaki JH, Cox JS, Daffe M, Brown EJ (2003) Requirement for kasB in Mycobacterium mycolic acid biosynthesis, cell wall impermeability and intracellular survival: implications for therapy. Mol Microbiol 49:1547–1563

Goldberg DE, Siliciano RF, Jacobs WR Jr (2012) Outwitting evolution: fighting drug-resistant TB, malaria, and HIV. Cell 148:1271–1283

Grzegorzewicz AE, Kordulakova J, Jones V, Born SE, Belardinelli JM, Vaquie A, Gundi VA, Madacki J, Slama N, Laval F et al (2012) A common mechanism of inhibition of the *Mycobacterium tuberculosis* mycolic acid biosynthetic pathway by isoxyl and thiacetazone. J Biol Chem 287:38434–38441

He L, Zhang L, Liu X, Li X, Zheng M, Li H, Yu K, Chen K, Shen X, Jiang H et al (2009) Discovering potent inhibitors against the beta-hydroxyacyl-acyl carrier protein dehydratase (FabZ) of *Helicobacter pylori*: structure-based design, synthesis, bioassay, and crystal structure determination. J Med Chem 52:2465–2481

Holm L, Rosenstrom P (2010) Dali server: conservation mapping in 3D. Nucleic Acids Res 38:W545–W549

Jackson M, McNeil MR, Brennan PJ (2013) Progress in targeting cell envelope biogenesis in *Mycobacterium tuberculosis*. Future Microbiol 8:855–875

Jenni S, Leibundgut M, Boehringer D, Frick C, Mikolasek B, Ban N (2007) Structure of fungal fatty acid synthase and implications for iterative substrate shuttling. Science 316:254–261

Jiang D, Zhang Q, Zheng Q, Zhou H, Jin J, Zhou W, Bartlam M, Rao Z (2014) Structural analysis of *Mycobacterium tuberculosis* ATP-binding cassette transporter subunit UgpB reveals specificity for glycerophosphocholine. FEBS J 281:331–341

Kimber MS, Martin F, Lu Y, Houston S, Vedadi M, Dharamsi A, Fiebig KM, Schmid M, Rock CO (2004) The structure of (3R)-hydroxyacyl-acyl carrier protein dehydratase (FabZ) from *Pseudomonas aeruginosa*. J Biol Chem 279:52593–52602

Kirkpatrick AS, Yokoyama T, Choi KJ, Yeo HJ (2009) *Campylobacter jejuni* fatty acid synthase II: structural and functional analysis of beta-hydroxyacyl-ACP dehydratase (FabZ). Biochem Biophys Res Commun 380:407–412

Kordulakova J, Janin YL, Liav A, Barilone N, Dos Vultos T, Rauzier J, Brennan PJ, Gicquel B, Jackson M (2007) Isoxyl activation is required for bacteriostatic activity against *Mycobacterium tuberculosis*. Antimicrob Agents Chemother 51:3824–3829

Koski MK, Haapalainen AM, Hiltunen JK, Glumoff T (2004) A two-domain structure of one subunit explains unique features of eukaryotic hydratase 2. J Biol Chem 279:24666–24672

Koski KM, Haapalainen AM, Hiltunen JK, Glumoff T (2005) Crystal structure of 2-enoyl-CoA hydratase 2 from human peroxisomal multifunctional enzyme type 2. J Mol Biol 345:1157–1169

Kostrewa D, Winkler FK, Folkers G, Scapozza L, Perozzo R (2005) The crystal structure of PfFabZ, the unique beta-hydroxyacyl-ACP dehydratase involved in fatty acid biosynthesis of *Plasmodium falciparum*. Protein Sci 14:1570–1580

Kremer L, Douglas JD, Baulard AR, Morehouse C, Guy MR, Alland D, Dover LG, Lakey JH, Jacobs WR Jr, Brennan PJ et al (2000) Thiolactomycin and related analogues as novel anti-mycobacterial agents targeting KasA and KasB condensing enzymes in *Mycobacterium tuberculosis*. J Biol Chem 275:16857–16864

Krissinel E, Henrick K (2007) Inference of macromolecular assemblies from crystalline state. J Mol Biol 372:774–797

Labonte JW, Townsend CA (2013) Active site comparisons and catalytic mechanisms of the hot dog superfamily. Chem Rev 113:2182–2204

Leesong M, Henderson BS, Gillig JR, Schwab JM, Smith JL (1996) Structure of a dehydratase-isomerase from the bacterial pathway for biosynthesis of unsaturated fatty acids: two catalytic activities in one active site. Structure 4:253–264

Leibundgut M, Jenni S, Frick C, Ban N (2007) Structural basis for substrate delivery by acyl carrier protein in the yeast fatty acid synthase. Science 316:288–290

Marrakchi H, Laneelle MA, Daffe M (2014) Mycolic acids: structures, biosynthesis, and beyond. Chem Biol 21:67–85

Matthews BW (1968) Solvent content of protein crystals. J Mol Biol 33:491–497

McCoy AJ, Grosse-Kunstleve RW, Adams PD, Winn MD, Storoni LC, Read RJ (2007) Phaser crystallographic software. J Appl Crystallogr 40:658–674

Mdluli K, Slayden RA, Zhu Y, Ramaswamy S, Pan X, Mead D, Crane DD, Musser JM, Barry CE 3rd (1998) Inhibition of a *Mycobacterium tuberculosis* beta-ketoacyl ACP synthase by isoniazid. Science 280:1607–1610

Moynie L, Leckie SM, McMahon SA, Duthie FG, Koehnke A, Taylor JW, Alphey MS, Brenk R, Smith AD, Naismith JH (2013) Structural insights into the mechanism and inhibition of the beta-hydroxydecanoyl-acyl carrier protein dehydratase from *Pseudomonas aeruginosa*. J Mol Biol 425:365–377

Nguyen C, Haushalter RW, Lee DJ, Markwick PR, Bruegger J, Caldara-Festin G, Finzel K, Jackson DR, Ishikawa F, O'Dowd B et al (2014) Trapping the dynamic acyl carrier protein in fatty acid biosynthesis. Nature 505:427–431

Nishida CR, Ortiz de Montellano PR (2011) Bioactivation of antituberculosis thioamide and thiourea prodrugs by bacterial and mammalian flavin monooxygenases. Chem Biol Interact 192:21–25

Otwinowski Z, Minor W (1997) Processing of X-ray diffraction data collected in oscillation mode. Academic Press, New York, pp 307–326

Potterton E, Briggs P, Turkenburg M, Dodson E (2003) A graphical user interface to the CCP4 program suite. Acta Crystallogr D Biol Crystallogr 59:1131–1137

Qian L, Ortiz de Montellano PR (2006) Oxidative activation of thiacetazone by the *Mycobacterium tuberculosis* flavin monooxygenase EtaA and human FMO1 and FMO3. Chem Res Toxicol 19:443–449

Robert X, Gouet P (2014) Deciphering key features in protein structures with the new ENDscript server. Nucleic Acids Res 42: W320–W324

Sacco E, Covarrubias AS, O'Hare HM, Carroll P, Eynard N, Jones TA, Parish T, Daffe M, Backbro K, Quemard A (2007) The missing piece of the type II fatty acid synthase system from *Mycobacterium tuberculosis*. Proc Natl Acad Sci USA 104:14628–14633

Schneider TR, Sheldrick GM (2002) Substructure solution with SHELXD. Acta Crystallogr D Biol Crystallogr 58:1772–1779

Schrodinger, LLC (2010). The PyMOL Molecular Graphics System, Version 1.3r1.

Sheldrick GM (2010) Experimental phasing with SHELXC/D/E: combining chain tracing with density modification. Acta Crystallogr D Biol Crystallogr 66:479–485

Swarnamukhi PL, Sharma SK, Bajaj P, Surolia N, Surolia A, Suguna K (2006) Crystal structure of dimeric FabZ of *Plasmodium falciparum* reveals conformational switching to active hexamers by peptide flips. FEBS Lett 580:2653–2660

Takayama K, Wang C, Besra GS (2005) Pathway to synthesis and processing of mycolic acids in *Mycobacterium tuberculosis*. Clin Microbiol Rev 18:81–101

Tasdemir D, Lack G, Brun R, Ruedi P, Scapozza L, Perozzo R (2006) Inhibition of *Plasmodium falciparum* fatty acid biosynthesis: evaluation of FabG, FabZ, and FabI as drug targets for flavonoids. J Med Chem 49:3345–3353

Trivedi OA, Arora P, Sridharan V, Tickoo R, Mohanty D, Gokhale RS (2004) Enzymic activation and transfer of fatty acids as acyl-adenylates in mycobacteria. Nature 428:441–445

Wang L, Li J, Wang X, Liu W, Zhang XC, Li X, Rao Z (2013) Structure analysis of the extracellular domain reveals disulfide bond forming-protein properties of *Mycobacterium tuberculosis* Rv2969c. Protein Cell 4:628–640

WHO (2014) Global tuberculosis report 2014 (Geneva, World Health Organization)

Wong HC, Liu G, Zhang YM, Rock CO, Zheng J (2002) The solution structure of acyl carrier protein from *Mycobacterium tuberculosis*. J Biol Chem 277:15874–15880

Zhang L, Kong Y, Wu D, Zhang H, Wu J, Chen J, Ding J, Hu L, Jiang H, Shen X (2008a) Three flavonoids targeting the beta-hydroxyacyl-acyl carrier protein dehydratase from *Helicobacter pylori*: crystal structure characterization with enzymatic inhibition assay. Protein Sci 17:1971–1978

Zhang L, Liu W, Hu T, Du L, Luo C, Chen K, Shen X, Jiang H (2008b) Structural basis for catalytic and inhibitory mechanisms of beta-hydroxyacyl-acyl carrier protein dehydratase (FabZ). J Biol Chem 283:5370–5379

Zumla A, George A, Sharma V, Herbert N, Masham Baroness, Ilton BM (2013) WHO's 2013 global report on tuberculosis: successes, threats, and opportunities. Lancet 382:1765–1767

18

RBP-J is required for M2 macrophage polarization in response to chitin and mediates expression of a subset of M2 genes

Julia Foldi[1], Yingli Shang[2], Baohong Zhao[3], Lionel B. Ivashkiv[1,3]✉, Xiaoyu Hu[2]✉

[1] Graduate Program in Immunology and Microbial Pathogenesis, Weill Cornell Graduate School of Medical Sciences, New York, NY 10021, USA
[2] School of Medicine and Institute for Immunology, Tsinghua University, Beijing 100084, China
[3] Arthritis and Tissue Degeneration Program, Hospital for Special Surgery, New York, NY 10021, USA
✉ Correspondence: ivashkivl@hss.edu (L. B. Ivashkiv), xiaoyuhu@tsinghua.edu.cn (X. Hu)

ABSTRACT

Development of alternatively activated (M2) macrophage phenotypes is a complex process that is coordinately regulated by a plethora of pathways and factors. Here, we report that RBP-J, a DNA-binding protein that integrates signals from multiple pathways including the Notch pathway, is critically involved in polarization of M2 macrophages. Mice deficient in RBP-J in the myeloid compartment exhibited impaired M2 phenotypes *in vivo* in a chitin-induced model of M2 polarization. Consistent with the *in vivo* findings, M2 polarization was partially compromised *in vitro* in *Rbpj*-deficient macrophages as demonstrated by reduced expression of a subset of M2 effector molecules including arginase 1. Functionally, myeloid *Rbpj* deficiency impaired M2 effector functions including recruitment of eosinophils and suppression of T cell proliferation. Collectively, we have identified RBP-J as an essential regulator of differentiation and function of alternatively activated macrophages.

KEYWORDS macrophages, RBP-J, M2, arginase, chitin

Lionel B. Ivashkiv and Xiaoyu Hu have contributed equally to this work.

INTRODUCTION

Macrophages are versatile cells with diverse functions in inflammation, tissue remodeling, angiogenesis, and tumor immunity. They respond to a wide variety of environmental cues to regulate immunity and inflammation by sensing microbial pathogens, secreting cytokines and inflammatory mediators, and presenting antigens to T cells. Depending on environmental signals, macrophages can display a spectrum of activation states ranging from classically activated, M1 inflammatory macrophages, to various alternatively activated M2 macrophages that are involved in immune regulation and tissue repair (Murray and Wynn, 2011; Ivashkiv, 2013). M1 macrophages are characterized by production of high levels of inflammatory mediators in response to stimulation by pathogen-associated molecular patterns and/or inflammatory cytokines. In contrast, M2 macrophages express less inflammatory mediators and play a key role in wound healing and resolution of inflammation (Mosser and Edwards, 2008; Murray et al., 2014). One mechanism by which M2 macrophages limit inflammation and restore homeostasis is suppression of T cell proliferation, which is mediated at least in part by a key M2 effector molecule arginase 1 (Arg1) (Chawla et al., 2011). Besides their suppressive functions, M2 macrophages are also involved in host defense against helminthes by recruiting eosinophils to the sites of infection. M2 polarization is driven by products of mast cells, T_H2 cells, and basophils including IL-4 that signals through the Jak-STAT signaling pathway to activate the latent transcription factor STAT6 (Chawla et al., 2011). IL-4-signaling promotes M2 polarization by upregulating expression of genes including those encoding Arg1, mannose receptor (MR), and Ym1, hallmarks of M2 phenotypes. In addition to IL-4-STAT6, a number of pathways and transcription factors such as PPARγ and IRF4 are also critically involved in driving M2 polarization (Odegaard et al., 2007; Satoh et al., 2010).

Recombinant recognition sequence binding protein at the Jκ site (RBP-J, also named CSL or CBF1) is the central nuclear mediator of canonical Notch signaling whose activation leads to transcription of Notch target genes. The evolutionarily conserved Notch signaling pathway regulates cell proliferation, apoptosis, and cell fate decisions during development and adult tissue homeostasis (Kopan and Ilagan, 2009). Recent studies using global expression analysis and chromatin immunoprecipitation deep-sequencing (ChIP-seq) have revealed genome-wide Notch-RBP-J targets in various systems including hematopoiesis (Hamidi et al., 2011), neural stem cell differentiation (Li et al., 2012), Epstein-Barr virus infection (Zhao et al., 2011), and T-lymhoblastic leukemia/lymphoma (Palomero et al., 2006; Wang et al., 2011). In the immune system, the most established functions for Notch signaling is regulation of development of lymphoid T and B cell lineages (Radtke et al., 2010; Yuan et al., 2010). Less well characterized, however, is the role of Notch signaling in myeloid cell differentiation and function. Notch-RBP-J signaling has been shown to control development and differentiation of multiple myeloid lineages including granulocyte/monocyte progenitors, dendritic cells, and osteoclasts (Klinakis et al., 2011; Lewis et al., 2011; Zhao et al., 2012). Although Notch signaling is not required for normal development of macrophage populations under homeostatic conditions, recently mounting evidence suggests that Notch signaling has profound effects on macrophage activation and polarization. Constitutive expression of Notch pathway components and constitutive activities of Notch signaling have been detected on macrophages of both human and mouse origin (Monsalve et al., 2006; Fung et al., 2007; Hu et al., 2008; Zhang et al., 2010). Notch signaling in macrophage regulates expression of pro- and anti-inflammatory mediators in response to macrophage activating signals such as LPS (Hu et al., 2008; Zhang et al., 2010; Zhang et al., 2012; Monsalve et al., 2009; Palaga et al., 2008; Tsao et al., 2011; Outtz et al. 2010). Moreover, we and others have shown that Notch-RBP-J is required for induction of a restricted subset of inflammatory M1 genes that includes Il12b (encodes p40 subunit shared by IL-12 and IL-23) and Nos2 (encodes iNOS). Thus, RBP-J promotes inflammatory M1 macrophage polarization in a focused manner that is important for processes such as anti-tumor responses and host defense against intracellular bacteria (Wang et al., 2010; Xu et al., 2012).

However, little is known about the role of the Notch-RBP-J pathway in M2 macrophages. As the M1 and M2 phenotypes are on extreme ends of the macrophage polarization spectrum and can oppose each other, we hypothesized that M2 responses may be augmented in RBP-J-deficient macrophages, which was supported by initial results that the M2-promoting factor JmjD3 is superinduced after TLR stimulation of RBP-J-deficient cells (Xu et al., 2012). In this study, we investigated the involvement of RBP-J in M2 macrophage polarization. Surprisingly, we found that RBP-J is critically involved in M2 polarization and function *in vivo* and *in vitro*, and regulates a restricted subset of key M2 effector molecules such as Arg1. Thus, RBP-J is the first transcription factor described to play a role in expression of transcriptional modules that are components of M1 and M2 phenotypes. This suggests that RBP-J may contribute to the complex 'mixed' M1-M2 phenotype that has been described *in vivo*. In addition, the common function of iNOS and Arg1 in suppressing T cell proliferation suggests an important role for RBP-J in restricting T cell proliferation in various settings of macrophage polarization.

RESULTS

RBP-J promotes M2 macrophage function *in vivo*

The Notch signaling pathway and RBP-J are generally thought to play a positive role in promoting classical M1 macrophage polarization (Wang et al., 2010; Xu et al., 2012). To investigate the functional roles of the Notch pathway in alternatively activated macrophage polarization, we generated mice conditionally deficient in *Rbpj* in their myeloid compartments (*Rbpj* cKO) by crossing *Rbpj*$^{flox/flox}$ animals to animals with a *Lyz2*-Cre as described previously (Hu et al., 2008). We then subjected *Rbpj* cKO mice and wildtype (WT) littermate control animals to an *in vivo* model of chitin-induced M2 polarization. Chitin is an *N*-acetyl-β-D-glucosamine polysaccharide and a major structural component of helminthes, fungi, and anthropods. It has been previously shown that intraperitoneal administration of chitin recruits macrophages with the M2 phenotype to the peritoneal cavity, which are important for the subsequent recruitment of eosinophils (Satoh et al., 2010; Reese et al., 2007). One day after intraperitoneal administration of chitin, we found that the total number of peritoneal exudate cells (PECs) did not significantly differ between WT control and RBP-J-deficient animals (data not shown). Next, we examined populations of macrophages in PECs by flow cytometry. Consistent with previous reports, CD11b$^+$F4/80$^+$ macrophages were present in the PECs from WT mice (Fig. 1A). Furthermore, the percentage of CD11b$^+$F4/80$^+$ macrophages in PECs was comparable between WT and *Rbpj* cKO mice after chitin administration (one representative experiment shown in Fig. 1A and cumulative data from three independent experiments shown in Fig. 1B), suggesting that *Rbpj* deficiency did not significantly alter the total macrophage population in response to an M2 stimulus *in vivo*. Next, we used a well established *in vivo* model of chitin-induced functional polarization of M2 macrophages, where an important M2 function is recruitment of eosinophils to the sites of inflammation (Satoh et al., 2010; Reese et al., 2007). As measured by percentage of CD45$^+$SiglecF$^+$ cells as previously described (Satoh et al., 2010), chitin-induced recruitment of eosinophils was significantly reduced in *Rbpj* cKO mice compared with WT controls (one representative experiment shown in Fig. 1C and cumulative data from 6 pairs of mice shown in Fig. 1D), suggesting that *Rbpj* deficiency in the myeloid

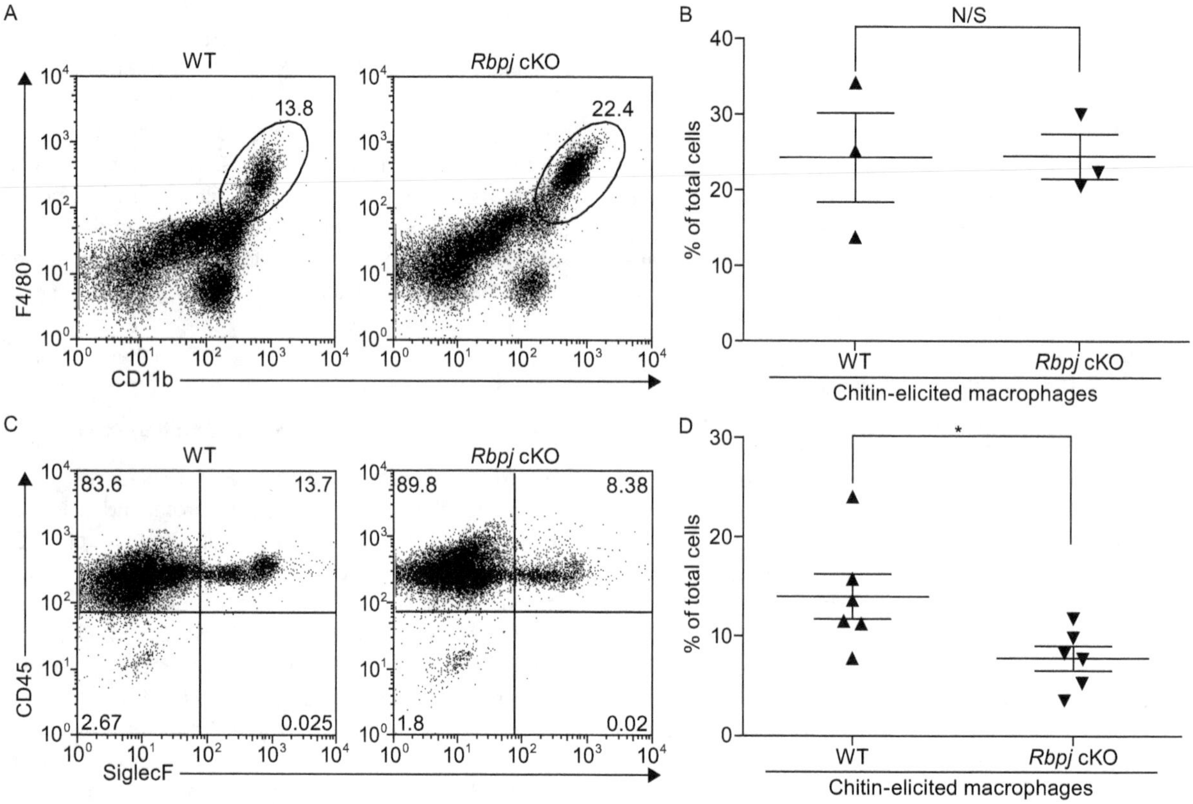

Figure 1. A role for *Rbpj* in eosinophil recruitment in response to chitin administration. (A) Expression of CD11b and F4/80 in peritoneal exudates cells (PECs) harvested from WT and *Rbpj* cKO mice 1 d after peritoneal injection with chitin. Circles and numbers indicate percentage of macrophages ($CD11b^+F4/80^+$) in total PECs. Results from one representative experiment are shown. (B) Cumulative data showing percentage of macrophages as in (A) from three independent experiments. (C) Expression of CD45 and SiglecF in PECs harvested from WT and *Rbpj* cKO mice 1 d after peritoneal injection with chitin. Quadrants and numbers indicate percentage of eosinophils ($CD45^+SiglecF^+$) in total PECs. (D) Cumulative data showing percentage of eosinophils as in (C) using six pairs of littermate mice with desired genotypes. Errors bars indicate s.d. $^*P < 0.05$ (two-tailed Student's *t*-test).

lineage impairs M2 macrophage function. Taken together, in the model of chitin-induced M2 polarization, recruitment of macrophages to peritoneal cavity does not require RBP-J, but RBP-J is indispensable for optimal functional polarization of M2 macrophages *in vivo*.

RBP-J promotes M2 polarization *in vivo*

Next, we wished to investigate the mechanisms underlying promotion of M2 function by RBP-J and examined phenotypes of chitin-elicited macrophages. A previous report has demonstrated that after chitin injection, classical M2 markers including arginase 1 (Arg1), mannose receptor (MR), and Ym1 are mainly expressed in macrophages but not in other PECs such as B cells or eosinophils (Satoh et al., 2010). Therefore, we assessed expression of M2 macrophage-associated genes in chitin-elicited PECs and found that expression of genes encoding Arg1 and MR was significantly lower in PECs isolated from *Rbpj* cKO mice compared with WT controls (Fig. 2A). In addition, expression of mRNA encoding another M2 marker, Ym1, also showed a trend towards decreased expression in *Rbpj* cKO mice (Fig. 2A), suggesting that *Rbpj* deficiency in the myeloid cells leads to impaired M2 polarization *in vivo* as manifested by reduced expression of a subset of M2-associated genes. However, RBP-J deficiency did not result in a global defect in M2 gene program as expression of *Marco* and *Jmjd3*, two prototypical M2-associated genes was not reduced in *Rbpj* cKO mice (Fig. 2B), indicating a restricted effect on M2 phenotype that is similar to the restricted effects of RBP-J deficiency on the M1 phenotype (Xu et al., 2012). In addition, the expression of an M1 macrophage-associated gene encoding IL-12p40 subunit was unchanged after chitin administration (Fig. 2C). Of note, we and others have shown that IL-12p40 upregulation following LPS treatment is attenuated in *Rbpj*-deficient bone-marrow derived macrophages (BMDMs) (Wang et al., 2010; Xu et al., 2012). Our results shown here are not in contradiction with previously published results regarding regulation of IL-12p40 expression because chitin administration did not upregulate IL-12p40 mRNA expression in

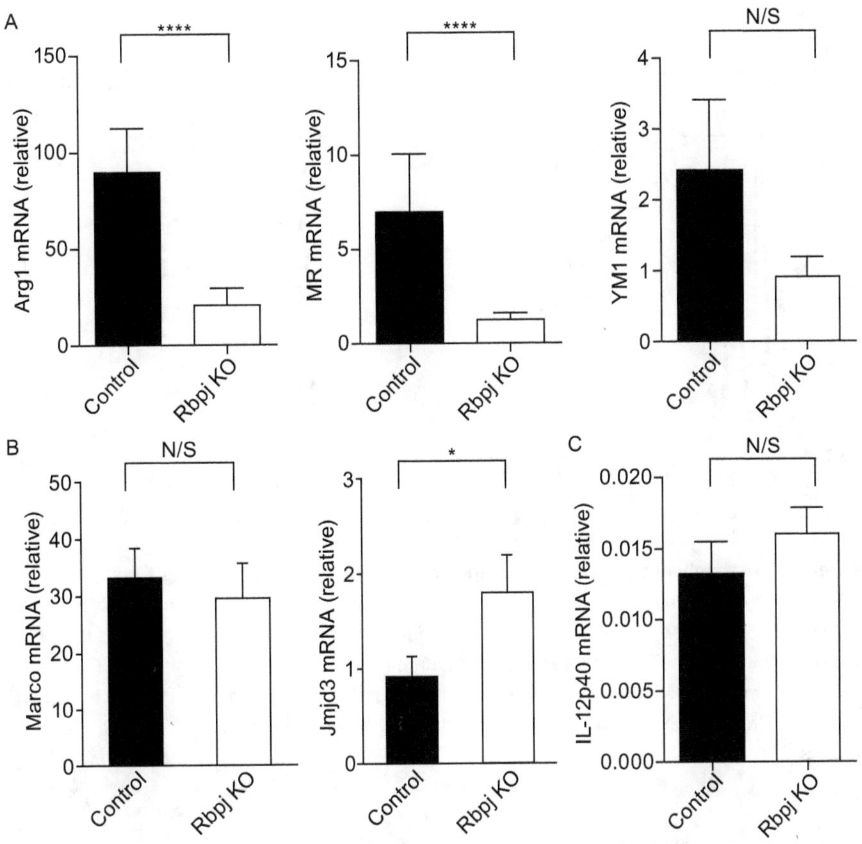

Figure 2. Crucial role for *Rbpj* in regulating M2 macrophage polarization in response to chitin administration *in vivo*. Total mRNA was prepared from PECs isolated from WT and *Rbpj* cKO mice 24 h after administration of chitin, and mRNA expression of Arg1, MR, Ym1, Marco, Jmjd3, and IL-12p40 (relative to GAPDH mRNA) was measured using quantitative real-time PCR. $^{****}P < 0.0001$ (two-tailed Student's *t*-test). $^{*}P < 0.05$ (two-tailed Student's *t*-test). Results are representative of four independent experiments, each with two-four pairs of littermate mice (analyzed individually).

PECs and IL-12p40 levels remain near baseline and low compared to the expression of the M2 genes encoding Arg1 and MR (data not shown). These results show that RBP-J is required for development of macrophages that express Arg1 *in vivo* in the chitin-induced model of M2 polarization.

RBP-J is required for M2-mediated suppression of T cell proliferation

One important function of M2 macrophages is the suppression of T cell proliferation and expression of Arg1 is thought to be critical for M2-mediated suppression of T cells (Pesce et al., 2009). As RBP-J promotes expression of M2-associated genes including *Arg1*, we hypothesized that RBP-J may be involved in inhibition of T cell responses by M2 macrophages and assessed the role of RBP-J in mediating this M2 function both *ex vivo* and *in vitro*. For *ex vivo* experiments, we co-cultured PECs isolated from *Rbpj* cKO mice and littermate control mice following intraperitoneal chitin administration with CFSE-labeled OT-II transgenic lymph node (LN) cells in the presence of cognate OVA peptide. Consistent with previous reports (Pesce et al., 2009), we found that compared to PECs isolated from PBS-injected WT mice, chitin-elicited PECs severely inhibited $CD4^+$ T cell proliferation in response to OVA (Fig. 3A, filled gray histogram versus black line), demonstrating a potent suppressive function of chitin-polarized M2 macrophages. However, when chitin-elicited PECs from *Rbpj* cKO mice were used in the co-culture, $CD4^+$ T cell proliferation was comparable to that mediated by control, PBS-treated, PECs (Fig. 3A, gray line), implicating a loss of suppressive function in *Rbpj*-deficient cells and thus suggesting that RBP-J plays a crucial role in the inhibition of T cell proliferation by chitin-elicited PECs. To confirm this result, we turned to an *in vitro* system of T cell suppression and used BMDMs isolated from *Rbpj* cKO and WT littermate control mice cultured in the presence of M-CSF either with or without IL-4 for 24 h to induce alternative activation before the addition of CFSE-labeled LN cells. Similar to our results with *ex vivo* PECs, we found that WT BMDMs treated with IL-4 strongly inhibited $CD4^+$ T cell proliferation as compared to BMDMs cultured without IL-4 (Fig. 3B, left panel). This inhibition was attenuated when *Rbpj*-deficient BMDMs were used in the co-cultures (Fig. 3B, right panel). Taken together, these results show that RBP-J

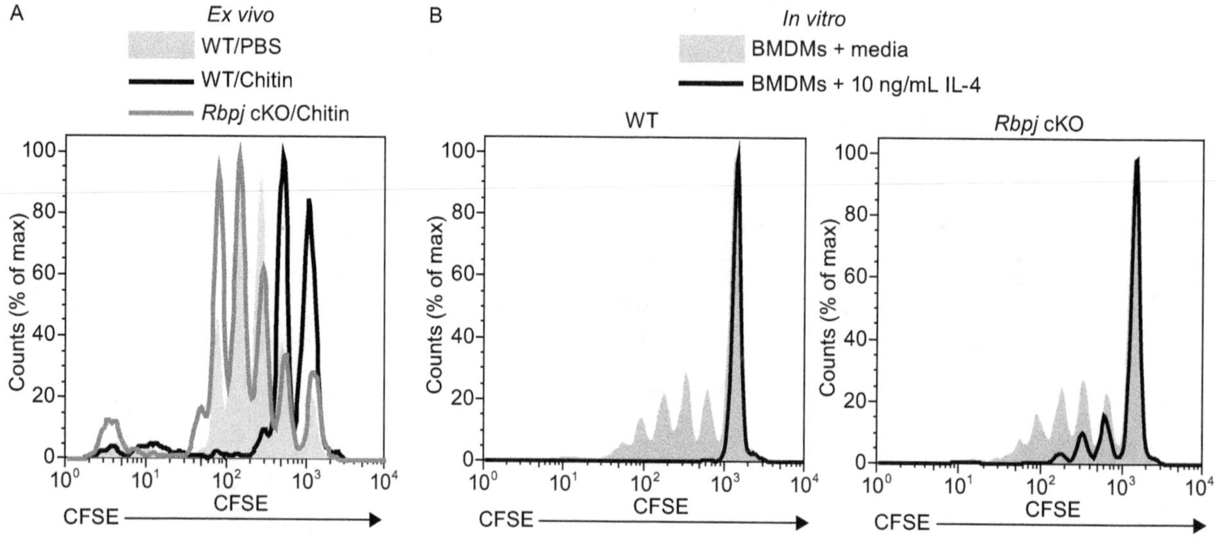

Figure 3. *Rbpj*-deficient alternatively activated macrophages are defective in suppressing T cell proliferation. (A) PECs from control, PBS-injected WT mice (filled gray histogram), and day-1 chitin-elicited PECs from WT mice (black line) and *Rbpj* cKO (grey line) were co-cultured with CFSE-labeled LN cells from OT-II transgenic mice and cognate OVA peptide. Proliferation of $CD4^+$ cells was examined at 96 h of co-culture. Results are representative of three independent experiments. (B) CFSE-labeled OT-II transgenic LN cells were co-cultured with control (filled gray histogram) or IL-4-treated (10 ng/mL, black line) BMDMs from WT or *Rbpj* cKO mice and cognate OVA peptide. Proliferation of $CD4^+$ cells was examined at 96 h of co-culture. Results are representative of two independent experiments.

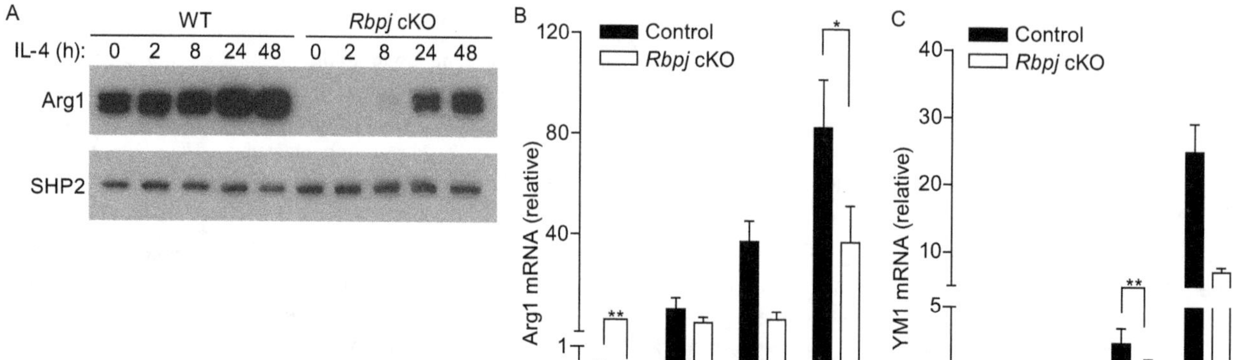

Figure 4. A crucial role for *Rbpj* in regulating expression of a subset of M2 macrophage genes. BMDMs from WT and *Rbpj* cKO mice were stimulated with IL-4 (10 ng/mL) for the indicated times. (A) Immunoblot analysis of whole cell lysates using antibodies recognizing Arg1. Levels of SHP2 served as loading controls. Results are representative of three independent experiments. (B and C) mRNA expression (relative to GAPDH mRNA) was measured using quantitative real-time PCR. Cumulative results from three independent experiments are shown (errors bars indicate s.d.). $P < 0.05$, $^{**}P < 0.01$ (two-tailed Student's *t*-test).

promotes M2 suppressive functions including inhibition of T cell proliferation, likely through regulation of expression of M2 effector molecules such as Arg1.

RBP-J promotes expression of M2-associated molecules

As we observed a requirement for RBP-J in maintaining high levels of M2-asscoiated genes *in vivo* in chitin-elicited PECs (Fig. 2A), we wished to further characterize RBP-J-mediated regulation of M2 genes. We utilized the *in vitro* system of M-CSF-differentiated BMDMs from *Rbpj* cKO and littermate control mice and treated the cells with IL-4 after 5 days of M-CSF-induced differentiation to induce M2 macrophage polarization. As expected, Arg1 protein and mRNA expression was induced by IL-4 in a time-dependent manner in control WT macrophages (Fig. 4A and 4B). Interestingly, *Rbpj*-deficient BMDMs expressed substantially less Arg1 protein and mRNA than BMDMs isolated from littermate controls prior to IL-4 treatment (Fig. 4A and 4B, 0 h time points). Nevertheless, *Rbpj*-deficient BMDMs were able to upregulate Arg1 mRNA and protein expression in response to IL-4 stimulation (Fig. 4A and 4B, lanes 6–10). However, due to the drastically reduced baseline expression, Arg1 expression was still markedly lower post IL-4 stimulation in *Rbpj*-deficient cells compared with Arg1 levels in control macrophages (Fig. 4A and 4B). These results implicate that RBP-J plays an essential role in maintaining baseline expression of Arg1 in macrophages but is dispensable for its IL-4-induced upregulation. In contrast, baseline mRNA expression of Ym1 was similar between WT and *Rbpj*-deficient BMDMs; however, IL-4-induced upregulation was attenuated in the absence of *Rbpj* (Fig. 4C). Taken together, our results indicate that RBP-J regulates expression of M2-associated molecules by either maintaining their basal levels in macrophages or promoting gene induction in response to M2-polarizing factors such as IL-4.

As Arg1 is a key M2 effector molecule whose basal expression is largely dependent on RBP-J in macrophages, we wished to further elucidate the mechanisms by which RBP-J regulates Arg1 expression. It has been previously established that the *Arg1* gene expression is controlled by several transcription factors such as STAT6 and CAAT/enhancer binding protein-β (C/EBPβ) (Chawla et al., 2011; El Kasmi et al., 2008). Therefore, we examined activation and expression of these transcription factors in *Rbpj*-deficient and control macrophages. We found that activation of STAT6 as assessed by its tyrosine phosphorylation was similar in WT and *Rbpj*-deficient macrophages after IL-4 treatment (Fig. 5A), showing that IL-4-induced STAT6 signaling remains intact in the absence of RBP-J. This result is consistent with the observation that IL-4-inducible Arg1 expression is not affected by *Rbpj*-deficiency (Fig. 4A and 4B), suggesting that RBP-J is dispensable for IL-4 signaling and STAT6 activation in macrophages. In addition, C/EBPβ protein expression was unchanged in *Rbpj*-deficient macrophages (data not shown), implying that neither STAT6 nor C/EBPβ is the target of RBP-J-mediated regulation. Recently, the transcription factor, interferon regulatory factor (IRF) 8, has also been implicated in regulating Arg1 expression under various experimental conditions (Pourcet et al., 2011). We found that both at baseline and following IL-4 stimulation, IRF8 protein expression did not significantly differ between WT and *Rbpj*-deficient macrophages (Fig. 5B), suggesting that IRF8 expression is not regulated by RBP-J under these experimental conditions. Furthermore, BMDMs from *Irf8*-null mice showed normal expression and induction of Arg1 mRNA (Fig. 5C) and protein (Fig. 5D) following IL-4 stimulation, confirming that IRF8 is dispensable for IL-4-induced expression of Arg1 in macrophages. Taken together, our results indicate that RBP-J determines basal macrophage Arg1 expression but does regulate transcription factors such as STAT6, C/EBPβ, and IRF8 that induce Arg1 expression in response to environmental cues.

DISCUSSION

In this study we demonstrated that functional polarization of M2 macrophages *in vitro* and *in vivo* is critically regulated by a transcription factor RBP-J that integrates signals from multiple input sources including the Notch signaling pathway. RBP-J expression in myeloid cells promoted M2-mediated immune effector functions including recruitment of eosinophils and suppression of T cell proliferation by regulating expression of a subset of M2-associated genes. Our results identified RBP-J as a key regulator of M2 polarization and suggest that targeting pathways that modulate RBP-J activities has the potential of altering macrophage polarization processes and subsequent immune effector responses.

One notable feature of regulation of the M2 gene program by RBP-J is its selectivity. In contrast to many previously described M2-driving transcription factors such as IRF4 and Stat6 that promote expression of a wide-range of M2-associated genes, RBP-J only selectively enhances expression of a subset of M2 genes including those encoding Arg1 and MR while imposing minimal or even inhibitory effects on other well-characterized M2 signature genes. The highly selective nature of RBP-J on the M2 gene program is reminiscent of its role in M1 polarization, in which only a subset of M1 signature genes are regulated by RBP-J. Thus, it emerges that instead of a so-called "master transcription regulator" by a conventional definition, RBP-J serves to focus the macrophage responses on certain modules appropriate for the current microenvironment and cellular context. For example, within the broad range of various M1 effector functions, RBP-J predominantly focuses on the "bacterial host defense" module by targeting IL-12 and iNOS with minimal impact on "acute inflammation" module by sparing IL-1β. Similarly, under M2 polarizing conditions, RBP-J mainly controls eosinophil recruitment and T cell responses by targeting Arg1 while leaving Stat6-dependent processes mostly untouched. Given the highly specific and focused nature of RBP-J-mediated regulation and the fact that RBP-J integrates signals from diverse environmental cues, we propose to conceptualize RBP-J as a modular regulator of M1-M2 macrophage polarization processes whose activities can be fine-tuned in a context-dependent manner.

Given the fact that RBP-J promotes certain components of canonical M1 as well as M2 responses, it is anticipated that RBP-J would possess complex and context-dependent biological functions *in vivo* that could not be attributed as simply promoting or suppressing inflammatory and immune responses. Indeed, this is often the case. While we and others have shown that RBP-J is essential for host defense against certain intracellular bacterial species, expression of RBP-J in tumor-associated macrophages (TAMs) is suppressive for T-cell mediated anti-tumor immune responses (Franklin et al., 2014). This notion is in line with our observations that RBP-J in macrophages is required for inhibition of T cell proliferation under both *ex vivo* and *in vitro* M2-

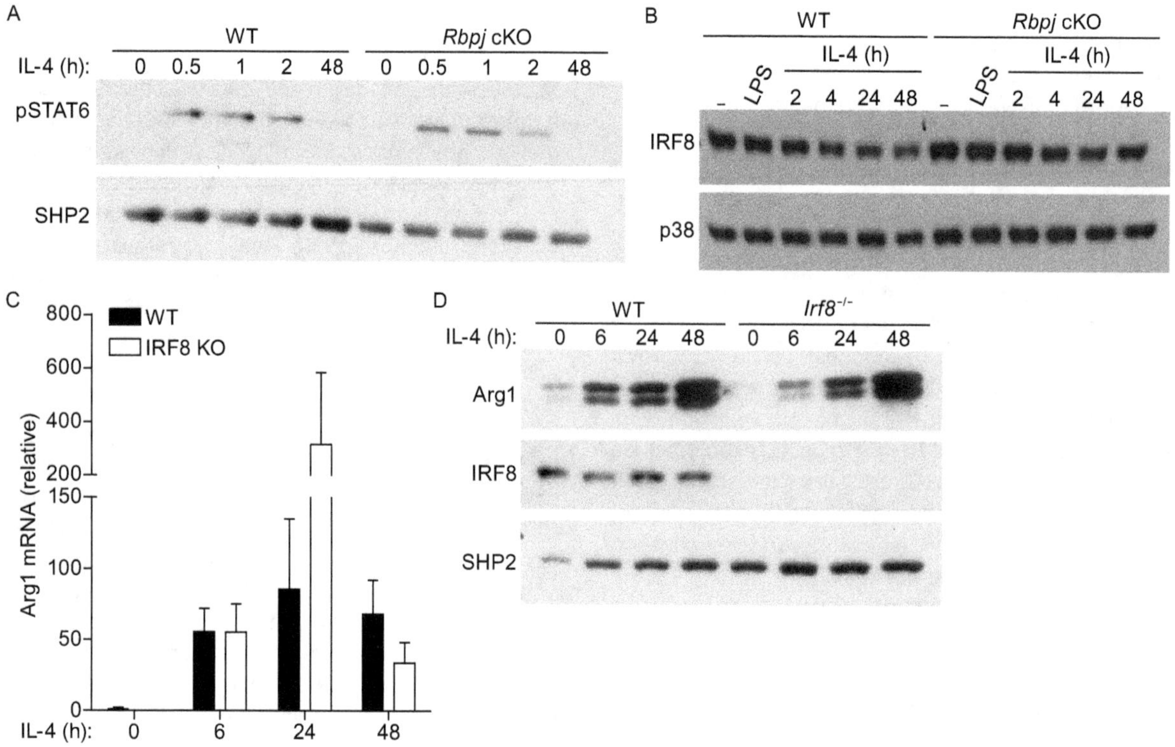

Figure 5. RBP-J regulates Arg1 expression in macrophages independently of STAT6 and IRF8. (A) BMDMs from WT and *Rbpj* cKO mice were stimulated with 100 ng/mL of IL-4 for the indicated times. Whole cell lysates were analyzed with immunoblotting using antibodies recognizing STAT6 phosphorylated on Tyr641 (pSTAT6). Levels of SHP2 served as loading controls. Results are representative of three independent experiments. (B) BMDMs from WT and *Rbpj* cKO mice were stimulated with 10 ng/mL of IL-4 and 1 ng/mL of LPS for the indicated times. Whole cell lysates were analyzed with immunoblotting using antibody recognizing IRF8. Levels of p38 served as loading controls. Results are representative of three independent experiments. (C) mRNA expression of Arg1 (relative to GAPDH mRNA) was measured in BMDMs from WT and *Irf8*$^{-/-}$ mice. Cumulative results from two independent experiments (errors bars indicate s.d.) are shown. (D) BMDMs from WT and *Irf8*$^{-/-}$ mice were stimulated with 10 ng/mL of IL-4 for the indicated times. Whole cell lysates were analyzed with immunoblotting using antibodies against Arg1 and IRF8. Levels of SHP2 served as loading controls. Results are representative of three independent experiments.

polarizing conditions (Fig. 3A and 3B) and suggests that antagonizing RBP-J activities in TAMs may represent a novel strategy of anti-tumor immune-therapy via boosting T cell responses.

In addition to biological consequences of RBP-J-mediated regulation, we also sought to investigate molecular mechanisms by which RBP-J regulated M2 gene expression especially expression of Arg1 given its role as a critical M2 effector molecule and its drastic regulation by RBP-J at the protein level (Fig. 4A). RBP-J appeared to maintain constitutive Arg1 expression in macrophages without significantly altering the capacity of macrophages to upregulating Arg1 expression in response to stimuli such as IL-4. These observations raised several interesting questions regarding mechanisms of Arg1 regulation. Firstly, is *Arg1* a direct Notch target gene or is its expression regulated by RBP-J via indirect means? These two possibilities can be distinguished by application of the Notch signaling inhibitors such as γ-secretase inhibitors or usage of Notch component-deficient animals in the future experiments. Secondly, the result that RBP-J did not regulate IL-4-activated Stat6 signaling suggests that regulation of Arg1 is Stat6-independent. Indeed, consistent with this notion, IL-4-Stat6 signaling is also dispensable for M2 effector functions in the chitin-induced model (Reese et al., 2007). Previous reports have shown that leukotriene B4 and its high-affinity receptor BLT1 are essential for chitin-induced M2 polarization (Reese et al., 2007) and it will be of great interest to investigate the involvement of leukotriene in RBP-J-mediated regulation of M2 phenotypes in the future.

MATERIALS AND METHODS

Mice

Rbpj$^{flox/flox}$ mice were kindly provided by Tasuku Honjo. Mice with a myeloid-specific deletion of *Rbpj* were generated by crossing *Rbpj*$^{flox/flox}$ animals to animals with a *Lyz2*-Cre on the C57/BL6 background as described previously (Hu et al., 2008). Mice with

Rbpj$^{flox/flox}$*Lyz2*-Cre genotypes were used for experiments. Gender-matched littermates with *Rbpj*$^{+/+}$*Lyz2*-Cre or *Rbpj*$^{flox/+}$*Lyz2*-Cre genotypes were used as controls. OT-II transgenic mice were obtained from the Jackson Laboratory. The experiments using mice were approved by the Hospital for Special Surgery and Tsinghua University Institutional Animal Care and Use Committees.

Chitin administration

Chitin was obtained from Sigma, washed three times in PBS, and sonicated on ice. The suspended solution was filtered through a 100 μmol/L strainer and diluted in 50 mL PBS. Approximately 800 ng of chitin was intraperitoneally injected to each mouse and peritoneal exudate cells (PECs) were collected 1 d after chitin injection for flow cytometry and gene expression analysis.

Flow cytometry

PECs were harvested and stained with antibodies specific to F4/80, CD11b, CD45, and/or SiglecF (BD). Cells were then washed three times with flow cytometry buffer. Data was acquired on a FACScan flow cytometer (BD) and analyzed using Cell Quest software (BD).

Cells and reagents

Murine BMDMs were obtained as described and maintained in DMEM supplemented with 10% FBS and 10% L929 cell supernatant as conditioned medium providing macrophage colony stimulating factor (M-CSF). After 5 days of culture, floating cells were discarded and attached macrophages were re-plated in 12-well plates overnight prior to stimulation. Recombinant mouse IL-4 was from Peprotech and used at 10 ng/mL unless otherwise noted.

mRNA isolation and real time PCR

RNA was extracted from whole cell lysates with an RNeasy Mini kit (Qiagen) and 0.5 μg of total RNA was reverse transcribed with a First Strand cDNA synthesis kit (Fermentas). Quantitative real time PCR (qRT-PCR) was performed in triplicate wells with an iCycler IQ thermal cycler and detection system (Biorad) using gene-specific primers. Threshold cycle numbers were normalized to triplicate samples amplified with primers specific for of glyceraldehyde-3-phosphate dehydrogenase (GAPDH).

Proliferation assay

CFSE-labeled OT-II transgenic LN cells were co-cultured with PECs or BMDMs from WT or *Rbpj*-deficient mice and cognate OVA peptide. Proliferation of CD4$^+$ cells was examined at 96 h of co-culture by assessing CFSE dilution using flow cytometry.

Immunoblotting analysis

Whole cell lysates were prepared by direct lysis in SDS loading buffer. For immunoblot analysis, lysates were separated by 7.5% SDS-PAGE and transferred to a PVDF membrane for probing with antibody. Polyclonal antibodies against arginase 1, IRF8, p38, and Shp2 were from Santa Cruz Biotechnology. pY-Stat6 antibody was from Cell Signaling Technology.

Statistical analysis

All statistical analyses were performed by using the Student's *t* test.

ACKNOWLEDGEMENTS

We thank Tasuko Honjo for providing *Rbpj*$^{flox/flox}$ mice, Keiko Ozato for providing *Ifr8*$^{-/-}$ mice, and Karmen Au for technical assistance. LBI and BZ are supported by the grants from NIH. XH is supported by the National Basic Research Program (973 Program) (No. 2015CB943201), National Natural Science Foundation of China Young Investigator Award 81422019, and funds from Peking-Tsinghua Center of Life Sciences.

AUTHOR CONTRIBUTIONS

Conceived and designed the experiments: JF LBI XH. Performed the experiments: JF YS BZ. Analyzed the data: JF LBI XH. Wrote the manuscript: JF LBI XH.

ABBREVIATIONS

Arg1, arginase 1; ChIP-seq, chromatin immunoprecipitation deep-sequencing; GAPDH, glyceraldehyde-3-phosphate dehydrogenase; M-CSF, macrophage colony stimulating factor; MR, mannose receptor; PECs, peritoneal exudate cells; qRT-PCR, quantitative real time PCR

COMPLIANCE WITH ETHICS GUIDELINES

Julia Foldi, Yingli Shang, Baohong Zhao, Lionel B. Ivashkiv, and Xiaoyu Hu declare that they have no conflict of interest.

All institutional and national guidelines for the care and use of laboratory animals were followed.

REFERENCES

Chawla A, Nguyen KD, Goh YP (2011) Macrophage-mediated inflammation in metabolic disease. Nat Rev Immunol 11:738–749

El Kasmi KC, Qualls JE, Pesce JT, Smith AM, Thompson RW et al (2008) Toll-like receptor-induced arginase 1 in macrophages thwarts effective immunity against intracellular pathogens. Nat Immunol 9:1399–1406

Franklin RA, Liao W, Sarkar A, Kim MV, Bivona MR et al (2014) The cellular and molecular origin of tumor-associated macrophages. Science 344:921–925

Fung E, Tang SM, Canner JP, Morishige K, Arboleda-Velasquez JF et al (2007) Delta-like 4 induces notch signaling in macrophages: implications for inflammation. Circulation 115:2948–2956

Hamidi H, Gustafason D, Pellegrini M, Gasson J (2011) Identification of novel targets of CSL-dependent Notch signaling in hematopoiesis. PLoS One 6:e20022

Hu X, Chung AY, Wu I, Foldi J, Chen J et al (2008) Integrated regulation of Toll-like receptor responses by Notch and interferon-gamma pathways. Immunity 29:691–703

Ivashkiv LB (2013) Epigenetic regulation of macrophage polarization and function. Trends Immunol 34:216–223

Klinakis A, Lobry C, Abdel-Wahab O, Oh P, Haeno H et al (2011) A novel tumour-suppressor function for the Notch pathway in myeloid leukaemia. Nature 473:230–233

Kopan R, Ilagan MX (2009) The canonical Notch signaling pathway: unfolding the activation mechanism. Cell 137:216–233

Lewis KL, Caton ML, Bogunovic M, Greter M, Grajkowska LT et al (2011) Notch2 receptor signaling controls functional differentiation of dendritic cells in the spleen and intestine. Immunity 35:780–791

Li Y, Hibbs MA, Gard AL, Shylo NA, Yun K (2012) Genome-wide analysis of N1ICD/RBPJ targets in vivo reveals direct transcriptional regulation of Wnt, SHH, and hippo pathway effectors by Notch1. Stem Cells 30(4):741–752

Monsalve E, Perez MA, Rubio A, Ruiz-Hidalgo MJ, Baladron V et al (2006) Notch-1 up-regulation and signaling following macrophage activation modulates gene expression patterns known to affect antigen-presenting capacity and cytotoxic activity. J Immunol 176:5362–5373

Monsalve E, Ruiz-Garcia A, Baladron V, Ruiz-Hidalgo MJ, Sanchez-Solana B et al (2009) Notch1 upregulates LPS-induced macrophage activation by increasing NF-kappaB activity. Eur J Immunol 39:2556–2570

Mosser DM, Edwards JP (2008) Exploring the full spectrum of macrophage activation. Nat Rev Immunol 8:958–969

Murray PJ, Wynn TA (2011) Protective and pathogenic functions of macrophage subsets. Nat Rev Immunol 11:723–737

Murray PJ, Allen JE, Biswas SK, Fisher EA, Gilroy DW et al (2014) Macrophage activation and polarization: nomenclature and experimental guidelines. Immunity 41:14–20

Odegaard JI, Ricardo-Gonzalez RR, Goforth MH, Morel CR, Subramanian V et al (2007) Macrophage-specific PPARgamma controls alternative activation and improves insulin resistance. Nature 447:1116–1120

Outtz HH, Wu JK, Wang X, Kitajewski J (2010) Notch1 deficiency results in decreased inflammation during wound healing and regulates vascular endothelial growth factor receptor-1 and inflammatory cytokine expression in macrophages. J Immunol 185:4363–4373

Palaga T, Buranaruk C, Rengpipat S, Fauq AH, Golde TE et al (2008) Notch signaling is activated by TLR stimulation and regulates macrophage functions. Eur J Immunol 38:174–183

Palomero T, Lim WK, Odom DT, Sulis ML, Real PJ et al (2006) NOTCH1 directly regulates c-MYC and activates a feed-forward-loop transcriptional network promoting leukemic cell growth. Proc Natl Acad Sci USA 103:18261–18266

Pesce JT, Ramalingam TR, Mentink-Kane MM, Wilson MS, El Kasmi KC et al (2009) Arginase-1-expressing macrophages suppress Th2 cytokine-driven inflammation and fibrosis. PLoS Pathog 5: e1000371

Pourcet B, Feig JE, Vengrenyuk Y, Hobbs AJ, Kepka-Lenhart D et al (2011) LXRalpha regulates macrophage arginase 1 through PU.1 and interferon regulatory factor 8. Circ Res 109:492–501

Radtke F, Fasnacht N, Macdonald HR (2010) Notch signaling in the immune system. Immunity 32:14–27

Reese TA, Liang HE, Tager AM, Luster AD, Van Rooijen N et al (2007) Chitin induces accumulation in tissue of innate immune cells associated with allergy. Nature 447:92–96

Palomero T, Lim WK, Odom DT, Sulis ML, Real PJ et al (2006) NOTCH1 directly regulates c-MYC and activates a feed-forward-loop transcriptional network promoting leukemic cell growth. Proc Natl Acad Sci USA 103:18261–18266

Pesce JT, Ramalingam TR, Mentink-Kane MM, Wilson MS, El Kasmi KC et al (2009) Arginase-1-expressing macrophages suppress Th2 cytokine-driven inflammation and fibrosis. PLoS Pathog 5: e1000371

Pourcet B, Feig JE, Vengrenyuk Y, Hobbs AJ, Kepka-Lenhart D et al (2011) LXRalpha regulates macrophage arginase 1 through PU.1 and interferon regulatory factor 8. Circ Res 109:492–501

Radtke F, Fasnacht N, Macdonald HR (2010) Notch signaling in the immune system. Immunity 32:14–27

Reese TA, Liang HE, Tager AM, Luster AD, Van Rooijen N et al (2007) Chitin induces accumulation in tissue of innate immune cells associated with allergy. Nature 447:92–96

19

Applications of RNA interference high-throughput screening technology in cancer biology and virology

Shan Gao[1✉], Chen Yang[2✉], Shan Jiang[3], Xiao-Ning Xu[4], Xin Lu[5], You-Wen He[3], Annie Cheung[6], Hui Wang[7,8✉]

[1] Department of Oncology, John Radcliffe Hospital, Weatherall Institute of Molecular Medicine, University of Oxford, Oxford OX3 9DS, UK
[2] Department of Molecular Genetics and Microbiology, Stony Brook University, Stony Brook, NY 11794, USA
[3] Department of Immunology, Duke University, Durham, NC 27708, USA
[4] Department of Immunology, Chelsea & Westminster Hospital, Imperial College of London, 369 Fulham Road, London SW10 9NH, UK
[5] Nuffield Department of Clinical Medicine, Ludwig Institute for Cancer Research Oxford Branch, University of Oxford, Old Road Campus Research Building, Oxford OX3 7DQ, UK
[6] Department of Pathology, Queen Mary Hospital, University of Hong Kong, Pokfulam Road, Hong Kong, China
[7] Natural Environment Research Council, Centre for Ecology & Hydrology, Benson Lane, Wallingford, Oxfordshire OX10 8BB, UK
[8] Department of Zoology, University of Oxford, South Parks Road, Oxford OX1 3PS, UK
✉ Correspondence: shan.gao@oncology.ox.ac.uk (S. Gao), chen.yang.1@stonybrook.edu (C. Yang), hui.wang@zoo.ox.ac.uk (H. Wang)

ABSTRACT

RNA interference (RNAi) is an ancient intra-cellular mechanism that regulates gene expression and cell function. Large-scale gene silencing using RNAi high-throughput screening (HTS) has opened an exciting frontier to systematically study gene function in mammalian cells. This approach enables researchers to identify gene function in a given biological context and will provide considerable novel insight. Here, we review RNAi HTS strategies and applications using case studies in cancer biology and virology.

KEYWORDS RNA interference (RNAi), short interfering RNA (siRNA), short hairpin RNA (shRNA), high-throughput screening, cancer, virology

INTRODUCTION

The systematic gene networks remain challenging, after many genomes from different species have been sequenced within the last two decades (Adams et al., 2000; Lander et al., 2001; Waterston et al., 2002; Yu et al., 2002). Although the functions of an increasing number of gene products have been revealed, a major challenge in the post-genomic era remains to clarify the precise roles of these genes in specific biological processes and ultimately to develop new strategies to fight diseases. RNA interference (RNAi) technology (Fire et al., 1998) is a powerful tool to study gene function by silencing transcription. Significant progress has been made using RNAi technology in the investigation of the molecular basis of cancer development. Such studies have identified several novel oncogenes, such as *sequence similarity 83, member B* (*FAM83B*) in breast cancer (Cipriano et al., 2012), *LIM-homeodomain-containing transcription factor 1B* (*Lmx1b*), *p21-activated kinase 4*, *inhibitor of apoptosis-stimulating protein of p53* (*iASPP*), and stem cell transcription factor *Nanog* in ovarian cancer (Siu et al., 2010; Jiang et al., 2011; He et al., 2013; Siu et al., 2013), and *abnormal spindle homologue, microcephaly associated* (*ASPM*) in glioblastoma (Horvath et al., 2006). Significant progress has also been made in the field of host immune responses against virus infections using RNAi technology. Indeed, many novel genes involved in virus entry and replication in the host have been uncovered (Brass et al., 2008; Brass et al., 2009).

With such technological advances, it is now feasible to interrogate phenotypes associated with the loss-of-function of many genes in mammalian cell culture systems. Furthermore, it becomes possible to construct genome-wide RNAi libraries that systematically target every individual gene in a given genome to perform high-throughput screening (HTS) of specific phenotypes of interests (Silva et al., 2008). The combination of RNAi and HTS in cell culture systems, known as cell-based RNAi HTS, is a new and exciting frontier in basic and applied biology. Inevitably, such a powerful methodology has led to significant progress in many areas of cancer biology, i.e., the crosstalk of biological signaling and the identification and validation of cancer therapeutic targets (Moffat and Sabatini, 2006; Iorns et al., 2007).

In this review, we provide an overview of RNAi and RNAi HTS in cell systems. Different technical strategies are described using case studies as examples. Finally, discussions are made in the context of the existing problems with these screens.

RNAi

RNAi is an evolutionarily conserved mechanism that regulates gene expression in eukaryotic cells. In this process, double-stranded RNAs (dsRNAs) can suppress the expression of target genes in a homology-dependent manner (Hannon, 2002; Almeida and Allshire, 2005). The effectors of RNAi are short interfering RNAs (siRNAs) produced from long dsRNA substrates by an RNAse III enzyme called Dicer. These Dicer products are 20- to 25-bp-long dsRNAs with a characteristic 2-nt overhang at the 3′-end (Fire et al., 1998; Bernstein et al., 2001). The siRNA duplexes are incorporated into the RNA-induced silencing complex (RISC) containing a core Argonaute (AGO) protein that destroys a "passenger strand" and keeps the "guiding strand" that hybridizes to the target mRNA by complementary homology. Such a homology-based association allows the AGO to degrade the mRNA transcript, or to inhibit the mRNA translation, resulting in post-transcriptional gene silencing (Martinez et al., 2002; Hock and Meister, 2008).

In mammalian cells, long dsRNAs induce interferon responses, which are the first line of defense against viral infections, and subsequently lead to a global shutdown of protein synthesis (Reynolds et al., 2006). However, vectors expressing short hairpin RNAs (shRNAs) and/or synthetic siRNAs designed to mimic endogenous 21-nt siRNAs can be manually introduced into mammalian cells to avoid the interferon response, thus mediating gene silencing without significant adverse effects (Elbashir et al., 2001; Kim and Rossi, 2007). With the availability of completely sequenced genomes, siRNA or vector-based shRNA libraries can be constructed with specific designs to maximize the probability of potent target gene silencing and to minimize the risk of off-target effects (Huesken et al., 2005; Turner et al., 2008). To date, RNAi has been successfully developed to become a powerful experimental tool (Huang et al., 2013)

RNAi HTS

Traditionally, functional genetic studies have been performed by so called forward genetics, in which random gene mutations are generated by induction (e.g., radiation, chemical treatments, and/or insertional mutagenesis), and then mutants with specific phenotypes are identified by breeding and segregation processes (Gao et al., 2008; Lawson and Wolfe, 2011). However, such procedures are time-consuming and not easily applicable to mammalian cell systems. Conversely, reverse genetic strategies focus on the mutants of a gene of interest by observing the phenotype so as to determine the gene function (Wang et al., 2011). Traditional gene knockouts, either in cell systems or living organisms, are expensive, time-consuming, and unsuitable for genome-scale screens. Using RNAi technology, a gene that is required for a certain function can be silenced by the introduction of siRNAs, and the corresponding phenotype can be determined by appropriate assays. This fascinating link between phenotype and RNAi-mediated gene silencing has promoted rapidly growing applications of RNAi in functional genomics, signal transduction, and drug target discovery. As such, RNAi represents a major technological advance for performing large-scale screenings in cell culture systems (Echeverri and Perrimon, 2006). HTS is a procedure that supports large-scale experiments, allowing a researcher to quickly conduct thousands or millions of chemical screens or genetic tests simultaneously (Krausz, 2007).

RNAi HTS METHODS

In the initial stage of an RNAi HTS experiment, the purpose of the screening should be clear, such as the identification of regulators for cell proliferation, size, morphology, adhesion, death, division, or the efficacy of viral infection. These questions enable researchers to develop corresponding phenotypic assays that should be clearly defined and easily scored with interpretations in relevant biological contexts (Boutros and Ahringer, 2008). Then, one can choose the appropriate type of RNAi library to screen (i.e., either genome-wide or for a subset of gene families of interest) and develop robust and reproducible assays for identification, validation, and characterization of candidate genes (Fig. 1A). Technical details that should be considered include: i) delivery methodology of the RNAi, ii) raw data collection, iii) appropriate positive and negative controls that can be used to optimize the discrepancy between genuine signals and background noise, iv) statistical methods that can define the initial hits, v) secondary validation assays to filter the primary hits, and vi) function determination of selected hits (Fig. 1B) (Sharma and Rao, 2009).

Currently, there are two major types of RNAi libraries that are widely used by researchers: siRNA and vector-based shRNA libraries. siRNA can be chemically synthesized or generated from cDNA templates by RNase III via a technique known as esiRNA, in which gene cDNA with RNA polymerase

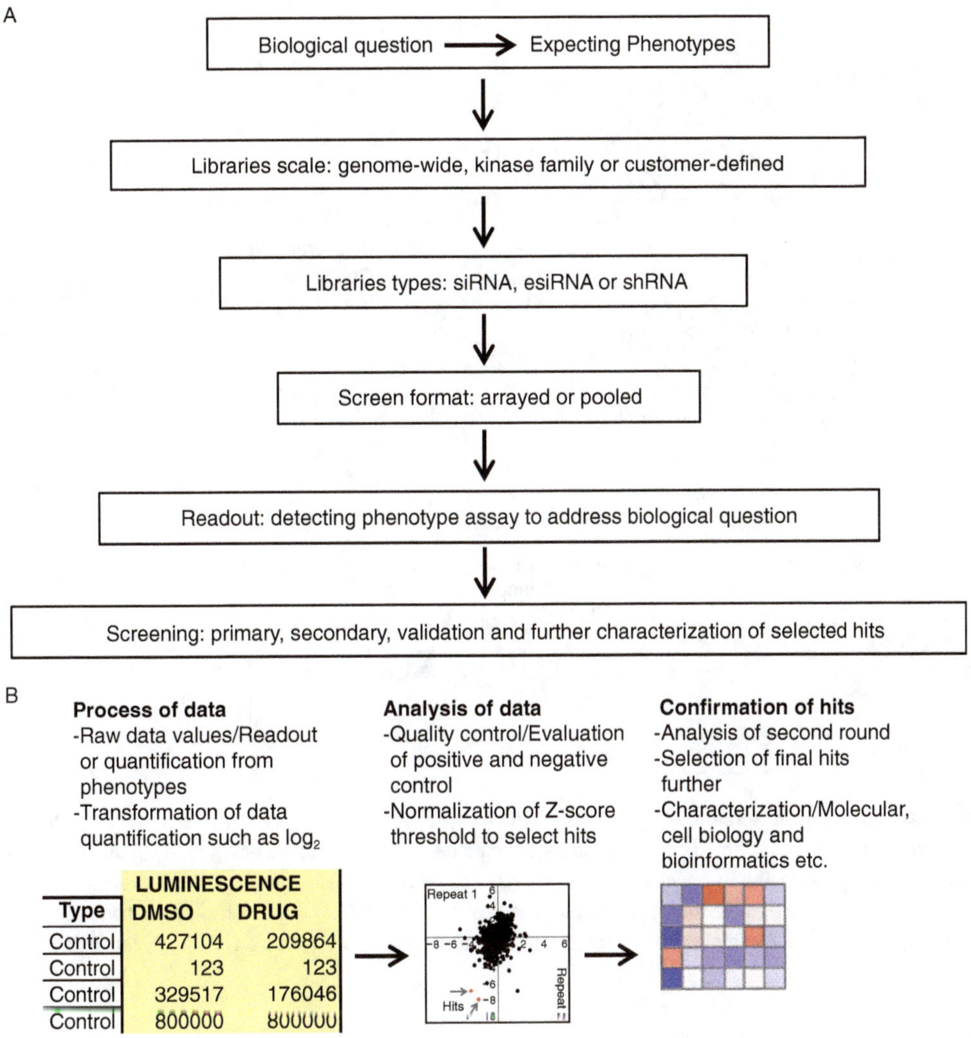

Figure 1. Flowchart of RNAi high-throughput screening. (A) The design and steps of a high-throughput screen. The choice of screen libraries such as scale, type, or format can be determined based on biological questions or the phenotypic assays developed. (B) Data analysis (Gao et al., unpublished data). Raw data can be analyzed according to your choice of statistical methods. Every experiment is subject to quality control (QC). If QC is passed, the primary hits are selected for a second round of screening. The final hits can be confirmed and further characterized using molecular biology, cell biology, or bioinformatics to determine their biological meanings.

promoter amplified by PCR is transcribed *in vitro*, and then its products is digested by RNase III to generate siRNAs that is similar to siRNAs generated *in vivo* by Dicer. These multiple silencing triggers result in higher effective gene silencing and lower off-target effects compared to single and pooled siRNAs (Yang et al., 2002; Kittler et al., 2004). Synthetic siRNA libraries are used for most situations for short-term gene silencing because these siRNA duplexes are not replicated and are progressively diluted as cells divide. Vector-based shRNA libraries are defined according to the type of viral vector used, e.g., retroviral, adenoviral, or lentiviral. Vector-based shRNA libraries are able to provide long-term and stable gene silencing because the vectors integrate into genomic DNA and are thus replicated. In particular, lentiviral-based shRNA libraries are quite useful in some cells, such as primary cells, that are difficult to transfect.

There are two distinguishable strategies used in RNAi screens. One is the array-based screen, and the other is pooled shRNA libraries coupled to next generation sequencing. In the array-based screen, both siRNA and shRNA can be used in this format. There are many factors to affect the generation of shRNAs, so it is hard to balance the concentration of every shRNA in HTS. However, the chemically synthesized siRNAs are easily to handle to titer their concentrations (Liu et al., 2010). Each gene of interest can be targeted by siRNA pools in one well, which is composed of three to six individual non-overlapping siRNAs, or by individual siRNAs in separated wells. After transfection (48–72 h), cells are divided into

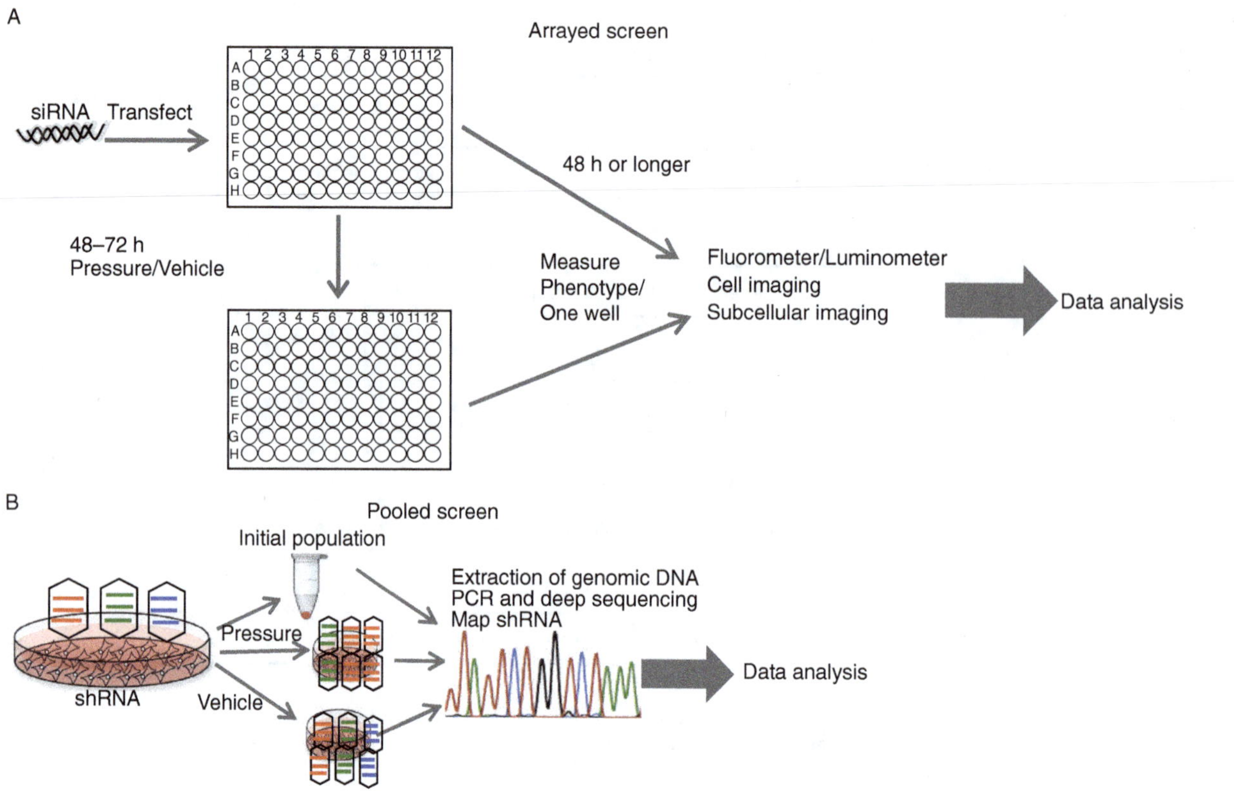

Figure 2. RNAi high-throughput screening approaches. (A) Arrayed screen using siRNA in a multiwell plate. siRNA can be reverse-transfected into cells for 48–72 h. Then, cells can be selected using some pressure or left for a longer time period to develop phenotypes, which can be determined using various readouts. (B) Pooled screen using pooled shRNA viral particles to infect target cells. After 48 h, cells can be divided into multiple groups. One group can be frozen as an initial population. Other groups can be treated with various pressures for some period. Then, relative enrichment of shRNAs can be analyzed using next generation sequencing.

different groups that can be treated with different selective pressures. For example, cells are treated with or without a drug for another 3–7 days and then examined for the phenotype of interest. In the case of using siRNA pools, screening is normally conducted in duplicate or triplicate in the first round, and then the pools are rescreened using the deconvoluted individual siRNAs to confirm final hits. In the pooled format, off-target effects of siRNAs are reduced as the concentration of each individual siRNA is decreased, while the total amount of siRNA molecules targeting the same mRNA species maintains the same as in the individual screening format.

In the individual siRNA screening format, in theory, different siRNAs targeting the same mRNA species should induce an identical phenotype. In reality, these siRNAs rarely lead to the same phenotype because different siRNAs are heterogeneous in inducing Dicer-mediated degradation of the target mRNA. Furthermore, a number of phenotypes only occur while the target is depleted to a certain level. Therefore, if any two individual siRNAs can confirm the same phenotype, they are considered as the final hits. Occasionally, a high proportion of a single siRNA hit may be ignored and subsequently lead to false negatives in the same assay (Fig. 2A) (Iorns et al., 2007).

In the pooled shRNA library coupled in next generation sequencing format, shRNAs are commonly cloned into specific vector backbones with unique barcodes. This approach is widely used for pooled competitive screening, so it is often called barcode screening and shows improvements in speed and scale compared to array-based screening. A large population of cells can be infected with a pool of vector-based shRNAs. After 48–72 h of infection, cells are then spilt into three or more groups. One group is immediately frozen as the initial population, the second group (or additional groups) is treated with specific selective pressure(s) for a longer period (14 days or longer), while the last group is treated with vehicle. After the selective pressure is applied, cells are harvested from the treatment groups, and genomic DNAs are extracted from these populations. Finally integrated shRNAs are recovered using PCR amplification followed by next generation sequencing.

If a given shRNA decreases cell viability, the relative abundance of that shRNA in the vehicle group will decrease when compared to the initial population. Similarly, the shRNA target gene under pressure will affect cell viability, so its relative abundance in the pressure group will correspondingly decrease when compared to the vehicle group, which is helpful to indicate the function or network connection of the target gene (Fig. 2B) (Sims et al., 2011; Corcoran et al., 2013). Alternatively, pooled shRNA can be used to infect different cell lines. After the selection period, PCR and next generation sequencing are employed to determine which shRNA reagents are under- or over-represented in these cell lines to define targets (Silva et al., 2008).

THE APPLICATION OF RNAi HTS TO CANCER BIOLOGY

The RNAi HTS approach is still at an early stage compared to many classical genetic screens, but it has already been used in a large number of studies that show some significant impact in a wide variety of fields, especially in cancer biology. Cancer cells acquire a set of mutated genes during carcinogenesis, and a vast amount of information about these mutated genes has been accumulated from whole exome sequencing and next generation sequencing (Berger et al., 2011; Banerji et al., 2012; Barbieri et al., 2012). However, that does not mean that all of these mutated genes prompt cancer development, rather than only a limited mutated genes causing the cancer (Ashworth et al., 2011). It is crucial to distinguish which mutated genes are a key driver for survival and proliferation of cancer cells. Moreover, developing potent and specific inhibitors targeting these key mutated genes represents a hot field in targeted cancer therapy, which has become an important part of many cancer treatment methods due to its precision in killing cancer cells with relatively few side effects as compared to traditional chemotherapies. However, precision often also means narrowness, which is an intrinsic drawback of targeted cancer therapy. Thus, these inhibitors often display less effective and limited activity in killing cancer cells and also allow cancer cells to develop drug resistance, one of the primary reasons for treatment failure (Guo and Wang, 2012). The major challenge and key step for cancer therapy is to identify the target that is essential for the survival and proliferation of the cancer cells, as well as biomarkers that are able to predict what types of cancers are sensitive or resistant to specific inhibitors. Therefore, RNAi HTS will be helpful in target selection and selective cancer patient treatment. Moreover, phenotypic assays to detect cell number and viability are well established. Thus, it is not surprising that RNAi HTS was applied to uncover mutant gene function or therapeutic target identification in neoplastic phenotypes in the earliest studies.

A genome-wide siRNA lethality screen has been performed to identify functional dependencies in two breast cell lines that were transformed using the same oncogene with different phenotypes. Proteasome genes were enriched among the 154 identified genes that show higher dependency in basal-like transformed cells than in myoepithelial-like transformed cells. Indeed, basal-like triple negative breast cancer (TNBC) lines are selectively sensitive to proteasome inhibitors compared to normal epithelial, luminal, and mesenchymal TNBC lines (Petrocca et al., 2013).

A genome-wide shRNA library in a pooled format has been applied to identify genes that regulate sensitivity to *RAS* mutant cancer cells in colorectal DLD-1 cells with and without a mutant form of the oncogene *KRAS*. A set of mitotic regulators, including ubiquitination, proteasome degradation of mitotic factors, and *PLK1*, increase the dependency of KRAS-mutant cells on mitotic checkpoints and progression. These KRAS-mutant cell lines are also more sensitive to treatment with a PLK1 inhibitor in both *in vivo* and xenograft models. PLK1 inhibitors (http://www.clinicaltrials.gov/ct2/results?term=PLK1&Search=Search) and siRNAs (http://www.tekmira.com/pipeline/tkm-plk1.php) are currently in clinical trials, and it will be very interesting to determine if RAS tumors display increased sensitivity in a clinical setting (Luo et al., 2009). MED12 was identified as a common determinant of drug resistance using 24,000 shRNAs targeting 8000 human genes in a lung cancer line harboring a translocation between *EML4* and the kinase *ALK*, which is sensitive to ALK inhibitors PF-02341066 (crizotinib) and NVP-TAE684. Furthermore *MED12* silencing causes resistance to various tyrosine receptor inhibitors by negatively regulating *TGF-βR2* via interaction with *TGF-βR2* (Huang et al., 2012).

Although several groups have made significant progress in the identification of key oncogenic events and cancer therapeutic targets using RNAi HTS, nearly all of these screens were conducted *in vitro*. Such *in vitro* screens result in some key and novel findings in the studies of cancer cells but are less able to recapitulate the complex interactions between tumors and their microenvironment, which is an important step toward understanding cancer cell growth in a more physiologic context, as it is not possible to design more rational treatments for cancers based on *in vitro* screens (Mbeunkui and Johann, 2009). Beronja and colleagues performed an *in vivo* genome-wide pooled lentiviral shRNA screen in normal embryonic epidermal tissue and a hyperproliferation of *Hras*^{G12V}-induced neoplasm in mice. A number of expected and unexpected genes that regulate embryonic epidermal growth were identified by the analysis of relative shRNA abundance. After eliminating genes that are essential under both conditions and regarded as housekeeping/viability genes, there are still ~250 candidates left that were defined as oncogene-specific growth regulators. They represent genes that could be targeted in cancer without causing any ill effect on normal tissue. Among the top *Hras*^{G12V}-dependent screen hits were the Wnt effector β-catenin and myeloid/lymphoid or mixed-lineage leukemia translocated to 6 (Mllt6). Silencing of either gene in *Hras*^{G12V}-expressing epidermal cells diminishes the formation of *Hras*^{G12V}-dependent squamous papillomas in mice and inhibits both the initiation and maintenance of human squamous cell carcinoma xenografts. These results demonstrate the feasibility of this *in vivo* screening approach and identify key regulators of oncogenic growth that may represent potential therapeutic targets (Beronja et al., 2013).

THE APPLICATION OF RNAi HTS IN VIROLOGY

Despite great effort in the antiviral drug development and vaccination research fields, viruses are still major threats to public health. Indeed, despite the availability of several different vaccines, influenza viruses infect up to one billion people globally, accounting for five million cases of severe disease and 250,000 to 500,000 deaths each year (Girard et al., 2005; Lambert and Fauci, 2010). In contrast, without a vaccine after 25 years of research, HIV has infected more than 70 million people to date and is responsible for 35 million deaths (WHO). Similarly, hepatitis C virus (HCV) infects 130–170 million people each year, causing acute and chronic infection (Madan et al., 2014). Dengue virus infects near 100 million people each year, resulting in half a million cases of hemorrhagic fever (Hussain and Asgari, 2014). In addition, viruses are also the major driving forces for certain types of tumors in humans. High risk types of human papillomaviruses (HPVs) are the causative agent of nearly all cervical cancers, which is one of the leading causes of mortality in women (Rositch et al., 2014). Similarly, Epstein-Barr virus (EBV), which infects 90% of adults worldwide, plays a key role in certain tumors, including Burkitt's lymphoma, Hodgkin's disease, and nasopharyngeal carcinomas (Pattle and Farrell, 2006).

The most effective way to prevent virus-related diseases is vaccines. However, in most cases, vaccines are not available or not always effective. Each year, the protection efficacy of trivalent inactivated influenza vaccines (TIVs) is only ~59% in the US adult population aged 18–65 years (Osterholm et al., 2012). Antiviral drugs are therefore critical in controlling viruses. The drugs currently in use mainly target virus proteins or genomes. However, it is very easy for viruses, especially RNA viruses, to develop resistance against such drugs due to the high mutation rates of the viral genomes. To that end, finding a more effective way to develop novel drugs has become a crucial research topic. It has been shown that pooled siRNAs targeting the Zaire Ebola virus (ZEBOV) key proteins including RNA polymerase L proteins, viral protein (VP) 24 and VP35 efficiently protect against ZEBOV in non-human primates, suggesting this will be useful treatment for other emerging viral infections in human (Geisbert et al., 2010).

Viruses rely on host cells to propagate. Therefore, understanding the involvement of host factors in virus infection may facilitate the discovery of potential drug candidates because host factors undergo much less mutagenic pressure than viral proteins and may have a universal function during different virus infections. With the development of HTS RNAi screens, it became possible to examine genome-wide interactions between viruses and host factors.

The Drosophila C virus (DCV), a picornavirus, was the first virus to undergo genome-wide RNAi screening to identify novel host factors required for internal ribosome entry site (IRES)-dependent viral translation (Cherry et al., 2005). The first virus in a mammalian system that was screened by genome wide RNAi is HCV in 2007. In an effort to discover novel druggable targets against HCV, a library of siRNAs targeting 4,000 human genes was used to identify genes that regulate HCV replication in Huh7-derived EN5-3 cells harboring an HCV subgenomic replicon (Fig. 3). Nine cellular genes, including members of the tumor necrosis factor/lymphotoxin signaling pathway, were identified (Table 1). Silencing these genes leads to inhibition of HCV replication, and the level of siRNA silencing of these host genes correlates well with the inhibition of HCV (Ng et al., 2007).

In 2008, three groups simultaneously reported genome-wide RNAi HTS for host factors required for HIV replication. Brass et al. developed a two-part siRNA screen to detect host factors involved in HIV infection (Brass et al., 2008). With 21,121 pools of siRNA, they identified 273 HIV-dependency factors in TZM-bl cells. Other than 36 host factors, including *CD4*, *CXCR4*, and components of *NF-κB* that were previously implicated in HIV pathogenesis, they revealed the involvement of the Golgi transport proteins Rab6 and Vps53 in HIV entry, TNPO3 in integration, and Med28 in transcription. Another group using a siRNA library composed of 22,329 pools of siRNA targeting 19,709 genes in HeLa P4/R5 cells detected 311 host factors, with an 18-gene overlap with Brass et al. (Zhou et al., 2008). They confirmed the involvement of the SP1/mediator complex and the *NF-κB* signaling pathway in HIV replication. Meanwhile, a study focusing on the early steps of HIV-1 infection revealed >200 genes in human 293T cells that may facilitate HIV infection. Among them, >40 genes specifically regulate the initiation of virus replication (Konig et al., 2008). Although each screen has an overlapping rate of approximately 6%, only three host factors were identified in all three screens, *MED6*, *MED7* (mediator complex), and *RelA* (*NF-κB* complex) (Friedel and Haas, 2011).

Due to the continuous outbreak of seasonal flu and the 2009 pandemic flu, influenza viruses have been vigorously studies using genome-wide RNAi screens. An initial attempt was performed in 2008, covering 90% of the Drosophila genome (Hao et al., 2008). Based on a *Renilla* luciferase reporter gene, >100 host factors were identified that alter influenza replication, including the cytochrome *c* oxidase subunit *COX6A1*, the ATPase *ATP6VoD1*, and the nuclear export factor *NXF1/TAP*. The interferon-inducible transmembrane proteins IFITM1, 2, and 3 were later discovered to restrict the early stage of influenza A virus replication via a siRNA screen in osteosarcoma cells (U2OS) (Brass et al., 2009). Further tests targeting 19,000 human genes were performed in human lung epithelial A549 cells in 2010 (Konig et al., 2010). Among the 295 host factors identified, 23 factors are necessary for influenza virus entry, and 10 factors are

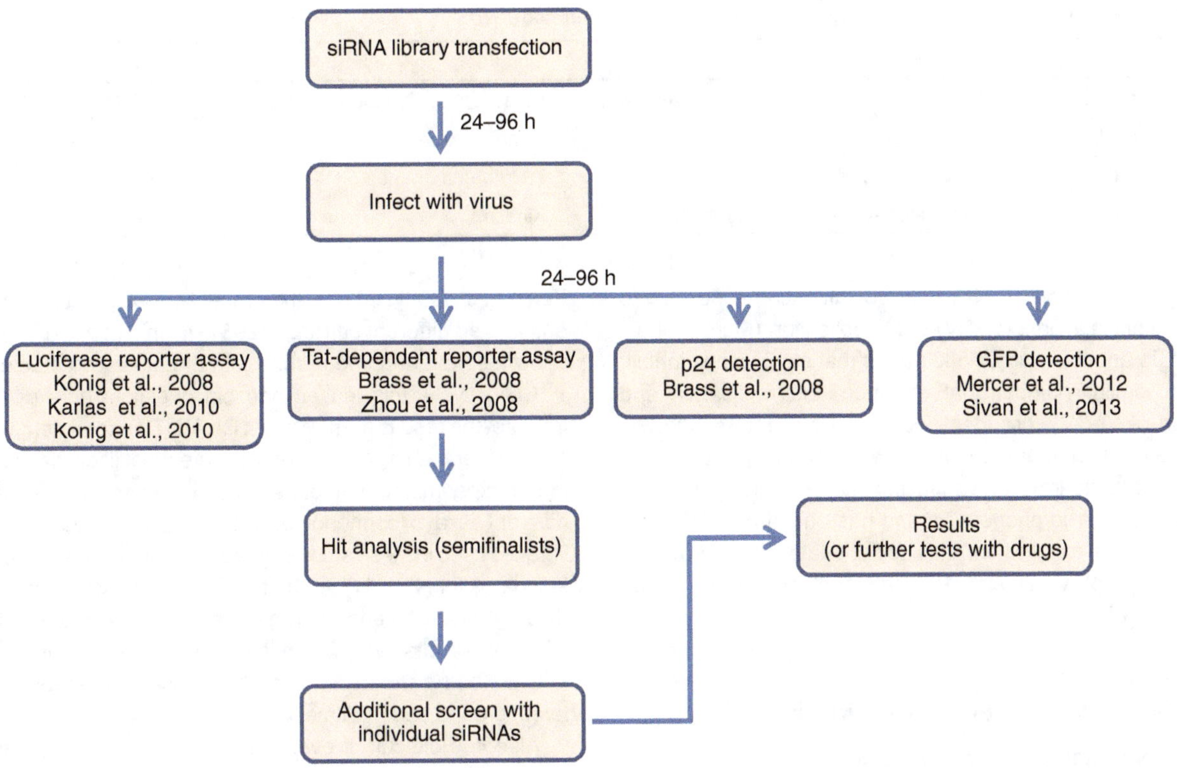

Figure 3. Work flow of genome wide siRNA screen for viral related host factor. Cells are initially transfected with different siRNA libraries, and then infected with viruses. Different assays can be utilized to detect screen results, including luciferase reporter assay, Tat-dependent reporter assay, p24 detection, as well as GFP detection. After first round of screen, hit analysis is performed and the candidates are subject to additional screen with individual siRNAs.

Table 1. Genome-wide siRNA screen for host factors related to virus infections

Virus	Group	Cells	Gene targeted	siRNAs per gene	Host factors identified	
HCV	Ng et al. (2007)	EN5-3 (Huh7)	4000	4	9	
HIV	Brass et al. (2008)	TZM-bl	21,121	4	273	3 overlap MED6, MED7, RelA.
	Zhou et al. (2008)	P4/R5 (HeLa)	19,709	3	311	
	Konig et al. (2008)	HEK-293T	19,628	6	295	
Influenza	Hao et al. (2008)	DL1	13,071	4	100	3 overlap ARCN1, ATP6AP1, COPG
	Brass et al. (2009)	U2OS	17,877	4	250	
	Konig et al. (2010)	A549	19,000	4	295	
	Karlas et al. (2010)	A549	22,843	4	168	
VACV	Mercer et al. (2012)	HeLa MZ	6979	3	188	23 overlap
	Sivan et al. (2013)	HeLa	21,566	4	500	

required for post-entry steps. Konig et al. confirm several of the factors using small molecule inhibitors, including that the vATPase and CAMK2B are indeed essential for influenza replication. In addition, a HTS RNAi study using pandemic swine-origin influenza virus identified 168 factors that inhibit virus infection, including the SON DNA binding protein (SON) that controls the trafficking of virions to late endosomes, as well as CDC-like kinase 1 (CLK1). In depth assays further uncovered the role of cyclin-dependent kinase inhibitor 1B (Cdkn1b) in influenza infection by using *p27*$^{-/-}$ mice (Karlas et al., 2010).

Careful design of RNAi screens also revealed detailed and specific information for poxvirus infection. A library was used to screen 7,000 druggable genes in HeLa cells and

Table 2. RNAi high-throughout screening for genes sensitive to mutated *KRAS*

Group	Cells	Libraries	Characterized genes
Barbie et al. (2009)	19 cell lines with wild-type or mutated *RAS*	Kinase, phosphatase and oncogenes	*TBK1*
Luo et al. (2009)	Isogenic DLD-1	Genome-wide	*PLK1*
Kumar et al. (2012)	28 NSCLC cell lines	7000 human genes	*GATA2*
Singh et al. (2012)	21 mutated *KRAS* colon cancer cell lines	Kinase	*TAK1*

clearly shows that Cullin3-based ubiquitination is needed to initiate vaccinia virus (VACV) DNA replication (Mercer et al., 2012). In another study, >500 genes that significantly inhibit, and a similar number that enhance, the replication and spread of VACV were identified from RNAi HTSs with two independent human genome-scale libraries. Functional studies demonstrate that silencing nucleoporin 62 strongly inhibits VACV morphogenesis and has only a modest effect on viral gene expression, thus recapitulating and providing insight into previous studies with enucleated cells (Sivan et al., 2013).

CONCLUSIONS AND FUTURE PERSPECTIVE

As demonstrated by the various studies described above, the applications of RNAi HTS have been successful for the identification of novel genes that regulated cancer cell growth (either *in vivo* or *in vitro*) and mediated the interactions between viruses and hosts. RNAi HTS has a more broad application as a powerful tool to identify gene networks in a given biological process. Although this approach has been successfully applied in many studies, challenges remain in understanding screen results and, particularly, determining the significance in clinical applications. For example, many studies related to KRAS-driven oncogenic events reveal different major determinants in different cancer cell lines. The non-canonical IκB kinase TBK1 was first identified as a major regulator of mutated KRAS in various cancer cell lines (Barbie et al., 2009). Subsequently, TAK1 and TATA2 have also been found to regulate mutated KRAS in colon cancer cell lines and non-small cell lung cancer cell lines, respectively (Kumar et al., 2012; Singh et al., 2012). However, a set of mitotic regulators including *PLK1* is more important in an isogenic colon cancer cell line (Luo et al., 2009) (Table 2). There are no major overlapping regulators in these studies. This may be caused by investigators using different cell lines with various genetic backgrounds, but these discrepancies may also be caused by unknown reasons.

In another example, for the identification of host regulators for HIV replication mentioned as above, overlaps were very limited among the three studies described (Table 1). The largest overlaps were observed between the studies of (Brass et al., 2008) and (Zhou et al., 2008). They use HeLa and HeLa-derived TZM-bl cells and focus on the entire virus life cycle as an interesting biological question. By contrast, König et al. analyzed only the processes subsequent to HIV-1 entry, used different cell lines (293T) (Konig et al., 2008), and show less overlapping results with the other two studies.

Collectively, this leaves one big question: how can one interpret the results from different RNAi HTS experiments? The reasons for the large discrepancies are most likely differences in the experimental setups, such as the cell culture systems, different assays for phenotype detecting, and the various RNAi libraries used in these studies. It will be crucial to carefully define standards for RNAi HTS. The standards will ensure and guide different research groups to generate RNAi HTS datasets as common annotation guidelines for disseminating data online. This will be helpful to compare the datasets generated by different groups and facilitate information sharing.

Also it is difficult, if ever possible to mimic many physiological phenotypes and micro-environments in the cell-based assays. Therefore, *in vivo* RNAi HTS assays as described in mice (Beronja et al., 2013; Fellmann and Lowe, 2014) are likely to become an essential technology advance that facilitates gene function identifications in physiological context. For example, in cancer biology, given the complex interactions of a tumor and its microenvironment, including the communications between tumor cells and surrounding cells (Hanahan and Weinberg, 2011), cell-based *in vitro* RNAi HTS may not able to discover the ideal therapeutic target. However, *in vivo* HTS will provide comprehensive information in much more physiologically relevant conditions, thus should support better design and execution for cancer therapy.

The combination of RNAi HTS and other genomic, transcriptomic, proteomic, and mice modeling techniques will lead to a systemic understanding of gene networks, which are more relevant in a physiological context. In the future, we anticipate that RNAi HTS will be applied as the very first step of many research endeavors, and the results of the screens will provide lead information to design defined validation experiments.

ACKNOWLEDGEMENTS

The authors would like to thank Ms. Yan Wu in the Institute of Microbiology, Chinese Academy of Sciences for help in preparation of this manuscript. HW is supported by NERC (UK) grants NE/I000593/1 and NE/L012863/1.

ABBREVIATIONS

EBV, Epstein-Barr virus; dsRNA, double-stranded RNA; HCV, hepatitis C virus; HPV, human papillomaviruses; HTS, high-

ABBREVIATIONS

EBV, Epstein-Barr virus; dsRNA, double-stranded RNA; HCV, hepatitis C virus; HPV, human papillomaviruses; HTS, high-throughput screening; RISC, RNA-induced silencing complex; RNAi, RNA interference; shRNA, short hairpin RNA; siRNA, short interfering RNA; TNBC, basal-like triple negative breast cancer.

COMPLIANCE WITH ETHICS GUIDELINES

Shan Gao, Chen Yang, Shan Jiang, Xiao-Ning Xu, Xin Lu, You-Wen He, Annie Cheung and Hui Wang declare they have no conflict of interest. This article does not contain any studies with human or animal subjects performed by the any of the authors.

REFERENCES

Adams MD, Celniker SE, Holt RA, Evans CA, Gocayne JD, Amanatides PG, Scherer SE, Li PW, Hoskins RA, Galle RF et al (2000) The genome sequence of *Drosophila melanogaster*. Science 287:2185–2195

Almeida R, Allshire RC (2005) RNA silencing and genome regulation. Trends Cell Biol 15:251–258

Ashworth A, Lord CJ, Reis-Filho JS (2011) Genetic interactions in cancer progression and treatment. Cell 145:30–38

Banerji S, Cibulskis K, Rangel-Escareno C, Brown KK, Carter SL, Frederick AM, Lawrence MS, Sivachenko AY, Sougnez C, Zou LH et al (2012) Sequence analysis of mutations and translocations across breast cancer subtypes. Nature 486: 405–409

Barbie DA, Tamayo P, Boehm JS, Kim SY, Moody SE, Dunn IF, Schinzel AC, Sandy P, Meylan E, Scholl C et al (2009) Systematic RNA interference reveals that oncogenic KRAS-driven cancers require TBK1. Nature 462:108–112

Barbieri CE, Baca SC, Lawrence MS, Demichelis F, Blattner M, Theurillat JP, White TA, Stojanov P, Van Allen E, Stransky N et al (2012) Exome sequencing identifies recurrent SPOP, FOXA1 and MED12 mutations in prostate cancer. Nat Genet 44:U107–U685

Berger MF, Lawrence MS, Demichelis F, Drier Y, Cibulskis K, Sivachenko AY, Sboner A, Esgueva R, Pflueger D, Sougnez C et al (2011) The genomic complexity of primary human prostate cancer. Nature 470:214–220

Bernstein E, Caudy AA, Hammond SM, Hannon GJ (2001) Role for a bidentate ribonuclease in the initiation step of RNA interference. Nature 409:363–366

Beronja S, Janki P, Heller E, Lien WH, Keyes BE, Oshimori N, Fuchs E (2013) RNAi screens in mice identify physiological regulators of oncogenic growth. Nature 501(7466):185–190

Boutros M, Ahringer J (2008) The art and design of genetic screens: RNA interference. Nat Rev Genet 9:554–566

Brass AL, Dykxhoorn DM, Benita Y, Yan N, Engelman A, Xavier RJ, Lieberman J, Elledge SJ (2008) Identification of host proteins required for HIV infection through a functional genomic screen. Science 319:921–926

Brass AL, Huang IC, Benita Y, John SP, Krishnan MN, Feeley EM, Ryan BJ, Weyer JL, van der Weyden L, Fikrig E et al (2009) The IFITM proteins mediate cellular resistance to influenza A H1N1 virus, West Nile virus, and dengue virus. Cell 139:1243–1254

Cherry S, Doukas T, Armknecht S, Whelan S, Wang H, Sarnow P, Perrimon N (2005) Genome-wide RNAi screen reveals a specific sensitivity of IRES-containing RNA viruses to host translation inhibition. Genes Dev 19:445–452

Cipriano R, Graham J, Miskimen KL, Bryson BL, Bruntz RC, Scott SA, Brown HA, Stark GR, Jackson MW (2012) FAM83B mediates EGFR- and RAS-driven oncogenic transformation. J Clin Invest 122:3197–3210

Corcoran RB, Cheng KA, Hata AN, Faber AC, Ebi H, Coffee EM, Greninger P, Brown RD, Godfrey JT, Cohoon TJ et al (2013) Synthetic lethal interaction of combined BCL-XL and MEK inhibition promotes tumor regressions in KRAS mutant cancer models. Cancer Cell 23:121–128

Echeverri CJ, Perrimon N (2006) High-throughput RNAi screening in cultured cells: a user's guide. Nat Rev Genet 7:373–384

Elbashir SM, Harborth J, Lendeckel W, Yalcin A, Weber K, Tuschl T (2001) Duplexes of 21-nucleotide RNAs mediate RNA interference in cultured mammalian cells. Nature 411:494–498

Fellmann C, Lowe SW (2014) Stable RNA interference rules for silencing. Nat Cell Biol 16:10–18

Fire A, Xu S, Montgomery MK, Kostas SA, Driver SE, Mello CC (1998) Potent and specific genetic interference by double-stranded RNA in *Caenorhabditis elegans*. Nature 391:806–811

Friedel CC, Haas J (2011) Virus–host interactomes and global models of virus-infected cells. Trends Microbiol 19:501–508

Gao S, Giansanti MG, Buttrick GJ, Ramasubramanyan S, Auton A, Gatti M, Wakefield JG (2008) Australin: a chromosomal passenger protein required specifically for *Drosophila melanogaster* male meiosis. J Cell Biol 180:521–535

Geisbert TW, Lee AC, Robbins M, Geisbert JB, Honko AN, Sood V, Johnson JC, de Jong S, Tavakoli I, Judge A et al (2010) Postexposure protection of non-human primates against a lethal Ebola virus challenge with RNA interference: a proof-of-concept study. Lancet 375:1896–1905

Girard MP, Cherian T, Pervikov Y, Kieny MP (2005) A review of vaccine research and development: human acute respiratory infections. Vaccine 23:5708–5724

Guo X, Wang XF (2012) A mediator lost in the war on cancer. Cell 151:927–929

Hanahan D, Weinberg RA (2011) Hallmarks of cancer: the next generation. Cell 144:646–674

Hannon GJ (2002) RNA interference. Nature 418:244–251

Hao L, Sakurai A, Watanabe T, Sorensen E, Nidom CA, Newton MA, Ahlquist P, Kawaoka Y (2008) Drosophila RNAi screen identifies host genes important for influenza virus replication. Nature 454:890–893

He L, Guo L, Vathipadiekal V, Sergent PA, Growdon WB, Engler DA, Rueda BR, Birrer MJ, Orsulic S, Mohapatra G (2013) Identification of LMX1B as a novel oncogene in human ovarian cancer. Oncogene 1–10. doi:10.1038/onc.2013.375

Hock J, Meister G (2008) The Argonaute protein family. Genome Biol 9:210

Horvath S, Zhang B, Carlson M, Lu KV, Zhu S, Felciano RM, Laurance MF, Zhao W, Qi S, Chen Z et al (2006) Analysis of oncogenic signaling networks in glioblastoma identifies ASPM as a molecular target. Proc Natl Acad Sci USA 103:17402–17407

Huang S, Holzel M, Knijnenburg T, Schlicker A, Roepman P, McDermott U, Garnett M, Gremrum W, Sun C, Prahallad A et al (2012) MED12 controls the response to multiple cancer drugs through regulation of TGF-beta receptor signaling. Cell 151:937–950

Huang L, Jin J, Deighan P, Kiner E, McReynolds L, Lieberman J (2013) Efficient and specific gene knockdown by small interfering RNAs produced in bacteria. Nat Biotechnol 31:350–356

Huesken D, Lange J, Mickanin C, Weiler J, Asselbergs F, Warner J, Meloon B, Engel S, Rosenberg A, Cohen D et al (2005) Design of a genome-wide siRNA library using an artificial neural network. Nat Biotechnol 23:995–1001

Hussain M, Asgari S (2014) MicroRNA-like viral small RNA from dengue virus 2 autoregulates its replication in mosquito cells. Proc Natl Acad Sci USA 111:2746–2751

Iorns E, Lord CJ, Turner N, Ashworth A (2007) Utilizing RNA interference to enhance cancer drug discovery. Nat Rev Drug Discov 6:556–568

Jiang L, Siu MK, Wong OG, Tam KF, Lu X, Lam EW, Ngan HY, Le XF, Wong ES, Monteiro LJ et al (2011) iASPP and chemoresistance in ovarian cancers: effects on paclitaxel-mediated mitotic catastrophe. Clin Cancer Res 17:6924–6933

Karlas A, Machuy N, Shin Y, Pleissner KP, Artarini A, Heuer D, Becker D, Khalil H, Ogilvie LA, Hess S et al (2010) Genome-wide RNAi screen identifies human host factors crucial for influenza virus replication. Nature 463:818–822

Kim DH, Rossi JJ (2007) Strategies for silencing human disease using RNA interference. Nat Rev Genet 8:173–184

Kittler R, Putz G, Pelletier L, Poser I, Heninger AK, Drechsel D, Fischer S, Konstantinova I, Habermann B, Grabner H et al (2004) An endoribonuclease-prepared siRNA screen in human cells identifies genes essential for cell division. Nature 432:1036–1040

Konig R, Zhou Y, Elleder D, Diamond TL, Bonamy GM, Irelan JT, Chiang CY, Tu BP, De Jesus PD, Lilley CE et al (2008) Global analysis of host-pathogen interactions that regulate early-stage HIV-1 replication. Cell 135:49–60

Konig R, Stertz S, Zhou Y, Inoue A, Hoffmann HH, Bhattacharyya S, Alamares JG, Tscherne DM, Ortigoza MB, Liang Y et al (2010) Human host factors required for influenza virus replication. Nature 463:813–817

Krausz E (2007) High-content siRNA screening. Mol Biosyst 3:232–240

Kumar MS, Hancock DC, Molina-Arcas M, Steckel M, East P, Diefenbacher M, Armenteros-Monterroso E, Lassailly F, Matthews N, Nye E et al (2012) The GATA2 transcriptional network is requisite for RAS oncogene-driven non-small cell lung cancer. Cell 149:642–655

Lambert LC, Fauci AS (2010) Influenza vaccines for the future. N Engl J Med 363:2036–2044

Lander ES, Consortium IHGS, Linton LM, Birren B, Nusbaum C, Zody MC, Baldwin J, Devon K, Dewar K, Doyle M et al (2001) Initial sequencing and analysis of the human genome. Nature 409:860–921

Lawson ND, Wolfe SA (2011) Forward and reverse genetic approaches for the analysis of vertebrate development in the zebrafish. Dev Cell 21:48–64

Liu YP, Vink MA, Westerink JT, Ramirez de Arellano E, Konstantinova P, Ter Brake O, Berkhout B (2010) Titers of lentiviral vectors encoding shRNAs and miRNAs are reduced by different mechanisms that require distinct repair strategies. RNA 16:1328–1339

Luo J, Emanuele MJ, Li D, Creighton CJ, Schlabach MR, Westbrook TF, Wong KK, Elledge SJ (2009) A genome-wide RNAi screen identifies multiple synthetic lethal interactions with the Ras oncogene. Cell 137:835–848

Madan V, Paul D, Lohmann V, Bartenschlager R (2014) Inhibition of HCV replication by cyclophilin antagonists is linked to replication fitness and occurs by inhibition of membranous web formation. Gastroenterology 146(5):1361-72.e1–1361-72.e9

Martinez J, Patkaniowska A, Urlaub H, Luhrmann R, Tuschl T (2002) Single-stranded antisense siRNAs guide target RNA cleavage in RNAi. Cell 110:563–574

Mbeunkui F, Johann DJ (2009) Cancer and the tumor microenvironment: a review of an essential relationship. Cancer Chemother Pharmacol 63:571–582

Mercer J, Snijder B, Sacher R, Burkard C, Bleck CK, Stahlberg H, Pelkmans L, Helenius A (2012) RNAi screening reveals proteasome- and Cullin3-dependent stages in vaccinia virus infection. Cell Rep 2:1036–1047

Moffat J, Sabatini DM (2006) Building mammalian signalling pathways with RNAi screens. Nat Rev Mol Cell Biol 7:177–187

Ng TI, Mo H, Pilot-Matias T, He Y, Koev G, Krishnan P, Mondal R, Pithawalla R, He W, Dekhtyar T et al (2007) Identification of host genes involved in hepatitis C virus replication by small interfering RNA technology. Hepatology 45:1413–1421

Osterholm MT, Kelley NS, Sommer A, Belongia EA (2012) Efficacy and effectiveness of influenza vaccines: a systematic review and meta-analysis. Lancet Infect Dis 12:36–44

Pattle SB, Farrell PJ (2006) The role of Epstein–Barr virus in cancer. Expert Opin Biol Ther 6:1193–1205

Petrocca F, Altschuler G, Tan SM, Mendillo ML, Yan HH, Jerry DJ, Kung AL, Hide W, Ince TA, Lieberman J (2013) A genome-wide siRNA screen identifies proteasome addiction as a vulnerability of basal-like triple-negative breast cancer cells. Cancer Cell 24:182–196

Reynolds A, Anderson EM, Vermeulen A, Fedorov Y, Robinson K, Leake D, Karpilow J, Marshall WS, Khvorova A (2006) Induction of the interferon response by siRNA is cell type- and duplex length-dependent. RNA 12:988–993

Rositch AF, Soeters HM, Offutt-Powell TN, Wheeler BS, Taylor SM, Smith JS (2014) The incidence of human papillomavirus infection following treatment for cervical neoplasia: a systematic review. Gynecol Oncol 132(3):767–779

Sharma S, Rao A (2009) RNAi screening: tips and techniques. Nat Immunol 10:799–804

Silva JM, Marran K, Parker JS, Silva J, Golding M, Schlabach MR, Elledge SJ, Hannon GJ, Chang K (2008) Profiling essential genes in human mammary cells by multiplex RNAi screening. Science 319:617–620

Sims D, Mendes-Pereira AM, Frankum J, Burgess D, Cerone MA, Lombardelli C, Mitsopoulos C, Hakas J, Murugaesu N, Isacke CM et al (2011) High-throughput RNA interference screening using pooled shRNA libraries and next generation sequencing. Genome Biol 12(10):104

Singh A, Sweeney MF, Yu M, Burger A, Greninger P, Benes C, Haber DA, Settleman J (2012) TAK1 inhibition promotes apoptosis in KRAS-dependent colon cancers. Cell 148:639–650

Siu MK, Chan HY, Kong DS, Wong ES, Wong OG, Ngan HY, Tam KF, Zhang H, Li Z, Chan QK et al (2010) p21-activated kinase 4 regulates ovarian cancer cell proliferation, migration, and invasion and contributes to poor prognosis in patients. Proc Natl Acad Sci USA 107:18622–18627

Siu MK, Wong ES, Kong DS, Chan HY, Jiang L, Wong OG, Lam EW, Chan KK, Ngan HY, Le XF et al (2013) Stem cell transcription factor NANOG controls cell migration and invasion via dysregulation of E-cadherin and FoxJ1 and contributes to adverse clinical outcome in ovarian cancers. Oncogene 32:3500–3509

Sivan G, Martin SE, Myers TG, Buehler E, Szymczyk KH, Ormanoglu P, Moss B (2013) Human genome-wide RNAi screen reveals a role for nuclear pore proteins in poxvirus morphogenesis. Proc Natl Acad Sci USA 110:3519–3524

Turner NC, Lord CJ, Iorns E, Brough R, Swift S, Elliott R, Rayter S, Tutt AN, Ashworth A (2008) A synthetic lethal siRNA screen identifying genes mediating sensitivity to a PARP inhibitor. EMBO J 27:1368–1377

Wang S, Huang X, Li Y, Lao H, Zhang Y, Dong H, Xu W, Li JL, Li M (2011) RN181 suppresses hepatocellular carcinoma growth by inhibition of the ERK/MAPK pathway. Hepatology 53:1932–1942

Waterston RH, Lindblad-Toh K, Birney E, Rogers J, Abril JF, Agarwal P, Agarwala R, Ainscough R, Alexandersson M, An P et al (2002) Initial sequencing and comparative analysis of the mouse genome. Nature 420:520–562

Yang D, Buchholz F, Huang ZD, Goga A, Chen CY, Brodsky FM, Bishop JM (2002) Short RNA duplexes produced by hydrolysis with *Escherichia coli* RNase III mediate effective RNA interference in mammalian cells. Proc Natl Acad Sci USA 99:9942–9947

Yu J, Hu SN, Wang J, Wong GKS, Li SG, Liu B, Deng YJ, Dai L, Zhou Y, Zhang XQ et al (2002) A draft sequence of the rice genome (*Oryza sativa* L. ssp indica). Science 296:79–92

Zhou H, Xu M, Huang Q, Gates AT, Zhang XD, Castle JC, Stec E, Ferrer M, Strulovici B, Hazuda DJ et al (2008) Genome-scale RNAi screen for host factors required for HIV replication. Cell Host Microbe 4:495–504

20

Phylogenomics of non-model ciliates based on transcriptomic analyses

Xiao Chen[1,2], Xiaolu Zhao[2], Xiaohui Liu[1], Alan Warren[3], Fangqing Zhao[4✉], Miao Miao[1✉]

[1] College of Life Sciences, University of Chinese Academy of Sciences, Beijing 100049, China
[2] Institute of Evolution & Marine Biodiversity, Ocean University of China, Qingdao 266003, China
[3] Department of Life Sciences, Natural History Museum, Cromwell Road, London SW7 5BD, UK
[4] Beijing Institutes of Life Science, Chinese Academy of Sciences, Beijing 100101, China
✉ Correspondence: zhfq@mail.biols.ac.cn (F. Zhao), miaomiao@ucas.ac.cn (M. Miao)

ABSTRACT

Ciliates are one of the oldest living eukaryotic unicellular organisms, widely distributed in the waters around the world. As a typical marine oligotrich ciliate, *Strombidium sulcatum* plays an important role in marine food webs and energy flow. Here we report the first deep sequencing and analyses of RNA-Seq data from *Strombidium sulcatum*. We generated 42,640 unigenes with an N50 of 1,451 bp after *de novo* assembly and removing rRNA, mitochondrial and bacteria contaminants. We employed SPOCS to detect orthologs from *S. sulcatum* and 17 other ciliates, and then carried out the phylogenomic reconstruction using 127 single copy orthologs. In phylogenomic analyses, concatenated trees have similar topological structures with concordance tree on the class level. Together with phylogenetic networks analysis, it aroused more doubts about the placement of *Protocruzia*, *Mesodinium* and *Myrionecta*. While epiplasmic proteins are known to be related to morphological characteristics, we found the potential relationship between gene expression of epiplasmic proteins and morphological characteristics. This work supports the use of high throughput approaches for phylogenomic analysis as well as correlation analysis between expression level of target genes and morphological characteristics.

Xiao Chen and Xiaolu Zhao have contributed equally to this work

KEYWORDS *Strombidium*, protozoa, transcriptome, non-model ciliate, phylogenomic analysis

INTRODUCTION

Ciliates are unique among unicellular organisms in having two types of nuclei thus separating the germline and somatic functions and are therefore commonly used to study genome structure and gene expression (Aury et al., 2006). Over the last ten years, high throughput sequencing has made ciliate genomic research available on a large scale, and the genomes of several species have been sequenced and analyzed, e.g. *Tetrahymena thermophila* (Eisen et al., 2006), *Paramecium tetraurelia* (Auryet al., 2006), *Ichthyophthirius multifiliis* (Coyne et al., 2011) and *Oxytricha trifallax* (Swart et al., 2013). In addition, the transcriptomes of 13 ciliate species representing 10 genera and five classes are now available (Table S1) including those of *Tetrahymena thermophila* (Xiong et al., 2012) and *Chilodonella uncinata* (Grant et al., 2012).

Strombidium sulcatum Claparède & Lachmann, 1859 (Fig. 1 A–F) is an oligotrich ciliate that plays an important role in food webs and energy flow in marine pelagic waters (Bernard & Rassoulzadegan, 1990, Montagnes et al., 1990). Since Claparède & Lachmann described *S. sulcatum* as the type species for the genus *Strombidium*, there have been numerous reports on its trophic status and diversity (Wiadnyana & Rassoulzadegan, 1989, e.g. Bernard & Rassoulzadegan, 1990, Allali et al., 1994, Christaki et al., 1998, Dolan et al., 2003). With the development of molecular sequencing technology, the taxonomic status of *S. sulcatum* and other species of *Strombidium* have been reassessed (Modeo et al., 2003, McManus et al., 2010, Li et al., 2013).

Ciliates are characterized by having basal bodies and associated appendages bounded to a submembrane cytoskeleton, the epiplasm, that lies beneath the inner alveolar membrane (Fauré-Fremiet, 1962). Two distinct classes of cytoskeletal protein have been identified in ciliates: articulins and epiplasmins (Nahon et al., 1993, Huttenlauch et al., 1995, Huttenlauch et al., 1998, Huttenlauch & Stick, 2003). These are homologous in protists and the presence of both has been established in the excavate flagellate *Euglena,* the dinoflagellate *Amphidinium* and the ciliates *Pseudomicrothorax, Paramecium, Tetrahymena* and *Euplotes* (Huttenlauch et al., 1998). Several novel families of articulins and epiplasmins have been identified including: plateins, which are articulin-related proteins in the ciliate *Euplotes* (Kloetzel et al., 2003a, b); EpiA, EpiB and EpiC, which are major cytoskeletal proteins in the cortex of *Tetrahymena* (Honts & Williams, 2003), and likewise Epi1-Epi51 in *Paramecium* (Pomel et al., 2006, Damaj et al., 2009, Aubusson-Fleury et al., 2013); alveolins, which are used to define the protist infrakingdom Alveolata (Gould et al., 2008). Most importantly, it had been shown that EpiC (Epc1p, gene: *EPC1*) influences both cell shape and the fidelity of cortical development (Huttenlauch et al., 1998).

Here we present Illumina Hiseq 2000 RNA-Seq data from *Strombidium sulcatum*. The transcriptome data are analyzed, including assembly and annotation. Phylogenomic reconstruction is carried out using 127 single copy orthologs, and evolutionary relationships are revealed by concatenated and concordance tree analyses. Thereafter, we investigate the transcriptional expression of epiplasmic proteins in *S. sulcatum* to determine whether there is a relationship between morphological characteristics and gene expression of epiplasmic proteins.

RESULTS

Assembly

A total of 24,958,136 paired-end reads (101 bp for each read) were produced for *S. sulcatum* (5,041,543,472 bases, % GC: 52%). After filtering low quality reads, 95.55% reads were retained. As rRNA had been depleted after extraction of the total RNA, it hardly existed in RNA-seq data (0.60%). The remaining rRNA sequences were removed for downstream analyses.

The remaining sequences were assembled with Trinity into 44,633 contigs (Max length: 18,235 bp, Min length: 201 bp, N50: 1498 bp). 94.33% of reads could be properly aligned to contigs in pairs. In order to eliminate assembly errors, 157 *trans* chemiras were detected and removed. Subsequently, the contigs were clustered by high similarity threshold (98%) using CD-HIT and one representative contig for each group was retained (43,437 contigs remained). After linking contigs whose terminals were perfectly matched, the contig number dropped to 43,009. These contigs were then BLASTed to *Oxytricha trifallax* whole genome, and BLAST hit results were contrasted to *O. trifallax* mitochondrial genomic peptides. This revealed that 18 contigs were related to the mitochondrial genome. Given that we used bacteria as the food source during the cultivation of *S. sulcatum*, the assembled contigs probably included a few contaminated sequences from bacteria. Based on BLAST results, we determined that 348 contigs are bacterial. After elimination of the contaminating rDNAs and bacterial sequences, raw reads from *S. sulcatum* were assembled into 42,640 unigenes. Of these unigenes, 27,965 had strong BLAST hits to eukaryotic sequences, and 16,668 had no strong affinity to any domain.

Gene content in transcriptomes

A variety of methods were used to assess the content of the *Strombidium sulcatum* transcriptome including determining closest relatives for ciliate proteins, identifying most highly expressed genes, CDS region prediction and protein annotation. *Strombidium sulcatum*'s transcriptome had a significant number of BLAST hits to those of *S. inclinatum* and *S. rassoulzadegani* and to the genome of *Oxytricha trifallax*. We identified highly expressed genes as contigs occupping 10% or more of sequence reads with high FPKM (Fragments Per Kilobase of transcript sequence per Millions base pairs sequenced). The functions of about 70% of the contigs with the highest read numbers are unclear. The majority of them have homologous contigs (identity > 90%) in the other 14 ciliates represented in the transcriptome dataset, the exceptions being comp7405_c0 and comp10361_c0, each of which had a blast top hit identity of only about 30%.

After CDS prediction, 40,746 CDS regions were identified from 42,640 unigenes with an average length of 471 aa. At the same time, 40,746 peptide sequences were predicted. With these predicted peptides, we conducted a one-way BLAST search to the GenBank ciliate database against all proteins contained therein; 25,676 of *S. sulcatum* predicted peptides had BLAST results and could be grouped into 16,553 gene groups, 9765 of which were found in the genomes of the other four ciliates (data not shown). The identity distribution of the top hits is shown in Fig. 1. The remaining *S. sulcatum* genes did not satisfy the pairing cut-off criteria (E-value $\leq 1 \times 10^{-5}$).

Annotation of predicted protein products were matched to models and Swiss-Prot accessions. 20,668 HSPs (Highly Scalable Parallel) were found when blasted to Swiss-Prot database, and each match 16,315 domains in Pfam-A database, 12,201 domains in Superfamily database, 8594 domains in TIGRFAMs database respectively. Assessment of the top BLAST hits' species distribution for *S. sulcatum* is shown in Fig. 2.

Ortholog detection and phylogenomics

A useful approach for surveying the protein-coding gene landscape of a newly sequenced genome is to group genes by orthology, which can provide guidance for functional annotation. Here, we grouped the proteins predicted from *S. sulcatum* transcriptome with four ciliate genomes (*Oxytricha trifallax, Tetrahymena thermophila, Ichthyophthirius*

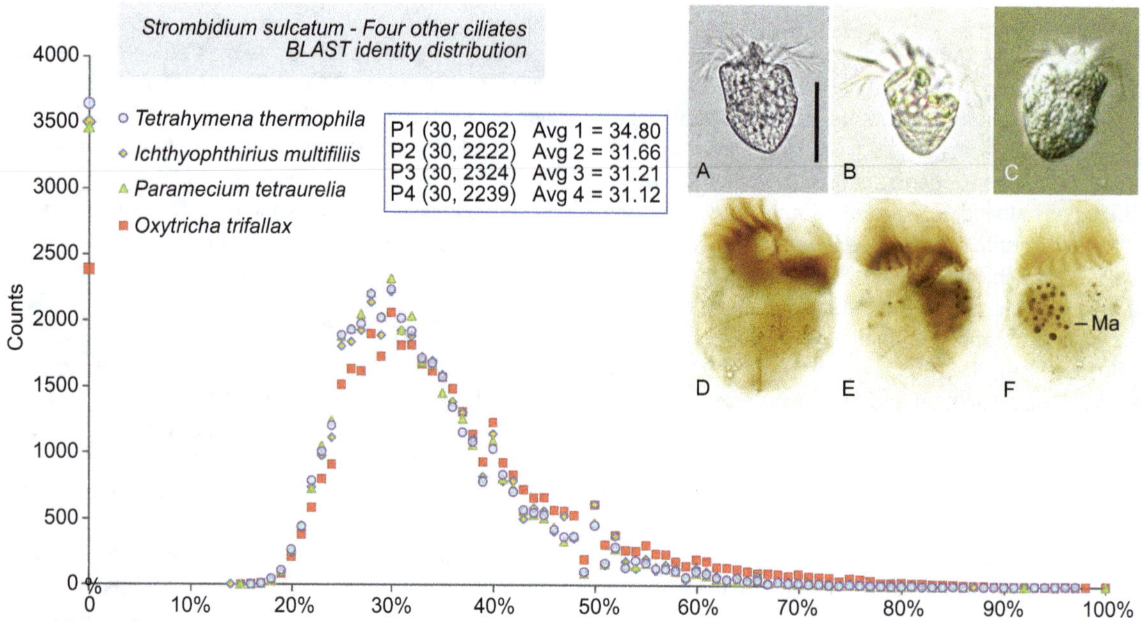

Figure 1. BLAST identity distribution between proteins of *Strombidium sulcatum* and that of four other ciliates (*Oxytricha trifallax, Tetrahymena thermophila, Ichthyophthirius multifiliis, Paramecium tetraurelia*). P(x,y), peaks (identity, peak value). Avg, average value. Photomicrographs of *Strombidium sulcatum in vivo* (A–C) and protargol-stained specimen (D–F). Ma, macronucleus. Scale bars are 25 μm.

multifiliis and *Paramecium tetraurelia*) using a reciprocal BLAST hit (RBH) approach. After removing short (<50 aa) and less-conserved (<30% identity) proteins, 684 orthologs were identified as shared by them all. *Strombidium sulcatum* shared most orthologs with *O. trifallax*, which is consistent with current systematic arrangements based on morphology and traditional molecular phylogeny (Fig. 3A). In addition, 1700 orthologs were identified in the three *Strombidium* spp. for which transcriptome data are available with *S. sulcatum* sharing most orthologs with *S. inclinatum* (Fig. 3B).

We detected further orthologs among the four ciliate genomes and the 14 ciliate transcriptomes. One hundred and twenty-seven orthologs shared by all 18 ciliates with BLAST identity ≥ 30% and match length ≥ 50 aa were retained to perform phylogenomic analysis. After the 127 ortholog concatenated alignment dataset was produced, and its model was selected as RtREV + I + G + F (gamma shape = 0.947; proportion of invariable sites = 0.063), a concatenated phylogenomic tree was calculated using both maximum likelihood and Bayesian inference methods (Fig. 4). According to the maximum likelihood tree based on 36,724 aa positions from 127 predicted peptide sequences, genera in the class Oligohymenophorea, which includes *Tetrahymena*, *Ichthyophthirius* and *Paramecium*, clustered as a clade with full support. The genus *Platyophrya* (class Colpodea) grouped with the class Oligohymenophorea with full support. The genus *Litonotus* (class Litostomatea) clustered with the class Colpodea and Oligohymenophorea with low support. The ambiguous genus *Protocruzia*, which is known to have an elevated rate of evolution, did not group within any of the major clades. However, most groups with morphological identity are recovered robustly.

The three species of *Strombidium* and *Oxytricha trifallax* clustered together with full support, and then grouped with *Euplotes*, *Strombidinopsis* and *Favella*. The *Mesodinium* (class Mesodiniea) did not group together with *Myrionecta* (class Mesodiniea). Instead, *Myrionecta* and *Tiarina* (class Prostomatea) formed a clade with moderate support that was sister to *Mesodinium* (class Mesodiniea) with strong support.

Coalescent analysis was also carried out and a concordance tree was estimated by consolidating all BI trees inferred from the 127 ortholog alignments (Fig. 4). The topology of the concordance tree was similar to that of the concatenated tree. However, *Protocruzia*'s systematic placement was even closer to the class Heterotrichea, while other Spirotrichs clustered as a monophyletic group.

Split decomposition analysis of phylogenetic networks was performed to uncover all possible relationships among the ciliate lineages (Fig. 5). The neighbor-net graphs were calculated based on 36,724 aa from 127 orthologs. The placement of most classes in the networks analysis was generally consistent with both the concatenated and the concordance analyses, and revealed that: (1) *Protocruzia* was in a position that is separate from all existing classes, and; (2) *Myrionecta* was grouped with *Tiarina* which represent the class Prostomatea, as same as the result indicated in maximum likelihood tree.

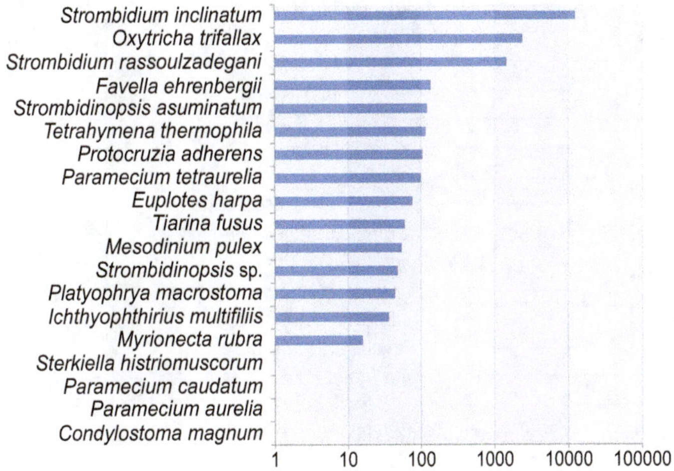

Figure 2. Species distribution of the most related homologous genes. The X-axis represents numbers of genes/contigs.

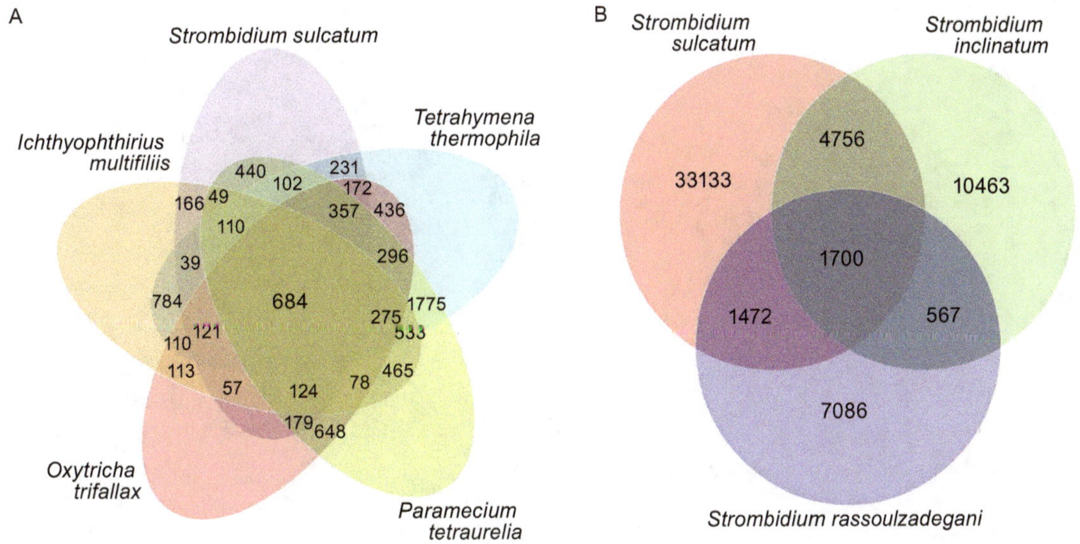

Figure 3. Comparisons of number of orthologs among *Strombidium sulcatum* and other ciliates. Venn diagram showing: (A) shared orthologs among *Strombidium sulcatum* (S) and four other ciliate species for which genomic data are available. Pairwise mutual best-hits by BLASTP were then identified as putative orthologs; (B) orthologs shared among *S. sulcatum*, *S. inclinatum* and *S. rassoulzadegani*. The numbers depict the total number of ortholog groups in each category.

Comparative transcriptomic analysis reveals the differential expression of epiplasmic proteins in ciliates

Strombidium is caharacterized by its pyriform body shape, well-developed adoral zone of membranelles, somatic cilia arranged as a girdle, and the presence of polysaccharide cortical plates in the posterior half of the cell (Fig. 1). In order to reveal the relationship between expression level difference and cellular geometry, morphological information of the thirteen ciliates (*S. sulcatum* plus other 12 ciliates whose RNA-seq transcriptome data were available from MMETSP, *Condylostoma magnum* was not involved because of its inadequate transcriptome data) is summarized in Table 1. Seven potential epiplasmic proteins related to morphology formation were

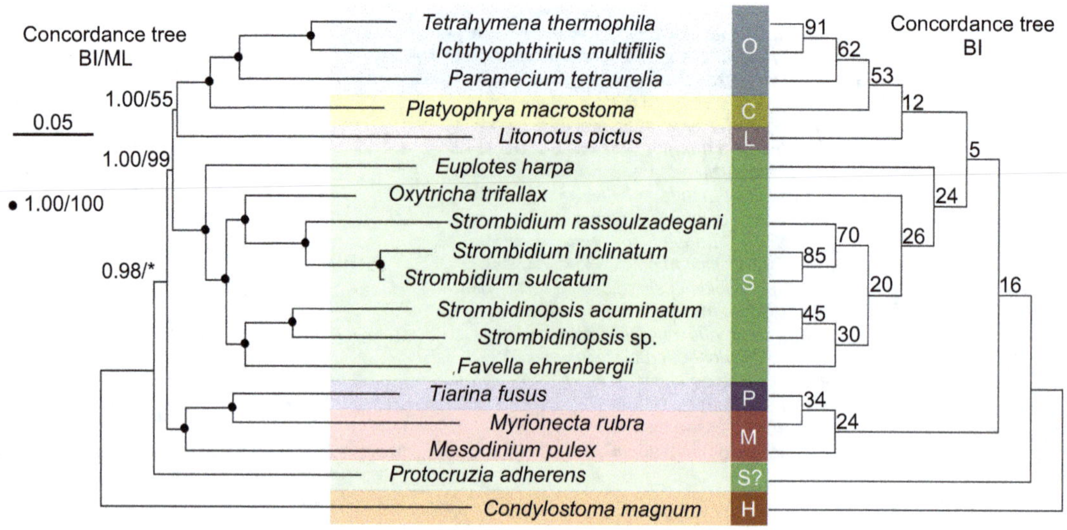

Figure 4. Comprehensive maximum likelihood phylogenomic tree based on 127 orthologs from 18 ciliates. Asterisks indicate bootstrap values less than 50% at a given node. The scale bar corresponds to five substitutions per one hundred nucleotide positions in the concatenated tree (left). The numbers depict the concordance factors in the concordance tree (right).

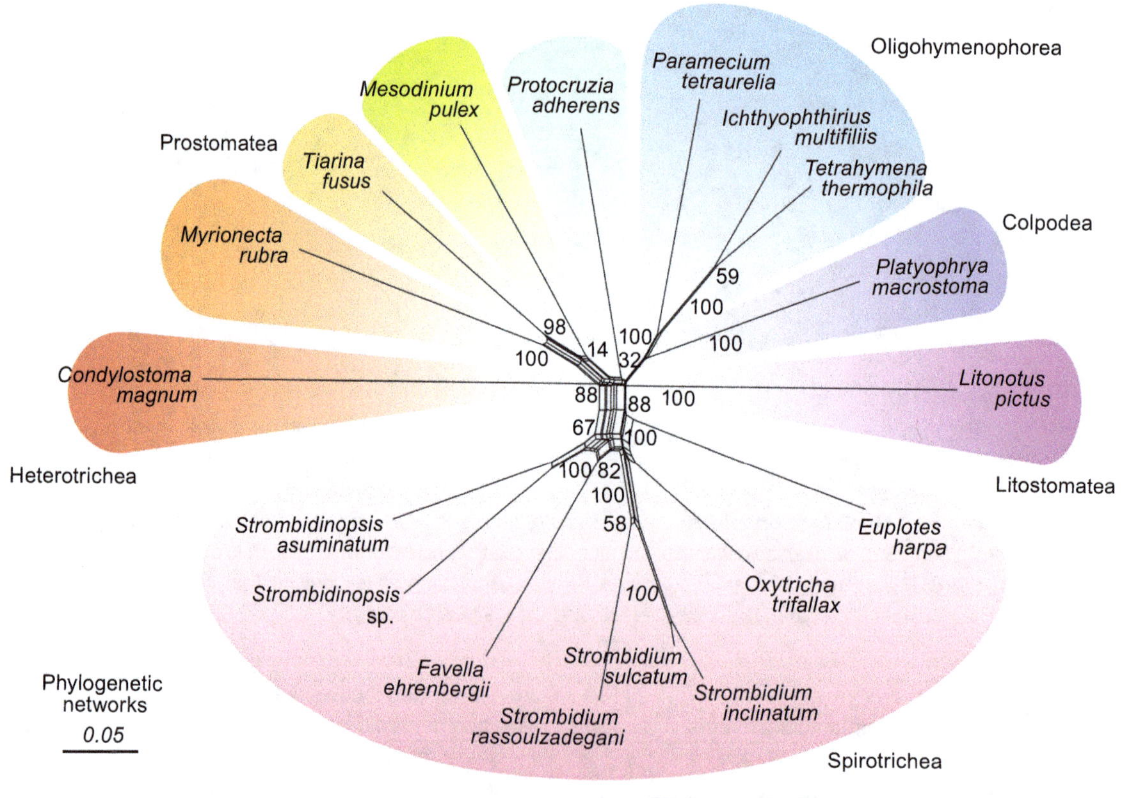

Figure 5. Phylogenetic network computed from the concatenated orthologs alignment dataset using the neighbornet algorithm and the uncorrected distances. Numbers along edges are bootstrap support values coming from 1000 replicates. Values <50% are not shown. The scale bar indicates five substitutions per one hundred nucleotide positions.

Table 1. Grouping according to each morphological characteristic criterion of 13 ciliate species

Class	SPIROTRICHEA							LITOSTOMATEA	COLPODEA	PROTOCRUZIIDEA	MESODINIEA		PROSTOMATEA
Subclass/ Order	Oligotrichia			Choreotrichia			Hypotrichia	Haptoria	Cyrtolophosidida				Prorodontida
Species	*Strombidium sulcatum*	*Strombidium inclinatum*	*Strombidium rassoulzadegani*	*Favella ehrenbergii*	*Strombidinopsis acuminatum*	*Strombidinopsis sp.*	*Euplotes harpa*	*Litonotus pictus*	*Platyophrya macrostoma*	*Protocruzia adherens*	*Mesodinium pulex*	*Myrionecta rubra*	*Tiarina fusus*
1. Classes	Group 1	Group 1	Group 1	Group 1	Group 1	Group 1	Group 1	Group 2	Group 3	Group 4	Group 5	Group 5	Group 6
2. Class Spirotrichea vs. other classes													
3. Strombidium vs other species													
4. Strombidium vs other species in class Spirotrichea													
5. Somatic kineties	Dikinetid	Dikinetid	Dikinetid	Dikinetid	Dikinetid	Dikinetid	Dikinetid	Monokinetid (Mostly)	Dikinetid	Dikinetid	Dikinetid	Dikinetid	Monokinetid
6. Circumoral kineties	Polykinetids	Polykinetids	Polykinetids	Polykinetids	Polykinetids	Polykinetids	Polykinetids	Dikinetid (Mostly)	Polykinetids	Polykinetids	-	-	Monokinetid
7. Cortical granules	-	-	-	-	-	-	-	Yellow to yellow-brown	-	Arranged along ciliary rows densely	-	-	-
8. Pellicular alveolus	-	-	-	-	-	-	Pellicular alveolus with plastein plates	-	-	-	-	-	Calcified polysaccharide plates in pelicular alveolus
9. Extrusomes	Rod-shaped, arround middle part of the cell, 12-15 µm in length, without grouping	In the posterior part of the cell	Rod-shaped, arround middle part of the cell, 14 µm in length	-	-	-	-	Rod-shaped, 5-8 µm in length, Ventral / Dorsal margin	-	-	Several spiky synapsis or toxicysts in anterior end of rostellum, 3-4 µm	Several spiky synapsis in anterior end of rostellum	-
10-1. Body size	Small, 30-50 x 25-40 µm	Small, 12.5-30µm x 12.5-21 µm	Small, 49-82 µm x 35-50 µm	Middle, 50-115 x 40-95µm	Middle, 70-124 µm x 29-50 µm	N/A	Middle, 120-260 µm	Larger, 250-350 µm	Small, 25-30 µm	Small, 20-50 µm	Small, 20-30 x 15-20 µm	Small, 10-100 µm, Usually 20-60 µm	Middle, 85-135 x 22-45 µm
10-2. Length-width ratio	1.2	1.2	1.2	1.2	2.5	N/A	1.5	4	1.5	1.5	1.2	1.2	4
11. Buccal area	Open buccal apparatus, with obvious protrusion in front of the buccal apparatus	Open buccal apparatus, without obvious protrusion in fornt of the buccal apparatus	Open buccal apparatus, with obvious protrusion and orange eyespots in fornt of the buccal apparatus.	Open buccal apparatus, broad	Closed buccal apparatus.Truncated.There are no obvious protrusion in form of the buccal apparatus	Closed buccal apparatus	Located in the front, well developed	Cytostome under the cell surface directly	Large buccal area occupies the whole oblique and truncated anterior end	Buccal cavity is 1/4-2/5 length of the body	Cytostome degenerated	Cytostome degenerated	Buccal at the top, concave

◀ Notes for Table 1. Group 1 members were shaded light orange, Group 2 members were shaded dark orange. (1) All 13 species were divided into 6 groups by classes (criterion 1); (2) all were divided into 2 groups as class Spirotrichea (Group 1) and other classes (Group 2) (criterion 2); (3) all were divided into 2 groups as *Strombidium* spp. (Group 1) and other species (Group 2) (criterion 3); (4) 7 species were divided into 2 groups as *Strombidium* spp. (Group 1) and other species in class Spirotrichea (Group 2) (criterion 4); (5) all 13 species were divided into 2 groups by somatic kineties characteristics, i.e. dikinetid (Group 1) or monokinetid (Group 2) (criterion 5); (6) all were divided into 2 groups by circumoral kineties characteristics, i.e. polykinetids (Group 1) or other types (dikinetid/monokinetid, Group 2) (criterion 6); (7) all were divided into 2 groups by cortical granules characteristics, i.e. cortical granules absent (Group 1) or cortical granules present (Group 2) (criterion 7); (8) all were divided into 2 groups by whether had obvious pellicular alveolus, i.e. not obvious (Group 1) or obvious (Group 2) (criterion 8); (9) all were divided into 2 groups by extrusome characteristics, i.e. extrusomes present (Group 1) or extrusomes absent (Group 2) (criterion 9); (10) all were divided into 2 groups by body size characteristics, i.e. either body length < 100 μm or length-width ratio < 2 (Group 1) or do not accord with that (Group 2) (criterion 10-1 & 10-2); (11) all were divided into 2 groups by buccal area characteristics, i.e. with well-developed buccal structure (Group 1) buccal structure not well-developed (Group 2) (criterion 11).

taken into account. Sequences from thirteen of all fourteen species' transcriptome data (*Condylostoma magnum* was omitted because of the great differences in its transcriptome compared with that of other ciliates) were BLASTed to the epiplasmic proteins dataset (BLASTP, E-value < 1×10^{-2}, identity > 30%) and homologs were identified in each species. These species, which represented six classes based on morphological characters, were always separated into two groups based on the distinction of one characteristic for each time Table 1. Then, rank-sum tests were performed to both expression level and number of types of epiplasmic protein under several conditions of different group divisions. Some notable results suggest: (1) *Strombidium* spp. have significantly higher expression levels of alveolin than other ciliates (Fig. 6A), including other spirotrichs (Fig. 6B); (2) ciliates that possess extrusomes (e.g. *Strombidium sulcatum*, see (Song et al., 2000) have significantly higher expression levels of alveolin than those that lack extrusomes (Fig. 6C); (3) spirotrichs have significantly lower expression levels of EPC than other ciliates (Fig. 6D), although EPC expression was significantly higher in *Strombidium* spp. than in other spirotrichs (Fig. 6E); (4) ciliates such as spirotrichs with a highly specialized AZM and well-developed cytostome structure, have significantly lower expression levels of EPC than those whose circumoral kinetids and cytostome structure were less well-developed or degenerated (Fig. 6F and 6G). On the other hand, the difference of epiplasmic protein homolog types does not have significant impact on morphological characteristics of these ciliates.

DISCUSSION

Non-model ciliate transcriptome assembly

A problem that has been largely ignored by previous *de novo* transcriptome analyses is the creation of chimeras (Yang & Smith, 2013). Chimeras can come from misassembly of short reads or PCR-induced recombination during library preparation. Chimeras may also be real biological products from gene fusion or transsplicing. Because most chimeras are *trans* chimera in Trinity assembly results, and *cis* chimera detection may lead to false identification according to Yang and Smith (2013), we simply removed all *trans* chimeras.

A second problem in transcriptome assembly of ciliates that cannot be grown axenically is that there may be mitochondrial as well as bacterial transcript-related reads mixed in RNA-seq data. In order to ensure that final unigenes covered only *S. sulcatum* nuclear transcripts, we BLASTed contigs to databases consisting of *Oxytricha trifallax* mitochrondrial genome and bacterial genomes because *Strombidium* is closely related to *Oxytricha*. Although mitochondrial genomes varied significantly among different classes, genes were poorly conserved even among closely related genera, and the gene content and gene order also varied greatly (Swart et al., 2012), only 18 contigs were identified as transcripts related to mitochondria.

Ortholog detection

The prediction of orthologs and paralogs is a basic step in transcriptome studies, and is usually used to identify the core and auxiliary genes among related organisms (Yang & Smith, 2013). When predicted peptides of four ciliate genomes and the *S. sulcatum* transcriptome are collapsed into ortholog sets, it is found that *S. sulcatum* and *O. trifallax* share more unique orthologs with each other (2005) than are shared between *S. sulcatum* and any of: *T. thermophila* (231), *P. tetraurelia* (440), *I. multifiliis* (166). Within the scope of ciliates, this finding provides direct evidence for the close evolutionary relationship between *S. sulcatum* and *O. trifallax*, and is much stronger than phylogenetic analyses based on sequence data for a single or a few genes.

Phylogenomic analysis

Strombidium sulcatum and two other *Strombidium* spp. clustered strongly in a group nesting deep within the Spirotrichea clade, and were the closest neighbors of *Oxytricha*.

The genus *Protocruzia* did not group within any of the major clades, as reported in a recent work (Gentekaki et al., 2014). The systematic placement of *Protocruzia* has been questioned when it was investigated phylogenetically, especially when based on multi-gene information (e.g. SSU rRNA gene, internal transcribed spacer 2 (ITS2) gene, histone H4 gene), as this usually results in *Protocruzia* being separated from an otherwise monophyletic spirotrichean assembalge (Li et al., 2010). The phylogenomic analyses

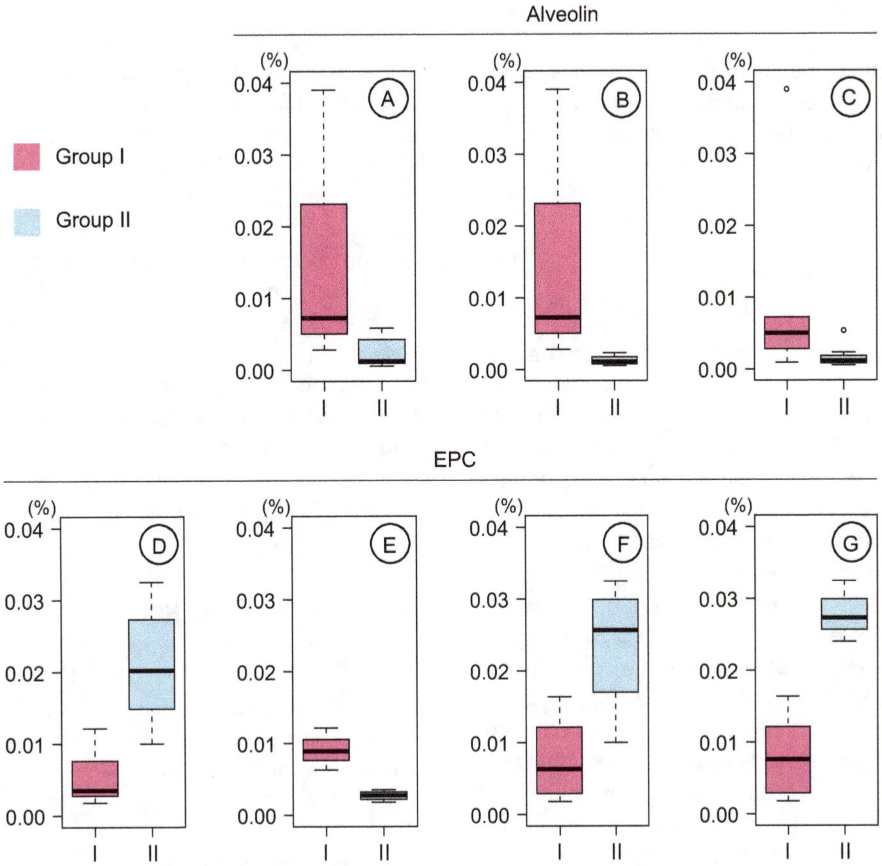

Figure 6. Proportion of transcripts of the epiplasmin gene family in 13 ciliates divided into groups referring to criteria shown in Table 1. The constituents of each group are as follows: (A) criterion 3: Group I [0, 11, 12], Group II [1–10]; (B) criterion 4: Group I [0, 11, 12], Group II [6, 7, 9, 10]; (C) criterion 9: Group I [0, 2–4, 11, 12], Group II [1, 5–10]; (D) criterion 2: Group I [0, 6, 7, 9–12], Group II [1–5, 8]; (E) criterion 4: Group I [0, 11, 12], Group II [6, 7, 9, 10]; (F) criterion 6: Group I [0, 1, 6–12], Group II [2–5]; (G) criterion 11: Group I [0–2, 6–12], Group II [3–5]. Numbers [0–12] in square brackets above are indexes of thirteen ciliates as: 0. *Strombidium sulcatum*; 1. *Platyophrya macrostoma*; 2. *Litonotus pictus*; 3. *Mesodinium pulex*; 4. *Myrionecta rubra*; 5. *Tiarina fusus*; 6. *Euplotes harpa*; 7. *Favella ehrenbergii*; 8. *Protocruzia adherens*; 9. *Strombidinopsis acuminatum*; 10. *Strombidinopsis* sp.; 11. *Strombidium inclinatum*; 12. *Strombidium rassoulzadegani*.

(both concatenated tree and concordance tree) and phylogenetic network in current study indicated similar findings that *Protocruzia* has a strong impact on the monophyly of the class Spirotrichea, and it might be an early branch of the subphylum Intramacronucleata in the evolutionary line from the class Heterotrichea (subphylum Postciliodesmatophora).

The similarity between the concatenated tree and concordance trees indicates that a robust phylogenetic topological structure could be predicted based on hundreds of molecular loci, even though these had markedly divergent rates of molecular evolution. Furthermore, according to Philippe et al. (2004), missing data in large alignments do not significantly affect inferred phylogeny when resolving relationships among eukaryotic groups. Therefore eukaryotic phylogenomic analysis at the rank of class or higher may be reliable when based on a large data set, even when taxon sampling is limited. This is in contrast to phylogenetic analyses based on a single gene such as the small subunit ribosome DNA (SSU rDNA) which is by far the most commonly used gene marker (Lynn, 2008, Adl et al., 2012). By analysing such a significantly expanded data set, phylogenomics is likely to result in reduced statistical errors and to reveal true evolutionary relationships more reliably.

On the other hand, a tree-like topological structure of relationships among species is assumed in traditional phylogenetic analyses, but recombination, hybridization, gene conversion, and gene transfer can all lead to evolutionary histories that are not adequately modelled by a single tree (Bryant & Moulton, 2004). It is quite likely that binary trees constructed in common analyses failed to reflect the complexity of true evolutionary histories of organisms (Morrison,

2011). Phylogenetic networks can break the restrictive supposition that the topological structure must be tree-like and can be used to visualize genealogical relationships among organisms (Huson et al., 2010). In addition, another advantage of phylogenetic network is to highlight conflicts for different taxa (Hall, 2011).

Epiplasmic proteins' transcriptional levels and morphological characteristics of ciliates

As the *Paramecium* epiplasm proteins (Epi1-Epi51) have no counterparts in other organisms (Aubusson-Fleury et al., 2013), we separated the epiplasmins into two types: EpiC (EPC) and Epi1-Epi51 (EPI). So epiplasmic proteins were divided into five classes: (1) the articulins (in various protists) (Marrs & Bouck, 1992, Huttenlauch et al., 1995, Huttenlauch et al., 1998); (2) articulin-like proteins (platein in *Euplotes*) (Kloetzel et al., 2003a); (3) EpiC (EPC, in *Tetrahymena*) (Honts & Williams, 2003); (4) Epi1-Epi51 (EPI, in *Paramecium*) (Damaj et al., 2009, Aubusson-Fleury et al., 2013); and (5) avleolins, perhaps homologues of alveolins and epiplasmins (Gould et al., 2008). Homologs of these five epiplasmic proteins types (articulins, platein, EPC, EPI, avleolins) in thirteen transcriptomes were investigated in this study. In order to test whether these five kinds of epiplasmic proteins' abundance difference in transcriptional level correlate with present systematic arrangement, we established four grouping criteria (Table 1, criteria 1–4). Rank sum test results suggest transcript abundances of only two epiplasmic protein groups, namely alveolins and EpiC, were significantly correlated to grouping criteria, specifically, criteria 3 (Fig. 6A) & 4 (Fig. 6B) to alveolins and criteria 2 (Fig. 6D) & 4 (Fig. 6E) to EpiC. Interestingly, both of these epiplamic protein groups show a strong correlation in criterion 4, i.e. *Strombidium* versus other species in the class Spirotrichea, which indicates that *Strombidium* spp. have unique cytoskeletal constituents within the class Spirotrichea. On the other hand, homologs of these five epiplasmic proteins do not show much correlation with grouping by present systematic arrangement.

Williams (2004) generated an epiplasm gene *EPC1* knockout construct that was successful in transforming *T. thermophila* cells, and the results suggested that the epiplasm plays a role in the control of either cell shape or cortical development. In this study, preliminary analyses revealed potential correlations between certain morphological characters and differences in the transcriptional levels of epiplasmic proteins. Here we identify seven morphological characteristics as potential grouping criteria for ciliates and which have potential correlations with epiplasmic proteins (Table 1, criteria 5–11). For example, the abundance of transcripts for alveolins is significantly (Sig. = 0.046 < 0.05, criterion 9 in Table 1) correlated with the presence of extrusomes, specifically, the abundance of transcripts for alveolins in species that have extrusomes is higher than that in which lack extrusomes (Fig. 6C). This supports the hypothesis that alveolins are the main constituents of the alveoli, which are part of the cortical epiplasm that includes the extrusomes (Gould et al., 2008). A second example is the correlation between EpiC (EPC) and the presence of circumoral kineties (criterion 6) and the size of the buccal area (criterion 11) (Fig. 6F and 6G). Williams (2004) reported that the most severe abnormality associated with *EPC1* knockout cells was the presence of branched and misaligned membranelles (clusters of cilia) found in the oral apparatus, indicating the importance of this group of epiplasmic proteins in the development of the oral ciliature. The results of the present study of EpiC support these hypotheses. Although present results indicate epiplasmic proteins' transcriptional levels are potentially correlated with morphological characteristics (such like circumoral kineties, extrusomes and buccal area) of ciliates, further experimental evidence is needed in order to confirm this finding.

MATERIALS AND METHODS

Cultures

Strombidium sulcatum was grown in 75 cm^2 plastic culture flasks in filtered marine water with a monoclonal population of *Escherichia coli* as the food source. Cultures were maintained at 25°C for 7–14 days until reaching over 10^6 cells in total.

RNA and Illumina Sequencing

Live species were harvested by centrifugation at 1700 *g* for 10 min and stored at −80°C until further treatment. The total RNA was extracted using the RNeasy kit (Qiagen, Hilden, Germany) and digested with DNase. The rRNA fraction was depleted using GeneRead rRNA Depletion Kit (Qiagen, Hilden, Germany). The quality of the remaining RNA was assayed using a BioAnalyzer (Agilent Technologies, Palo Alto, CA, USA). In total, 250 ng of the RNA was used to synthesize cDNA using Affymetrix 30 IVT Express Kit (Affymetrix Inc., Santa Clara, CA, USA). The resulting high quality cDNAs were used to construct a library for paired-end sequencing.

Assembly

The sequencing adapter was trimmed and low quality reads were filtered using FASTX-Toolkit (Gordon & Hannon, 2010). The remaining rRNA reads were removed by mapping to *S. sulcatum*'s 18S rRNAs [GenBank accession: DQ777745] and 5.8S rRNAs [GenBank accession: DQ811089] employing Bowtie v1.0.0 (Langmead et al., 2009). Trinity (r2013-02-25) (Grabherr et al., 2011) was run on a 2.66 GHz, 64 core Intel Xeon 7500 high-performance computing clusters with 512 GB RAM in order to assemble reads from the fastq format paired-end read files from the Illumina Hiseq2000 sequencer. Standard settings for Illumina sequence data and *de novo* transcriptome assembly (with default parameters) were used. After assembly, paired-end reads were aligned to the assembled transcripts for quality assessment using PERL scripts in the trinity package based on Bowtie v1.0.0 (Langmead et al., 2009), following the protocol of Trinity. RNA-seq reads were mapped back to the assembled contigs and were visualized using the inGAP package (Qi et al., 2010, Qi & Zhao, 2011).

The ciliate protein dataset consisted of: (1) peptides of four ciliate genomes (*Ichthyophthirius multifiliis* [Assembly ID: GCA_000220395.1] (Coyne et al., 2011), *Oxytricha trifallax* [Assembly ID: GCA_000295675.1] (Swart et al., 2013), *Paramecium tetraurelia* [Assembly ID: GCA_000165425.1] (Aury et al., 2006), and *Tetrahymena thermophila* [Assembly ID: GCA_000189635.1] (Eisen et al., 2006)) from GenBank; (2) peptides of thirteen transcriptomes (*Platyophrya macrostoma*, *Condylostoma magnum*, *Litonotus pictus*, *Mesodinium pulex*, *Myrionecta rubra*, *Tiarina fusus*, *Euplotes harpa*, *Favella ehrenbergii*, *Protocruzia adherens*, *Strombidinopsis acuminatum*, *Strombidinopsis* sp., *Strombidium inclinatum*, *Strombidium rassoulzadegani*, MMETSP ID see Table S1) from the Marine Microbial Eukaryote Transcriptome Sequencing Project (Resources, 2014); and (3) all ciliate proteins available on GenBank.

Trans chimeras were detected and removed following the protocol of Yang and Smith (2013). Assembled sequences were BLASTXed against the ciliate genome/transcriptome protein dataset. Redundancy of contigs was eliminated by CD-HIT v4.6.1 (Fu et al., 2012) (CD-HIT-EST, with 98% sequence identity threshold). Non-redundant contigs were passed to custom PERL scripts that used a BLAST-based strategy to link contigs with others whose terminals were a perfect match (100%). *Oxytricha trifallax* mitochondrial genomic peptides and bacterial genomes database were downloaded from GenBank as a mapping reference to remove reads which were probably contaminated by mitochondria or bacteria.

Gene prediction and annotation in transcriptomes

Abundance estimation of contigs was analysed using RSEM v1.2.3 (Li & Dewey, 2011) and PERL scripts in the Trinity package based on Bowtie, following the protocol of Trinity. CDS regions were predicted using ESTScan v2.2.1 (Iseli et al., 1999), and only predicted sequences longer than 50 bp were retained. Predicted protein products were blasted to the Swiss-Prot database, and were restricted to the top five HSP bitscores with E-value $\leq 1.0 \times 10^{-20}$. Predicted proteins were matched to HMM models (Pfam-A, Superfamily and TIGRFAMs) using HMMER3 (Eddy, 2009), and were restricted to E-value $\leq 1.0 \times 10^{-5}$ with the top five hits reported.

Orthologs among *S. sulcatum* and other ciliates

SPOCS (Curtis et al., 2013) (-H -M 2) was employed to identify shared orthologs among *S. sulcatum* and genome peptide data of the other four ciliates. Transcriptome peptide data of 13 ciliates were BLASTed to the *S. sulcatum* predicted peptide dataset by BLASTP (E-value $\leq 1 \times 10^{-2}$, identity ≥ 30%, length ≥ 50 aa), and vice versa. Pairwise mutual best-hits were identified as putative orthologs.

Phylogenomics

One hundred and twenty-seven orthologs shared by all 18 ciliates were retained and aligned using MUSCLE (Edgar, 2004). Sequence alignments were trimmed by Gblocks v0.91b (Castresana, 2000) (Parameters were set as: Maximum number of contiguous nonconserved positions = 8; Minimum length of a block = 10; Allowed gap positions = with half). A concatenated alignment dataset was produced by putting each ortholog alignment in series using Bioedit v7.2.0 (Hall, 1999). ProtTest v3.3 (Darriba et al., 2011) was employed to select the best model. The Maximum Likelihood (ML) tree was produced by PhyML v3.0 (Guindon et al., 2010). The Bayesian Inference (BI) tree was predicted by MrBayes v3.2.2 (Ronquist & Huelsenbeck, 2003). Coalescent analysis was proformed using Bucky v1.4.2 (Larget et al., 2010) by consolidating 127 BI trees based on 127 ortholog alignments respectively. Clades on concordance trees were annotated with their concordance factor (CF) (Baum, 2007), i.e. the proportion of the genome for which the clade is true. Bayesian concordance analysis was used to obtain point estimates and credibility intervals on CF's (Ane et al., 2007).

In order to visualize all available phylogenetic signals (36,724 aa in total) in the 127 ortholog alignments, split decomposition analyses were calculated with the computer program SplitsTree ver. 4 (Huson, 1998, Huson & Bryant, 2006). Since we were interested in the ambiguity of relationships of taxa at class level, phylogenetic networks were generated for the concatenated dataset of 127 ortholog alignments using the neighbornet algorithm with uncorrected distances (Bryant & Moulton, 2004). To assess the reliability of the phylogenetic networks, bootstrap analyses with 1000 replicates were carried out.

Epiplasmic protein expression analyses based on thirteen ciliates' transcriptome data

Sequences of the major epiplasmic proteins (EPI, Articulin, EPC, Platein, Alveolin) were acquired from GenBank and ParameciumDB (http://paramecium.cgm.cnrs-gif.fr/) and formed the epiplasmic proteins dataset (Acc. No. see Table S2). This epiplasmic proteins dataset were used as queries to BLASTed (E-value $\leq 1.0 \times 10^{-2}$, identity ≥ 30%) aganist the transcriptome data of these ciliate species in order to find homologs of epiplasmic proteins. Relative transcript abundance (number of reads mapping to the gene/total number of all reads of this species) and homolog type numbers of these genes were acquired from their transcriptome mapping data (Resources, 2014).

Morphologically similar groups were determined based on four taxonomic arrangements (different classes; the class Spirotrichea versus other classes; *Strombidium* spp. versus other ciliates; *Strombidium* spp. versus other species in the class Spirotrichea) and seven morphological character states concerning: somatic kineties, circumoral kineties, cortical granules, pellicular alveolus, extrusomes, cell size and shape, and the buccal area (Agatha, 2004, Foissner et al., 2007). Morphological character states were collected from morphological descriptions of each species as follows: *Strombidium sulcatum* Claparède & Lachmann, 1858 (Song et al., 2000), *Strombidium inclinatum* Montagnes, 1990 (Montagnes et al., 1990), *Strombidium rassoulzadegani* McManus, 2010 (McManus et al., 2010), *Favella ehrenbergii* (Claparède and Lachmann, 1858) Jörgensen, 1924 (Kim et al., 2010), *Strombidinopsis acuminatum* Faure-Fremiet, 1924 (Lynn et al., 1991, Dale & Lynn, 1998, Kim et al., 2010), *Euplotes harpa* Stein, 1859 (Dragesco et al., 1986), *Litonotus pictus* Gruber, 1884 (Kim & Min, 2009), *Platyophrya macrostoma* Foissner, 1980 (Foissner, 1993), *Protocruzia adherens* (Mansfeld, 1923) Kahl, 1930 (Song & Wilbert, 1997), *Mesodinium pulex* Claparède & Lachmann, 1858 (Tamar, 1992), *Myrionecta rubra* (Lohmann 1908) Jankowski, 1976 (Lynn, 2008) and *Tiarina fusus* (Claparède and Lachmann, 1859) Bergh, 1882 (Corliss, 1979). SPSS v16.0 (Inc, 2007) was employed to perform rank sum test among groups.

ACKNOWLEDGEMENTS

This work was supported by the Natural Science Foundation of China (Grant Nos. 31430077, 91131013, and 31272285). Many thanks are given to Prof. Weibo Song (Ocean University of China), for his indispensable advice during the preparation of the manuscript.

COMPLIANCE WITH ETHICS GUIDELINES

Xiao Chen, Xiaolu Zhao, Xiaohui Liu, Alan Warren, Fangqing Zhao and Miao Miao declare that they have no conflict of interest.

This article does not contain any studies with human or animal subjects performed by the any of the authors.

REFERENCES

Adl SM, Simpson AG, Lane CE et al (2012) The revised classification of eukaryotes. J Eukaryot Microbiol 59:429–493

Agatha S (2004) A cladistic approach for the classification of oligotrichid ciliates (Ciliophora: Spirotricha). Acta Protozool 43:201

Allali K, Dolan J, Rassoulzadegan F (1994) Culture characteristics and orthophosphate excretion of a marine oligotrich ciliate, *Strombidium sulcatum*, fed heat-killed bacteria. Mar Ecol Prog Ser 105:159

Ane C, Larget B, Baum DA, Smith SD, Rokas A (2007) Bayesian estimation of concordance among gene trees. Mol Biol Evol 24:1575

Aubusson-Fleury A, Bricheux G, Damaj R et al (2013) Epiplasmins and epiplasm in *Paramecium*: the building of a submembraneous cytoskeleton. Protist 164:451–469

Aury J-M, Jaillon O, Duret L et al (2006) Global trends of whole-genome duplications revealed by the ciliate *Paramecium tetraurelia*. Nature 444:171–178

Baum DA (2007) Concordance trees, concordance factors, and the exploration of reticulate genealogy. Taxon 56:417–426

Bernard C, Rassoulzadegan F (1990) Bacteria or microflagellates as a major food source for marine ciliates: possible implications for the microzooplankton. Mar Ecol Prog Ser 64:147–155

Bryant D, Moulton V (2004) Neighbor-net: an agglomerative method for the construction of phylogenetic networks. Mol Biol Evol 21:255–265

Castresana J (2000) Selection of conserved blocks from multiple alignments for their use in phylogenetic analysis. Mol Biol Evol 17:540–552

Christaki U, Dolan JR, Pelegri S, Rassoulzadegan F (1998) Consumption of picoplankton-size particles by marine ciliates: effects of physiological state of the ciliate and particle quality. Limnol Oceanogr 43:458–464

Corliss JO (1979) The ciliated protozoa. Characterization, classification and guide to the literature. Pergamon, Oxford

Coyne RS, Hannick L, Shanmugam D et al (2011) Comparative genomics of the pathogenic ciliate *Ichthyophthirius multifiliis*, its free-living relatives and a host species provide insights into adoption of a parasitic lifestyle and prospects for disease control. Genome Biol 12:R100

Curtis DS, Phillips AR, Callister SJ, Conlan S, McCue LA (2013) SPOCS: software for predicting and visualizing orthology/paralogy relationships among genomes. Bioinformatics 29:2641–2642

Dale T, Lynn DH (1998) Stomatogenesis of the ciliate genus *Strombidinopsis* with an improved description of *S. spiniferum* and *S. acuminatum*. J Eukaryot Microbiol 45:210–217

Damaj R, Pomel S, Bricheux G, Coffe G, Viguès B, Ravet V, Bouchard P (2009) Cross-study analysis of genomic data defines the ciliate multigenic epiplasmin family: strategies for functional analysis in *Paramecium tetraurelia*. BMC Evol Biol 9:125

Darriba D, Taboada GL, Doallo R, Posada D (2011) ProtTest 3: fast selection of best-fit models of protein evolution. Bioinformatics 27:1164–1165

Dolan JR, Sall N, Metcalfe A, Gasser B (2003) Effects of turbulence on the feeding and growth of a marine oligotrich ciliate. Aquat Microbial Ecol 31:183–192

Dragesco J, Dragesco-Kernéis A, Fryd-Versavel G (1986) Ciliés libres de l'Afrique intertropicale: introduction à la connaissance et à l'étude des Ciliés. IRD Editions

Eddy SR (2009) A new generation of homology search tools based on probabilistic inference. Genome Inform 23:205–211

Edgar RC (2004) MUSCLE: multiple sequence alignment with high accuracy and high throughput. Nucleic Acids Res 32:1792–1797

Eisen JA, Coyne RS, Wu M et al (2006) Macronuclear genome sequence of the ciliate *Tetrahymena thermophila*, a model eukaryote. PLoS Biol 4:e286

Fauré-Fremiet E (1962) Le genre Paranassula Kahl (Ciliata, Cyrtophorina). Cah Biol Mar 3:61–77

Foissner W (1993) Colpodea (Ciliophora). In: Balogh scientific books. Gustav Fischer Verlag, Stuttgart, p 798

Foissner W, Müller H, Agatha S (2007) A comparative fine structural and phylogenetic analysis of resting cysts in oligotrich and hypotrich Spirotrichea (Ciliophora). Eur J Protistol 43:295–314

Fu L, Niu B, Zhu Z, Wu S, Li W (2012) CD-HIT: accelerated for clustering the next-generation sequencing data. Bioinformatics 28:3150–3152

Gentekaki E, Kolisko M, Boscaro V et al (2014) Large-scale phylogenomic analysis reveals the phylogenetic position of the problematic taxon *Protocruzia* and unravels the deep phylogenetic affinities of the ciliate lineages. Mol Phylogenet Evol 78:36–42

Gordon A, Hannon G (2010) Fastx-toolkit. FASTQ/A short-reads pre-processing tools (unpublished). http://hannonlab.cshl.edu/fastx_toolkit

Gould SB, Tham W-H, Cowman AF, McFadden GI, Waller RF (2008) Alveolins, a new family of cortical proteins that define the protist infrakingdom Alveolata. Mol Biol Evol 25:1219–1230

Grabherr MG, Haas BJ, Yassour M et al (2011) Full-length transcriptome assembly from RNA-Seq data without a reference genome. Nat Biotechnol 29:644–652

Grant JR, Lahr DJ, Rey FE et al (2012) Gene discovery from a pilot study of the transcriptomes from three diverse microbial eukaryotes: *Corallomyxa tenera*, *Chilodonella uncinata*, and *Subulatomonas tetraspora*. Protist Genomics 1:3–18

Guindon S, Dufayard JF, Lefort V, Anisimova M, Hordijk W, Gascuel O (2010) New algorithms and methods to estimate maximum-likelihood phylogenies: assessing the performance of PhyML 3.0. Syst Biol 59:307–321

Hall TA (1999) BioEdit: a user-friendly biological sequence alignment editor and analysis program for Windows 95/98/NT. Nucl Acids Symp Ser 41:95–98

Hall BG (2011) Phylogenetic trees made easy. A how-to manual. Sinauer, Sunderland

Honts JE, Williams NE (2003) Novel cytoskeletal proteins in the cortex of Tetrahymena. J Eukaryot Microbiol 50:9–14

Huson DH (1998) SplitsTree: analyzing and visualizing evolutionary data. Bioinformatics 14:68–73

Huson DH, Bryant D (2006) Application of phylogenetic networks in evolutionary studies. Mol Biol Evol 23:254–267

Huson DH, Rupp R, Scornavacca C (2010) Phylogenetic networks. Concepts, algorithms and applications. Cambridge University Press, Cambridge

Huttenlauch I, Stick R (2003) Occurrence of articulins and epiplasmins in protists. J Eukaryot Microbiol 50:15–18

Huttenlauch I, Geisler N, Plessmann U, Peck RK, Weber K, Stick R (1995) Major epiplasmic proteins of ciliates are articulins: cloning, recombinant expression, and structural characterization. J Cell Biol 130:1401–1412

Huttenlauch I, Peck RK, Stick R (1998) Articulins and epiplasmins: two distinct classes of cytoskeletal proteins of the membrane skeleton in protists. J Cell Sci 111:3367–3378

Inc S (2007) SPSS 16.0 for Windows. SPSS Inc, Chicago

Iseli C, Jongeneel CV, Bucher P (1999) ESTScan: a program for detecting, evaluating, and reconstructing potential coding regions in EST sequences. Proc Int Conf Intell Syst Mol Biol 99:138–148

Kim S-J, Min G-S (2009) Taxonomic study of poorly-known marine pleurostomatid ciliates of *Litonotus paracygnus* and *L. pictus* (Ciliophora: Pleurostomatida) from Korea. Anim Syst Evol Divers 25.167–178

Kim SY, Yang EJ, Gong J, Choi JK (2010a) Redescription of *Favella ehrenbergii* (Claparède and Lachmann, 1858) Jörgensen, 1924 (Ciliophora: Choreotrichia), with phylogenetic analyses based on small subunit rRNA gene sequences. J Eukaryot Microbiol 57:460–467

Kim YO, Kim SY, Lee WJ, Choi JK (2010b) New Observations on the Choreotrich Ciliate Strombidinopsis acuminata Fauré-Fremiet 1924, and Comparison with Strombidinopsis jeokjo Jeong et al., 2004. J Eukaryot Microbiol 57:48–55

Kloetzel JA, Baroin-Tourancheau A, Miceli C, Barcetta S, Farmar J, Banerjee D, Fleury-aubusson A (2003a) Plateins: a novel family of signal peptide-containing articulins in euplotid ciliates. J Eukaryot Microbiol 50:19–33

Kloetzel JA, Baroin-Tourancheau A, Miceli C, Barchetta S, Farmar J, Banerjee D, Fleury-Aubusson A (2003b) Cytoskeletal proteins with N-terminal signal peptides: plateins in the ciliate *Euplotes* define a new family of articulins. J Cell Sc 116:1291–1303

Langmead B, Trapnell C, Pop M, Salzberg SL (2009) Ultrafast and memory-efficient alignment of short DNA sequences to the human genome. Genome Biol 10:R25

Larget BR, Kotha SK, Dewey CN, Ane C (2010) BUCKy: gene tree/species tree reconciliation with Bayesian concordance analysis. Bioinformatics 26:2910–2911

Li B, Dewey CN (2011) RSEM: accurate transcript quantification from RNA-Seq data with or without a reference genome. BMC Bioinform 12:323

Li L, Thorsten S, Kyoon SM, Al-Rasheid Khaled AS, Al-Khedhairy Bdulaziz A, Song W (2010) *Protocruzia*, a highly ambiguous ciliate (Protozoa; Ciliophora): very likely an ancestral form for Heterotrichea, Colpodea or Spirotrichea? With reevaluation of its evolutionary position based on multigene analyses. Sci China Life Sci 53:131–138

Li J, Liu W, Gao S, Warren A, Song W (2013) Multigene-based analyses of the phylogenetic evolution of oligotrich ciliates, with consideration of the internal transcribed spacer 2 secondary structure of three systematically ambiguous genera. Eukaryot Cell 12:430–437

Lynn DH (2008) Ciliated protozoa: characterization, classification, and guide to the literature. Springer, Dordrecht

Lynn DH, Montagnes DJS, Dale T, Gilron GL, Strom SL (1991) A reassessment of the genus *Strombidinopsis* (Ciliophora, Choreotrichida) with descriptions of four new planktonic species and remarks in its taxonomy and phylogeny. J Mar Biol Assoc UK 71:597–612

Marrs JA, Bouck GB (1992) The two major membrane skeletal proteins (articulins) of *Euglena gracilis* define a novel class of cytoskeletal proteins. J Cell Biol 118:1465–1475

McManus GB, Xu D, Costas BA, Katz LA (2010) Genetic identities of cryptic species in the *Strombidium stylifer/apolatum/oculatum* cluster, including a description of *Strombidium rassoulzadegani* n. sp. J Eukaryot Microbiol 57:369–378

Modeo L, Petroni G, Rosati G, Montagnes DJ (2003) A multidisciplinary approach to describe protists: redescriptions of *Novistrombidium testaceum* Anigstein 1914 and *Strombidium inclinatum* Montagnes, Taylor, and Lynn 1990 (Ciliophora, Oligotrichia). J Eukaryot Microbiol 50:175–189

Montagnes DJ, Taylor F, Lynn D (1990) *Strombidium inclinatum* n. sp. and a reassessment of *Strombidium sulcatum* Claparède and Lachmann (Ciliophora). J Protozool 37:318–323

Morrison DA (2011) An Introduction to phylogenetic networks. RJR Productions, Uppsala

Nahon P, Coffe G, Guyader H, Darmanaden-Delorme J, Jeanmaire-Wolf R, Clerot J-C, Adoutte A (1993) Identification of the epiplasmins, a new set of cortical proteins of the membrane cytoskeleton in *Paramecium*. J Cell Sci 104:975–990

Philippe H, Snell EA, Bapteste E, Lopez P, Holland PW, Casane D (2004) Phylogenomics of eukaryotes: impact of missing data on large alignments. Mol Biol Evol 21:1740–1752

Pomel S, Diogon M, Bouchard P, Pradel L, Ravet V, Coffe G, Viguès B (2006) The membrane skeleton in *Paramecium*: molecular characterization of a novel epiplasmin family and preliminary GFP expression results. Protist 157:61–75

Qi J, Zhao F (2011) inGAP-sv: A novel scheme to identify and visualize structural variation from paired end mapping data. Nucleic Acids Res 39:W567–W575

Qi J, Zhao F, Buboltz A, Schuster SC (2010) inGAP: An integrated next-generation genome analysis pipeline. Bioinformatics 26:127–129

Resources NCfG (2014) Marine Microbial Eukaryote Transcriptome Sequencing Project

Ronquist F, Huelsenbeck JP (2003) MrBayes 3: Bayesian phylogenetic inference under mixed models. Bioinformatics 19:1572–1574

Song W, Wilbert N (1997) Morphological investigations on some free living ciliates (Protozoa, Ciliophora) from China Sea with description of a new hypotrichous genus Hemigastrostyla nov. gen. Archiv für Protistenkunde 148:413–444

Song W, Wang M, Warren A (2000) Redescriptions of three marine ciliates, *Strombidium elegans* Florentin, 1901, *Strombidium sulcatum* Claparède & Lachmann, 1859 and *Heterostrombidium paracalkinsi* Lei, Xu & Song, 1999 (Ciliophora, Oligotrichida). Eur J Protistol 36:327–342

Swart EC, Nowacki M, Shum J et al (2012) The *Oxytricha trifallax* mitochondrial genome. Genome Biol Evol 4:136–154

Swart EC, Bracht JR, Magrini V et al (2013) The *Oxytricha trifallax* macronuclear genome: a complex eukaryotic genome with 16,000 tiny chromosomes. PLoS Biol 11:e1001473

Tamar H (1992) Four marine species of *Mesodinium* (Ciliophora: Mesodiniidae) II. *Mesodinium pulex* Clap. & Lachm., 1858. Archiv für Protistenkunde 141:284–303

Wiadnyana NN, Rassoulzadegan F (1989) Selective feeding of *Acartia clausi* and *Centropages typicus* on microzooplankton. Mar Ecol Prog Ser 53:37–45

Williams NE (2004) The epiplasm gene EPC1 influences cell shape and cortical pattern in *Tetrahymena thermophila*. J Eukaryot Microbiol 51:201–206

Xiong J, Lu X, Zhou Z et al (2012) Transcriptome analysis of the model protozoan, *Tetrahymena thermophila*, using deep RNA sequencing. PLoS One 7:e30630

Yang Y, Smith SA (2013) Optimizing de novo assembly of short-read RNA-seq data for phylogenomics. BMC Genomics 14:328

21

Microvesicle-delivery miR-150 promotes tumorigenesis by up-regulating VEGF and the neutralization of miR-150 attenuate tumor development

Yuchen Liu[1], Luming Zhao[1], Dameng Li[1], Yuan Yin[2], Chen-Yu Zhang[1✉], Jing Li[1✉], Yujing Zhang[1✉]

[1] Jiangsu Engineering Research Center for MicroRNA Biology and Biotechnology, State Key Laboratory of Pharmaceutical Biotechnology, School of Life Sciences, Nanjing University, Nanjing 210093, China
[2] Oncology Institute, Fourth Affiliated Hospital of Suzhou University, Wuxi 214062, China
✉ Correspondence: yjzhang@nju.edu.cn (Y. Zhang), jingli220@nju.edu.cn (J. Li), cyzhang@nju.edu.cn (C.-Y. Zhang)

ABSTRACT

Tumor-associated macrophages (TAMs) mostly exhibit M2-like (alternatively activated) properties and play positive roles in angiogenesis and tumorigenesis. Vascular endothelial growth factor (VEGF) is a key angiogenic factor. During tumor development, TAMs secrete VEGF and other factors to promote angiogenesis; thus, anti-treatment against TAMs and VEGF can repress cancer development, which has been demonstrated in clinical trials and on an experimental level. In the present work, we show that miR-150 is an oncomir because of its promotional effect on VEGF. MiR-150 targets TAMs to up-regulate their secretion of VEGF *in vitro*. With the utilization of cell-derived vesicles, named microvesicles (MVs), we transferred antisense RNA targeted to miR-150 into mice and found that the neutralization of miR-150 down-regulates miR-150 and VEGF levels *in vivo* and attenuates angiogenesis. Therefore, we proposed the therapeutic potential of neutralizing miR-150 to treat cancer and demonstrated a novel, natural, microvesicle-based method for the transfer of nucleic acids.

KEYWORDS microvesicle, miR-150, tumorigenesis, VEGF, neutralization, attenuation

INTRODUCTION

Tumors have a complex cellular ecology. Previous studies have shown that tumors consist not only of malignant cells but also of various stromal cell types (Lewis et al., 1995; Hanahan and Coussens, 2012). The stromal cells in cancer can be divided into three subtypes: angiogenic vascular cells (AVCS), infiltrating immune cells (IICs), and cancer-associated fibroblastic cells (CAFs) (Hanahan and Coussens, 2012). Tumor growth is clearly dependent on angiogenesis, as tumors need to establish a blood vessel network to access oxygen and nutrients (Millan Nunez-Cortes, 1991; Cristofanilli et al., 2002; Folkman, 2007). The contributions of stromal cells include resisting cell death, avoiding immune destruction, activating invasion, inducing angiogenesis, and sustaining proliferative signaling (Hanahan and Coussens, 2012).

The macrophages found in tumors are usually termed tumor-associated macrophages (TAMs) (Lewis and Pollard, 2006). The general understanding is that macrophages are subdivided into M1 (classically activated) and M2 (alternatively activated) phenotypes (Murdoch et al., 2008). M1 macrophages show a pro-inflammatory character, while M2 macrophages exhibit an immunosuppressive phenotype. TAMs often express many genes similar to those found in macrophages with the M2 phenotype and are thought to be M2-like macrophages (Murdoch et al., 2008). A number of studies have shown that TAMs influence tumor angiogenesis (Qian and Pollard, 2010; Fukuda et al., 2012). In human tumors, TAMs accumulate in poorly vascularized, necrotic areas and are also found surrounding blood vessels (Ferrara, 2010). A large number of publications have indicated that high numbers of TAMs are a poor prognostic indicator in cancer (Bingle et al., 2002). Lin and colleagues showed that decreasing macrophage infiltration in tumors resulted in the inhibition of tumor angiogenesis and delayed tumor development (Lin and Pollard, 2007). Further studies have found that TAMs express a broad range of pro-angiogenic factors that positively regulate angiogenesis, including vascular endothelial growth factor (VEGF) (Bolat et al., 2006), basic fibroblast growth factors (bFGF) (Bolat et al., 2006), tumor necrosis factor (TNF-α), and transforming growth factor-β (TGF-β) (Luo et al., 2006). These TAM-secreted factors explain the role of TAMs in tumor development. Because of their pivotal role in tumor growth, TAMs and their factors have been considered to be anticancer therapeutic targets. The blockade of colony stimulating factor-1 (CSF-1) signaling resulted in macrophage depletion in mammary tumors, reduced vascular density and improved responses to chemotherapy (Lin et al., 2001). Histidine-rich glycoprotein (HRG) can induce the reprogramming of TAMs, which has been shown to normalize vascular formations, resulting in a decrease in torturous vasculature, better pericyte coverage, less erratic blood flow, and improved responses to chemotherapy (Rolny et al., 2011).

VEGF is an angiogenic factor that can bind to receptors to induce endothelial cell migration and form new blood vessels (Ferrara et al., 2003). TAM-derived VEGF acts significantly on angiogenesis and tumor development, whereas down-regulating VEGF levels induces antitumorigenesis (Bolat et al., 2006). The deletion of the VEGF gene in macrophages decreases tumor angiogenesis (Stockmann et al., 2008). Anti-bodies against VEGF exert a large inhibitory effect on tumor growth (Kim et al., 1993). Cancer patients can currently receive anti-VEGF treatment, which results in a greater anti-tumor effect.

miRNAs are a class of small, non-coding RNAs that regulate gene expression at the post-transcriptional level (Bartel, 2004). Previous studies have demonstrated that miRNAs are relevant to cancers (Esquela-Kerscher and Slack, 2006), and anti-miRNA treatments have demonstrated anti-tumor effects on an experimental level (Kota et al., 2009). Previous study has found that miRNAs can be packaged into microvesicles (MVs), which are membranous vesicles that are secreted from most cell types, and transferred to neighboring or distant cells, where these miRNAs can regulate the functions of the recipient cells (Zhang et al., 2010). MVs have been utilized to transfer therapeutic RNAi to treat diseases (Alvarez-Erviti et al., 2011). In our present study, we demonstrated that miR-150 serves as an oncomir by regulating the VEGF secretion of TAMs. Then, MVs were utilized as the means for delivering antisense RNA targeted to miR-150 into mice to treat tumors and decrease tumor development by influencing the level of VEGFs secreted by TAMs. Our present study proposed a new anti-tumor molecule and a new method for the transfer of antineoplastic drugs.

RESULTS

Microvesicle-delivered miR-150 promotes tumor development *in vivo*

MiR-150 is an immune-related miRNA that is up-regulated in the plasma with the occurrence of several types of cancers (Fig. 1A). Thus, we hypothesized that miR-150 plays a role in tumorigenesis. To verify our hypothesis, we conducted the following experiments. We delivered miR-150 into mice using microvesicles to assess the effect of miR-150 on tumorigenesis. The human embryo kidney cell line 293T and its MVs contain a very low level of miR-150 (Zhang et al., 2010); thus, 293T MVs were chosen to serve as the carrier for miR-150. MiR-150 was over-expressed in the 293T cell line by transfecting mimic-miR-150 RNA; non-sense RNAs were also transfected into 293T cells to serve as the negative control. After transfection, MVs from 293T cells transfected with mimic-miR-150 RNA contained much higher levels of miR-150 compared with MVs from 293T cells transfected with ncRNA (Fig. 1B). Mice were subcutaneously implanted with sarcoma cells (S-180) to create a tumor model, and 293T MVs with or without miR-150 were intravenously injected into the tumor mice models as described in Fig. 2A. The results showed that the injection of 293T MVs containing miR-150 increased the plasma miR-150 levels (Fig. 2B). Tumor weight also dramatically increased after treatment with miR-150 (293T MV/miR-150) when compared with the saline group and the 293T/ncRNA group (Fig. 2C and 2D), suggesting that 293T MV-delivered exogenous miR-150 promotes tumor development *in vivo*.

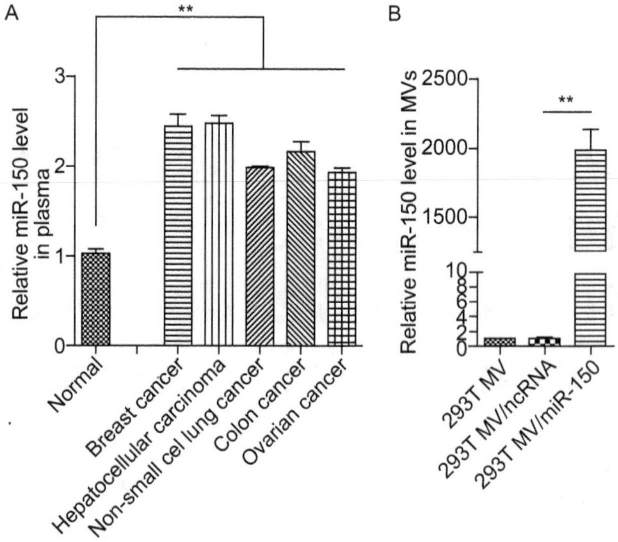

Figure 1. Microvesicle-delivered miR-150 promotes tumor development *in vivo*. (A) qRT-PCR analysis of the miR-150 levels in plasma from normal controls and patients with breast cancer, hepatocellular carcinoma, non-small cell lung cancer, colon cancer or ovarian cancer ($n = 4$). (B) qRT-PCR analysis of the miR-150 levels in MVs derived from 293T cells, 293T cells transfected with miRNA mimic negative control and 293T cells transfected with miR-150 mimic ($n = 4$). *, $P < 0.05$; **, $P < 0.01$.

VEGF is a very crucial factor in the process of tumor development. Tumors secrete VEGF to induce endothelial cell migration and the formation of new blood vessels. Serum VEGF levels were also assessed after the injection of miR-150. The ELISA results showed that VEGF levels increased in

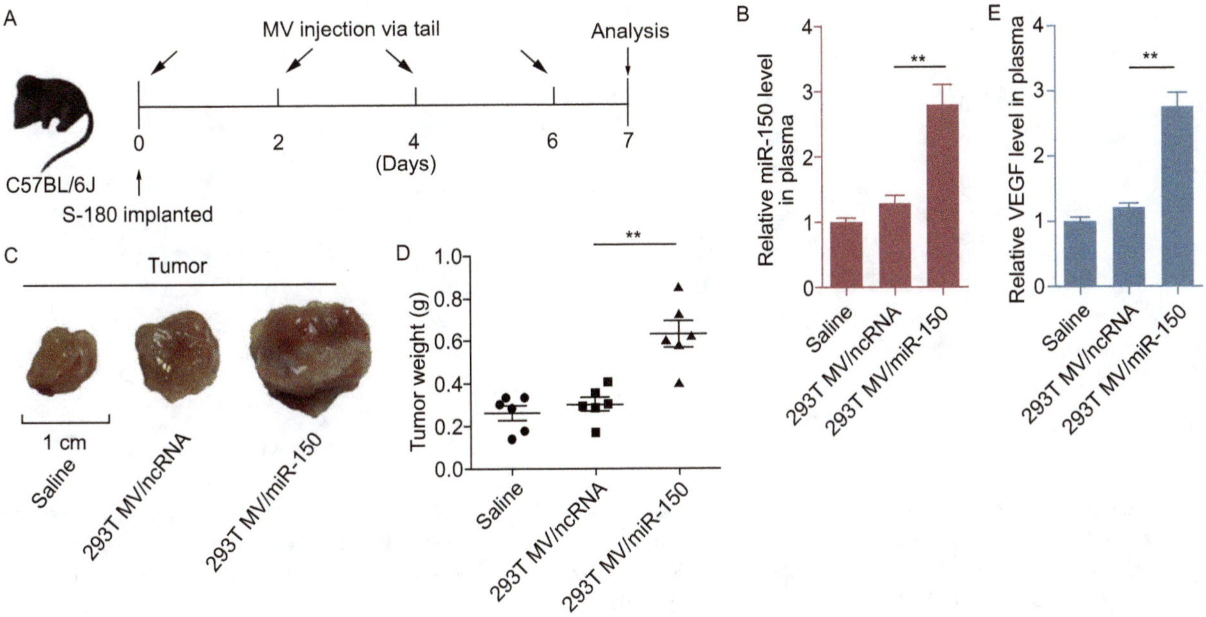

Figure 2. Microvesicle-delivered miR-150 promotes tumor development *in vivo*. (A) Flow chart depicting the experimental design. The C57BL/6J mice were first implanted with S-180 sarcoma cells and injected via the tail vein with saline (control), normal or modified 293T MVs (named 293T MV/miR-150) once every two days. The modified 293T MVs were harvested by transfecting the cells with the miR-150 mimic. Seven days later, the tumor tissues were analyzed and tumor sizes were measured. (B) qRT-PCR analysis of the plasma miR-150 levels in tumor-implanted mice that were injected with saline (control), normal 293T MVs or 293T MV/miR-150 ($n = 8$). (C) Morphology of the tumor tissues excised from tumor-implanted mice that were injected with saline (control), normal 293T MVs or 293T MV/miR-150. (D) Weight of the tumor tissues excised from tumor-implanted mice that were injected with saline (control), normal 293T MVs or 293T MV/miR-150. (E) ELISA analysis of the plasma VEGF levels in tumor-implanted mice that were injected with saline (control), normal 293T MVs or 293T MV/miR-150 ($n = 8$). *, $P < 0.05$; **, $P < 0.01$.

the plasma following the injection of miR-150 (Fig. 2E). Taken together, these results demonstrate that miR-150 promoted tumor development and suggest that the promotion of tumor development may be mediated by VEGF.

MiR-150 targets TAMs and promotes VEGF secretion by TAMs

To further assess whether miR-150 has a positive effect on VEGF secretion, we performed an *in vitro* assay. Considering that most TAMs are thought to be M2-like and that TAMs are the major origin of VEGF during tumor development, we hypothesized that miR-150 may target TAMs and consequently alter the levels of VEGF secreted. To validate our hypothesis, the effect of miR-150 on TAMs was assessed. TAMs were isolated from mice sarcoma tissue. To verify whether TAMs from the sarcoma were M2-like macrophages, bone marrow cells were isolated from the femur and were polarized into M1 and M2 macrophages, which were used as the reference cells for TAMs. Markers that can efficiently distinguish between M1 and M2 macrophages were tested. As we expected, the expression of these markers in TAMs was similar to M2 macrophages and differed from those observed in M1 macrophages, suggesting that the sarcoma-derived TAMs were M2-like macrophages (Fig. 3A). MiR-150 is abundantly expressed in MVs derived from THP-1 human monocyte cells following AGE treatment (Fig. S1A). Thus, to examine the effect of miR-150 on TAMs, AGE-treated THP-1 MVs that were rich in miR-150 were incubated with TAMs for 8 h, after which the medium was collected for use in an ELISA assay. The ELISA results showed that THP-1 MVs (AGE) dramatically enhanced VEGF secretion from TAMs (Fig. 3B). Although miR-150 is abundantly expressed in THP-1 MVs, other contents in the MVs may be responsible for the promotion of VEGF secretion. To solve this problem and confirm the role of miR-150, the following experiment was conducted. We knocked down miR-150 in THP-1 MVs by transfecting THP-1 cells with antisense oligonucleotides against miR-150 (anti-miR-150); ncRNAs were also transfected into THP-1 cells as a negative control. The transfected THP-1 cells were treated with AGE before medium collection. The results showed that miR-150 expression was reduced in the modified THP-1 MVs (Fig. S1B). TAMs were incubated with these miR-150-deficient MVs. After an 8 h incubation, the results revealed that the miR-150-deficient MVs (THP-1 MV/miR-150^{def}) lost the ability to enhance VEGF secretion when compared to THP-1 MVs loaded with ncRNAs (THP-1 MV/anti-ncRNA), suggesting that the promotion of VEGF secretion was mediated by miR-150 but not by other contents in the MVs (Fig. 3C). Taken together, these results indicated that miR-150

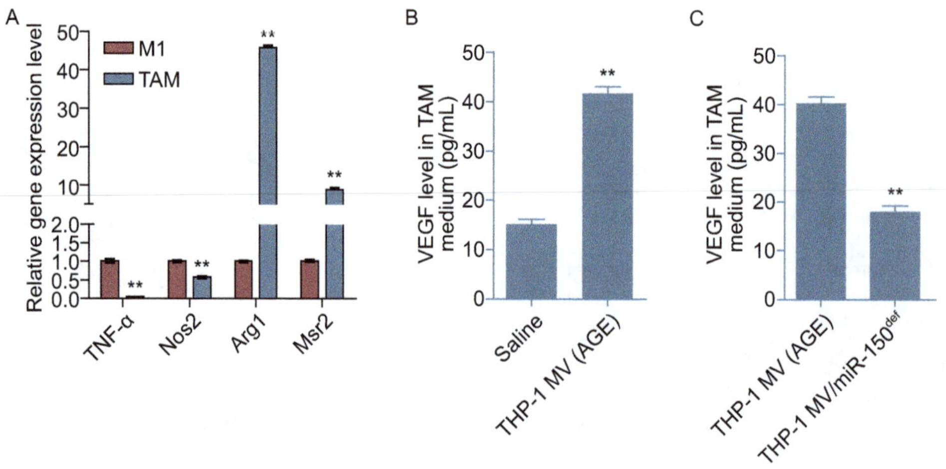

Figure 3. MiR-150 targets TAMs and promotes VEGF secretion by TAMs. (A) qRT-PCR analysis of the relative gene expression levels of TNF-α, NOS1, Arg2, and Msr2 of M1 macrophages or TAMs. (B) ELISA analysis of the VEGF levels in the medium of TAMs treated with saline (control) or THP-1 MV (AGE treated). (C) ELISA analysis of the VEGF levels in the medium of TAMs treated with normal THP-1 MV or miR-150 deficient MV (THP-1 MV/miR-150^{def}). *, $P < 0.05$; **, $P < 0.01$.

increased the VEGF secretion of TAMs, resulting in a positive effect on tumor development.

Microvesicle-delivered miR-150 inhibitor abolishes tumorigenesis and is proposed to be a potential therapeutic molecule against tumors

Because high levels of miR-150 promote tumor development, neutralizing miR-150 could potentially attenuate tumorigenesis. Taking advantage of the low level of miR-150 in 293T cells and 293T MVs, 293T MVs could be used as a carrier to transfer antisense RNA against miR-150 (anti-miR-150) into tumor models. We transfected 293T cells with antisense miR-150 RNA to load the MVs with the miR-150 inhibitor (293T MV/anti-miR-150). Mice were subcutaneously implanted with S-180 sarcoma cells to create a tumor model, and we treated this model with 293T MVs loaded with the miR-150 inhibitor or ncRNA, as depicted in Fig. 4A. The plasma miR-150 levels in mice treated with anti-miR-150 (293T MV/anti-miR-150) decreased significantly compared with those from the ncRNA group and the control group (Fig. 4B), indicating that the miR-150 inhibitor effectively decreased the plasma miR-150 levels of the tumor models. The VEGF levels in the plasma and tumor tissue also decreased when mice were treated with miR-150 antisense RNA (Fig. 4C). Subsequently, the tumor weights from the 293T MV/anti-miR-150 group dramatically decreased (Fig. 4D and 4E), suggesting that tumor development was strongly restrained. Consistent with the previous result, the protein levels of VEGF in xenograft decreased dramatically with the treatment of miR-150 antisense RNA (Fig. 4F). The results of the HE staining of paraffin sections showed that tumors treated with the miR-150 inhibitor presented large areas of necrosis because of fewer palingenetic blood vessels (Fig. 4G), further suggesting that angiogenesis and tumor development had been inhibited by the miR-150 inhibitor injection. The results of the immunohistochemical staining of paraffin sections revealed that the expression levels of the endothelial cell marker CD31 were dramatically reduced after miR-150 inhibitor treatment (Fig. 4H), suggesting that the migration of endothelial cells into the tumor was inhibited and that angiogenesis was inhibited in the tumor. Taken together, these results demonstrated that the down-regulation of miR-150 in a tumor model efficiently inhibited VEGF secretion and consequently restrained tumor development, suggesting that miR-150 could be a potential therapeutic target in cancers and that MV-delivered anti-miR-150 could be a novel strategy to treat cancers.

Microvesicle-delivered miR-150 inhibitor prevents tumorigenesis

We also found that the inhibitor of miR-150 played a long-term function in regulating tumorigenesis *in vivo*. The experiment was conducted as depicted in Fig. 5A. The MVs containing the inhibitor of miR-150 were injected into mice via the tail vein prior to the mice being implanted with sarcoma cells. After 7 days, the plasma miR-150 and VEGF levels decreased dramatically in mice receiving miR-150 inhibitor treatment (293T MV/anti-miR-150) compared with mice treated with saline or ncRNA (293T MV/anti-ncRNA) (Fig. 5B and 5C). The tumor weights from the anti-miR-150 injection group (293T MV/anti-miR-150) were lower than those from mice that were injected with 293T MV/ncRNA or saline (Fig. 5D and 5E). The result of the HE staining of paraffin sections showed a decrease in angiogenesis in tumors from mice treated with the miR-150 inhibitor (293T MV/anti-miR-150) compared with those from the other two groups (Fig. 5F). The immunohistochemical staining of paraffin sections showed that CD31 expression was reduced in tumors from mice treated with the inhibitor of miR-150 (293T MV/anti-

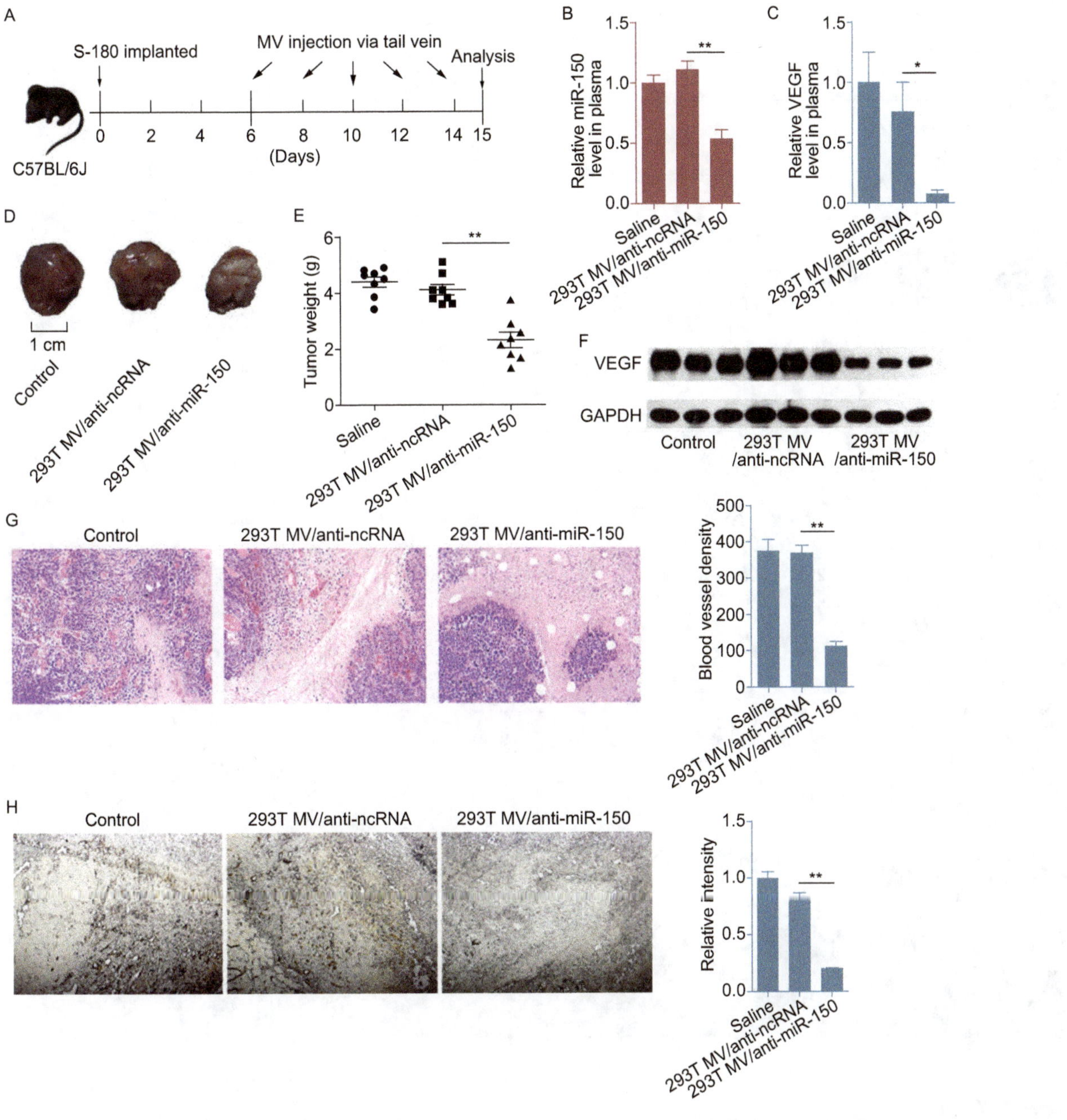

Figure 4. Microvesicle-delivered miR-150 inhibitor abolishes tumorigenesis and is proposed to be a potential therapeutic molecule for tumor treatment. (A) Flow chart depicting the experimental design. The C57BL/6J mice were first implanted with S-180 sarcoma cells. After 6 days, the mice were injected via the tail vein with saline (control) or with "modified" 293T MVs (named 293T MV/ncRNA or 293T MV/anti-miR-150) once every two days. The "modified" 293T MVs were harvested by transfecting the cells with negative control RNA or anti-miR-150. Nine days later, the tumor tissues were analyzed and tumor sizes were measured. (B) qRT-PCR analysis of the plasma miR-150 levels in tumor-implanted mice that were injected with saline (control), 293T MV/ncRNA or 293T MV/anti-miR-150 (n = 8). (C) ELISA analysis of the plasma VEGF levels in tumor-implanted mice that were injected with saline (control), 293T MV/ncRNA or 293T MV/anti-miR-150 (n = 8). (D) Morphology of the tumor tissues excised from tumor-implanted mice that were injected with saline (control), 293T MV/ncRNA or 293T MV/anti-miR-150. (E) Weight of the tumor tissues excised from tumor-implanted mice that were injected with saline (control), 293T MV/ncRNA or 293T MV/anti-miR-150. (F) Representative image of Western blot analysis of xenograft VEGF from mice that were untreated or treated with 293T MV/ncRNA or 293T MV/anti-miR-150 (n = 8). (G) Left panel: H&E stained sections from the tumor tissues shown in panel C, with arrows pointing to the luminal structures containing red blood cells. Right panel: Quantitative analysis of the numbers of newly formed vessels in tumor sections (n = 8). (H) Left panel: Immunohistochemistry of the paraffin-embedded tumor tissues shown in Fig. 4C using a CD31 antibody. Right panel: Quantitative analysis of the intensity of CD31 in tumor sections (n = 8). *, $P < 0.05$; **, $P < 0.01$.

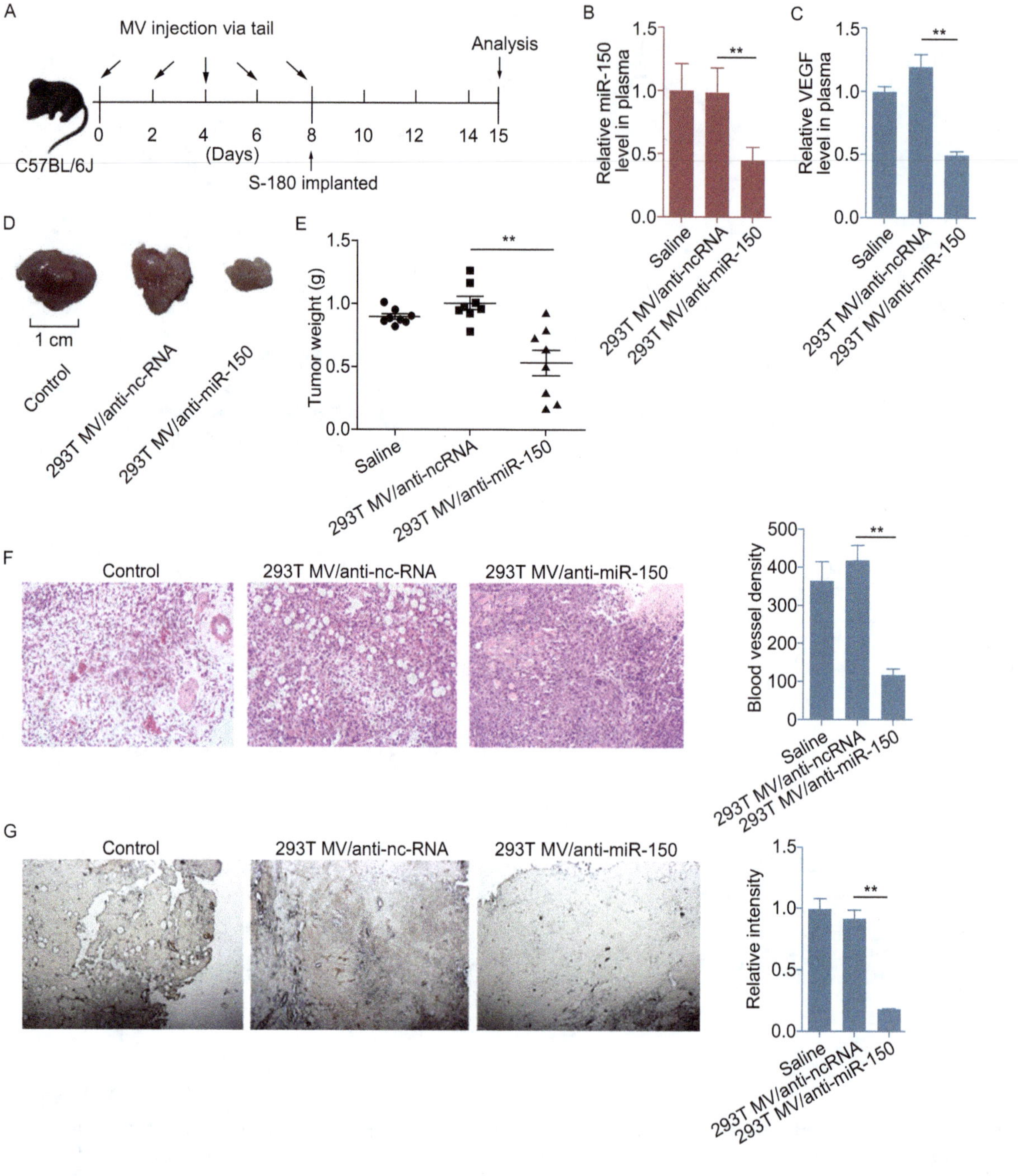

Figure 5. Microvesicle-delivered miR-150 inhibitor prevents tumorigenesis. (A) Flow chart depicting the experimental design. The C57BL/6J mice were first injected via the tail vein with saline (control) or "modified" 293T MVs (named 293T MV/ncRNA or 293T MV/anti-miR-150) once every two days. The "modified" 293T MVs were harvested by transfecting the cells with negative control RNA or anti-miR-150. After 8 days, the mice were implanted with S-180 sarcoma cells. Seven days later, the tumor tissues were analyzed and tumor sizes were measured. (B) qRT-PCR analysis of the plasma miR-150 levels in tumor-implanted mice that were injected with saline (control), 293T MV/ncRNA or 293T MV/anti-miR-150 (n = 8). (C) ELISA analysis of the plasma VEGF levels in tumor-implanted mice that were injected with saline (control), 293T MV/ncRNA or 293T MV/anti-miR-150 (n = 8). (D) Morphology of the tumor tissues excised from tumor-implanted mice that were injected with saline (control), 293T MV/ncRNA or 293T MV/anti-miR-150. (E) Weight of the tumor tissues excised from tumor-implanted mice that were injected with saline (control), 293T MV/ncRNA or 293T MV/anti-miR-150. (F) Left panel: H&E stained sections from the tumor tissues shown in panel C, with arrows pointing to the luminal structures containing red blood cells. Right panel: Quantitative analysis of the numbers of newly formed vessels in tumor sections (n = 8). (G) Left panel: Immunohistochemistry of the paraffin-embedded tumor tissues shown in Fig. 5C using a CD31 antibody. Right panel: Quantitative analysis of the intensity of CD31 in tumor sections (n = 8). *, $P < 0.05$; **, $P < 0.01$.

miR-150), suggesting that MV-delivered anti-miR-150 inhibited endothelial cell migration and angiogenesis in the tumors (Fig. 5G). Taken together, these results demonstrated that the preemptive neutralization of miR-150 may alter the microenvironment of tumors and prevent subsequent tumorigenesis. The inhibitor of miR-150, protected by MVs, may survive and contribute to the regulation of tumor development over a long-term period.

DISCUSSION

During tumor development, tumor cells secrete chemoattractants to ensure the ongoing recruitment of monocytes into tumors, including some types of growth factors, such as VEGF, placenta growth factor (PIGF), and CSF-1 (Coffelt et al., 2009). TAMs exhibit several characteristics of M2-polarized macrophages and promote tumor development by promoting angiogenesis and lymphangiogenesis, sustaining proliferative signaling and resisting cell death (Fukuda et al., 2012). Our present study focused on the role of TAMs in the regulation of angiogenesis, largely via their production of VEGF. MiRNAs are a class of classic molecule that regulates tumor development. The term oncomir is specifically used to reference miRNAs that have critical functions in tumor development (Goga and Benz, 2007). MiR-150 is an immune-related miRNA, and, for the first time, the present study has found that miR-150 promotes tumorigenesis (Fig. 2C). MiR-150 also promotes increased levels of VEGF in plasma (Fig. 2E). Considering the large role VEGF plays in the induction of angiogenesis and tumor development, the significant promotion of tumor growth by miR-150 may be mediated by increased VEGF levels. TAMs are a major source of VEGF, which binds to receptors on endothelial cells to promote migration and budding (Dulat et al., 2000). Thus, we proposed the hypothesis that TAMs may be the target cell of miR-150 during tumor development. In the present study, our results showed that exogenous miR-150 can increase the VEGF secretion by TAMs *in vitro*, revealing that miR-150 acts as a promotion factor in tumor development by regulating the VEGF secretion of TAMs (Fig. 3B and 3C). VEGF is not the direct target of miR-150, and miR-150 may target other upstream proteins involved in VEGF regulation. We predicted the target genes of miR-150 using Targetscan (http://www.targetscan.org/) and found that the inhibitor of growth family member 4 (ING4) is a very perfect target. It is well known that ING4 is a tumor suppressor for it intervenes the activity of hypoxia inducible factor (HIF) which has an important positive role in angiogenesis and tumorigenesis. HIF promote tumor angiogenesis by inducing VEGF and other factors (Ozer et al., 2005; Colla et al., 2007). Thus, the up-regulation of VEGF by miR-150 may be mediated by ING4 (Brahimi-Horn and Pouyssegur, 2006). Another predicted target of miR-150 by miRnada (http://www.microma.org/) is OS9, which is an essential component of a multiprotein complex that regulates HIF1A levels in an oxygen-dependent manner (Baek et al., 2005). It is reported that knockdown of OS9 by RNA interference increased HIF1A protein levels, HIF1A transcriptional activity, and VEGF mRNA levels under nonhypoxic conditions (Baek et al., 2005). We also found that THP-1-derived MVs contained VEGF, which raised the issue of whether the VEGF in the MVs or miR-150 was responsible for the increased levels of VEGF. Through calculation, we found that the proportion of VEGF contained in the MVs was only 1.5% of the increased VEGF in medium, and the other 98.5% of increased VEGF was contributed by the effects of exogenous miR-150 (Fig. S1C).

Great progress has been made in developing therapeutic strategies that target TAMs and VEGF to treat cancer (Kim et al., 1993). Several studies have revealed that TAMs are a very critical cell group in tumor tissue and that TAMs contribute greatly to tumorigenesis and the protection of tumors from therapy (McDonnell et al., 2003). VEGF inhibition represses tumor angiogenesis, and this result has led to the clinical utilization of VEGF inhibitors to treat cancer (Warren et al., 1995; Presta et al., 1997). In our present study, we used a microvesicle-delivered antagonucleotide against miR-150 to treat tumors. After the injection, VEGF levels decreased because of the down-regulation of miR-150 (Fig. 4C), and tumor growth was abolished by anti-miR-150 treatment (Fig. 4D). Through the angiogenic morphogenesis results, we found that angiogenesis was attenuated after antisense injection (Fig. 4F and 4G). Thus, we concluded that the miR-150 antisense caused the anesis of tumor development by down-regulating VEGF. This method provides a new, potential target to treat cancer, and the function of miRNAs in cancer development should not be ignored.

Microvesicles, a class of membrane vesicles with a diameter of 30–1000 nm, contain shedding vesicle and exosome (Chen et al., 2012). They are secreted by most cell types *in vitro* (Valadi et al., 2007). MVs have been utilized for the delivery of therapeutic RNAi and are considered to be a more effective, advantageous method than other options. Compared with the delivery strategies of viruses, lipid nanoparticles and polymeric nanoparticles, MVs present some major advantages. MVs are a natural carrier and are not subject to attack by antibodies, complements or opsonins in circulation. Other methods are prone to be cleared or trigger unwanted immune responses (van den Boorn et al., 2011). In our present study, MVs delivered the antagonucleotide for miR-150 into tumors efficiently and repressed tumor development, further suggesting that MVs are the perfect carrier for transferring therapeutic small RNA. Furthermore, we found that the function of the injected antisense continued for quite a long time. Seven days after the injection, miR-150 was maintained at a low level in the plasma, and VEGF levels and tumor development were also repressed (Fig. 5B–D), demonstrating that the antisense was still functional *in vivo*. In this process, MVs may act as a protector for the inner antisense and allow the antisense to exist and act for a long time.

An increasing body of research has revealed the critical role of miRNAs in the process of tumor development and demonstrated that miRNAs can serve as therapeutic targets. Our present study found that anti-miRNA treatment efficiently triggered the attenuation of tumor growth and suggests that MVs may be a potential carrier for the delivery of small nucleic acids. The utilization of MV-delivered small RNA may be a prospective therapeutic strategy.

MATERIALS AND METHODS

Cell lines

The human embryonic kidney 293T cell line and the human acute monocytic leukemia cell line THP-1 were purchased from the Shanghai Institute of Cell Biology, Chinese Academy of Sciences (Shanghai, China). The 293T cells were cultured in high-glucose DMEM medium, supplemented with 10% fetal bovine serum (FBS) (Gibco, CA, USA). The THP-1 cells were cultured in standard RPMI 1640 medium, supplemented with 10% FBS (Gibco).

Animals, tumor model generation, and primary cell isolation

C57BL/6J (male, 8-week-old) mice were purchased from the Model Animal Research Center (MARC) of Nanjing University (Nanjing, China) and were maintained in pathogen-free conditions. The animals received humane care according to the guidelines prepared by the National Institutes of Health's Guide for the Care and Use of Laboratory Animals and approved by the Institutional Review Board of Nanjing University, Nanjing, China.

To generate the xenograft tumor models, S-180 mouse sarcoma cells (10^6 cells per mouse) were subcutaneously injected into the left armpits of the animals. The tumor sizes were measured using a caliper and the tumor weights were determined after the tumors were separated from the animals.

Bone marrows were isolated from 8-week-old mice by flushing their femurs. The cells were suspended and maintained in RPMI 1640 media, supplemented with 10% FBS (Gibco) and 800 pg/mL mice M-CSF. Bone marrow cells were differentiated into non-polarized macrophages (M0) after 7 days. M0 cells were treated with 1 μg/mL LPS and 20 ng/mL IFN-γ to polarize them into classically activated macrophages (M1), or treated with 20 ng/mL IL-4 to polarize them into alternatively activated macrophages (M2).

Primary tumor-associated macrophage cells were isolated from the S-180 mouse sarcomas. Briefly, solid tumors were shredded and disaggregated by exposure to 0.3% collagenase IV (Invitrogen, CA, USA) for 40 min at 37°C; 10^6 cells/mL were seeded in 60-mm dishes (Corning, Tewksbury, MA). After 12 h of incubation, non-adherent cells were washed away, and adherent cells were maintained.

MV isolation

MVs were isolated from cell culture medium by differential centrifugation, according to previous publications (Valadi et al., 2007; Skog et al., 2008). After removing cells and other debris by centrifugation at 300 *g* and 3000 *g*, the supernatant was centrifuged at 110,000 *g* for 70 min (all steps were performed at 4°C). MVs were collected from the pellet and resuspended in PBS.

Cell transfection with scrambled negative control RNA, mimic-miR-150 RNA, and anti-miR-150 RNA

Synthetic RNA molecules, including mimic-miR-150, anti-miR-150, and scrambled negative control oligonucleotides (mimic-ncRNA or anti-ncRNA), were purchased from Ambion (Austin, TX). THP-1 cells and 293T cells were seeded in 75 cm^2 flasks (Corning) and were transfected the following day using Lipofectamine 2000 (Invitrogen), according to the manufacturer's instructions. To over-express miR-150 in 293T cells, 750 pmol of mimic-miR-150 RNA or scrambled negative control RNA was used. To knockdown miR-150, 750 pmol of anti-miR-150 RNA or scrambled negative control anti-miRNA (anti-ncRNA) was used. Cells and media were harvested 48 h after transfection.

RNA isolation and quantitative RT-PCR

The total RNA of cells was extracted using TRIzol Reagent (Invitrogen), according to the manufacturer's recommended protocols. Quantitative RT-PCR of mature miRNA was performed on an ABI 7900 (Applied Biosystems; Foster City, CA) instrument using the Taqman miRNA probes (Applied Biosystems). Gene Quantitative RT-PCR was performed on an ABI 7900 machine, and the primers used were purchased from Invitrogen. The primers used for macrophage marker genes were as follows: TNF-α: 5′-ACGGCATGGATCTCAAAGAC-3′ (forward primer) and 5′-AGATAGCAAATCGGCTGACG-3′ (reverse primer); Nos2: 5′-CCAAGCCCTCACCTACTTCC-3′ (forward primer) and 5′-CTCTGAGGGCTGACACA-3′ (reverse primer); Arg1: 5′-CTCCAAGCCAAAGTCCTTAGAG-3′ (forward primer) and 5′-AGGAGCTGTCATTAGGGACATC-3′ (reverse primer); and Msr2: 5′-AAAGAAAGCCCGAGTCCC-3′ (forward primer) and 5′-TGCCCAAGGAGATAGCAAGA-3′ (reverse primer).

ELISA assay

The VEGF levels in cell supernatants were measured using specific enzyme-linked immunosorbent assays. Briefly, 2 mL cell-free supernatant (each sample) from TAMs was harvested from 10^5 cells after 8-hour incubations, and VEGF levels were determined using an ELISA kit (Invitrogen), following the manufacturer's instructions. Data were representative of 3 independent experiments.

Western blotting

The levels of VEGF protein in xenograft were quantified by Western blot analysis using antibodies against VEGF (147, Santa Cruz). Normalization was conducted by blotting the same samples with an antibody against GAPDH (6C5, Santa Cruz).

Hematoxylin and eosin staining

The tumors were fixed in 4% paraformaldehyde, embedded in paraffin, sectioned, and then stained with hematoxylin and eosin. To perform a quantitative analysis, the numbers of erythrocyte-filled blood vessels (capillary density) were counted in 5 independent fields from three sections.

Immunohistochemistry

The tumors were fixed in 4% paraformaldehyde, embedded in paraffin, sectioned, and then stained with DBE-conjugated anti-CD31 antibodies (Abcam, MA, USA). Quantitative analysis was conducted by quantifying the fluorescence intensity from five sections.

Statistical analysis

Statistical analysis was performed using a *T*-test. Data are presented as the means ± SEM of at least three independent experiments. Differences were considered statistically significant at $P < 0.05$.

ACKNOWLEDGMENTS

This work was supported by grants from the National Basic Research Program (973 Program) (No. 2014CB542300), the National Natural Science Foundation of China (Grant Nos. 31200969, 31200874, 81101330, 31271378, and 81250044), the National Science Fund for Distinguished Young Scholars (81025019), the Natural Science Foundation of Jiangsu Province (BK2011013 and BK2012014), and the Research Fund for the Doctoral Program of Higher Education of China (20120091120044).

ABBREVIATIONS

AVCS, angiogenic vascular cells; CAFs, cancer-associated fibroblastic cells; CSF-1, colony stimulating factor-1; HIF, hypoxia inducible factor; HRG, histidine-rich glycoprotein; IICs, infiltrating immune cells; MVs, microvesicles; TAMs, tumor-associated macrophages; TGF-β, transforming growth factor-β; TNF-α, tumor necrosis factor; VEGF, vascular endothelial growth factor

COMPLIANCE WITH ETHICS GUIDELINES

Yuchen Liu, Luming Zhao, Dameng Li, Yuan Yin, Chen-Yu, Jing Li, Yujing Zhang declare that they have no conflict of interest. Additional informed consent was obtained from all patients for which identifying information is included in this article. All institutional and national guidelines for the care and use of laboratory animals were followed.

REFERENCES

Alvarez-Erviti, L., Seow, Y., Yin, H., Betts, C., Lakhal, S., and Wood, M.J. (2011). Delivery of siRNA to the mouse brain by systemic injection of targeted exosomes. Nat Biotechnol 29, 341–345.

Baek, J.H., Mahon, P.C., Oh, J., Kelly, B., Krishnamachary, B., Pearson, M., Chan, D.A., Giaccia, A.J., and Semenza, G.L. (2005). OS-9 interacts with hypoxia-inducible factor 1alpha and prolyl hydroxylases to promote oxygen-dependent degradation of HIF-1alpha. Mol Cell 17, 503–512.

Bartel, D.P. (2004). MicroRNAs: genomics, biogenesis, mechanism, and function. Cell 116, 281–297.

Bingle, L., Brown, N.J., and Lewis, C.E. (2002). The role of tumour-associated macrophages in tumour progression: implications for new anticancer therapies. J Pathol 196, 254–265.

Bolat, F., Kayaselcuk, F., Nursal, T.Z., Yagmurdur, M.C., Bal, N., and Demirhan, B. (2006). Microvessel density, VEGF expression, and tumor-associated macrophages in breast tumors: correlations with prognostic parameters. J Exp Clin Cancer Res 25, 365–372.

Brahimi-Horn, C., and Pouyssegur, J. (2006). The role of the hypoxia-inducible factor in tumor metabolism growth and invasion. Bulletin du cancer 93, E73–80.

Chen, X., Liang, H., Zhang, J., Zen, K., and Zhang, C.Y. (2012). Secreted microRNAs: a new form of i ntercellular communication. Trends Cell Biol 22, 125–132.

Coffelt, S.B., Hughes, R., and Lewis, C.E. (2009). Tumor-associated macrophages: effectors of angiogenesis and tumor progression. Biochim Biophys Acta 1796, 11–18.

Colla, S., Tagliaferri, S., Morandi, F., Lunghi, P., Donofrio, G., Martorana, D., Mancini, C., Lazzaretti, M., Mazzera, L., Ravanetti, L., et al. (2007). The new tumor-suppressor gene inhibitor of growth family member 4 (ING4) regulates the production of proangiogenic molecules by myeloma cells and suppresses hypoxia-inducible factor-1 alpha (HIF-1alpha) activity: involvement in myeloma-induced angiogenesis. Blood 110, 4464–4475.

Cristofanilli, M., Charnsangavej, C., and Hortobagyi, G.N. (2002). Angiogenesis modulation in cancer research: novel clinical approaches. Nat Rev Drug Discov 1, 415–426.

Esquela-Kerscher, A., and Slack, F.J. (2006). Oncomirs-microRNAs with a role in cancer. Nat Rev Cancer 6, 259–269.

Ferrara, N. (2010). Pathways mediating VEGF-independent tumor angiogenesis. Cytokine Growth Factor Rev 21, 21–26.

Ferrara, N., Gerber, H.P., and LeCouter, J. (2003). The biology of VEGF and its receptors. Nat Med 9, 669–676.

Folkman, J. (2007). Angiogenesis: an organizing principle for drug discovery? Nat Rev Drug Discov 6, 273–286.

Fukuda, K., Kobayashi, A., and Watabe, K. (2012). The role of tumor-associated macrophage in tumor progression. Front Biosci (Schol Ed) 4, 787–798.

Goga, A., and Benz, C. (2007). Anti-oncomir suppression of tumor phenotypes. Mol Interv 7, 199–202, 180.

Hanahan, D., and Coussens, L.M. (2012). Accessories to the crime: functions of cells recruited to the tumor microenvironment. Cancer Cell 21, 309–322.

Kim, K.J., Li, B., Winer, J., Armanini, M., Gillett, N., Phillips, H.S., and Ferrara, N. (1993). Inhibition of vascular endothelial growth factor-induced angiogenesis suppresses tumour growth in vivo. Nature 362, 841–844.

Kota, J., Chivukula, R.R., O'Donnell, K.A., Wentzel, E.A., Montgomery, C.L., Hwang, H.W., Chang, T.C., Vivekanandan, P., Torbenson, M., Clark, K.R., et al. (2009). Therapeutic microRNA delivery suppresses tumorigenesis in a murine liver cancer model. Cell 137, 1005–1017.

Lewis, C.E., Leek, R., Harris, A., and McGee, J.O. (1995). Cytokine regulation of angiogenesis in breast cancer: the role of tumor-associated macrophages. J Leukoc Biol 57, 747–751.

Lewis, C.E., and Pollard, J.W. (2006). Distinct role of macrophages in different tumor microenvironments. Cancer Res 66, 605–612.

Lin, E.Y., Nguyen, A.V., Russell, R.G., and Pollard, J.W. (2001). Colony-stimulating factor 1 promotes progression of mammary tumors to malignancy. J Exp Med 193, 727–740.

Lin, E.Y., and Pollard, J.W. (2007). Tumor-associated macrophages press the angiogenic switch in breast cancer. Cancer Res 67, 5064–5066.

Luo, Y., Zhou, H., Krueger, J., Kaplan, C., Lee, S.H., Dolman, C., Markowitz, D., Wu, W., Liu, C., Reisfeld, R.A., et al. (2006). Targeting tumor-associated macrophages as a novel strategy against breast cancer. J Clin Invest 116, 2132–2141.

McDonnell, C.O., Bouchier-Hayes, D.J., Toomey, D., Foley, D., Kay, E.W., Leen, E., and Walsh, T.N. (2003). Effect of neoadjuvant chemoradiotherapy on angiogenesis in oesophageal cancer. Br J Surg 90, 1373–1378.

Millan Nunez-Cortes, J. (1991). Angiogenesis: a crucial element in tumor development. An Med Interna 8, 369–371.

Murdoch, C., Muthana, M., Coffelt, S.B., and Lewis, C.E. (2008). The role of myeloid cells in the promotion of tumour angiogenesis. Nat Rev Cancer 8, 618–631.

Ozer, A., Wu, L.C., and Bruick, R.K. (2005). The candidate tumor suppressor ING4 represses activation of the hypoxia inducible factor (HIF). Proc Natl Acad Sci U S A 102, 7481–7486.

Presta, L.G., Chen, H., O'Connor, S.J., Chisholm, V., Meng, Y.G., Krummen, L., Winkler, M., and Ferrara, N. (1997). Humanization of an anti-vascular endothelial growth factor monoclonal antibody for the therapy of solid tumors and other disorders. Cancer Res 57, 4593–4599.

Qian, B.Z., and Pollard, J.W. (2010). Macrophage diversity enhances tumor progression and metastasis. Cell 141, 39–51.

Rolny, C., Mazzone, M., Tugues, S., Laoui, D., Johansson, I., Coulon, C., Squadrito, M.L., Segura, I., Li, X., Knevels, E., et al. (2011). HRG inhibits tumor growth and metastasis by inducing macrophage polarization and vessel normalization through downregulation of PIGF. Cancer Cell 19, 31–44.

Stockmann, C., Doedens, A., Weidemann, A., Zhang, N., Takeda, N., Greenberg, J.I., Cheresh, D.A., and Johnson, R.S. (2008). Deletion of vascular endothelial growth factor in myeloid cells accelerates tumorigenesis. Nature 456, 814–818.

Valadi, H., Ekstrom, K., Bossios, A., Sjostrand, M., Lee, J.J., and Lotvall, J.O. (2007). Exosome-mediated transfer of mRNAs and microRNAs is a novel mechanism of genetic exchange between cells. Nat Cell Biol 9, 654–659.

van den Boorn, J.G., Schlee, M., Coch, C., and Hartmann, G. (2011). SiRNA delivery with exosome nanoparticles. Nat Biotechnol 29, 325–326.

Warren, R.S., Yuan, H., Matli, M.R., Gillett, N.A., and Ferrara, N. (1995). Regulation by vascular endothelial growth factor of human colon cancer tumorigenesis in a mouse model of experimental liver metastasis. J Clin Invest 95, 1789–1797.

Zhang, Y., Liu, D., Chen, X., Li, J., Li, L., Bian, Z., Sun, F., Lu, J., Yin, Y., Cai, X., et al. (2010). Secreted monocytic miR-150 enhances targeted endothelial cell migration. Mol Cell 39, 133–144.

Permissions

All chapters in this book were first published in P&C, by Springer; hereby published with permission under the Creative Commons Attribution License or equivalent. Every chapter published in this book has been scrutinized by our experts. Their significance has been extensively debated. The topics covered herein carry significant findings which will fuel the growth of the discipline. They may even be implemented as practical applications or may be referred to as a beginning point for another development.

The contributors of this book come from diverse backgrounds, making this book a truly international effort. This book will bring forth new frontiers with its revolutionizing research information and detailed analysis of the nascent developments around the world.

We would like to thank all the contributing authors for lending their expertise to make the book truly unique. They have played a crucial role in the development of this book. Without their invaluable contributions this book wouldn't have been possible. They have made vital efforts to compile up to date information on the varied aspects of this subject to make this book a valuable addition to the collection of many professionals and students.

This book was conceptualized with the vision of imparting up-to-date information and advanced data in this field. To ensure the same, a matchless editorial board was set up. Every individual on the board went through rigorous rounds of assessment to prove their worth. After which they invested a large part of their time researching and compiling the most relevant data for our readers.

The editorial board has been involved in producing this book since its inception. They have spent rigorous hours researching and exploring the diverse topics which have resulted in the successful publishing of this book. They have passed on their knowledge of decades through this book. To expedite this challenging task, the publisher supported the team at every step. A small team of assistant editors was also appointed to further simplify the editing procedure and attain best results for the readers.

Apart from the editorial board, the designing team has also invested a significant amount of their time in understanding the subject and creating the most relevant covers. They scrutinized every image to scout for the most suitable representation of the subject and create an appropriate cover for the book.

The publishing team has been an ardent support to the editorial, designing and production team. Their endless efforts to recruit the best for this project, has resulted in the accomplishment of this book. They are a veteran in the field of academics and their pool of knowledge is as vast as their experience in printing. Their expertise and guidance has proved useful at every step. Their uncompromising quality standards have made this book an exceptional effort. Their encouragement from time to time has been an inspiration for everyone.

The publisher and the editorial board hope that this book will prove to be a valuable piece of knowledge for researchers, students, practitioners and scholars across the globe.

List of Contributors

Kelei Bi, Yueting Zheng, Feng G"ao, Jianshu Dong, Jiangyun Wang, Yi Wang and Weimin Gong
Laboratory of Non-coding RNA, Institute of Biophysics, Chinese Academy of Sciences, Beijing 100101, China

Haijin He, Marlyn Gonzalez and Fei Li
Department of Biology, New York University, New York, NY 10003, USA

Fan Zhang
Clinical Translational Research Center, Shanghai Pulmonary Hospital, Tongji University School of Medicine, Shanghai 200433, China

Changfei Li, Jun Hu, Junli Hao, Bao Zhao, Bo Wu, Lu Sun, Shanxin Peng, George F. Gao and Songdong Meng
CAS Key Laboratory of Pathogenic Microbiology and Immunology, Institute of Microbiology, Chinese Academy of Sciences (CAS), Beijing 100101, China

Hanqin Li, Susu Mao, Ke Zen, Chenyu Zhang and Liang Li
Jiangsu Engineering Research Center for MicroRNA Biology and Biotechnology, State Key Laboratory of Pharmaceutical Biotechnology, Nanjing University School of Life Sciences, Nanjing 210093, China

Haitao Wang
Department of Pharmacy, 306th Hospital of PLA, Beijing 100101, China

Hong Wu and Ying Huang
State Key Laboratory of Molecular Biology, National Center for Protein Science Shanghai, Shanghai Science Research Center, Shanghai Key Laboratory of Molecular Andrology, Institute of Biochemistry and Cell Biology, Shanghai Institutes for Biological Sciences, Chinese Academy of Sciences, Shanghai 200031, China

Yanchao Liu
State Key Laboratory of Molecular Biology, National Center for Protein Science Shanghai, Shanghai Science Research Center, Shanghai Key Laboratory of Molecular Andrology, Institute of Biochemistry and Cell Biology, Shanghai Institutes for Biological Sciences, Chinese Academy of Sciences, Shanghai 200031, China
University of Chinese Academy of Sciences, Beijing 100049, China

Yu Yu
State Key Laboratory of Genetic Engineering, Collaborative Innovation Center of Genetics and Development Institute of Plant Biology, School of Life Science, Fudan University, Shanghai 200433, China

Yang Yang, Ran Zhao and Chen Xu
Department of Physiology and Pathophysiology, School of Basic Medical Sciences, Fudan University, Shanghai 200032, China

Chao Lu and Ruizhe Qian
Department of Physiology and Pathophysiology, School of Basic Medical Sciences, Fudan University, Shanghai 200032, China
Research Center on Aging and Medicine, Fudan University, Shanghai 200032, China

Ning Sun
Department of Physiology and Pathophysiology, School of Basic Medical Sciences, Fudan University, Shanghai 200032, China
Research Center on Aging and Medicine, Fudan University, Shanghai 200032, China
State Key Laboratory of Medical Neurobiology, Fudan University, Shanghai 200032, China

Bingxuan Hua and Zuoqin Yan
Department of Orthopedics, Zhongshan Hospital, Fudan University, Shanghai 200032, China

Jun-Yu Ma and Wei Shen
College of Animal Science and Technology, Qingdao Agricultural University, Qingdao 266109, China

Teng Zhang and Qing Yuan Sun
College of Animal Science and Technology, Qingdao Agricultural University, Qingdao 266109, China
State Key Laboratory of Reproductive Biology, Institute of Zoology, Chinese Academy of Sciences, Beijing 100101, China

Heide Schatten
Department of Veterinary Pathobiology, University of Missouri, Columbia, MO 65211, USA

Susu Mao, Hui Xiao, Chenyu Zhang and Liang Li
State Key Laboratory of Pharmaceutical Biotechnology, Nanjing Advanced Institute for Life Sciences (NAILS), Nanjing University School of Life Sciences, Nanjing 210093, China
Jiangsu Engineering Research Center for microRNA Biology and Biotechnology, Nanjing 210093, China

Qi Sun
State Key Laboratory of Pharmaceutical Biotechnology, Nanjing Advanced Institute for Life Sciences (NAILS), Nanjing University School of Life Sciences, Nanjing 210093, China
Jiangsu Engineering Research Center for microRNA Biology and Biotechnology, Nanjing 210093, China
Center of Kidney Disease, 2nd Affiliated Hospital, Nanjing Medical University, Nanjing 210003, China

Zhanping Shi, Yanan Geng, Jiping Liu, Huina Zhang, Liqiang Zhou, Juehua Yu, Kunshan Zhang, Jie Liu, Xinpei Gao, Chunxue Zhang, Yinan Yao and Chong Zhang
Stem Cell Translational Research Center, Tongji Hospital, Tongji University School of Medicine, Shanghai 200065, China
Department of Psychiatry and Biobehavioral Sciences and Intellectual Development and Disabilities Research Center, David Geffen School of Medicine, University of California, Los Angeles, CA 90095, USA
Collaborative Innovation Center for Brain Science, Tongji University, Shanghai 200092, China

Quan Lin
Stem Cell Translational Research Center, Tongji Hospital, Tongji University School of Medicine, Shanghai 200065, China
Department of Psychiatry and Biobehavioral Sciences and Intellectual Development and Disabilities Research Center, David Geffen School of Medicine, University of California, Los Angeles, CA 90095, USA

Yi E. Sun
Stem Cell Translational Research Center, Tongji Hospital, Tongji University School of Medicine, Shanghai 200065, China
Department of Psychiatry and Biobehavioral Sciences and Intellectual Development and Disabilities Research Center, David Geffen School of Medicine, University of California, Los Angeles, CA 90095, USA
Collaborative Innovation Center for Brain Science, Tongji University, Shanghai 200092, China

Tang Hai, Ying Wang, Runfa Guo, Wei Li, Liu Wang and Qi Zhou
State Key Laboratory of Reproductive Biology, Institute of Zoology, Chinese Academy of Sciences, Beijing 100101, China

Yuran Song
State Key Laboratory of Reproductive Biology, Institute of Zoology, Chinese Academy of Sciences, Beijing 100101, China
Graduate University of the Chinese Academy of Sciences, Beijing 100049, China

Changhui Sun, Dan Chen, Pingrong Wang and Xiaojian Deng
Rice Research Institute, Sichuan Agricultural University, Chengdu 611130, China

Jun Fang and Chengcai Chu
State Key Laboratory of Plant Genomics and Center for Plant Gene Research (Beijing), Institute of Genetics and Developmental Biology, Chinese Academy of Sciences, Beijing 100101, China

Qianlan Xu and Xiaoyuan Song
Hefei National Laboratory for Physical Sciences at the Microscale, University of Science and Technology of China, Hefei 230026, China
CAS Center for Excellence in Molecular Cell Science, Shanghai 200031, China School of Life Sciences, University of Science and Technology of China, Hefei 230071, China

Ruoyu Wang and A. R. Ghanam
School of Life Sciences, University of Science and Technology of China, Hefei 230071, China
Anatomy and Embryology Department, Suez Canal University, Ismailia 41522, Egypt

Wei Miao
Key Laboratory of Aquatic Biodiversity and Conservation, Institute of Hydrobiology, Chinese Academy of Sciences, Wuhan 430072, China

Guanxiong Yan
Key Laboratory of Aquatic Biodiversity and Conservation, Institute of Hydrobiology, Chinese Academy of Sciences, Wuhan 430072, China
University of Chinese Academy of Sciences, Beijing 100049, China

Zhixiu Yang, Qiang Guo, Yuling Chen, Ningning Li, Kaige Yan, Yixiao Zhang, Haiteng Deng, Jianlin Lei and Ning Gao
Ministry of Education Key Laboratory of Protein Sciences, Center for Structural Biology, School of Life Sciences, Tsinghua University, Beijing 100084, China

Simon Goto, Akira Muto and Hyouta Himeno
Department of Biochemistry and Molecular Biology, Faculty of Agriculture and Life Science, Hirosaki University, Hirosaki 036-8561, Japan

Kai Yu, Yang Yu, Xiaoyan Tang, Huimin Chen and Xiao-Dong Su
Biodynamic Optical Imaging Center, School of Life Science, Peking University, Beijing 100871, China
State Key Laboratory of Protein and Plant Gene Research, Peking University, Beijing 100871, China

Junyu Xiao
State Key Laboratory of Protein and Plant Gene Research, Peking University, Beijing 100871, China
Peking-Tsinghua Center for Life Sciences, Peking University, Beijing 100871, China

Yongchang Zhang
College of Life Sciences, Nanjing Normal University, Nanjing 210023, China
State Key Laboratory for Diagnosis and Treatment of Infectious Diseases, Collaborative Innovation Center for Diagnosis and Treatment of Infectious Diseases, First Affiliated Hospital, Zhejiang University School of Medicine, Hangzhou 310003, China
Department of Medical Microbiology and Parasitology, Center for Infection and Immunity, Zhejiang University School of Medicine, Hangzhou 310058, China

Rongsui Gao, Qingjing Wang and Youjun Feng
State Key Laboratory for Diagnosis and Treatment of Infectious Diseases, Collaborative Innovation Center for Diagnosis and Treatment of Infectious Diseases, First Affiliated Hospital, Zhejiang University School of Medicine, Hangzhou 310003, China
Department of Medical Microbiology and Parasitology, Center for Infection and Immunity, Zhejiang University School of Medicine, Hangzhou 310058, China

Huiyan Ye
State Key Laboratory for Diagnosis and Treatment of Infectious Diseases, Collaborative Innovation Center for Diagnosis and Treatment of Infectious Diseases, First Affiliated Hospital, Zhejiang University School of Medicine, Hangzhou 310003, China
Department of Medical Microbiology and Parasitology, Center for Infection and Immunity, Zhejiang University School of Medicine, Hangzhou 310058, China
College of Life Science and Technology, Guangxi University, Nanning 530004, China

Huimin Zhang and Rongsui Gao
Department of Medical Microbiology & Parasitology, Zhejiang University School of Medicine, Hangzhou 310058, China

Beiwen Zheng and Youjun Feng
Department of Medical Microbiology & Parasitology, Zhejiang University School of Medicine, Hangzhou 310058, China
State Key Laboratory for Diagnosis and Treatment of Infectious Diseases, Zhejiang University School of Medicine, Hangzhou 310058, China

Yuna Sun, Xuemei Li and Jun Li
National Laboratory of Biomacromolecules, Institute of Biophysics, Chinese Academy of Sciences, Beijing 100101, China
University of Chinese Academy of Sciences, Beijing 100049, China

Yu Dong and Xiaodi Qiu
National Laboratory of Biomacromolecules, Institute of Biophysics, Chinese Academy of Sciences, Beijing 100101, China
University of Chinese Academy of Sciences, Beijing 100049, China

Neil Shaw
National Laboratory of Biomacromolecules, Institute of Biophysics, Chinese Academy of Sciences, Beijing 100101, China
Structure Biology Laboratory, Tsinghua University, Beijing 100084, China

Zihe Ra
National Laboratory of Biomacromolecules, Institute of Biophysics, Chinese Academy of Sciences, Beijing 100101, China
Structure Biology Laboratory, Tsinghua University, Beijing 100084, China
Tianjin Key Laboratory of Protein Science, College of Life Sciences, Nankai University, Tianjin 300071, China

Yueyang Xu
Tianjin Key Laboratory of Protein Science, College of Life Sciences, Nankai University, Tianjin 300071, China

Julia Foldi
Graduate Program in Immunology and Microbial Pathogenesis, Weill Cornell Graduate School of Medical Sciences, New York, NY 10021, USA

Lionel B. Ivashkiv
Graduate Program in Immunology and Microbial Pathogenesis, Weill Cornell Graduate School of Medical Sciences, New York, NY 10021, USA
Arthritis and Tissue Degeneration Program, Hospital for Special Surgery, New York, NY 10021, USA

Yingli Shang and Xiaoyu Hu
School of Medicine and Institute for Immunology, Tsinghua University, Beijing 100084, China

Baohong Zhao
Arthritis and Tissue Degeneration Program, Hospital for Special Surgery, New York, NY 10021, USA

Shan Gao
Department of Oncology, John Radcliffe Hospital, Weatherall Institute of Molecular Medicine, University of Oxford, Oxford OX3 9DS, UK

Chen Yang
Department of Molecular Genetics and Microbiology, Stony Brook University, Stony Brook, NY 11794, USA

Shan Jiang and You-Wen He
Department of Immunology, Duke University, Durham, NC 27708, USA

Xiao-Ning Xu
Department of Immunology, Chelsea & Westminster Hospital, Imperial College of London, 369 Fulham Road, London SW10 9NH, UK

Xin Lu
Nuffield Department of Clinical Medicine, Ludwig Institute for Cancer Research Oxford Branch, University of Oxford, Old Road Campus Research Building, Oxford OX3 7DQ, UK

Annie Cheung
Department of Pathology, Queen Mary Hospital, University of Hong Kong, Pokfulam Road, Hong Kong, China

Hui Wang
Natural Environment Research Council, Centre for Ecology & Hydrology, Benson Lane, Wallingford, Oxfordshire OX10 8BB, UK
Department of Zoology, University of Oxford, South Parks Road, Oxford OX1 3PS, UK

Xiaohui Liu and Miao Miao
College of Life Sciences, University of Chinese Academy of Sciences, Beijing 100049, China

Xiao Chen
College of Life Sciences, University of Chinese Academy of Sciences, Beijing 100049, China
Institute of Evolution & Marine Biodiversity, Ocean University of China, Qingdao 266003, China

Xiaolu Zhao
Institute of Evolution & Marine Biodiversity, Ocean University of China, Qingdao 266003, China

Alan Warren
Department of Life Sciences, Natural History Museum, Cromwell Road, London SW7 5BD, UK

Fangqing Zhao
Beijing Institutes of Life Science, Chinese Academy of Sciences, Beijing 100101, China

Yuchen Liu, Luming Zha, Dameng Li, Chen-Yu Zhang, Jing Li and Yujing Zhang
Jiangsu Engineering Research Center for MicroRNA Biology and Biotechnology, State Key Laboratory of Pharmaceutical Biotechnology, School of Life Sciences, Nanjing University, Nanjing 210093, China

Yuan Yin
Oncology Institute, Fourth Affiliated Hospital of Suzhou University, Wuxi 214062, China

Di-Jing Shi and Xu Cao
Department of Neurobiology, Neuroscience Research Institute, Peking University Health Science Center, Beijing 100191, China

Sheng Ye and Rongguang Zhang
National Laboratory of Biomacromolecules, Institute of Biophysics, Chinese Academy of Sciences, Beijing 100101, China

KeWei Wang
Department of Neurobiology, Neuroscience Research Institute, Peking University Health Science Center, Beijing 100191, China
Department of Molecular and Cellular Pharmacology, State Key Laboratory of Natural and Biomimetic Drugs, Peking University School of Pharmaceutical Sciences, Beijing 100191, China
PKU-IDG/McGovern Institute for Brain Research, Peking University, Beijing 100871, China

Index

www.ingramcontent.com/pod-product-compliance
Lightning Source LLC
La Vergne TN
LVHW080948210926
848485LV00004B/805